Oceanology of the
Antarctic Continental Shelf

American Geophysical Union | ANTARCTIC RESEARCH SERIES

Physical Sciences

ANTARCTIC OCEANOLOGY
　Joseph L. Reid, *Editor*
ANTARCTIC OCEANOLOGY II: THE AUSTRALIAN-NEW ZEALAND SECTOR
　Dennis E. Hayes, *Editor*

ANTARCTIC SNOW AND ICE STUDIES
　Malcolm Mellor, *Editor*
ANTARCTIC SNOW AND ICE STUDIES II
　A. P. Crary, *Editor*

ANTARCTIC SOILS AND SOIL FORMING PROCESSES
　J. C. F. Tedrow, *Editor*
DRY VALLEY DRILLING PROJECT
　L. D. McGinnis, *Editor*
GEOLOGY AND PALEONTOLOGY OF THE ANTARCTIC
　Jarvis B. Hadley, *Editor*
GEOLOGY OF THE CENTRAL TRANSANTARCTIC MOUNTAINS
　Mort D. Turner and John F. Splettstoesser, *Editors*
GEOMAGNETISM AND AERONOMY
　A. H. Waynick, *Editor*
METEOROLOGICAL STUDIES AT PLATEAU STATION, ANTARCTICA
　Joost A. Businger, *Editor*
OCEANOLOGY OF THE ANTARCTIC CONTINENTAL SHELF
　Stanley S. Jacobs, *Editor*
STUDIES IN ANTARCTIC METEOROLOGY
　Morton J. Rubin, *Editor*
UPPER ATMOSPHERE RESEARCH IN ANTARCTICA
　L. J. Lanzerotti and C. G. Park, *Editors*
THE ROSS ICE SHELF: GLACIOLOGY AND GEOPHYSICS
　C. R. Bentley and D. E. Hayes, *Editors*

Biological and Life Sciences

BIOLOGY OF THE ANTARCTIC SEAS
　Milton O. Lee, *Editor*
BIOLOGY OF THE ANTARCTIC SEAS II
　George A. Llano, *Editor*
BIOLOGY OF THE ANTARCTIC SEAS III
　George A. Llano and Waldo L. Schmitt, *Editors*
BIOLOGY OF THE ANTARCTIC SEAS IV
　George A. Llano and I. Eugene Wallen, *Editors*
BIOLOGY OF THE ANTARCTIC SEAS V
　David L. Pawson, *Editor*
BIOLOGY OF THE ANTARCTIC SEAS VI
　David L. Pawson, *Editor*

BIOLOGY OF THE ANTARCTIC SEAS VII
　David L. Pawson, *Editor*
BIOLOGY OF THE ANTARCTIC SEAS VIII
　David L. Pawson and Louis S. Kornicker, *Editors*
BIOLOGY OF THE ANTARCTIC SEAS IX
　Louis S. Kornicker, *Editor*
BIOLOGY OF THE ANTARCTIC SEAS X
　Louis S. Kornicker, *Editor*
BIOLOGY OF THE ANTARCTIC SEAS XI
　Louis S. Kornicker, *Editor*
BIOLOGY OF THE ANTARCTIC SEAS XII
　David L. Pawson, *Editor*
BIOLOGY OF THE ANTARCTIC SEAS XIII
　Louis S. Kornicker, *Editor*
BIOLOGY OF THE ANTARCTIC SEAS XIV
　Louis S. Kornicker, *Editor*
BIOLOGY OF THE ANTARCTIC SEAS XV
　Louis S. Kornicker, *Editor*
BIOLOGY OF THE ANTARCTIC SEAS XVI
　Louis S. Kornicker, *Editor*

ANTARCTIC TERRESTRIAL BIOLOGY
　George A. Llano, *Editor*
TERRESTRIAL BIOLOGY II
　Bruce Parker, *Editor*
TERRESTRIAL BIOLOGY III
　Bruce Parker, *Editor*

ANTARCTIC ASCIDIACEA
　Patricia Kott
ANTARCTIC BIRD STUDIES
　Oliver L. Austin, Jr., *Editor*
ANTARCTIC PINNIPEDIA
　William Henry Burt, *Editor*
ANTARCTIC CIRRIPEDIA
　William A. Newman and Arnold Ross
BIRDS OF THE ANTARCTIC AND SUB-ANTARCTIC
　George E. Watson
ENTOMOLOGY OF ANTARCTICA
　J. Linsley Gressitt, *Editor*
HUMAN ADAPTABILITY TO ANTARCTIC CONDITIONS
　E. K. Eric Gunderson, *Editor*
POLYCHAETA ERRANTIA OF ANTARCTICA
　Olga Hartman
POLYCHAETA MYZOSTOMIDAE AND SEDENTARIA OF ANTARCTICA
　Olga Hartman
RECENT ANTARCTIC AND SUBANTARCTIC BRACHIOPODS
　Merrill W. Foster

Volume 43 | ANTARCTIC RESEARCH SERIES

Oceanology of the Antarctic Continental Shelf

Stanley S. Jacobs, Editor

American Geophysical Union
Washington, D.C.
1985

| Volume 43 | ANTARCTIC RESEARCH SERIES |

OCEANOLOGY OF THE ANTARCTIC CONTINENTAL SHELF

STANLEY S. JACOBS, *Editor*

Published under the aegis of the
Board of Associate Editors, Antarctic Research Series
Charles R. Bentley, Chairman
Samuel C. Colbeck, Robert H. Eather, David Elliot,
Dennis Hayes, Louis S. Kornicker, Heinz Lettau, and Bruce Parker

Library of Congress Cataloging in Publication Data
Main entry under title:

Oceanology of the Antarctic continental shelf.

(Antarctic research series, ISSN 0066-4634; v. 43)
1. Continental shelf—Antarctic regions.
I. Jacobs, Stanley S. II. Series.
GC85.2.A57024   1985   551.4   85-19980
ISBN 0-87590-196-4
ISSN 0066-4634

Copyright 1985 by the American Geophysical Union
2000 Florida Avenue, N.W.
Washington, D.C. 20009

Figures, tables, and short excerpts may be reprinted in scientific books and journals
if the source is properly cited.

Authorization to photocopy items for internal or personal use, or the internal or personal
use of specific clients, is granted by the American Geophysical Union for libraries and
other users registered with the Copyright Clearance Center (CCC) Transactional Reporting
Service, provided that the base fee of $1.00 per copy, plus $0.10 per page is paid directly to CCC,
21 Congress St., Salem, MA 01970. 0066- 4634/85/$01.00 + 0.10.
  This consent does not extend to other kinds of copying, such as copying for creating new
collective works or for resale. The reproduction of multiple copies and the use of full
articles or the use of extracts, including figures and tables, for commercial purposes
requires specific permission from AGU.

Published by
AMERICAN GEOPHYSICAL UNION
With the aid of grant DPP-80-19997 from the
National Science Foundation

Printed in the United States of America

CONTENTS

The Antarctic Research Series: Statement of Objectives   vii
   Board of Associate Editors

Preface   ix

GEBCO Bathymetric Sheet 5.18 (Circum-Antarctic)   1
   J. R. Vanney and G. L. Johnson

Circulation and Water Masses on the Southern Weddell Sea Shelf   5
   A. Foldvik, T. Gammelsrød, and T. Tørresen

Bottom Currents Near the Continental Shelf Break in the Weddell Sea   21
   Arne Foldvik, Thor Kvinge, and Tor Tørresen

Interaction Between Ice Shelf and Ocean in George VI Sound, Antarctica   35
   J. R. Potter and J. G. Paren

Origin and Evolution of Water Masses Near the Antarctic Continental Margin: Evidence From $H_2^{18}O/H_2^{16}O$ Ratios in Seawater   59
   Stanley S. Jacobs, Richard G. Fairbanks, and Yoshio Horibe

Preliminary Observations From Long-Term Current Meter Moorings Near the Ross Ice Shelf, Antarctica   87
   R. Dale Pillsbury and Stanley S. Jacobs

Tidal Rectification Below the Ross Ice Shelf, Antarctica   109
   D. R. MacAyeal

Evolution of Tidally Triggered Meltwater Plumes Below Ice Shelves   133
   D. R. MacAyeal

The Winter Oceanography of McMurdo Sound, Antarctica   145
   E. L. Lewis and R. G. Perkin

Observations in the Boundary Layer Under the Sea Ice in McMurdo Sound   167
   W. M. Mitchell and J. A. T. Bye

A Recurring, Atmospherically Forced Polynya in Terra Nova Bay   177
   Dennis D. Kurtz and David H. Bromwich

Antarctic Offshore Leads and Polynyas and Oceanographic Effects   203
   H. Jay Zwally, J. C. Comiso, and A. L. Gordon

A Passive Microwave Study of Polynyas Along the Antarctic Wilkes Land Coast   227
   Donald J. Cavalieri and Seelye Martin

Some Effects of Ocean Currents and Wave Motion on the Dynamics of Floating Glacier Tongues   253
   G. Holdsworth

Tidal Measurements Along the Antarctic Coastline   273
   J. R. E. Lutjeharms, C. C. Stavropoulos, and K. P. Koltermann

Oceanographic Influences on Sedimentation Along the Antarctic Continental Shelf   291
   Robert B. Dunbar, John B. Anderson, Eugene W. Domack, and Stanley S. Jacobs

THE ANTARCTIC RESEARCH SERIES:
STATEMENT OF OBJECTIVES

The Antarctic Research Series, an outgrowth of research done in the Antarctic during the International Geophysical Year, was begun early in 1963 with a grant from the National Science Foundation to AGU. It is a book series designed to serve scientists and graduate students actively engaged in Antarctic or closely related research and others versed in the biological or physical sciences. It provides a continuing, authoritative medium for the presentation of extensive and detailed scientific research results from Antarctica, particularly the results of the United States Antarctic Research Program.

Most Antarctic research results are, and will continue to be, published in the standard disciplinary journals. However, the difficulty and expense of conducting experiments in Antarctica make it prudent to publish as fully as possible the methods, data, and results of Antarctic research projects so that the scientific community has maximum opportunity to evaluate these projects and so that full information is permanently and readily available. Thus the coverage of the subjects is expected to be more extensive than is possible in the journal literature.

The series is designed to complement Antarctic field work, much of which is in cooperative, interdisciplinary projects. The Antarctic Research Series encourages the collection of papers on specific geographic areas (such as the East Antarctic Plateau or the Weddell Sea). On the other hand, many volumes focus on particular disciplines, including marine biology, oceanology, meteorology, upper atmosphere physics, terrestrial biology, snow and ice, human adaptability, and geology.

Priorities for publication are set by the Board of Associate Editors. Preference is given to research projects funded by U.S. agencies, long manuscripts, and manuscripts that are not readily publishable elsewhere in journals that reach a suitable reading audience. The series serves to emphasize the U.S. Antarctic Research Program, thus performing much the same function as the more formal expedition reports of most of the other countries with national Antarctic research programs.

The standards of scientific excellence expected for the series are maintained by the review criteria established for the AGU publications program. The Board of Associate Editors works with the individual editors of each volume to assure that the objectives of the series are met, that the best possible papers are presented, and that publication is achieved in a timely manner. Each paper is critically reviewed by two or more expert referees.

The format of the series, which breaks with the traditional hard-cover book design, provides for rapid publication as the results become available while still maintaining identification with specific topical volumes. Approved manuscripts are assigned to a volume according to the subject matter covered; the individual manuscript (or group of short manuscripts) is produced as a soft cover 'minibook' as soon as it is ready. Each minibook is numbered as part of a specific volume. When the last paper in a volume is released, the appropriate title pages, table of contents, and other prefatory matter are printed and sent to those who have standing orders to the series. The minibook series is more useful to researchers, and more satisfying to authors, than a volume that could be delayed for years waiting for all the papers to be assembled. The Board of Associate Editors can publish an entire volume at one time in hard cover when availability of all manuscripts within a short time can be guaranteed.

BOARD OF ASSOCIATE EDITORS
ANTARCTIC RESEARCH SERIES

PREFACE

This is the first oceanology volume of the Antarctic Research Series to be devoted to the continental shelf. That region is of special interest because of its great depth and its climatic role in the production of sea ice, ventilation of the deep ocean, and wastage of the Antarctic ice sheet. Sea ice persists along much of the continental shelf during the austral summer, so that shipboard observations there have often been difficult and sometimes dangerous. Fortunately, time series measurements from bottom-moored instruments and analyses of satellite imagery are beginning to supplement the data that can be collected on summer expeditions. Geochemical tracers, high-resolution vertical profiling instruments and computer modeling are also providing new insights into the continental shelf circulation.

Antarctic oceanology is both interdisciplinary and international, as attested by the authors' diverse interests and affiliations. Their contributions are not lacking in jargon peculiar to the various trades, nor in adopted terminology with subtly different meanings. For example, a geologist's meltwater is evidenced in shelf sediment patterns, and may have been generated by pressure melting or frictional heat where the ice sheet is grounded. An oceanographer's meltwater is identified by water column tracers and has mostly been derived from oceanic or atmospheric heat flux into the glacial ice or sea ice. Oceanology is also in a transitional stage between classical, discrete shipboard measurements and continuous remote sensing. That evolution is emphasized in this volume by literature reviews, early results from an ongoing experiment, and an updated description of some pioneering long-term current and temperature measurements. Several authors took advantage of the opportunity to exchange critical reviews and cross-reference their work to other papers in the volume. The separate reference lists may together serve as a useful bibliography for the continental shelf regime.

The accompanying GEBCO circum-Antarctic chart effectively portrays the continental shelf in relation to the glaciated continent and surrounding deep ocean. It also displays several ephemeral bathymetric features and extensive areas that are relatively unknown. Some of the hypotheses and interpretations in this volume may also prove to be short-lived, particularly where the data bases are incomplete. Better information about the thickness distribution and drift of sea ice, the calving and attrition rate of icebergs, and the details of processes at shelf water boundaries could improve the heat, salt, and other budgets. Perhaps the work reported here will suggest other leads to be followed and stimulate synergistic collaborations.

Many anonymous reviewers made valuable contributions to the papers in this volume. The GEBCO 5.18 sheet was made available with the assistance of the Canadian Hydrographic Service. Copyediting was carried out under the direction of the staff of the American Geophysical Union. Camera-ready word processing was done by D. Criscione at Lamont-Doherty Geological Observatory, in part with support from the Department of Energy. Many of the field projects and laboratory studies were supported by the U.S. Antarctic Research Program of the National Science Foundation.

Stanley S. Jacobs

GEBCO BATHYMETRIC SHEET 5.18 (CIRCUM-ANTARCTIC)

J. R. Vanney

Universite Pierre et Marie Curie and Universite de Paris-Sorbonne, Paris, France

G. L. Johnson

Office of Naval Research, Arlington, Virginia 22217

Introduction

The large Antarctic map accompanying this volume represents the Antarctic chart of a global series of bathymetric charts that was completed in 1982. Sheet 5.18 is one of the 18 charts of the fifth edition of the General Bathymetric Chart of the Oceans (GEBCO). This series of bathymetric charts originated at the Seventh International Geological Congress in Berlin in 1899. The first edition became a reality when Prince Albert I of Monaco assembled a small group of scientists to begin work on the first edition in 1903. A project of 24 large sheets to cover the world on a scale of 1:10 million was developed and issued. A second edition was brought out by the scientific group between 1912 and 1927.

With the invention of continuous sounding, however, the flood of data became so great that the International Hydrographic Bureau (IHB), Monaco, was asked to take over the program. Between 1932 and 1955 the IHB produced the third edition which consisted of 18 sheets based on 1001 plotting sheets on a scale 1:1 million. In 1974, once again the workload became so heavy that the IHB, which had changed its name in the meantime to the International Hydrographic Organization (IHO), agreed to cooperate on this project with the Intergovernmental Oceanographic Commission (IOC) of UNESCO. Following a study by the Scientific Committee on Oceanic Research (SCOR), it was decided to develop an entirely new form of presentation for the series, to be known as the fifth edition. For the fifth edition the IHO was responsible, in conjunction with 19 volunteering Hydrographic Offices in its member states, for graphic advice and supervision of the final product. The IOC in conjunction with SCOR, the International Association for the Physical Sciences of the Ocean, and the Commission for Marine Geology accepted responsibility for all scientific input into the project, including contouring of the bathymetric data and compilation on the final artwork for each sheet. Cartographic production of the sheets was undertaken by the Canadian Hydrographic Service in Ottawa, Canada.

Sheets of the fifth edition of GEBCO can be obtained thoughout the world through agents for the sale of Canadian Nautical Charts-- $5.00 (Can.) plus handling charges, or from:

Hydrographic Chart Distribution Office
Department of Fisheries and the Environment
1675 Russell Road, P.O. Box 8080
Ottawa, Ontario, Canada, K1G 3H6

Sheet 5.18

This bathymetric chart is the result of a series of geomorphological studies initiated by the authors of the circum-Antarctic seas. All of the original charts were completed on mercator projection and published in the open literature progressing westward from the Antarctic Peninsula. Figure 1 shows the limit of each study and is keyed to the references.

In constructing Sheet 5.18, all available sounding data were used from all available sources as is indicated in the chart legend. As is readily noted, there are large areas with few or no data. Since publication of this chart, it has come to the attention of the authors that the Islas Orcadas Seamounts in the northern Weddell Sea are probably "phantoms" based on a single faulty echo sounding line. Likewise, the small seamounts just southeast of the Kerguelen Plateau (Figure 1) have recently been shown to be non-existent by Quilty et al. [1984]. It is likely that dense accumulations of biological scatterers were mistaken for the true bottom in these cases.

The Antarctic continental shelf as revealed by the GEBCO sheet exhibits typical high-latitude morphology. It has been deeply incised by glacial activity with both coast parallel and normal shelf troughs (dark on Figure 1). The Antarctic shelf is deep, 500-900 m, which probably is a reflection of depression by the thick inland ice sheet. A recent morphologic synthesis appears in Johnson et al.

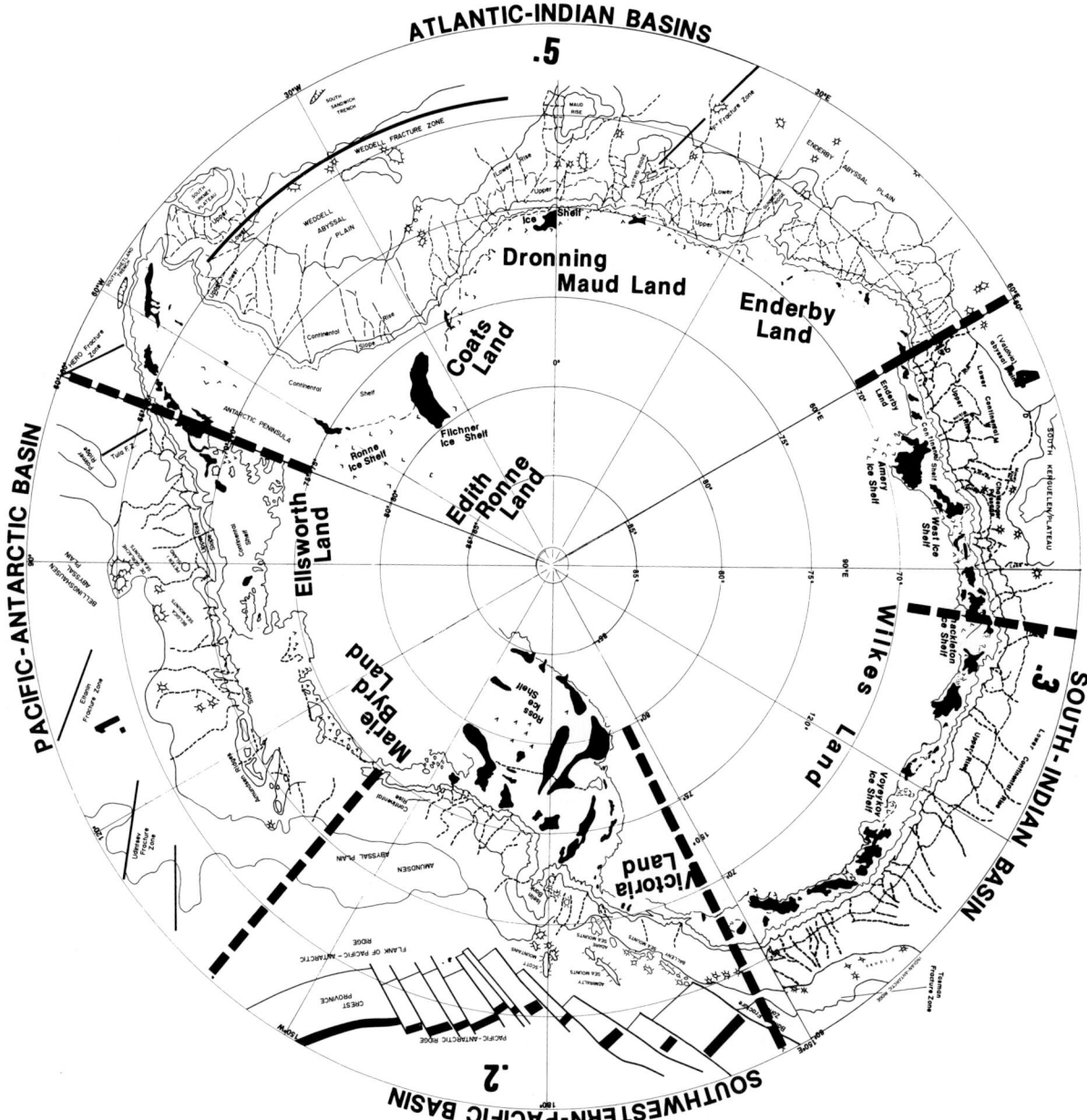

Fig. 1. Physiographic provinces of circum-Antarctic seas based on references: (1) Vanney and Johnson (1976), (2) Vanney et al. (1981), (3) Vanney and Johnson (1980), (4) Vanney and Johnson (1982), and (5) Johnson et al. (1981). Reference numbers and heavy dashed lines denote the regions covered by each study. Dark areas on continental shelves are troughs. Stippled region is continental shelf, thin dashed line denotes approximate edge of ice shelf.

[1982] and specific areas are discussed in detail in the references of Figure 1.

Acknowledgments. The authors are deeply indebted to Dave Monahan of the Canadian Hydrographic Service for his skilled assistance and counsel during all phases of construction of Sheet 5.18.

References

Johnson, G. L., J. R. Vanney, A. Elverhoi, and J. La Brecque, Morphology of the Weddell Sea and southwest Indian Ocean, Dtsch. Hydrogr. Z., 34, 263-272, 1981.

Johnson, G. L., J. R. Vanney, and D. Hayes, The Antarctic Continental Shelf in Antarctic Geoscience, edited by C. Craddock, pp. 995-1002, University of Wisconsin Press, Madison, 1982.

Quilty, P. G., R. J. Thwaites, and R. M. Burbury, Does Gribb Bank Exist?, Polar Rec., 138, 319-324, 1984.

Vanney, J. R., and G. L. Johnson, The Bellingshausen-Amundsen basins (southeastern Pacific): Major sea-floor units and problems, Mar. Geol., 22, 71-101, 1976.

Vanney, J. R., and G. L. Johnson, Wilkes Land continental margin physiography, East Antarctica, Polarforschung, 49, 20-29, 1980.

Vanney, J. R., and G. L. Johnson, Marine geomorphology of the Kerguelen-Antarctica Passage, in B. C. Heezen Memorial Volume, edited by R. Scruton and M. Talwani, pp. 237-254, John Wiley, New York, 1982.

Vanney, J. R., R. K. H. Falconer, and G. L. Johnson, Geomorphology of the Ross Sea and adjacent oceanic provinces, Mar. Geol., 41, 73-102, 1981.

(Received February 20, 1984; accepted April 27, 1984.)

CIRCULATION AND WATER MASSES ON THE SOUTHERN WEDDELL SEA SHELF

A. Foldvik, T. Gammelsrød, and T. Tørresen

Geophysical Institute, University of Bergen, Bergen, Norway

Abstract. Circulation and water masses on the southern Weddell Sea shelf are discussed, based upon observations from three summer expeditions. The circulation is dominated by two cyclonic gyres, one in the Filchner Depression and one off Ronne Ice Shelf. In both areas a relatively warm (T ~ -1.3°C) southward flow of Modified Weddell Deep Water and a cold (T ≤ -1.9°C) northward flow of Ice Shelf Water are observed. Ice Shelf Water spills over the sill of the Filchner Depression and is observed on the continental slope as a narrow bottom-trapped current. Based on current meter observations at the sill, the overflow is estimated to be 10^6 m^3s^{-1}, with no appreciable seasonal variation. Weddell Sea Bottom Water forms by mixing between Ice Shelf Water and Weddell Deep Water on the slope.

Introduction

In this paper we present some results from three expeditions to the southern Weddell Sea with the Norwegian icebreaker Polarsirkel, (Figure 1): the Norwegian Antarctic Research Expeditions of 1977 and 1979, and the Federal Republic of Germany Expedition of 1980. Detailed discussions of the oceanographic observations from these cruises are given by Foldvik et al. [1985a, b] and Foldvik et al. [1985], hereafter referred to as Pol-77, Pol-79, and Pol-80. The data are available through the Norwegian Oceanographic Data Center, Bergen.

The Weddell Sea has been regarded as the primary source of Antarctic Bottom Water ever since the pioneering work of Brennecke [1921], Mosby [1934], and Deacon [1937]. This bottom water has been shown to spread northwards into the Atlantic, Pacific, and Indian oceans [Deacon, 1937; Wüst, 1938; Reid and Lynn, 1971]. It is generally accepted that bottom water formation involves the cold, saline water masses formed on the wide Antarctic continental shelves, particularly in the southern Weddell Sea. It is therefore essential to understand the processes of water mass formation and circulation on the shelf.

General Circulation and Water Masses

The circulation of the deep Weddell Sea is dominated by the cyclonic Weddell Gyre [Carmack and Foster, 1975a, 1977; Deacon, 1979; Gordon et al., 1981]. The cyclonic circulation is believed to extend to the bottom, but the mean speed in the abyss is rather weak, ~1 cm s^{-1}, below 4000 m [Foster and Middleton, 1979]. In the southeastern part, the cold and fresh Antarctic coastal current flows along the shelf break towards the west [Deacon, 1937; Gill, 1973] until about 27°W where it splits, one branch continuing along the shelf break and another flowing southwards on the eastern side of the Filchner Depression [Carmack and Foster, 1977]. Typical mean velocities near bottom at the shelf break near 40°W are ~7 cm s^{-1} along the isobaths [Foldvik et al., this volume].

The different water masses encountered in the Weddell Sea are shown schematically in a θ-S diagram in Figure 2. Near the bottom in the deep ocean we find cold (θ ≤ -0.7°C) Weddell Sea Bottom Water (WSBW), which can be distinguished from the slightly warmer Antarctic Bottom Water (AABW) by a break in the θ-S curve [Carmack and Foster, 1975a]. Above the bottom water resides the Weddell Deep Water (WDW) which is characterized by potential temperatures between 0.0°C and 0.8°C and salinities from 34.64 to 34.72 [Foster and Carmack, 1976b; Gordon, 1982]. Overlying the WDW we find Winter Water (WW) with temperatures near freezing and salinities in the range 34.36 to 34.52 [Foster and Carmack, 1976a], but modified in summer by heating and melting. The mixing of warm WDW and WW produces the Modified Weddell Deep Water (MWDW), termed Modified Warm Deep Water by Foster and Carmack [1976a].

On the continental shelf additional water masses appear such as the Eastern Shelf Water (ESW) which is also referred to as Low Salinity Shelf Water [Carmack, 1977; Jacobs et al., this volume]. The coldest water mass is the Ice Shelf Water (ISW), defined as water with potential temperature below the surface freezing point. The most saline and dense

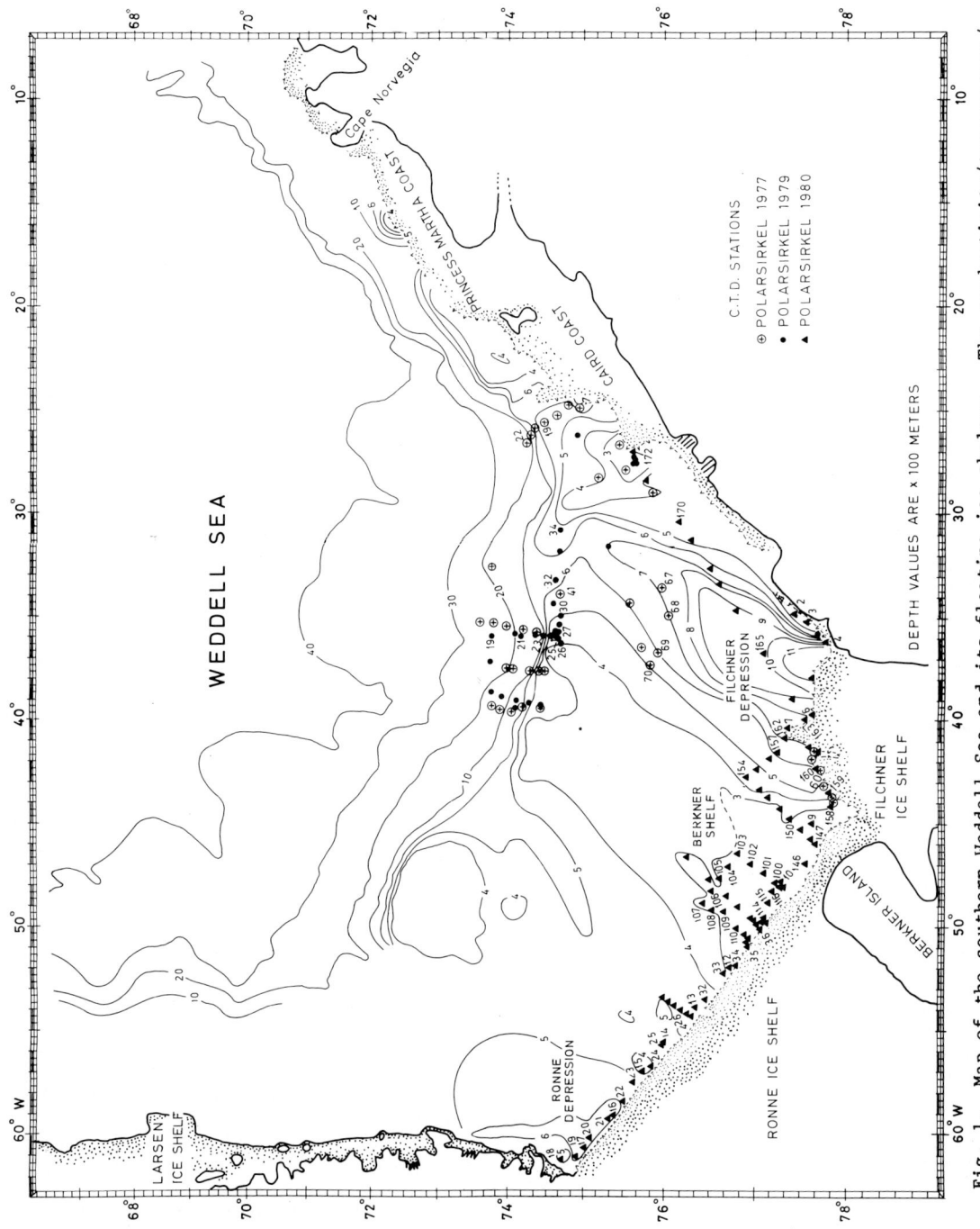

Fig. 1. Map of the southern Weddell Sea and its floating ice shelves. The conductivity/temperature/depth (CTD) stations obtained from Polarsirkel in 1977, 1979, and 1980 are marked with different symbols, and selected station numbers are indicated.

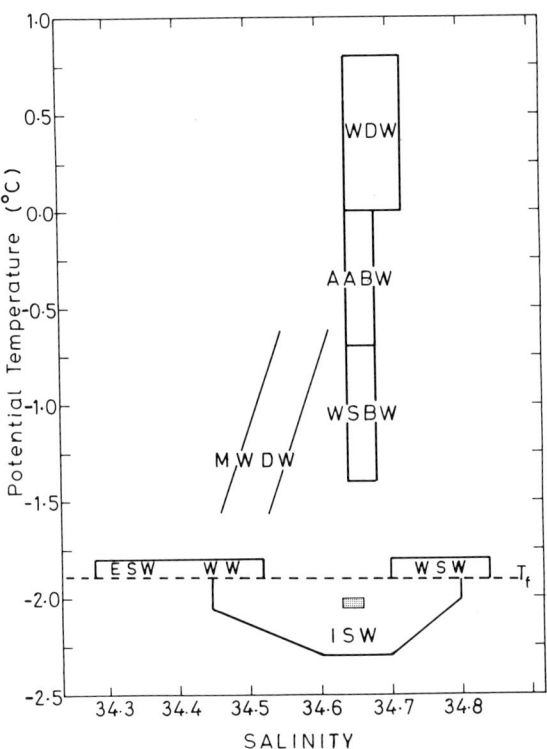

Fig. 2. Schematic representation of potential temperature (θ) versus salinity (S) for the major Weddell Sea water masses: Weddell Deep Water (WDW), Antarctic Bottom Water (AABW), Weddell Sea Bottom Water (WSBW), Eastern Shelf Water (ESW), Winter Water (WW), Ice Shelf Water (ISW), Western Shelf Water (WSW) and Modified Weddell Deep Water (MWDW). The figure is based on Carmack and Foster [1975a, b], Foster and Carmack [1976a,b], Carmack and Foster [1977], Gordon [1982], Pol-77, Pol-79, and Pol-80. The broken line represents the surface freezing point. The shaded rectangle within the ISW sector indicates thermohaline characteristics of the overflow at the sill of the Filchner Depression.

water mass is the Western Shelf Water (WSW) or High Salinity Shelf Water [Carmack, 1977; Jacobs et al., this volume] with temperatures near freezing and salinities above 34.70. The latter two water masses appear to be essential for the formation of bottom water.

Water Masses on the Shelf

Western Shelf Water

Western Shelf Water (WSW) is formed by brine rejection during ice freezing, most efficiently on the shallow shelves in the southern Weddell Sea, [Mosby, 1934; Foster, 1972]. The systematic transport of newly formed ice away from the barrier (seaward extent of the ice shelf) due to prevailing offshore winds makes this area especially favorable for freezing [Gill, 1973]. This process is further enhanced by periodic ice divergence due to strong tides [Gammelsrød and Slotsvik, 1981].

In a hydrographic section taken along the Filchner and Ronne Ice Shelves (Figure 3) WSW appears as a local salinity and density maximum on the shallow shelf north of Berkner Island, hereafter referred to as the Berkner Shelf. The most saline water, $S \geq 34.70$, is found in the deep part of the shelf near the Antarctic Peninsula, which will be referred to as the Ronne Depression. This water is slightly supercooled with respect to surface pressure (i.e. potentially supercooled) and has thus been modified by circulation under the ice shelf. Because of its high salinity we treat it here as WSW. Similarly, we observe a lens of potentially supercooled saline water ($S \geq 34.70$) near the bottom on the eastern slope of the Filchner Depression (Figures 3 and 4).

Sections taken normal to the ice shelf show that maximum salinities need not be located at the barrier in summer. In the "box" shown folded out in Figure 5 (see Figure 1 for location) it can be observed that the salinity of WSW increases with distance from the ice shelf on the eastern leg (stations 97-103).

An E-W transect further north on the western slope of the Filchner Depression (Figure 6) shows saline ($S \geq 34.70$) water in a 100-m thick layer near the bottom. On the sill of the Filchner Depression WSW is not observed (see Figure 7). This is an indication that the WSW is recirculated and trapped in the depression (see discussion of Circulation of WSW), as indicated by Carmack and Foster [1975b].

Ice Shelf Water

Ice Shelf Water is cooled to the in situ freezing point at the underside of the floating ice shelves [Sverdrup, 1940; Lusquinos, 1963]. The freezing point decreases with depth (about 0.075°C/100 m) [Millero, 1978] and temperatures down to -2.3°C have been observed in the Filchner Depression where the maximum draught of the ice shelf is as much as 400 m at the barrier.

Most ISW is probably formed from WSW which is the most dense water on the continental shelf and therefore may flow underneath the ice shelf [Carmack and Foster, 1975b]. The cooling at the underside of the ice shelf is associated with net melting of glacier ice

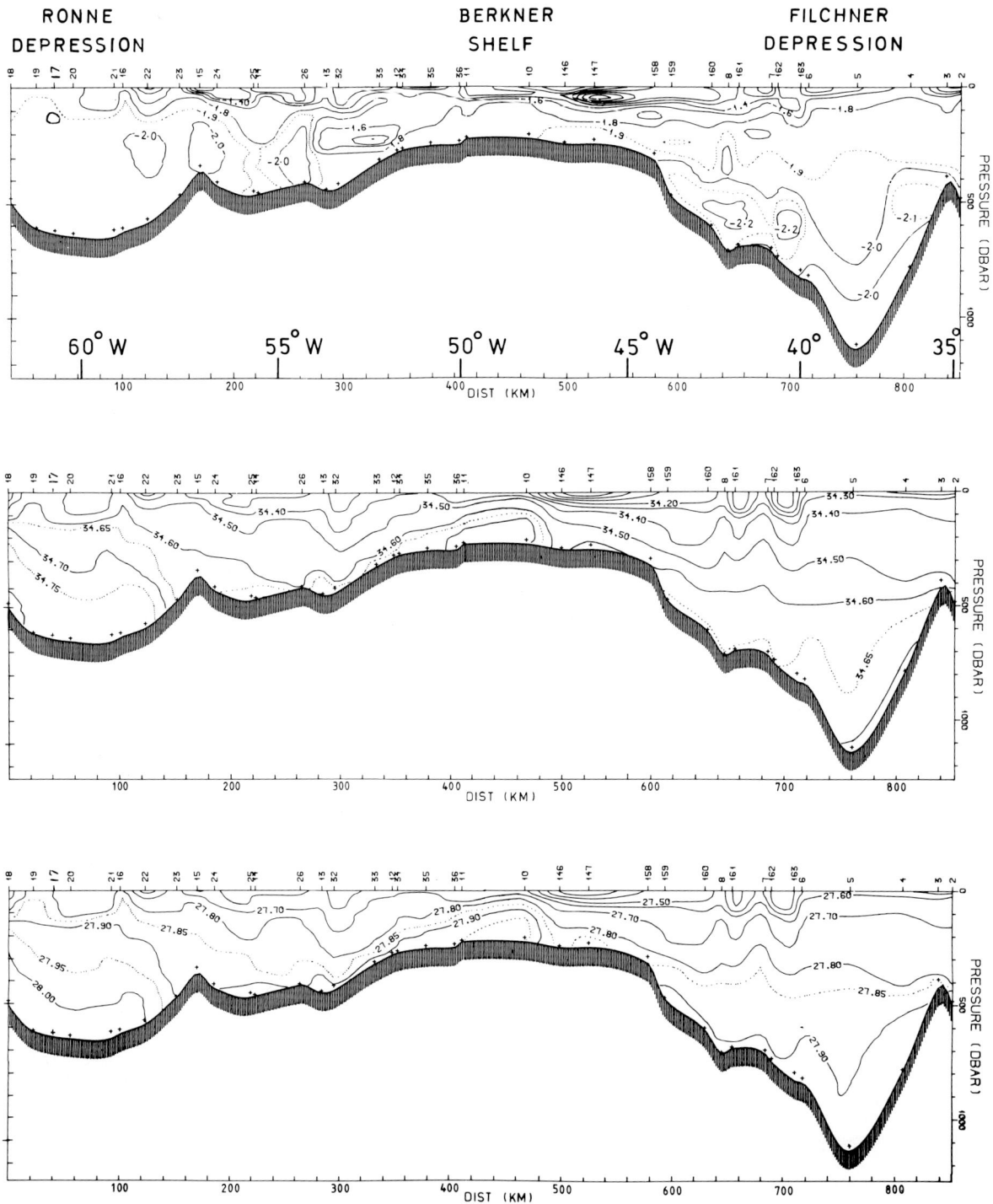

Fig. 3. Section from Polarsirkel 1980 stations along the barrier in Figure 1. Crosses denote deepest observations. (Top) Potential temperature (°C), (Middle) salinity and (Bottom) potential density.

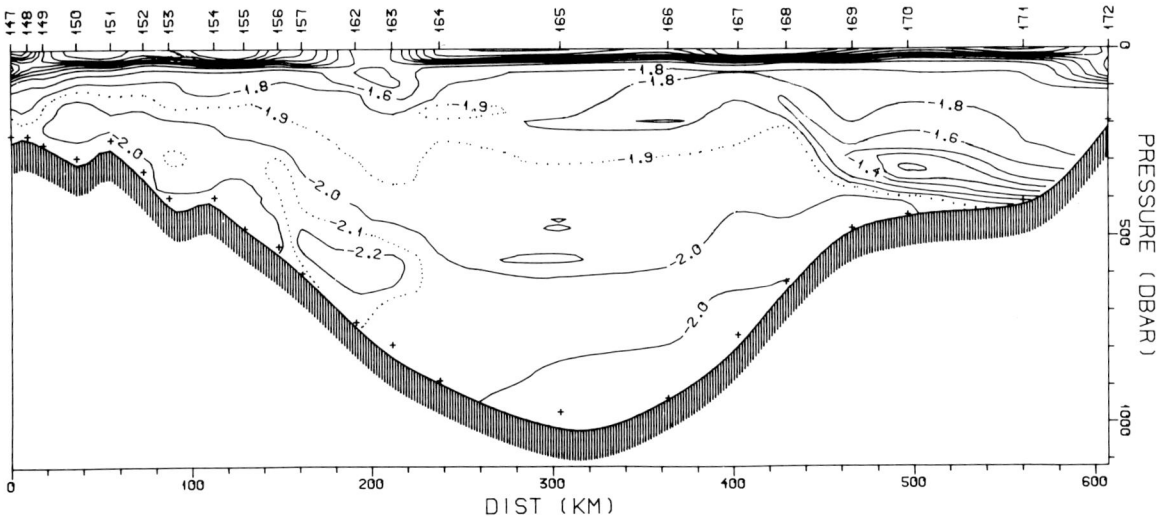

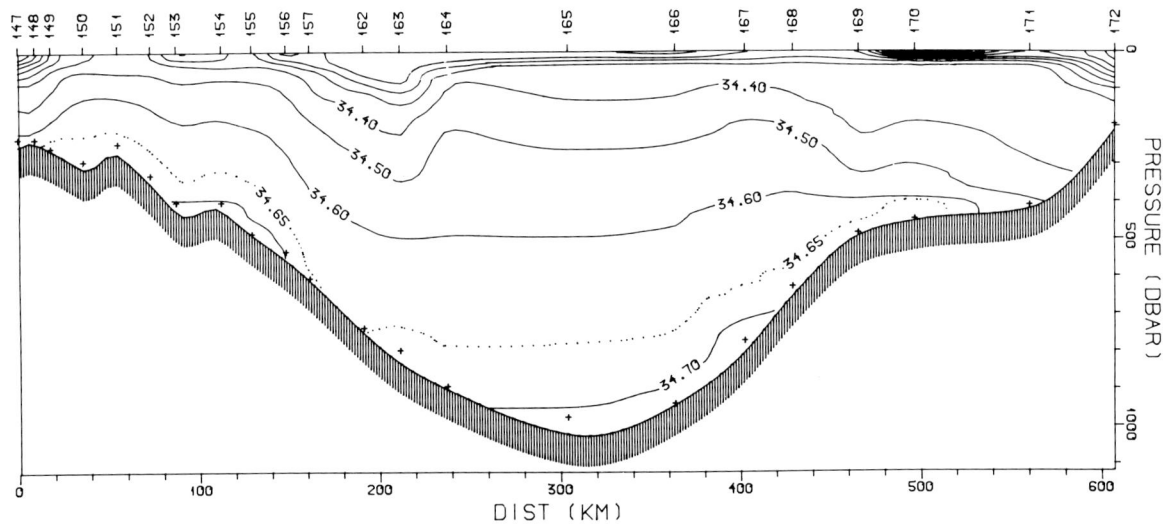

Fig. 4. Zig-zag Pol-80 section with one leg northeast from Berkner Island (stations 147-154), one leg crossing the western slope of the Filchner Depression (stations 154-163) and one leg from the Filchner Ice Shelf northeast across the Filchner Depression (stations 163-172). See Figure 1 for location. (Top) Potential temperature (°C) and (Bottom) salinity.

over large areas [Robin et al., 1983]. ISW is therefore less saline and less dense than WSW. Intermediate stages exist between the ISW and WSW end members depending on the residence time of WSW under the ice shelf.

The draught of the Ronne Ice Shelf is around 200 m at the barrier [Robin et al., 1983]. Below this depth there is a temperature minimum layer in the Ronne Depression with four separate cores of ISW with potential temperature below -2.0°C (Figure 3). Two of these cores were observed on both traverses along the barrier. Similar cores were observed from Glacier in 1968 (Figure 12) and from Polarstern in 1984 [Rohardt, 1984] indicating that these are relatively permanent features. On the western slope of the Berkner Shelf the ISW is missing. It appears again near the bottom at station 146 on the Berkner Shelf (Figure 3), while further offshore it

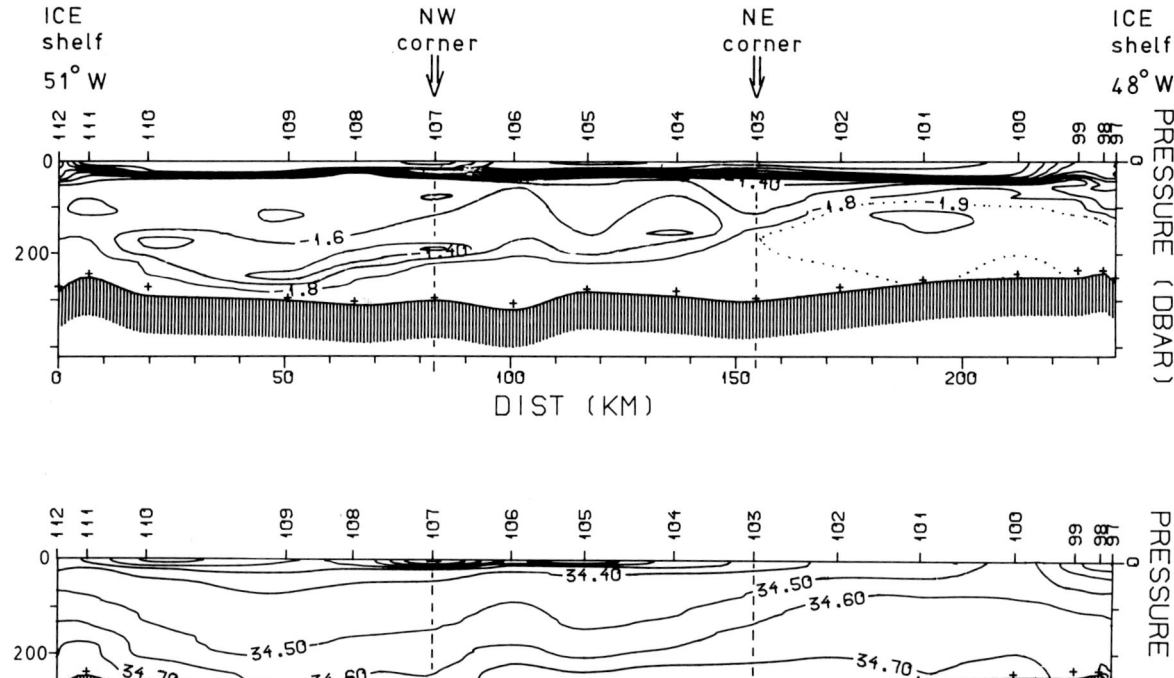

Fig. 5. Pol-80 section on the Berkner Shelf forming a "box" with the ice shelf as one of the sides. Stations 97-103 and 107-112 are the eastern and western sides respectively, and stations 103-107 are parallel to the barrier. (Figure 1). The section is shown folded out with the NW and NE corners indicated. (Top) Potential temperature (°C) and (Bottom) salinity.

seems to override the WSW (stations 100-103, Figure 5).

In the Filchner Depression (Figures 3 and 4) temperature minima occur at 500 to 600 m depth at both slopes, with the most extreme temperatures ($\theta \leq -2.2°C$) at the western slope [see also Carmack and Foster 1975b]. Further north on the western slope, the core of ISW is found some 150-200 m above bottom (Figure 6), but at the sill (Figure 7) minimum temperatures occur near the bottom. Some of the ISW crosses the sill and in the section normal to the shelf break at 36°W (Figure 8) ISW is found near bottom at 1500-m depth. In transects parallel to and west of the 36°W section (Figure 1), ISW was not observed, probably because these sections did not extend to sufficient depths. However, Foster and Carmack [1976a] observed newly formed bottom water at 2000 to 3500 m depth on the slope near 40°W. We believe that to be the continuation of the cold flow of ISW.

The Influence of Weddell Deep Water on the Shelf

Mixing between Weddell Deep Water and the overlying Winter Water, producing Modified Weddell Deep Water (MWDW), becomes especially effective at the shelf break, where the MWDW shows a tendency to invade the shelf at preferred locations. Foster and Carmack [1976a] noted this on their 40°W section and we observed evidence of the inflow in the Pol-77 and Pol-79 sections (Figure 8). The section along the shelf break across the sill of the Filchner Depression also shows MWDW at about 400-m depth on the eastern slope at 32°W (Figure 7).

Figure 9 shows the distribution of MWDW on the shelf. This map was obtained by examining all vertical temperature profiles and extracting any distinct deep temperature maximum. Our data base (Pol-77, Pol-79, Pol-80) was supple-

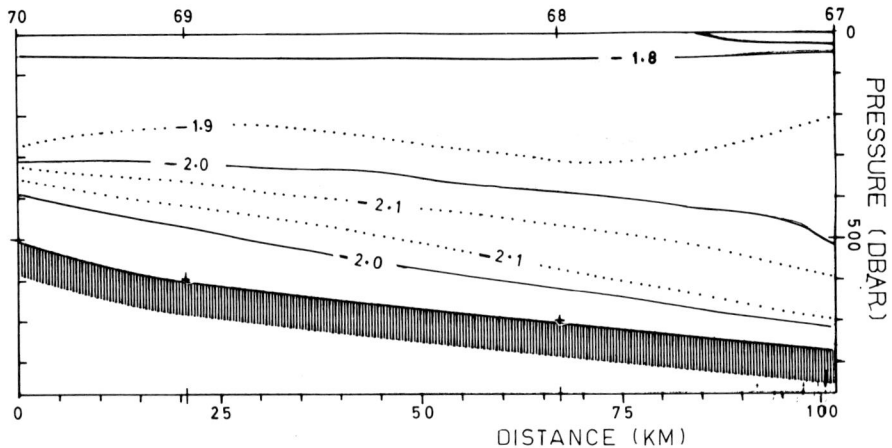

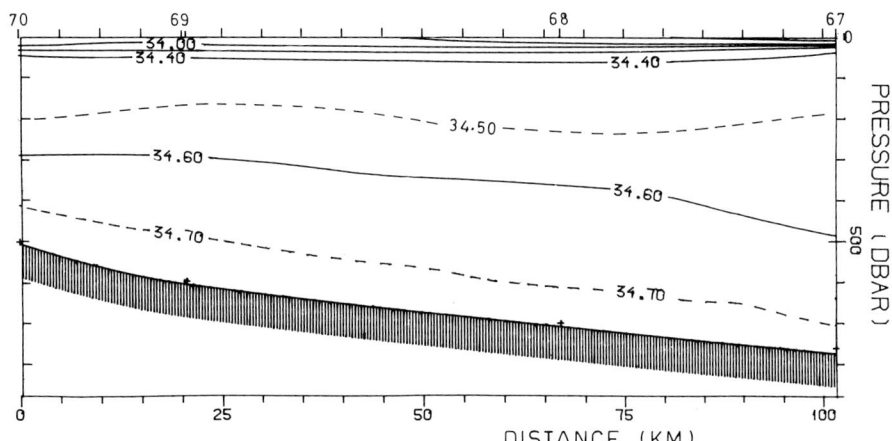

Fig. 6. Pol-77 section across the western slope of the Filchner Depression at 76°S (Figure 1). (Top) Potential temperature (°C) and (Bottom) salinity.

mented with selected stations from other expeditions to cover areas not occupied by us. Nearby stations from different years are sometimes incompatible and in such cases the highest value was chosen. In Figure 9 the -1.8°C isotherm indicates the maximum extent of MWDW while the -1.3° isotherm represents its core on the shelf.

Figure 9 demonstrates that the MWDW, which forms the deep part of the coastal current, divides at about 30°W, with one branch following the 400-m isobath on the eastern side of the Filchner Depression. This branch is readily seen in the section across the Filchner Depression around station 170 (Figure 4). It can be tracked south to the Filchner Ice Shelf where it was found in 1979, but not in 1980 nor in 1969, which indicates that this current is intermittent.

Similarly, there is a western branch, probably steered by the 400-m isobath west of the Berkner Shelf, indicating a dividing of the coastal current at about 40°W. This western branch is observed to touch the northwest corner of the "box" (station 107, Figure 5) as a warm ($\theta \sim -1.1°C$) layer around 200-m depth. The core of the MWDW appears to flow under the ice shelf at station 33 (Figure 3).

Circulation on the Shelf

Circulation of Western Shelf Water

Since WSW is the most dense water mass on the shelf, the horizontal distribution of bottom salinities (Figure 10) is useful for a

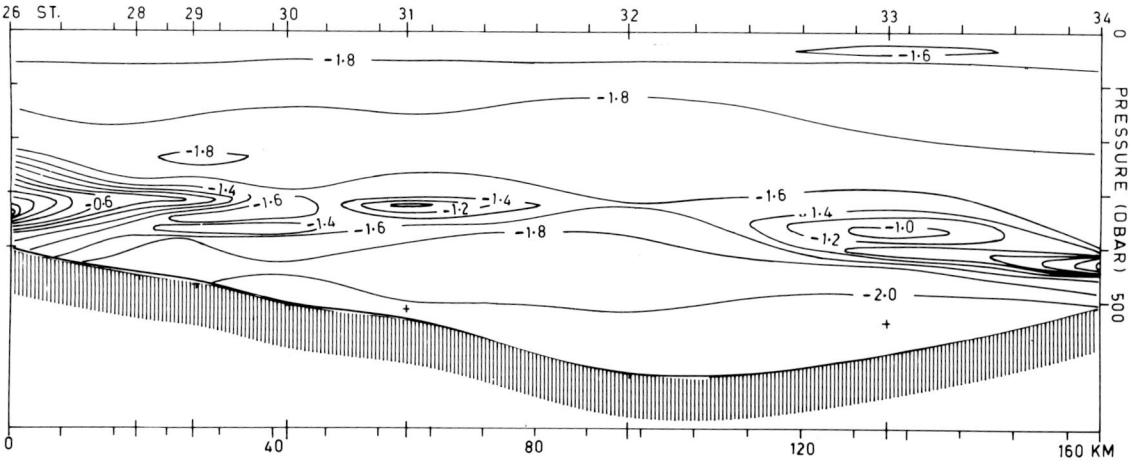

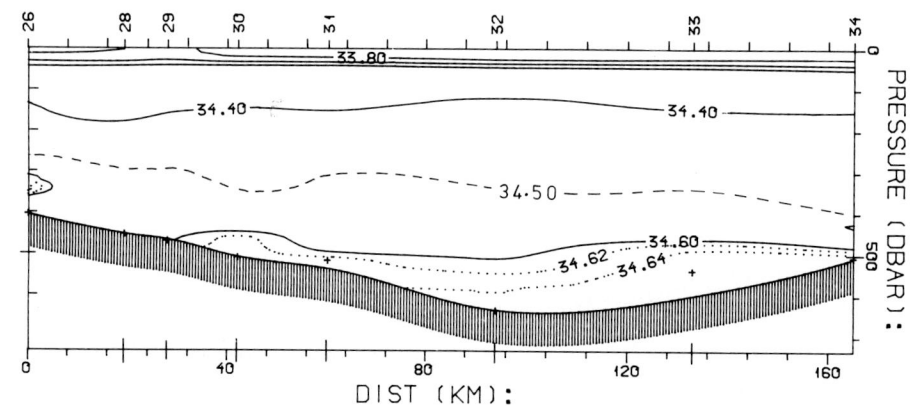

Fig. 7. Pol-79 section along the sill of the Filchner Depression from west to east (Figure 1). (Top) Potential temperature (°C) and (Bottom) salinity.

discussion of the WSW flow. Again our data were supplemented with results from other expeditions, and when nearby stations did not compare the highest value was selected.

Figure 10 shows that on the Berkner Shelf maximum salinities occur some distance north of the barrier. It was argued above that WSW production is most efficient close to the Barrier. However, since all our observations are taken in the summer season when WSW production has ceased, weak offshore currents may explain the observed salinity distribution. To our knowledge, the only current measurement from the area is a short series obtained in 1980 (Pol-80), which indicated a mean current of ~1 cm/s, but the direction was onshore rather than offshore. This series was much too short (4 days), however, to draw any conclusion about the mean current. The same series revealed a rather strong tidal current (~20 cm/s). Tidal mixing utilizing the fresh water supply underneath the shelf could also explain the offshore salinity maximum.

From the Berkner Shelf the WSW seems to flow into the deep Filchner Depression (Figure 10) taking part in a clockwise circulation there [Carmack and Foster, 1975b]. It would thus be moving south where observed on the eastern slope (Figures 3 and 4). Figures 3 and 10 also indicate that some of the WSW spreads westward from the Berkner Shelf along the barrier.

Below about 300-m depth in the Ronne Depression the water is very saline (Figure 3). Since the temperature is slightly below the surface freezing point, this water has been modified by the ice shelf. The most dense water is found on the western slope (Figure 3, Bottom), indicating a deep flow away from the ice. However, since the topography to the

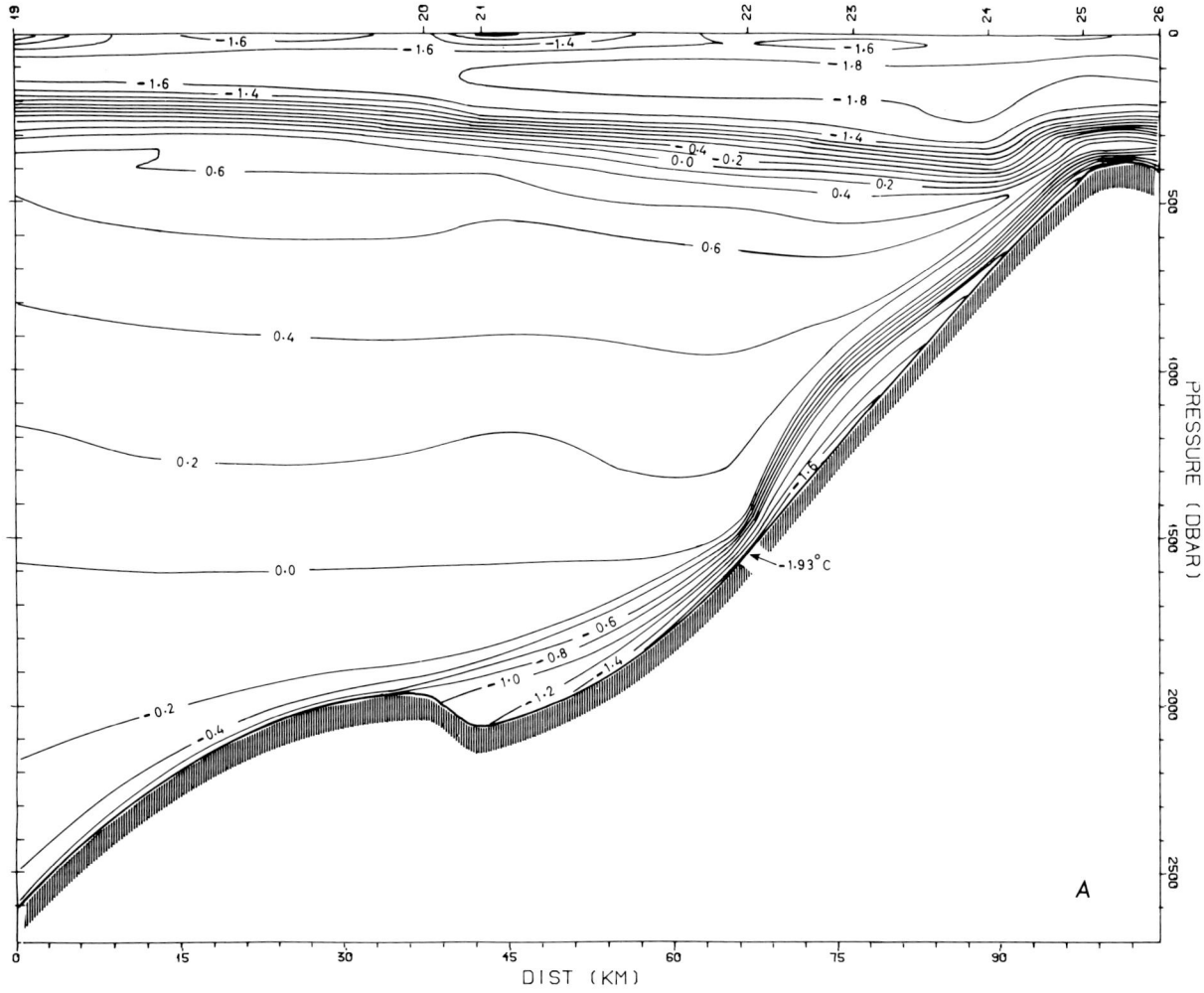

Fig. 8. Pol-79 section normal to the shelf break at 36°W (Figure 1). (a) Potential temperature (°C) and (b) salinity.

north of Ronne Depression is not known we cannot infer how much WSW is actually leaving the region.

Circulation of Ice Shelf Water

A major outflow of ISW from beneath the Filchner Ice Shelf is related to the extreme temperatures ($\theta \sim -2.2°C$) observed on the western slope of the Filchner Depression (Figures 3 and 4). This ISW can be traced as it follows the isobaths of the western slope of the depression (Figure 6) until it flows over the sill (Figure 7). Current measurements from the sill (Figure 11) show that this flow is significant throughout the year (Pol-79).

On its way down the slope towards the deep ocean, ISW is deflected westward because of the rotation of the earth, and passes 36°W at about 1500-m depth (Figure 8). This is also demonstrated in Figure 12 where we have plotted the ISW core (minimum temperature) in the same manner as on the MWDW map (Figure 9). The -1.9°C isotherm defines the maximum extent of ISW. In addition we have plotted the -1°C isotherm near bottom north of the shelf break to demonstrate the elongation of ISW flow as a cold tongue towards the deep ocean.

Some ISW is recirculated in the Filchner Depression [Carmack and Foster, 1975b]. This return flow of ISW is slightly warmer and about 100 m shallower than the outflow, (see Figures 3 and 4).

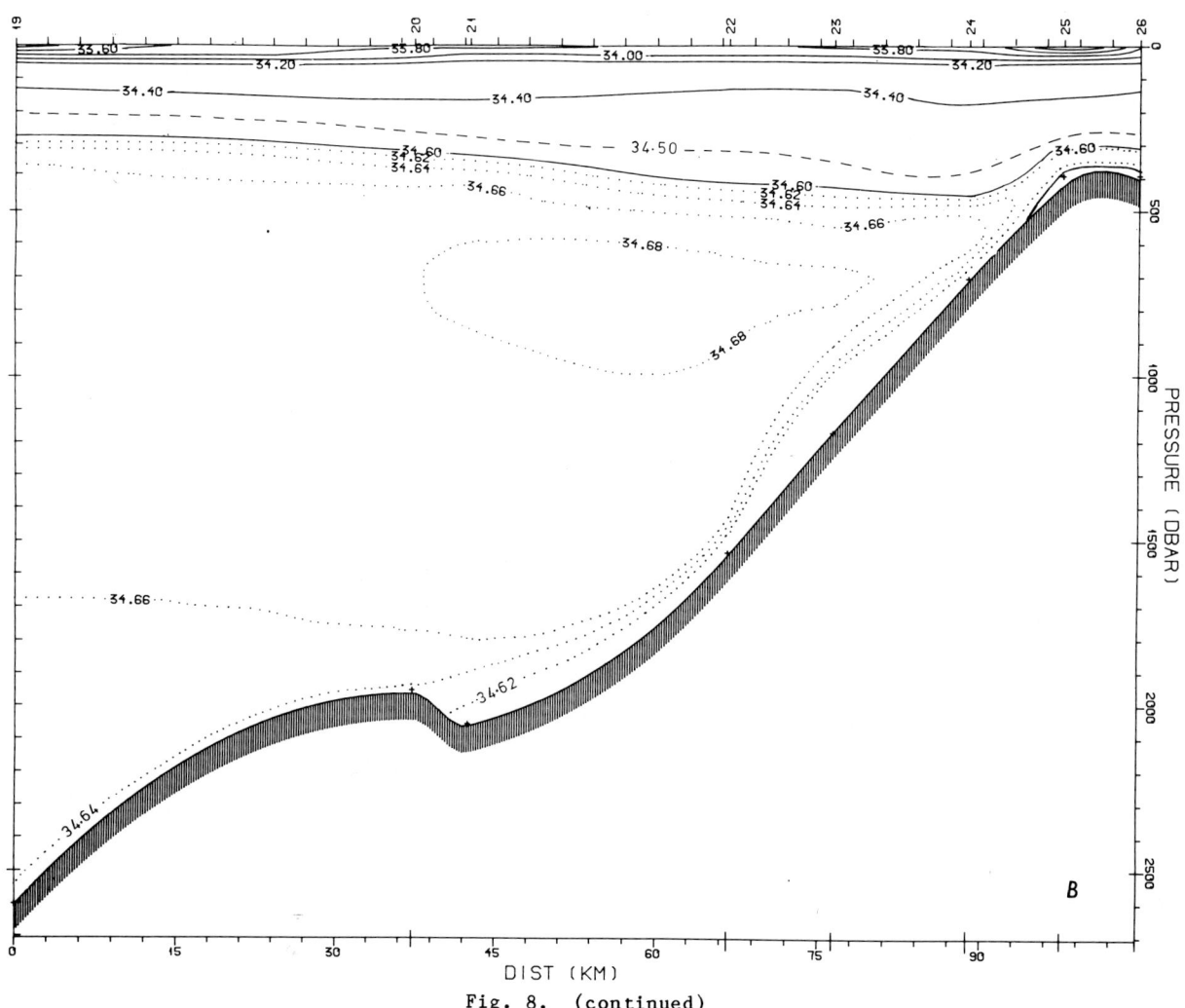

Fig. 8. (continued)

On the western slope of the Berkner Shelf, ISW is missing because of the intruding MWDW. West of about 54°W, ISW is again present (see Figures 3 and 12), indicating an offshore current as also noted by Seabrooke et al. [1971]. Since the two easternmost of the four cold cores are relatively fresh (S ~ 34.60) these are probably from a different origin than the western cores where the salinity is higher (S ~ 34.70).

Circulation Beneath the Floating Ice Shelves

Filchner Ice Shelf

The major outflow of ISW on the western slope of the Filchner Depression also appears in the bottom salinity map (Figure 10) as a local minimum around station 162. However, the ISW higher up on the shelf is fresher, indicating that there are branches of ISW of somewhat different origin. The draught of the eastern part of the Filchner Ice Shelf is between 300 and 400 m [Robin et al., 1983], which means that even some of the ESW there (Figure 3) may move under the ice shelf. The ice draught increases southwards by some 100 m/50 km [Robin et al., 1983]. Therefore, after a relatively short journey south, ESW would be deflected and forced to recirculate as the fresh ISW observed at the western slope of the Filchner Depression. The underlying more saline ISW/WSW may penetrate much farther south under the ice to form the coldest ISW. Vertical temperature and salinity profiles at station 60 (Pol-77; Figure 13) provide an example of two different cores of ISW in the same water column. The upper core may be

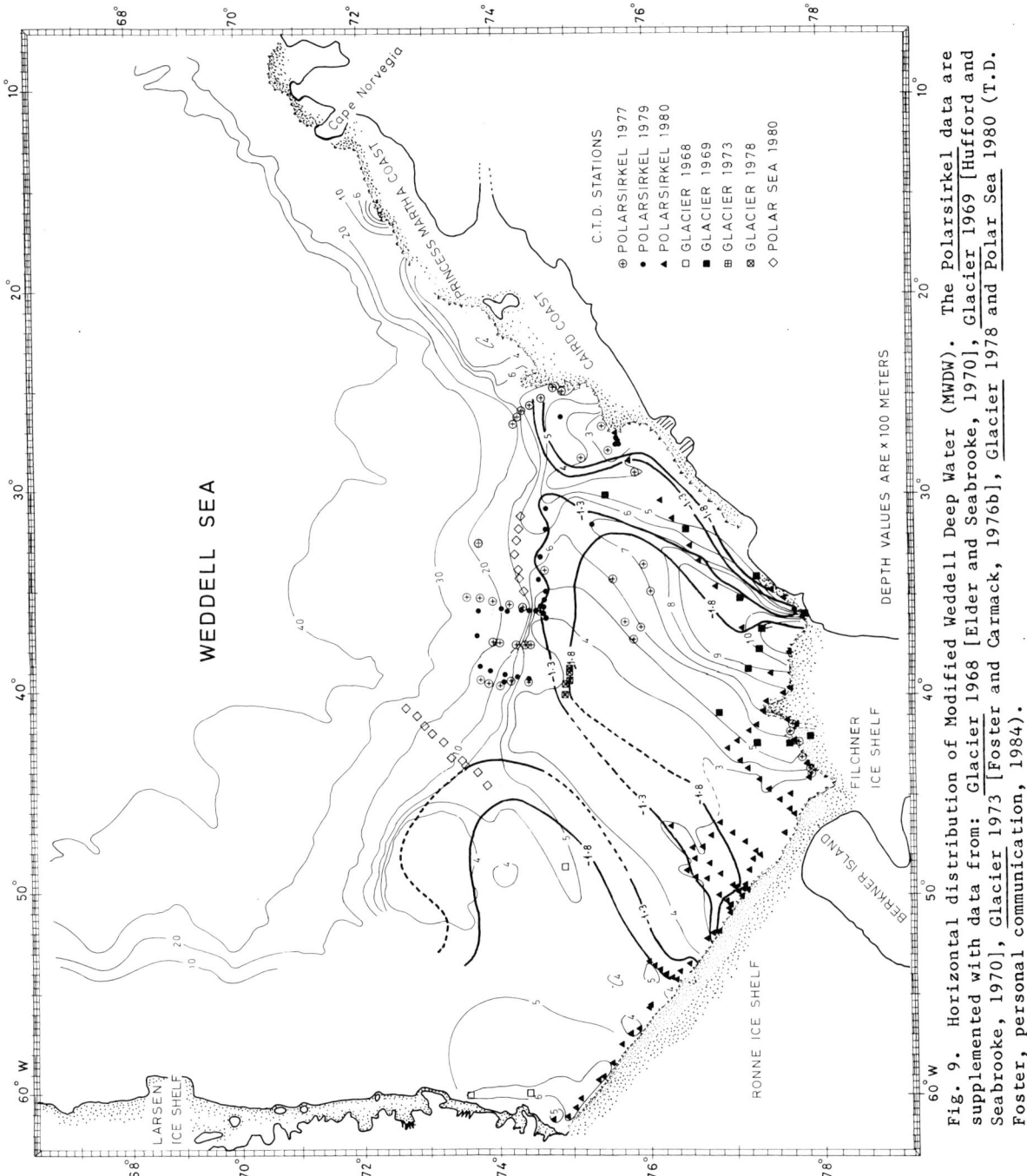

Fig. 9. Horizontal distribution of Modified Weddell Deep Water (MWDW). The Polarsirkel data are supplemented with data from: Glacier 1968 [Elder and Seabrooke, 1970], Glacier 1969 [Hufford and Seabrooke, 1970], Glacier 1973 [Foster and Carmack, 1976b], Glacier 1978 and Polar Sea 1980 (T.D. Foster, personal communication, 1984).

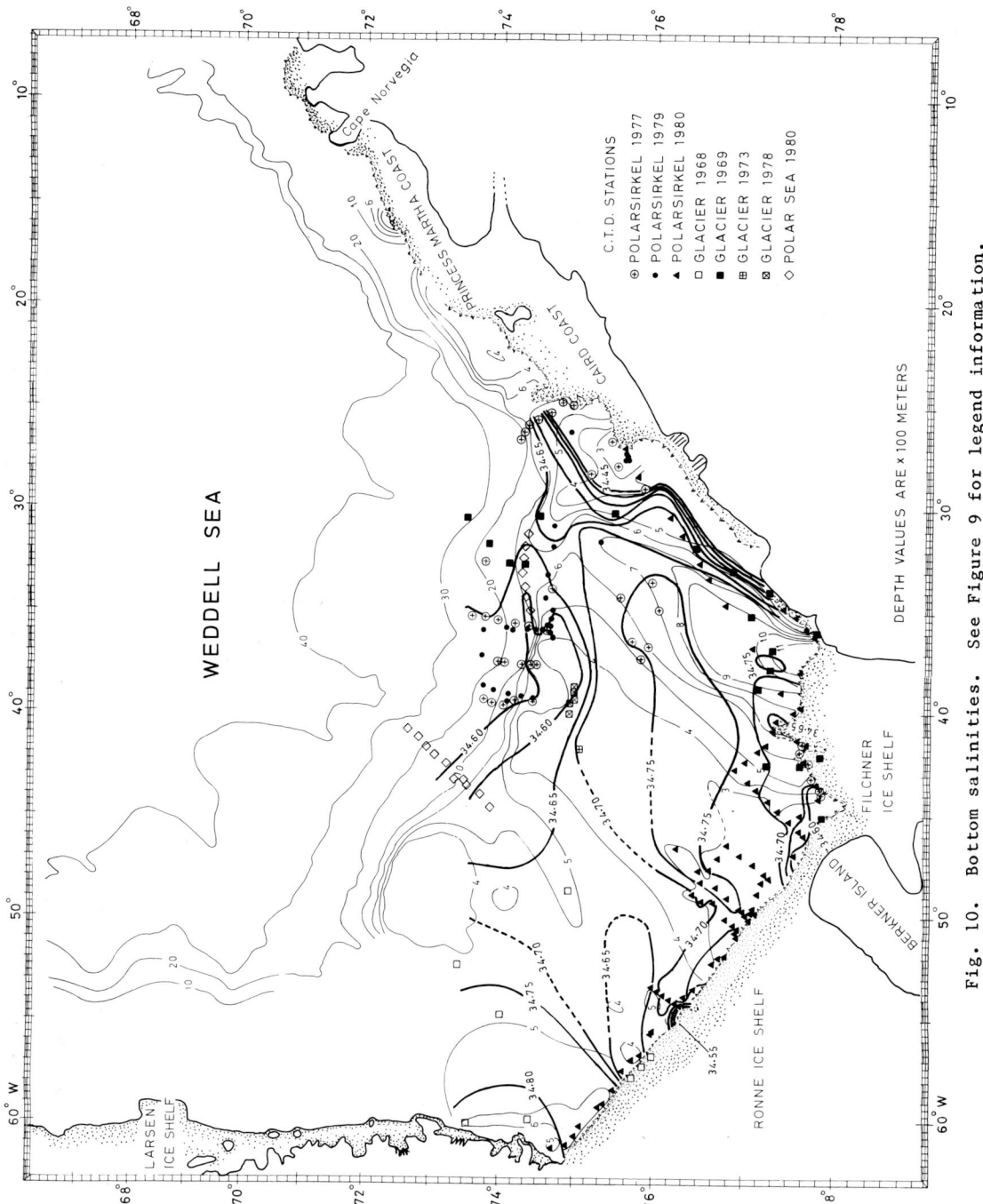

Fig. 10. Bottom salinities. See Figure 9 for legend information.

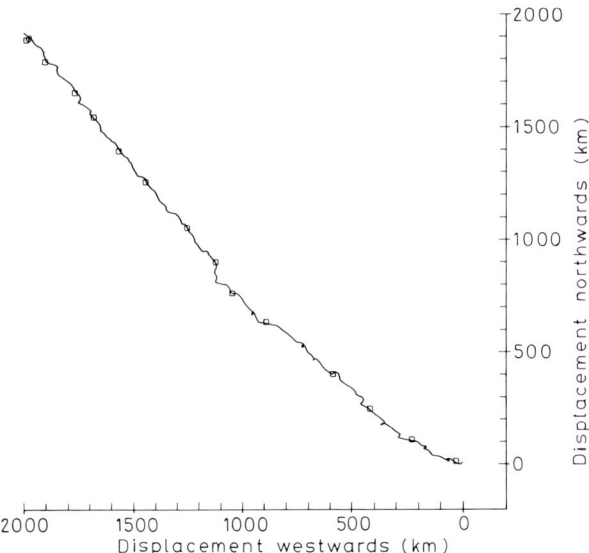

Fig. 11. Progressive vector diagram showing the flow of Ice Shelf Water 25 m above the bottom near station 41 (1977) from January 1977 to March 1978. The average temperature was -2.05°C, and the average speed was 8.1 cm/s. The beginning of each month is indicated with a square, starting with February 1, 1977, in the lower right corner.

should then occur at the western side of the Ronne Depression, whereas we observe three cold ISW cores on the eastern side (Figure 3). The two easternmost and freshest (S ~ 34.60) cores may be formed from MWDW flowing underneath the ice where it is thinnest [Robin et al., 1983], following the ice isopleths in a small cyclonic gyre before emerging again.

Discussion and Conclusions

Based on observations from three recent summer expeditions we have discussed the circulation on the shelf in the southern Weddell Sea. Cyclonic (clockwise) flow patterns in the Filchner Depression and under the floating Filchner Ice Shelf [Carmack and Foster, 1975b; Robin et al., 1983] agree qualitatively with our interpretation of CTD observations. A branch of Modified Weddell Deep Water enters the region at 30°W flowing south towards the southeastern part of the Filchner Depression. Ice Shelf Water appears at the Filchner Ice Shelf Barrier and follows the western slope of the Filchner Depression as a concentrated flow towards the north. Most of this ISW has been formed from WSW under the ice shelf by cooling and adding glacier meltwater.

Some ISW flows over the sill of the Filchner Depression. Due to its density and extremely low temperature (i.e. high compressibility) the ISW flows down-slope towards the deep ocean basin as a narrow bottom-trapped current (Figure 8). From Figure 7 the area occupied by ISW on the shelf break is roughly estimated at $10^7 m^2$. If this water leaves the region with the mean observed speed ~10 cm s^{-1} (Figure 11) then the ISW overflow becomes of the order 10^6 m^3 s^{-1}. The temperature/salinity characteristics of the overflowing ISW delineated in Figure 2 illustrate that Weddell Sea Bottom Water may readily be formed by mixing ISW with the overlying WDW at the slope. If ISW mixes with core WDW at + 0.5°C, then approximately equal parts of these water types are needed to produce WSBW at -0.8°C, yielding 2×10^6 m^3 s^{-1} of bottom water. However, as seen from the slope section (Figure 8), most of the mixing takes place below the core of WDW, i.e. at lower temperatures, and thus the net bottom water production may be considerably higher than the above estimate. Our estimate of WSBW production due to mixing of ISW and WDW is comparable to the 2-5 x 10^6 m^3 s^{-1} estimated by Carmack and Foster [1975a].

On the relatively shallow (~300 m) Berkner Shelf, high-salinity WSW (S ≥ 34.7) forms due to freezing of seawater in the winter season. Presumably this process is most effective close to the barrier, and the WSW product ap-

formed by cooling of ESW and the lower core by cooling of WSW [see also Jacobs et al., this volume]. The observed in situ supercooling in Figure 13 may be due to subsequent upward displacement of the water column [Foldvik and Kvinge, 1977].

This picture of cyclonic circulation cells beneath the Filchner Ice Shelf agrees qualitatively with the circulation scheme proposed by Robin et al. [1983]. They interpreted variations in the strength of radio echo signals in terms of basal melting and freezing, and related freezing to flow toward regions of thinner ice.

Ronne Ice Shelf

The high salinities observed in the Ronne Depression may be produced locally by freezing and subsequent trapping by a sill to the north, presumably at 400 to 500-m depth. Another possible explanation is that this water originated at the Berkner Shelf where maximum salinities are expected during the freezing season, and has circulated in a large cyclonic gyre underneath the Ronne Ice Shelf [see Robin et al., 1983]. However, the coldest ISW

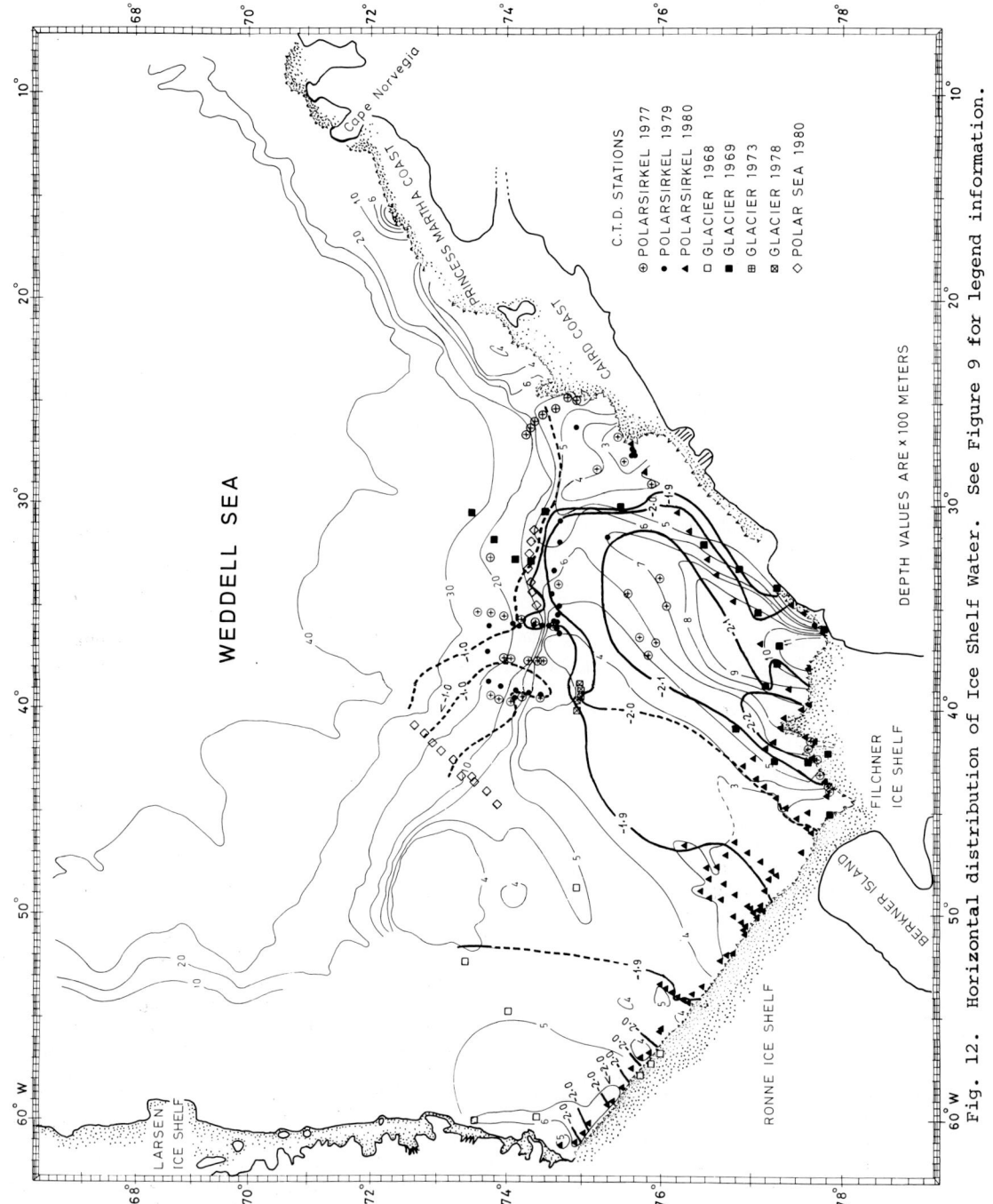

Fig. 12. Horizontal distribution of Ice Shelf Water. See Figure 9 for legend information.

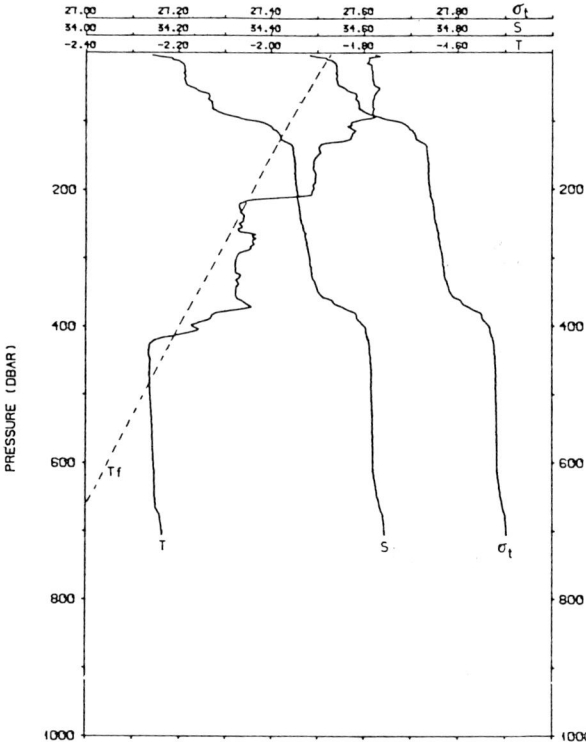

Fig. 13. Vertical profiles of temperature, salinity, and σ_t in the Filchner Depression at Pol-77 station 60. The broken line represents the in situ freezing temperature [Millero, 1978].

pears to spill into both the Filchner and Ronne Depressions.

The topography of the Ronne Depression is poorly known and most of our observations were located along the barrier. However, considerable similarities exist between the physical conditions in the Filchner and Ronne Depressions. Another branch of relatively warm MWDW flows into this region, steered by the bottom contours east of the Ronne Depression. This warm water (T ~ -1.3°C) appears to flow under the floating Ronne Ice Shelf.

The temperature of the ISW is higher in the Ronne Depression (~ -2.0°C) than in the Filchner Depression (~ -2.2°C). This reflects the difference in ice shelf draught in the two areas. The salinity is markedly higher on the west side of Ronne depression, but some of this cold, saline water may be trapped by a sill to the north.

Acknowledgments. We especially wish to acknowledge the generous invitation by the Alfred-Wegener-Institut für Polarforschung to the Geophysical Institute, University of Bergen, to carry out the physical oceanography program during the Federal Republic of Germany 1980 Expedition. We also gratefully acknowledge the assistance of the Norwegian Research Council for Sciences and the Humanities (NAVF) through the research grant supporting one of us (Tor Tørresen) and for the support of the field program including the loan of the CTD equipment. Our sincere thanks are due to Captain M. Aklestad and his crew onboard the Polarsirkel for efficient and professional assistance throughout all three cruises. We also want to thank our colleagues R. Bø, H.G. Gade and N. Slotsvik whose participation in different cruises was highly appreciated. Our sincere thanks are also due to B. Biskopshavn, A. Naess, and A. Revheim for all help and support in technical matters.

References

Brennecke, W., Die ozeanographischen Arbeiten der Deutschen Antarktischen Expedition 1911-1912, Ark. Dtsch. Seewarte, 39, 214 pp., 1921.

Carmack, E.C., Water characteristics of the Southern Ocean south of the Polar Front, in A Voyage of Discovery, edited by M. Angel, pp. 15-42, Pergamon, New York, 1977.

Carmack, E.C., and T.D. Foster, On the flow of water out of the Weddell Sea, Deep Sea Res., 22, 711-724, 1975a.

Carmack, E.C., and T.D. Foster, Circulation and distribution of oceanographic properties near the Filchner Ice Shelf, Deep Sea Res., 22, 77-90, 1975b.

Carmack, E.C., and T.D. Foster, Water masses and circulation in the Weddell Sea, in Polar Oceans, edited by M. Dunbar, pp. 151-164, Arctic Institute of North America, Calgary, Alberta, Canada, 1977.

Deacon, G.E.R., The hydrography of the Southern Ocean, Discovery Reports, 15, 124 pp., 1937.

Deacon, G.E.R., The Weddell Gyre, Deep Sea Res., 26, 981-995, 1979.

Elder, R.B., and J.M. Seabrooke, Oceanography of the Weddell Sea, U.S. Coast Guard Oceanogr. Rep., 30, 98 pp., 1970.

Foldvik, A., T. Kvinge, and T. Tørresen, Bottom currents near the continental shelf break in the Weddell Sea, this volume.

Foldvik, A., and T. Kvinge, Thermohaline convection in the vicinity of an ice shelf, in Polar Oceans, edited by M. Dunbar, pp. 247-255, Arctic Institute of North America, Calgary, Alberta, Canada, 1977.

Foldvik, A., T. Gammelsrød, and T. Tørresen, Hydrographic observations from the Weddell Sea during the Norwegian Antarctic Research

Expedition 1976/77, Polar Res., 3(2), in press, 1985a.

Foldvik, A., T. Gammelsrød, and T. Tørresen, Physical oceanography studies in the Weddell Sea during the Norwegian Antarctic Research Expedition 1978/79, Polar Res., 3(2), in press, 1985b.

Foldvik, A., T. Gammelsrød, N. Slotsvik, and T. Tørresen, Oceanographic conditions during the German Antarctic Expedition 1979/80 to the Weddell Sea, Polar Res., 3(2), in press, 1985.

Foster, T.D., Haline convection in leads and polynyas, J. Phys. Oceanogr., 2, 462-469, 1972.

Foster, T.D., and E.C. Carmack, Frontal zone mixing and Antarctic Bottom Water formation in the southern Weddell Sea, Deep Sea Res., 23, 301-317, 1976a.

Foster, T.D., and E.C. Carmack, Temperature and salinity structure in the Weddell Sea, J. Phys. Oceanogr., 6, 36-44, 1976b.

Foster, T.D., and J.H. Middleton, Variability in the bottom water of the Weddell Sea, Deep Sea Res., 26, 743-762, 1979.

Gammelsrød, T., and N. Slotsvik, Hydrographic and current measurements in the southern Weddell Sea 1979/80, Polarforschung, 51, 101-111, 1981.

Gill, A.E., Circulation and bottom water formation in the Weddell Sea, Deep Sea Res., 20, 111-140, 1973.

Gordon, A.L., Weddell Deep Water variability, J. Mar. Res., 40, suppl., 199-217, 1982.

Gordon, A.L., D.G. Martinson, and H.W. Taylor, The wind-driven circulation in the Weddell-Enderby Basin, Deep Sea Res., 28, 151-163, 1981.

Hufford, G.L., and J.M. Seabrooke, Oceanography of the Weddell Sea in 1969 (IWSOE), U.S. Coast Guard Oceanogr. Rep., 31, 32 pp., 1970.

Jacobs, S.S., R.G. Fairbanks, and Y. Horibe, Origin and evolution of water masses near the Antarctic continental margin: Evidence from $H_2^{18}O/H_2^{16}O$ ratios in seawater, this volume.

Lusquinos, A.J., Extreme temperatures in the Weddell Sea, Arbok for Universitetet i Bergen, Mat. Naturv. Ser., 23, 1, 1963.

Millero, F.J., Freezing point of sea water, Eighth Report of the Joint Panel on Oceanographic Tables and Standards, Appendix 6, UNESCO Tech. Pap. Mar. Sci. 28, pp. 29-35, 1978.

Mosby, H., The waters of the Atlantic Antarctic Ocean, Sci. Results Norw. Antarct. Exped. 1927-1928, 1, 131 pp., 1934.

Reid, J.L., and R.J. Lynn, On the influence of the Norwegian-Greenland and Weddell seas upon the bottom waters of the Indian and Pacific oceans, Deep Sea Res., 18, 1063-1088, 1971.

Robin, G. deQ., C.S.M. Doake, H. Kohnen, R.D. Crabtree, S.R. Jordan, and D. Möllner, Regime of the Filchner-Ronne ice shelves, Antarctica, Nature, 302, 582-586, 1983.

Rohardt, G., Hydrographische Untersuchungen am Rand des Filchner Schelfeises, Berichte zur Polarforschung, 19, 137-143, 1984.

Seabrooke, J.M., G.L. Hufford, and R.B. Elder, Formation of Antarctic Bottom Water in the Weddell Sea, J. Geophys. Res., 76, 2164-2178, 1971.

Sverdrup, H.U., Hydrology, Section 2, Discussion, B.A.N.Z. Antarct. Res. Exped. 1921-31, Reports, Ser. A. 3, Oceanography, Pt. 2, 88-126, 1940.

Wüst, G. Bodentemperatur und Bodenstrom in der Atlantischen, Indischen und Pazifischen Tiefsee, Gerlands Beitr. Geophys., 54, 1-8, 1938.

(Received May 22, 1984; accepted October 10, 1984.)

BOTTOM CURRENTS NEAR THE CONTINENTAL SHELF BREAK IN THE WEDDELL SEA

Arne Foldvik

Geophysical Institute, University of Bergen, Bergen, Norway

Thor Kvinge

Christian Michelsens Institute, Bergen, Norway

Tor Tørresen

Geophysical Institute, University of Bergen, Bergen, Norway

Abstract. The results of year-long current meter moorings during 1968-1969 and 1977-1978 near the continental shelf break in the southern Weddell Sea are discussed. Mooring design and materials for the first moorings are evaluated in terms of exposure to the Antarctic ocean environment over the 5 years between deployment and recovery. All mean currents 20-100 m above the seafloor and about 630 m below the sea surface are in the direction of the local isobaths with a small (~5°) cross-isobath component. Mean speeds are 6-7 cm/s, and mean (15 day) temperatures vary between -0.5°C in summer and -1.6°C in winter. Individual temperature records never exceed 0°C. The records are dominated by tidal currents, with the diurnal component K_1 about three times larger than the semidiurnal component M_2. A nearby tide gauge record shows a dominant semidiurnal tide with K_1 roughly two-thirds the amplitude of M_2. In the winter of 1968 the diurnal tidal current component breaks down. This phenomenon is attributed to a weakening of the local stratification and the subsequent breakdown of the baroclinic tidal forcing. The energy in the subtidal frequency bands shows marked seasonal effects, with a general decrease of the energy level in winter.

Introduction

In 1968, the National Science Foundation sponsored the International Weddell Sea Oceanographic Expedition (IWSOE-68). The Geophysical Institute at the University of Bergen, Norway, was invited to participate with a program to study the formation mechanism of Weddell Sea Bottom Water. Four current meter moorings were emplanted on the upper continental slope in the southern Weddell Sea, and two of these were finally recovered 5 years later during IWSOE-73. Data and preliminary results from that experiment have appeared in a technical report by Foldvik and Kvinge [1974].

During the Norwegian Antarctic research expeditions in 1977 and 1979 another mooring was set and recovered from a location very close to the 1968 mooring site (Figure 1). This paper presents a somewhat condensed version of the Foldvik and Kvinge [1974] report with the new material added and discussed.

Experience Related to Long-Term Moorings in Antarctic Waters, 1968-1969

The Buoy Stations

The 1968 program, although very simple in principle, confronted us with some rather difficult technical problems for which there were no prior solutions. It was necessary to construct an instrument mooring that would remain in place and operate correctly under the ice cover for at least 1 year. A location had to be found for the buoy station that would satisfy the logistics and scientific requirements. It would be necessary to locate and recover the instruments after they had wintered over beneath the sea ice. Allowance had to be made for possible battery failure and dredging to recover the instruments. It was decided to base the observations on the standard Aanderaa current meter RCM-4 [Aanderaa, 1964], which was designed for recordings of low-frequency oscillations.

The project was considered rather hazardous, and so it was decided to make use of four nearly identical and closely spaced stations to increase the probability of success. A detailed description of the instruments, the buoy station and the launching operation is given by Kvinge [1968]. The construction of the buoy rig is shown in Figure 2.

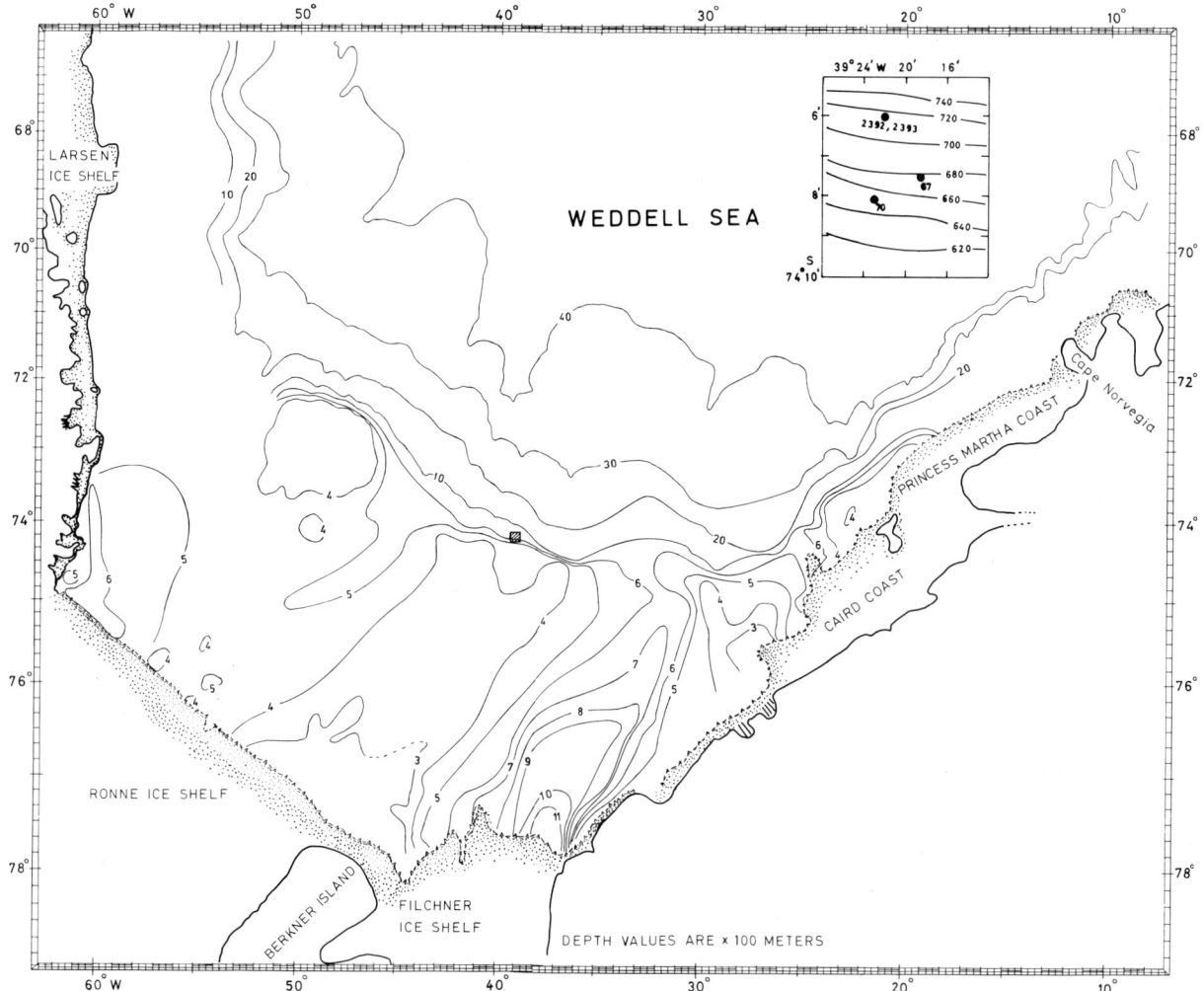

Fig. 1. Map of the Weddell Sea showing location of the buoy sites and the relative position of the instruments. The moorings are marked with the current meter numbers.

Technical State of Equipment After Recovery

The recovery. Unfavorable ice conditions made the buoy site inaccessible until February 1973. After 5 years the batteries in the acoustic release units were presumably exhausted, and the recovery had to be based upon dredging for the ground line shown in Figure 2. Station 2 (74°07.6'S, 39°18.5'W) and station 4 (74°08.1'S, 39°23.0'W) were snagged during the same haul and brought to the surface heavily entangled, apparently having been dragged along the bottom for quite some distance. It is reasonable to assume that this rough recovery caused the loss of an attached Braincon Watersampler [Kvinge, 1968].

The buoy rig. Two types of floats were used. Station 2 was equipped with seven 16" glass floats suspended in a nylon net, whereas on station 4, the buoyancy was provided by four 22-inch steel floats shackled together in series. Neither the glass floats nor the nylon net showed signs of destruction or heavy wear. The steel floats had been protected by several layers of anticorrosive paint. This paint was mostly intact, with only a few superficial and insignificant rust spots on the floats.

The shackles showed minor mechanical wear, estimated to be less than 5%. Both float systems were covered with a thin layer of biological growth which was hardly noticeable and of no significance to the behavior of the rig. The buoy wires, 6-mm-diameter stainless steel 6 x 19 + 1, were spliced over polyvinyl chloride (PVC) thimbles in order to protect the wire from corrosion and wear. The wires showed no signs of either corrosion or wear, and a breaking test proved that the wire was fully intact.

It was assumed that the anchor, a 350-kg railway wheel, would sink deep into the mud

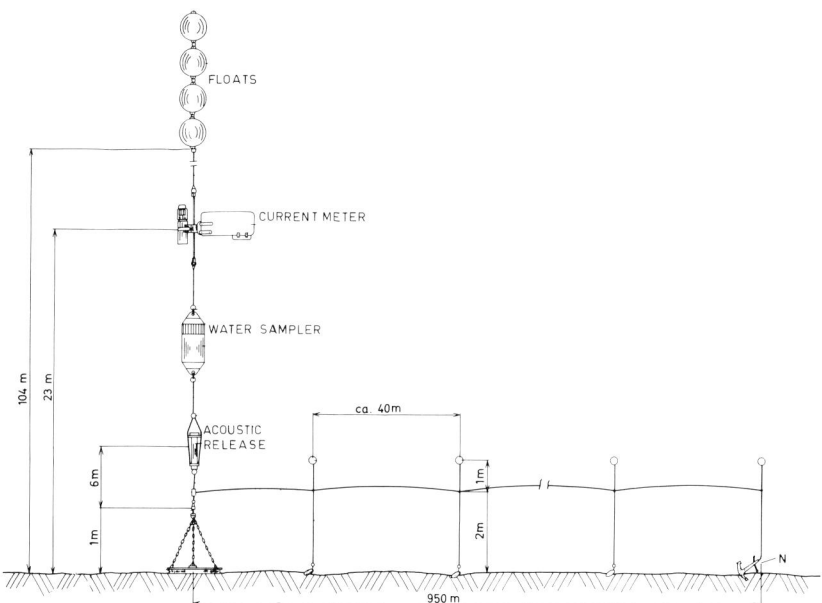

Fig. 2. Design of the 1968 buoy stations.

and get stuck. For this reason, a weak link with a breaking strength of 1000 kg was inserted just above the triple anchor chain. The penetration and resistance to movement was less than expected, as the anchor let go before the weak link broke and showed signs of having been immersed only about 5 cm into the mud. The anchor and triple chain were moderately covered by rust, and wear on the chains was estimated to be about 15%. The 16-mm-diameter polypropylene ground line, connected to the buoy wire by a triple PVC thimble, was recovered in very good condition. The ground line showed no signs of wear and very little biological growth. Breaking strength tests proved that the ropes were fully intact.

The current meters. The Aanderaa RCM-4 current meters were suspended from standard current meter brackets. Electrolytic corrosion was anticipated to be one of the main problems, so the instrument containers were insulated from the brackets and straps by several layers of vinyl tape and sprayed with an anticorrosive compound. Prior to launching, the instruments had been prepared in a cold room in order to avoid condensation when exposed to cold water.

The instruments were recovered in remarkably good condition. The suspension rod suffered pit corrosion below a plastic ring that had been clamped to the rod to prevent the instrument bracket from sliding down in case the pivot pins should break. Screws holding the rotor bracket and rivets were rusty on the surface.

One Savonious rotor was missing and probably lost during recovery. The other rotor was fully intact, but the rotor bearings had been worn oval, apparently without affecting the calibration of the speed recordings. The tape and the anticorrosive spray were still sticky and in good condition. The original chromium-nickel plating showed no signs of corrosion or destruction. Like the other equipment, the instruments were only slightly fouled on the exterior, and it is not likely that this biological growth had any effect on the recordings. Both current meters were dry and clean when opened after recovery.

Instrument 70 had apparently worked until the main batteries were exhausted, using all available tape. Instrument 67 had stopped, obviously due to clock failure, after about half of the tape was used. The instruments were sent to the manufacturer for recalibration and a thorough technical examination. This examination showed that the electronics and the sensors were in excellent condition. Recalibration of the rotor on instrument 67 (the rotor on instrument 70 was lost) and the temperature sensors showed no significant deviation from the original calibration. A recalibration of the clocks was done and is described below. For more details about the technical examination, see the original report by Foldvik and Kvinge [1974]. The current meter data are archived at the Geophysical Institute, University of Bergen.

Summary of experience related to long-term mooring. It was demonstrated that long-term moorings are technically possible in very cold water ($\leq 0°C$). In fact, the technical state of the moorings was such that the rigs would most likely have held up for another couple of years. The most serious deterioration was due to pit corrosion on the stainless steel parts. This type of corrosion can be hampered or even avoided by careful passivation of the metal surfaces and allowing for free exposure to the ambient water. Electrolytic corrosion can in practice be eliminated by avoiding direct metallic contact between metals of different electrode potentials. Several layers of anticorrosive paints seem to provide adequate protection for ordinary steel, and a protective spray like Tectyl has a good long-lasting effect.

It is believed that the low rate of corrosion was related to the low ambient water temperature, short mooring lines, and small vertical temperature and salinity gradients. The instruments are adequately designed for operation for periods of a year or more, depending on the sampling rate. The mechanical clocks were a weak point in the 1968 moorings. Newer instruments were equipped with crystal clocks that are more accurate and reliable.

Alternative recovery systems such as ground lines have also proved to be worthwhile, particularly in waters that may be inaccessible for long periods of time.

General Remarks on the Quality of the Data

The two 1968 current meters, instruments 67 and 70, were located in an area of even bottom topography at about 650 m depth, 23 m above the bottom. The horizontal separation was only 2.5 km, so one might expect high coherence between the two sets of observations, in particular with regard to barotropic disturbances, but little useful information about the spatial variation. On the other hand, these adjacent moorings provide an excellent means for verifying conspicuous events and checking doubtful recordings.

Current meters 2392 and 2393 were mounted in 1977 on the same mooring at 720 m depth, about 3.5 km north of the 1968 moorings (74°06'S, 39°22'W; Figure 1). The mooring depths were 25 m and 100 m above the bottom for instruments 2392 and 2393, respectively. The current meters were suspended in taut and relatively short subsurface moorings which permitted a limited degree of freedom only. The consequent mooring effect on the observations is presumably small and has been neglected. The duration of the time series varied from 460 days (instrument 70) to 264 days (instrument 67) and 257 days (instrument 2393). Instrument 2392 gave current data only for the first 6 days, but provided a temperature series of 564 days.

Since the clocks had stopped when the instruments were recovered, there existed no means for a direct check of the time reference for the records. This calibration has, therefore, been based on the following indirect methods:

1. After recovery the mechanical clocks were calibrated at 20°C by the manufacturer with the following results: instrument 67 lost 44.6 s per day, whereas instrument 70 gained 12.4 s per day. This relative time difference adds up to 4 1/2 hours during 270 days, so the recordings should show a phase difference of the dominant diurnal tide of approximately 67°. An examination of the recordings, however, showed that the actual phase difference was far less than this estimate. It is reasonable to assume that the inconsistency may have been related to the drying out of lubrication in the clocks after 5 years, and to an excessive ambient temperature during recalibration. This calibration was, therefore, considered irrelevant and was not taken into account.

2. The observed and expected phase angle shifts for the principal constituents, based on harmonic analysis of subseries of the current components were compared. The deviation and the long time stability of the instrument clocks have been estimated by means of the following analysis. Synchronized series of current components were divided into subseries of 311 measurements corresponding to 311 hours or 13 days. Each series was then analyzed separately and the phase angles of the twelfth, thirteenth, twenty-fifth, and twenty-sixth harmonics relative to the phase angles of the tidal constituents O_1, K_1, M_2, and S_2 were plotted as a function of time (Figure 3). The observed systematic decrease of the phase angles with time for instrument 70 indicates that this clock had regressed an average of 18.5 s per day (Table 1, last column). This seemed to be a reasonable deviation, in agreement with previous experience with similar clocks. The average deviation of the clock in instrument 67 was apparently negligible. Instrument 2393 was fitted with a quartz clock and showed no sign of deviation.

3. The observed and theoretical frequencies of the tidal constituents, based on harmonic analysis of long series were compared. The absolute clock deviation of instrument 70 was also estimated independently (Figure 4). The apparent periods of the tidal constituents O_1, K_1, M_2, and S_2 were determined by harmonic analyses of series consisting of $(8760 \pm n)$ hours, with n=1, 2, ..., 20. These observed periods were then compared with the theoretical periods for the corresponding tidal constituents. The difference can be resolved by assuming that the clock in instrument 70 regressed at an average rate of 18.5 s per day, or nearly 2 hours per year. The results are listed in Table 1.

We are not able to give a more accurate estimate of the clock deviation, but it is

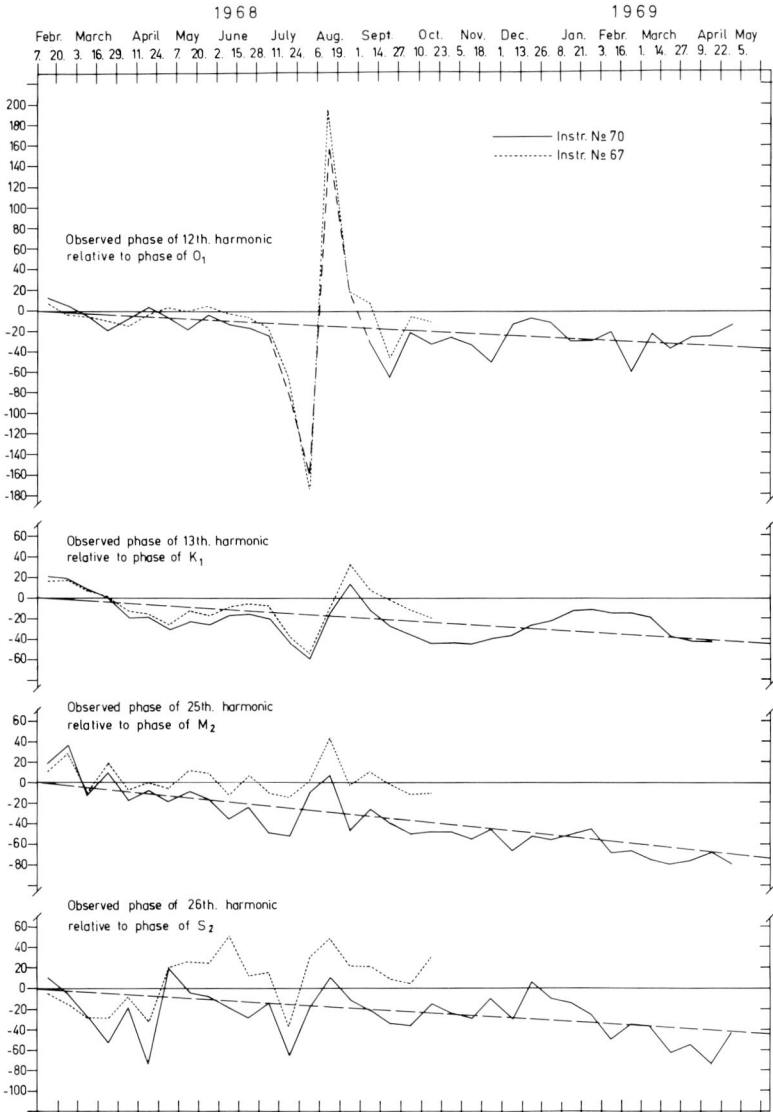

Fig. 3. The observed phase of selected harmonics relative to the phase of associated tidal constituents. The harmonic analyses are based on 35 consecutive subseries, each of 311 hours. The slanting broken lines represent the best linear fit to the observations for instrument 70. The zero point is arbitrary.

quite evident that the deviations are small and apparently linear. By the use of an interpolation routine we have adjusted instrument 70, assuming a retardation of 18.5 s per day. The other instruments were assumed correct.

In the Aanderaa RCM-4 instrument, the current speed S_i is recorded as the integrated flow in the time interval Δt between two consecutive recordings t'_i, whereas the current direction $\vec{D}'_i$ (unit vector) is obtained from the instantaneous compass reading at the recording time t'_i. We have constructed centered current data (Figure 5) by defining the direction $\vec{D}_i$ at the centered time $t_i = t'_i - \Delta t/2$ as follows:

$$\vec{D}_i = \frac{\vec{V}'_{i-1} + \vec{V}'_i}{|\vec{V}'_{i-1} + \vec{V}'_i|}$$

where

$$\vec{V}'_i = \frac{S_i + S_{i+1}}{2} \vec{D}'_i$$

The velocity vector then becomes

TABLE 1. Check of Clock Deviation for Instrument 70

Tidal Constituent	T_{obs}, hours	T_{theor}, hours	$T_{theor} - T_{obs}$, hours	Deviation, s/d	Deviation (From Figure 3) s/d
O_1	25.8136	25.8193	0.0057	19	19
K_1	23.9284	23.9345	0.0061	22	22
M_2	12.4177	12.4206	0.0029	20	19
S_2	11.9982	12.0000	0.0018	13	14

T_{obs} denotes the period of maximum amplitude response in the harmonic analysis (Figure 4). T_{theor} denotes the theoretical period of the tidal constituents. The last column gives the deviation obtained from Figure 3.

$$\vec{V}_i = S_i \vec{D}_i \qquad \text{at } t = t_i$$

The manufacturer used a linear calibration formula for the current direction, but the compass calibrations of instruments 67 and 70 showed a deviation of up to 15°. This may cause systematic errors, especially in the residual currents in areas with strong tidal currents [Forbes and Church, 1980]. We have therefore fit a fifth-order polynomial to the calibration data and used this as our calibration formula.

The magnetic deviation at the buoy site is about 6° east, and the direction of magnetic north is almost normal to the local isobaths. We therefore decided not to correct the original data for magnetic deviation. The directions refer to magnetic north, and the north current component thus approximately represents the cross-isobath component.

Current and Temperature Conditions

The Mean Currents

The recordings of the velocity components shown in Figures 6, 7, and 8 are, with few exceptions, dominated by pronounced periods of apparent tidal origin. The mean flow is demonstrated by the progressive vector diagrams shown in Figure 9. The mean speed given by instrument 70 is 7.2 cm/s in the direction 277° (magnetic). Instrument 67 shows a systematic 8% lower mean speed, and its direction deviates slightly to the south relative to instrument 70. Apart from this discrepancy, which is within the accuracy of the instruments, the two progressive vector diagrams coincide quite well. Instrument 2393, which is north of instruments 67 and 70 and higher above the seafloor, shows a slightly lower mean speed (6.2 cm/s) in the direction 282° (magnetic).

Variations in mean current speed have no apparent relationship to the mean current directions, which are remarkably stable. Near the end of August 1968, there was a small deviation in current direction toward the south. The variations in mean current speed are substantial, with monthly winter averages twice the values for the summer season, see Figures 9 and 10. The local isobaths are approximately oriented in the direction 275°-281° (Figure 1). The true direction of the mean current

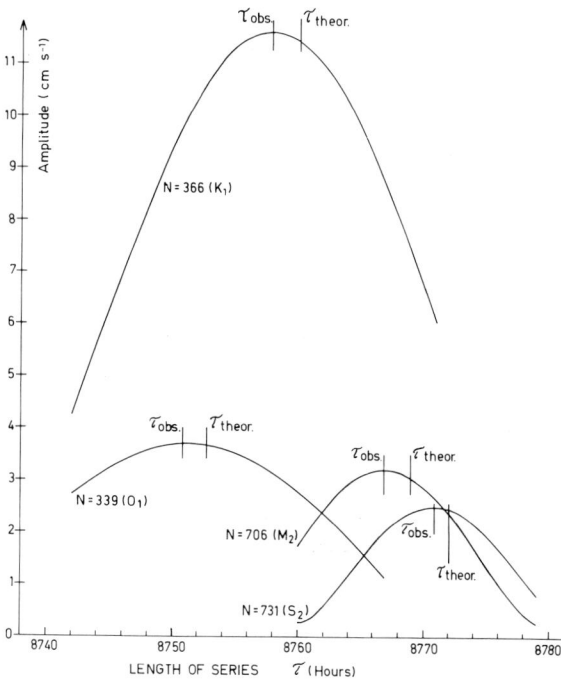

Fig. 4. The amplitude of selected harmonics N as a function of τ, the length of the time series used in the harmonic analysis. These harmonics represent the tidal constituents O_1, K_1, M_2, and S_2. The observed tidal periods T_{obs} are defined by $T_{obs} = \tau_{obs}/N$, where τ_{obs} denotes the length of the time series giving maximum amplitude response. The theoretical length of the time series τ_{theor} is accordingly given by $T_{theor} \cdot N$, where T_{theor} is the period of the tidal constituent in question.

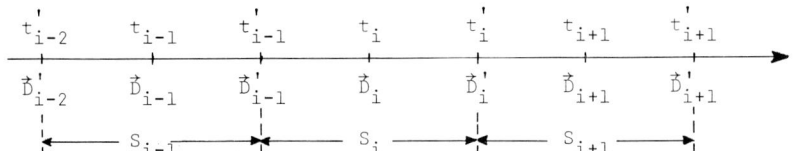

Fig. 5. Schematic representation of sampling times and the computed centered times for an Aanderaa current meter: t'_i is observation time, t_i is centered time, $\vec{D}'_i$ is observed direction (unit vector), $\vec{D}_i$ is computed centered direction (unit vector), and S_i is observed current speed.

varies from 281° (instrument 67) to 288° (instrument 2393). It thus appears that there is a systematic deviation in the mean currents of about 5°-6° to the right of the direction of the isobaths. It is tempting to attribute this cross-isobath flow to frictional boundary effects, but some of it might be accounted for by instrumental uncertainty.

Characteristic Periods

The conspicuous feature of the recordings of velocity components is the diurnal oscillation. A direct comparison of the recordings from instruments 67 and 70 reveals a remarkable similarity, both in major events (discussed in the next section) and in minor details (Figures 6 and 7). This coherence is also pronounced in the autospectra for these two stations (Figures 11 and 12). The spectra are based on 6300 observations from the same time interval, and a direct comparison is therefore permissible. The autospectra for the total series for instruments 70 and 2393 are shown in Figures 13 and 14. The spectra show high energy densities in two narrow bands of diurnal and semidiurnal period. These bands are obviously related to the tides, and the major tidal components O_1, K_1, M_2, and S_2 are resolved. The diurnal frequency band carries by far the highest energy density, and here the v components (cross-isobath) have more energy than the u components (along-isobath).

In the low-frequency band (periods longer

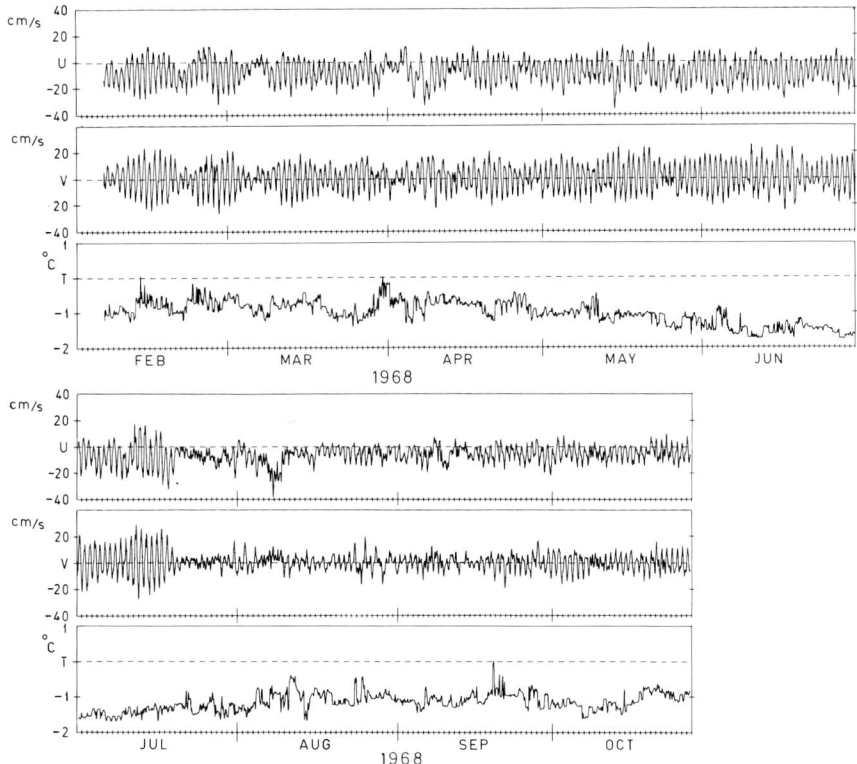

Fig. 6. Observed hourly values of current components and temperature for instrument 67.

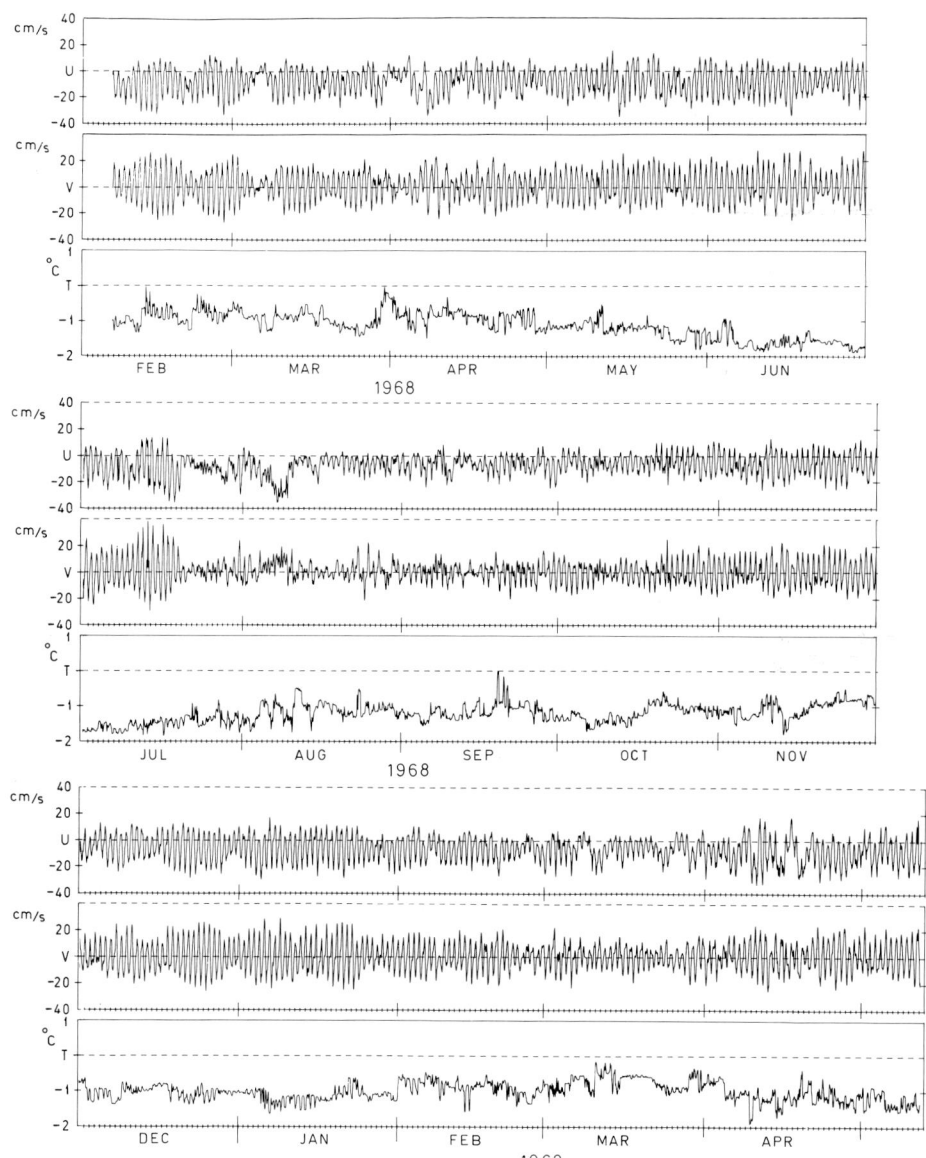

Fig. 7. Observed hourly values of current components and temperature for instrument 70.

than a day) the motion is predominantly zonal (along-isobath) for periods above about 80 hours, whereas the meridional (cross-isobath) component dominates for periods between 80 and 50 hours. Low-frequency oscillations in the area are described by Middleton et al. [1982], based in part upon the records from instrument 2393. They conclude that over the continental slope the energy in long-period fluctuations is mostly due to baroclinic modes of motion, while barotropic modes dominate on the shelf. Furthermore, fluctuations on time scales longer than about 5 days are mainly due to wind forcing, whereas continental shelf waves are responsible for fluctuations with period between about 2 and 5 days. The measurements described here are from the upper part of the continental slope and should thus contain some energy from baroclinic motion. The instruments from 1968 and 1977 are separated by only 3.5 km, and the auto-spectra from these instruments show the same pattern. This indicates that the features described by Middleton et al. [1982] are relatively permanent.

At the latitudes of the current meters the inertial periods are 12.4415 hours, 12.4410 hours, and 12.4433 hours for instruments 67, 70, and 2393, respectively. Consequently, power spectra analysis cannot separate inertial oscillations from the semidiurnal tidal

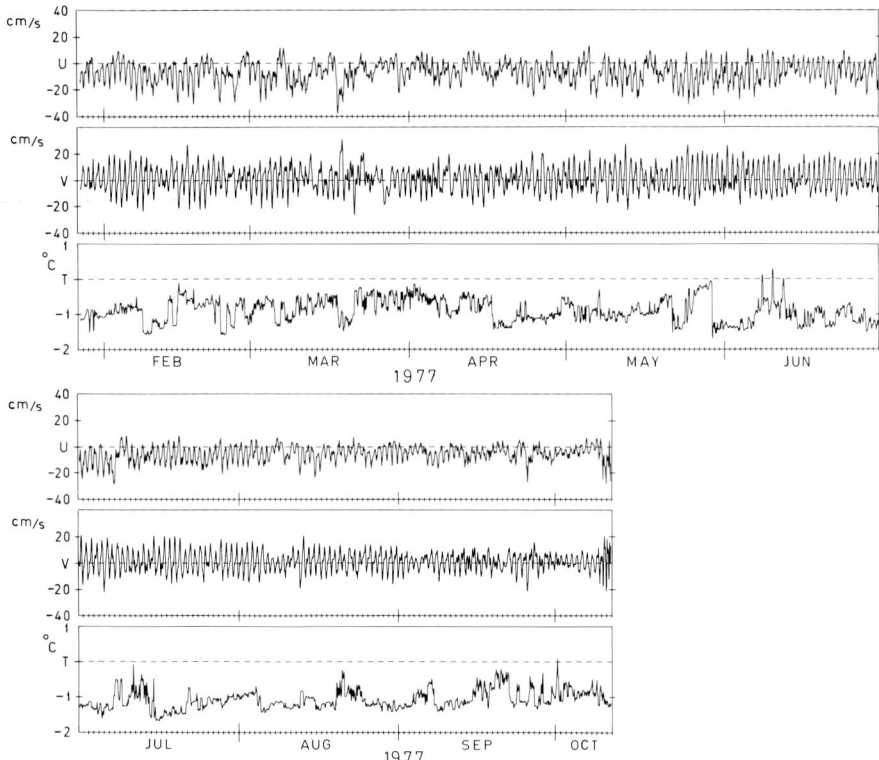

Fig. 8. Observed hourly values of current components and temperature for instrument 2393.

constituent M_2 even with the length of the present data series. Further, the drift of the inertial frequency with the "age" of the disturbance will broaden the inertial frequency band and make a separation from the M_2 component even more difficult. In the frequency range between the inertial and the Nyquist frequency (2 hours) the energy density E falls off with frequency σ as described by the power law $E \sim \sigma^{-c}$, where the slope c is close to 2.8 (Figures 11 and 12). There is no tendency for the spectrum to level out at high frequencies, which indicates that the instrument functioned correctly in that range.

Figures 15 and 16 show the time variation of the autospectra from instruments 70 and 2393. The time series are divided into subseries of 1024 hours (43 days) with an overlap of 512 hours. The autospectrum for each subseries is then computed and plotted along a time axis (see Figure 15 caption). These plots give a picture of the time variation of energy in the low-frequency band. They confirm what is seen in the autospectra in Figures 11-14, namely, that the cross-isobath component has the most energy between 1.5 and 3 days, whereas the along-isobath component exhibits the most energy on periods longer

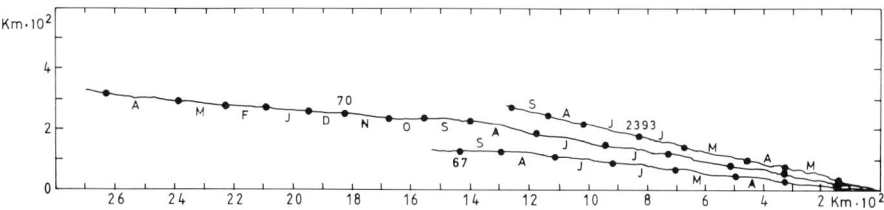

Fig. 9. Progressive vector diagrams for instruments 67 and 70 (start February 6, 1968) and 2393 (start February 6, 1977). The series are filtered with a Butterworth low-pass filter with 40-hour cutoff. Solid circles denote the beginning of each month, and the months are indicated with capital letters. The axes are relative to magnetic west and north.

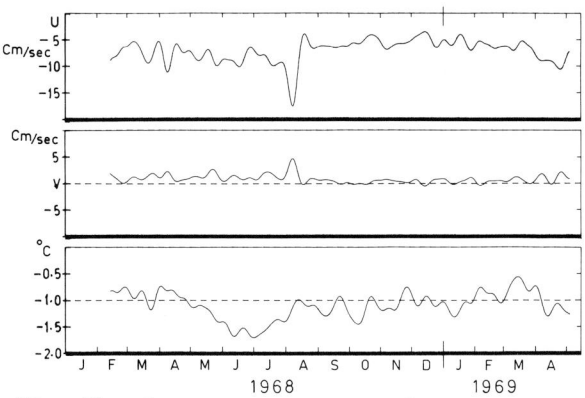

Fig. 10. Current components and temperature from instrument 70, filtered with a Butterworth low-pass filter with 15-day cutoff and resampled every 12 hours.

than 2.5 days. These phenomena are not stationary. For instrument 70 the low-frequency energy vanished in September (1968) and built up again in January, and for Instrument 2393 the low frequency energy disappears earlier in

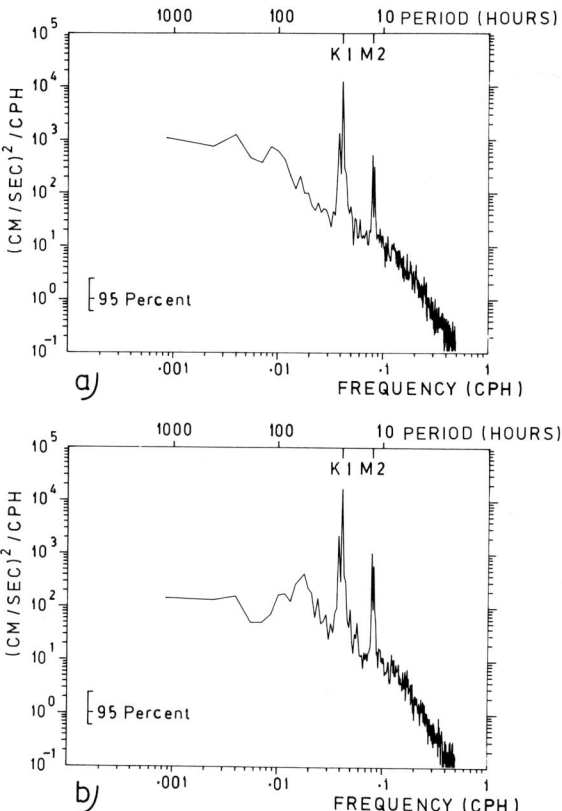

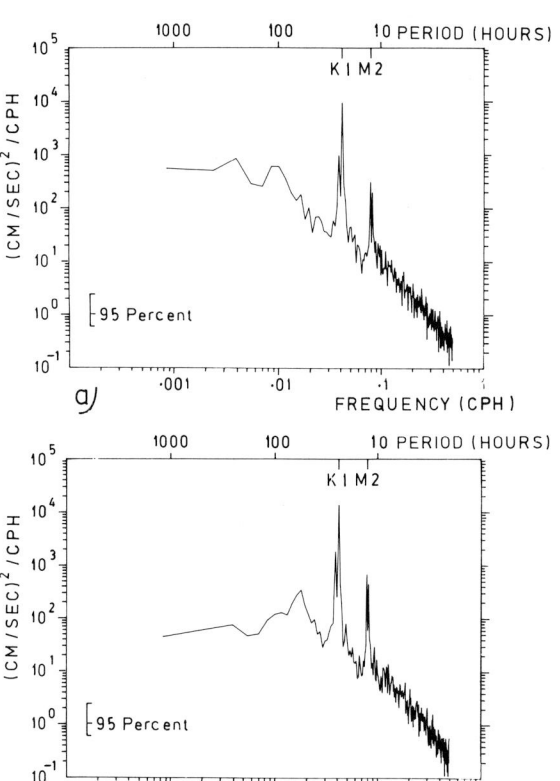

Fig. 11. Autospectra of current components for instrument 67 based on 6300 hours (262 days): (a) u (east) component; (b) v (north) component.

Fig. 12. Autospectra of current components for instrument 70 based on 6300 hours (262 days): (a) u (east) component; (b) v (north) component.

the season (July 1977). These variations appear as a seasonal effect with very little energy in the low-frequency band when the ice cover reaches its maximum extent.

If the energy in the low-frequency band is derived from the along-isobath component of the wind stress [Middleton et al., 1982], then the lack of energy in late winter might be caused by low-pressure systems passing farther north in winter than in summer, reduced input of wind energy to the sea due to extensive ice cover, or reduction of static stability in winter.

Amplitude Variations of Selected Modes

Changes in the currents apparent in the recordings of current components and in the power spectra have been studied in more detail by focusing on discrete frequency bands. We selected the diurnal and semidiurnal tidal components O_1, K_1, M_2, and S_2 for a study of the time variations of these energy-containing components of the current. We divided the time series from instruments 70 and 2393 into subseries of 30 days, with an overlap of 15

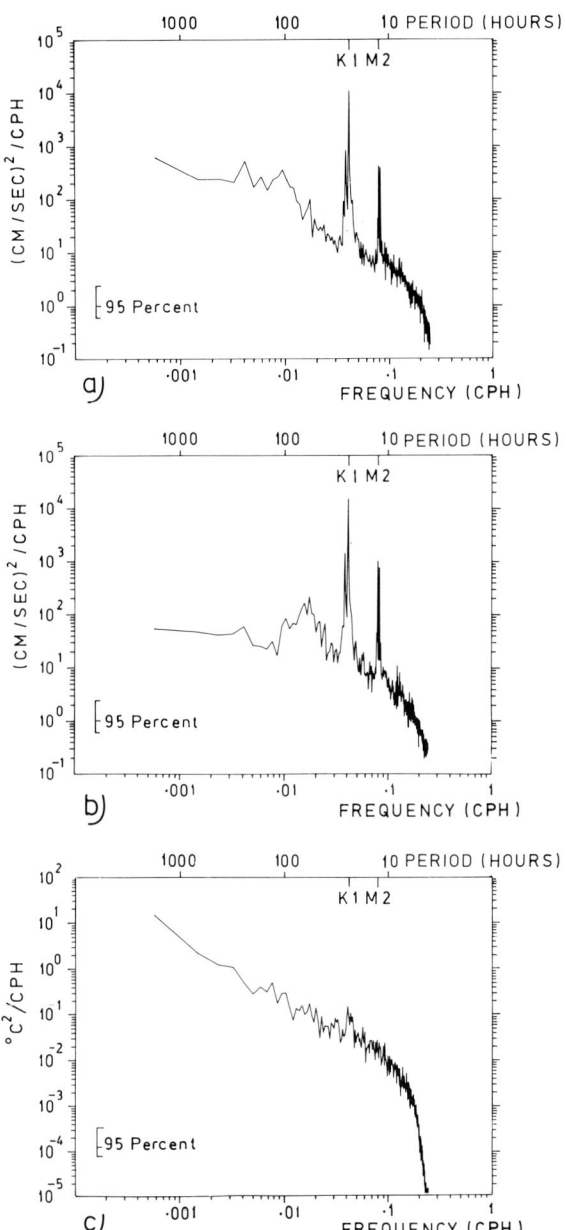

Fig. 13. Autospectra of current component and temperature for instrument 70 based on 10,976 hours (457) days. (a) u (east) component; (b) v (north) component; (c) temperature.

This is especially important with P_1 and K_2, which will cause oscillation in the amplitude and phase of K_1 and S_2 with a period of half a year. We can reduce this interaction by inferring the interacting constituent from the analyzed component, assuming a constant amplitude ratio and phase difference between the neighboring constituents (found from analyzing the whole series). This was done for the consti-

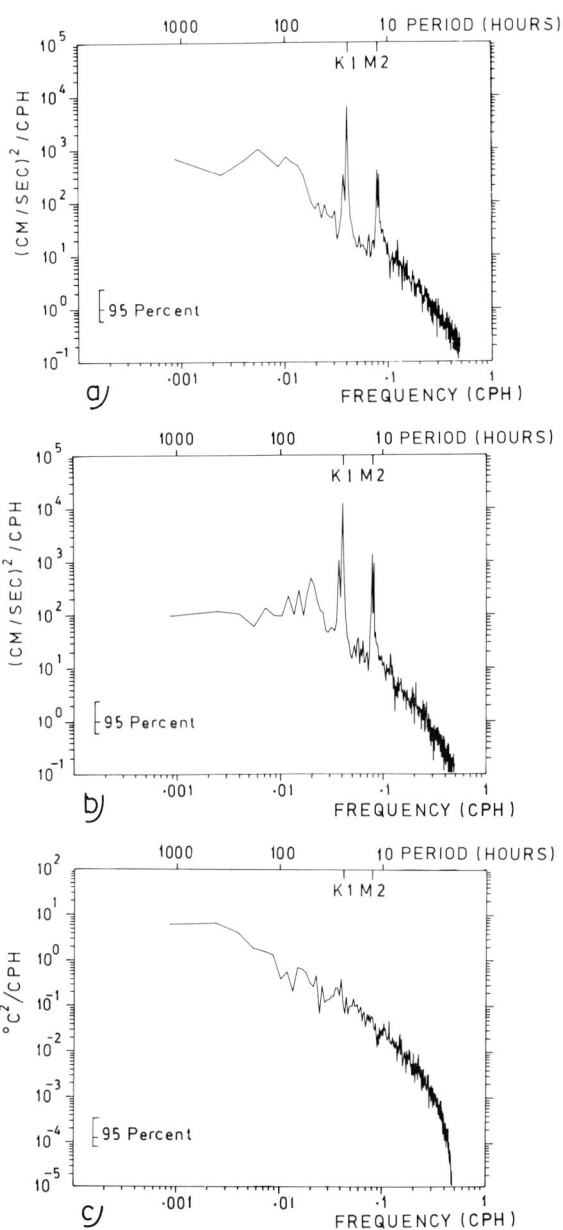

Fig. 14. Autospectra of current components and temperature for instrument 2393 based on 6174 hours (257 days): (a) u (east) component; (b) v (north) component; (c) temperature.

days, and made a tidal analysis of each subseries using a program by Foreman [1978]. A problem with tidal analysis of short time series is that constituents close in frequency to the analyzed components result in a frequency difference corresponding to a period longer than the time series analyzed. These constituents will interact with the analyzed components, resulting in a periodic behavior.

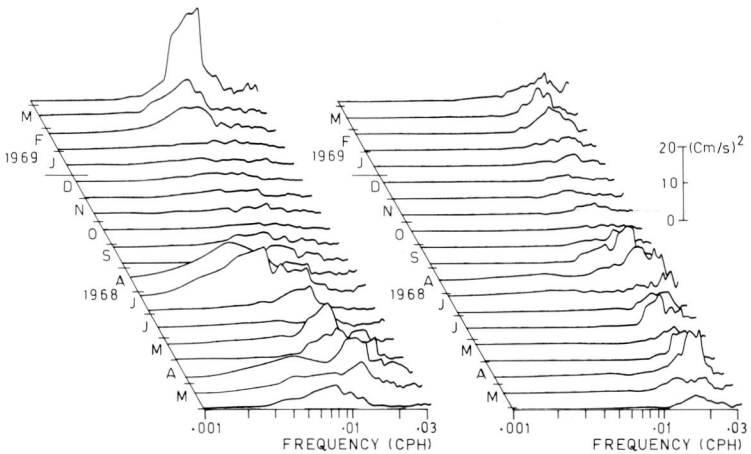

Fig. 15. Time variation of energy in the low-frequency band from instrument 70. The time series is divided into subseries of 1024 hours with an overlap of 512 hours, and the autospectrum for each subseries is computed and plotted along a time axis. The spectra are plotted in energy-conserving form (equal areas under a spectral curve represent equal energies): (a) u (east) component; (b) v (north) component.

tuents K_1, P_1 and S_2, K_2, removing artificial periods of half a year.

The results of the harmonic analyses are shown in Figure 17. The amplitude of the oscillation, defined as the long axis of the current ellipse, is shown for the four constituents. The conspicuous feature of the diagram is the different rate of variation in time of the diurnal, especially K_1, and the semidiurnal modes. The diurnal tides appear to have a seasonal period with minimum amplitude in winter. The amplitudes of O_1 and K_1 from instruments 70 and 2393 are very similar from February until the middle of July, when K_1 from instrument 70 suddenly drops in accordance with the breakdown of the diurnal tides seen in Figure 7. These variations are not reflected in the semidiurnal constituents.

In Table 2 the four principal tidal components are shown from a tide gauge deployed at the shelf break about 60 km southeast of instrument 2393. We assume that these are also representative for the current meter site. For barotropic tidal currents one would expect the ratio between the tidal current constituents to be about the same as the corresponding ratio between the tidal water level constituents, but this is clearly not the case. The diurnal tidal currents (Figure 17) are 3 to 4 times stronger than the semidiurnal tidal currents, while the tide gauge data (Table 2) show the diurnal components to be smaller than the semidiurnal ones.

Middleton et al. [1982] presented a dispersion diagram for barotropic shelf waves, using a representative shelf profile for the area. The diagram shows a high-frequency cutoff for the lowest-order barotropic mode between the diurnal and semidiurnal tidal bands. All the higher-order barotropic modes have periods of more than 3 days. When stratification is added, these modes are shifted toward higher frequencies as shown by Huthnance [1978], but the high-frequency cutoff for the lowest-order mode will still be located between the diurnal and semidiurnal frequency bands. Thus, there exists the possibility of baroclinic resonant forcing at the diurnal tidal frequency band for at least the lowest-order mode. Such forcing of the diurnal frequency may qualitatively explain the major feature in Figure 17. The breakdown of the diurnal tides in July

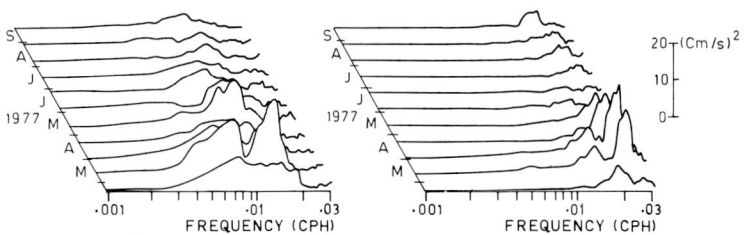

Fig. 16. Time variation of energy in the low-frequency band from instrument 2393. Legend as for Figure 15: (a) u (east) component; (b) v (north) component.

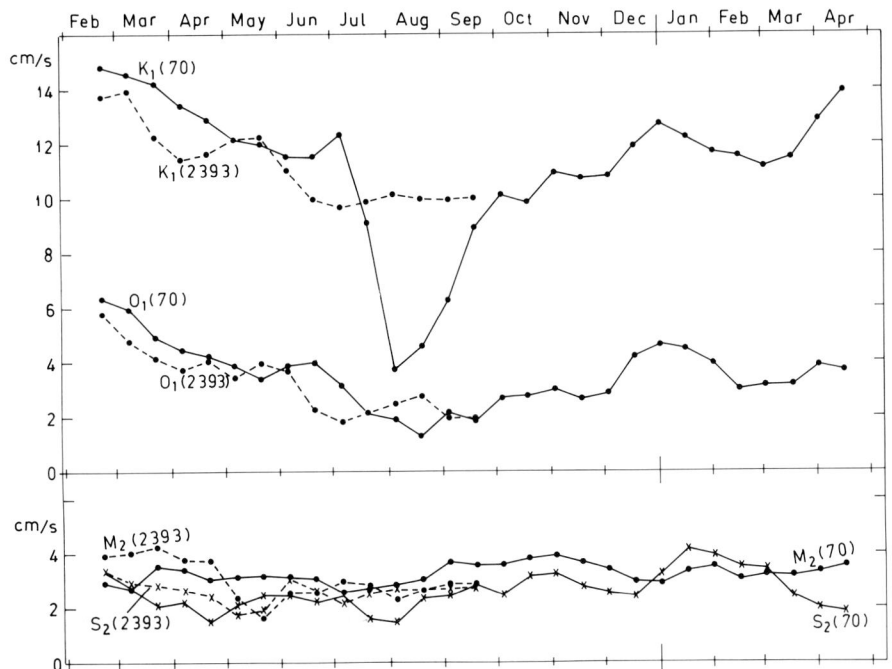

Fig. 17. Time variation of the tidal currents constituents O_1, K_1, M_2, and S_2 for instrument 70 (1968-1969) and instrument 2393 (1977). The long axes of the tidal current ellipses are plotted.

1968 may then be attributed to a corresponding breakdown of the stratification in the area.

The measurements from 1977 do not show a midwinter breakdown of the diurnal tide, but it is likely that the stratification was quite different during the winters of 1968 and 1977. The temperature records from instrument 70 (1968) and instrument 2393 (1977) (Figures 10 and 18) both show a slight seasonal trend, but the winter minimum was much lower in 1968 than in 1977. This could have been caused by advection of cold, homogeneous shelf water into the buoy site.

TABLE 2. Amplitudes and Greenwich Phases of Principal Tidal Components From the Shelf Break at 74°23'S, 37°39'W

Constituent	O_1	K_1	M_2	S_2
Amplitude, cm	30.3	36.4	57.6	38.3
Greenwich phase G°	5	16	242	264

The analysis is based upon pressure recordings with instrument TG3A-220 during the period from January 25, 1977, to July 31, 1977 (Norwegian Antarctic Research Expeditions 1977 and 1979).

Temperature Variations

According to Foster and Carmack [1977] the shelf break in the southern Weddell Sea is a region where cold shelf water (Western Shelf Water, $\theta \sim -1.9°C$) meets the warmer Modified Warm Deep Water ($\theta \sim -0.7°C$). The latter water mass is a mixture of Warm Deep Water ($\theta \sim 0.5°C$) found between 200 m and 1500 m in the deep part of the Weddell Sea and Winter Water ($\theta \sim -1.8°C$) from the upper 200 m. The mixing processes in this region give rise to temperature step structures on a wide range of horizontal and vertical scales. These complex temperature structures are advected with the local currents. Ignoring local diffusion, the observed temperature fluctuations are related to advection through the equation

$$\frac{\partial T}{\partial t} = -U \frac{\partial T}{\partial x} - V_b \frac{\partial T}{\partial y_b}$$

where T denotes temperature, U the velocity component in the x (east) direction, and V_b the velocity component in the direction of y_b (north) parallel to the sloping bottom. The velocity component normal to the sloping bottom has been ignored, probably a valid approximation due to the proximity of the bottom. The first term defines local temperature variations due to advection along the isobaths. The second term defines local tempera-

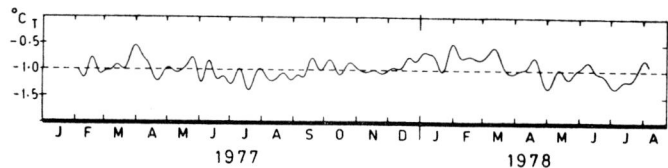

Fig. 18. Temperature from instrument 2392, filtered with a Butterworth-low pass filter with 15-day cutoff and resampled every 12 hours.

ture variations by cross-isobath motion due to temperature gradients in the direction of the sloping bottom. These temperature gradients may alternate in sign depending upon the relative position at the shelf break of the surrounding water masses.

From the general remarks above, it is not obvious that high correlations will exist between temperature variations and currents, and an inspection of the original data in Figures 6, 7, and 8 confirms this suspicion. The temperature variations are more irregular and abrupt than the corresponding variations in the current speed. Often the temperature variations take the form of steps, as might be expected for passage of thermal fronts. Other small-scale temperature variations may be interpreted as advection of smaller volumes of water of different temperature and/or oscillations of frontal zones.

Temperature oscillations near diurnal periods can be identified in the autospectra of the temperature from instrument 70. These would result from the dominant diurnal cross-isobath motion combined with temperature gradients. The main contribution to the variance is found in the low-frequency range and is associated with advection along the isobaths. The very small contribution in the diurnal frequency range indicates the absence of appreciable systematic temperature gradients along the bottom at the buoy site.

The 15-day average temperatures range from about $-1.6°C$ in winter to about $-0.5°C$ in summer (Figures 10 and 18). Hourly values (Figures 6, 7, and 8) range from $-1.9°C$ to $0.0°C$. Apparently, the Warm Deep Water does not come within 100 m of the seafloor at this site on the upper continental slope. The temperature data do suggest that both cold shelf water and Modified Warm Deep Water can be found at this site, downslope from their primary locations on the continental shelf [Foldvik et al., 1985].

Acknowledgments. The 1968 current meter project was initiated by Hakon Mosby, Geophysical Institute, University of Bergen, whose encouraging support we gratefully acknowledge. Thanks are also due to Odd Dahl, Jan Strømme, and Reidar Bø at the Christian Michelsens Institute, Bergen, for their cooperation. We are also gratefully indebted to the captains, officers, crew, and scientists on board the USCGC Glacier and the R/V Polarsirkel for their valuable assistance. We also would like to thank the editor of this volume for the invitation to publish this paper and for constructive remarks. This project has been supported by the National Science Foundation, the Norwegian Polar Institute, and the Norwegian Research Council for Science and the Humanities (NAVF).

References

Aanderaa I., A recording and telemetering instrument, NATO Subcommittee on Oceanographic Research, Tech. Rep. 16, Christian Michelsens Inst, Bergen, Norway, 1964.

Foldvik, A. and T. Kvinge, Bottom currents in the Weddell Sea, Results of Long Time Current Meter Moorings at 74°S, 40°W During IWSOE 1968-1973, Rep. 37, Geophys. Inst. Div. A., Univ. of Bergen, Bergen, Norway, 43 pp., 1974.

Foldvik, A., T. Gammelsrød, and T. Tørresen, Circulation and water masses on the southern Weddell Sea shelf, this volume.

Forbes, A.M.G., and J.A. Church, The effects of compass calibration on Aanderaa current meter records, Tech. Rep. 121, Aust. CSIRO Div. Fish. Oceanogr., 1980.

Foreman, M.G.G., Manual for tidal currents analysis and prediction, Pac. Mar. Sci. Rep., 78-6, 1978.

Foster, T., and E.C. Carmack, Antarctic bottom water formation in the Weddell Sea, in Polar Oceans, edited by M. Dunbar, pp. 167-177, Arctic Inst. of North America, Calgary, 1977.

Huthnance, J.M., On coastal trapped waves: Analysis and numerical calculation by inverse iteration, J. Phys. Oceanogr., 8, 74-92, 1978.

Kvinge, T., Technical report of project to measure currents related to the formation of Antarctic bottom water in the Weddell Sea. Geophys. Inst., Univ. of Bergen, Bergen, Norway, 19 pp., 1968.

Middleton, J.H., T.D. Foster, and A. Foldvik, Low-frequency currents and continental shelf waves in the southern Weddell Sea, J. Phys. Oceanogr., 12, 618-634, 1982.

(Received August 27, 1984;
accepted January 3, 1985.)

INTERACTION BETWEEN ICE SHELF AND OCEAN IN GEORGE VI SOUND, ANTARCTICA

J. R. Potter and J. G. Paren

British Antarctic Survey, Natural Environment Research Council, Cambridge, CB3 OET, England

Abstract. George VI Ice Shelf floats on warmer water than any other ice shelf in the Antarctic. Profiles of temperature (T) and salinity (S) taken in the vicinity of the northern ice front show a linear T/S dependence confirming a thermodynamic model of ice melting in Circumpolar Deep Water and indicate that thermohaline convection is the principal mixing process. Oxygen isotope (δ) profiles demonstrate that the melting ice has a δ value of -20°/oo with respect to Standard Mean Ocean Water (SMOW). An integration of accumulation and isotope data over the ice catchment confirms that this is the mean isotope ratio of present-day accumulation. Since the basal ice is formed from accumulation over several millennia, it is unlikely that there has been any significant net climatic change in the Antarctic Peninsula over this period. Both summer and long-term measurements show that currents are weak except at the western margin of the northern ice front where a northward jet conveys some 0.05×10^6 m^3 s^{-1} of water into Marguerite Bay. This leads to a simple circulation model for the northern part of George VI Sound; Circumpolar Deep Water is advected under the ice shelf at depth, upwells transferring heat which melts the ice and then collects in a northward outflow gathered to the west by Coriolis force. The circulation is driven by the melting process which causes the upwelling of warmer water from greater depths. A salt and energy balance shows that the outflow conveys some 16 km^3 yr^{-1} of ice melt. It is inferred from T/S profiles that the northern circulation penetrates at least 160 km south but not so far as the southern ice front in Ronne Entrance. These geographical limits constrain the basal melt to values between 1.1 and 3.6 m yr^{-1}. A calculation balancing accumulation over the catchment with ice losses from the ice shelf predicts a basal melt of 2.1 m yr^{-1}. If the ice shelf is in equilibrium it alone supplies 53 km^3 yr^{-1} of ice melt or about one-sixth of the total for Antarctica. Tidal height and current measurements both show a highly suppressed M_2 tide and the presence of "shallow water" constituents. A nonlinear ice shelf reponse to tidal forcing is suspected.

Introduction

In order to understand the state of stability of the Antarctic ice sheet and its effect on sea level it is essential to understand the mechanism of heat transfer and melting of the ice shelves which fringe its coast. Oceanographic calculations are needed to determine whether ice loss due to iceberg calving and melting is less or more than is required to balance accumulation over the ice sheet. Any positive difference between mass accumulation and depletion rates must result in a thickening or extension of either the ice sheet or ice shelf. Thus both glaciological studies of ice accumulation and oceanographic studies to infer ice loss are required to determine the true state of balance. This has been the objective of our work on George VI Ice Shelf.

Setting

The west coast of the Antarctic Peninsula is the warmest part of mainland Antarctica. Along this coast, ice shelves do not exist further north than 67°S. George VI Ice Shelf covers an area of 25,000 km^2 and is the largest ice shelf on the west coast of the peninsula. Even so, by continental standards it is small, representing only 1.5% of the total area of Antarctic ice shelves. Its importance is that it floats on warmer water than any other ice shelf in the Antarctic and can be presumed to have a high rate of basal melting.

George VI Sound is a channel extending for about 500 km between roughly parallel flanks of land from Cape Jeremy in Marguerite Bay in the north to Ronne Entrance in the Bellingshausen Sea (Figure 1). The sound is probably part of a faulted valley system [Crabtree et al., in press]. It is 25 km wide at its northern end, comprising two lateral troughs up to 800 m deep separated by a central ridge only 400 m below sea level. The channel incises into the continental shelf and extends north into Marguerite Bay [Vanney and Johnson, this volume]. The thickness of George VI Ice Shelf is now well known as a result of extensive airborne radio echo sounding. A contour map of ice shelf thickness has recently been

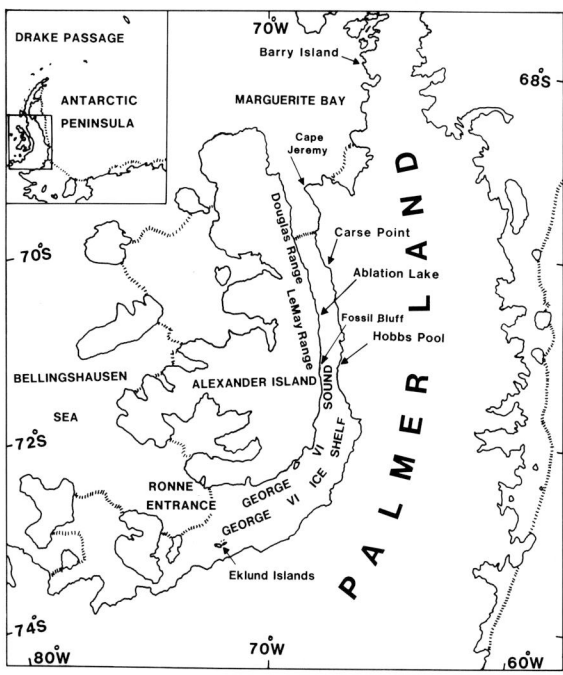

Fig. 1. Map of the Antarctic Peninsula showing George VI Sound and its location in the Antarctic Peninsula.

published [British Antarctic Territory, 1983]. Bedrock topography, however, is only known along four lines across the sound. The first was determined at the northern ice front by plumb line in 1976-1977. The remaining lines were determined by seismic surveying during the austral summers of 1980-1981 and 1983-1984. The ice thickness along the central line of the sound and the known bedrock topography are shown in Figure 2. George VI Ice Shelf occupies most of George VI Sound and its thickness varies from 100 m near the northern ice front to a maximum of 600 m at 72°50'S, 67°50'W. The thickest region of the ice shelf which occupies the whole width of the channel is on a line between the coast of Alexander Island at 72°38'S, 70°10'W and the peninsula coast at 73°06'S, 69°10'W. The minimum ice shelf thickness here is 470 ± 10 m. The ice shelf thins toward both ice fronts from this line, making it an inverted sill at the base of the ice shelf. Ronne Entrance is usually ice-free in the summer but the southern ice front generally abuts fast ice. The northern ice front has one small polynya (approximately 15 km^2) at its western margin during the summer months but is otherwise icebound.

Theodolite surveys of glacier movement and ice strain have been made on the shelf [Bishop and Walton, 1981]. The data were analyzed using steady state theory and the results predicted high rates of basal melting (up to 8 m yr^{-1}); a simple average of the melt rates gives 2 m yr^{-1}. The pattern of basal melting derived from the surveys bears no simple relationship to obvious parameters, such as the time for which the ice has been afloat or distance from the gounding line. Doake [1982] argues that there is no reason to suppose that oceanographic conditions are sufficiently variable to account for the apparent melting pattern, leading to the conclusion that parts of the ice shelf must be thickening or thinning at several meters per year.

Temperature (T), salinity (S), and oxygen isotope ratio profiles have been taken for several summers at sites through the ice shelf. Data from five sites within 13 km of the northern ice front have been published [Lennon et al., 1982; Potter et al., 1984]. More recent data have not been fully processed. The latter include profiles taken 40 km south of the northern ice front at Carse Point (70°15'S, 68°12'W), profiles from three adjacent sites 120 km further south near Hobbs Pool (71°18'S, 67°35'W), and profiles taken at nine sites between the southern ice front and Eklund Islands (73°14'S, 72°00'W).

Tidal height observations have been made at a total of five sites on the ice shelf. Measurements have been made at three lake sites; Hobbs Pool, Ablation Lake (70°49'S, 68°25'W), and an adjacent site. These were analyzed by Cartwright [1980]. Recent data obtained near each ice front using an Aanderaa WLR5 pressure-sensitive tide gauge are of excellent quality. The data are presently being analyzed. An Aanderaa RCM4 current meter has provided a 5-month record from February to July 1980, from a site in the center of the sound 5 km south of the northern ice front at the fixed depth of 156 m [Loynes et al., 1984]. In addition, two RCM4 meters have recorded 13 months of simultaneous data from 116 and 201 m depth at the same northern site and presently two RCM4s are operating beneath the ice shelf near the southern ice front.

All the oceanographic measurements have been taken by field parties traveling over the ice shelf and gaining access to the underlying seawater through natural rifts in the ice shelf or through fast ice immediately adjacent to an ice front. The rifts are fissures which penetrate the entire depth of the ice shelf and are common near ice fronts where they define future calving lines and inland where ice flow diverges (Figure 3). The rifts used to collect data from the northern part of George VI Ice Shelf are typically 400 m wide and floored with snow overlying sea ice. The ice cliffs separating the edge of the rifts from the sea ice are frequently rounded and drifted with snow which causes the rifts to appear as shallow valleys in the ice shelf. Close to the rifts, the ice shelf is thinner than some few hundred meters back. The transition from ice shelf to sea ice is therefore very gentle

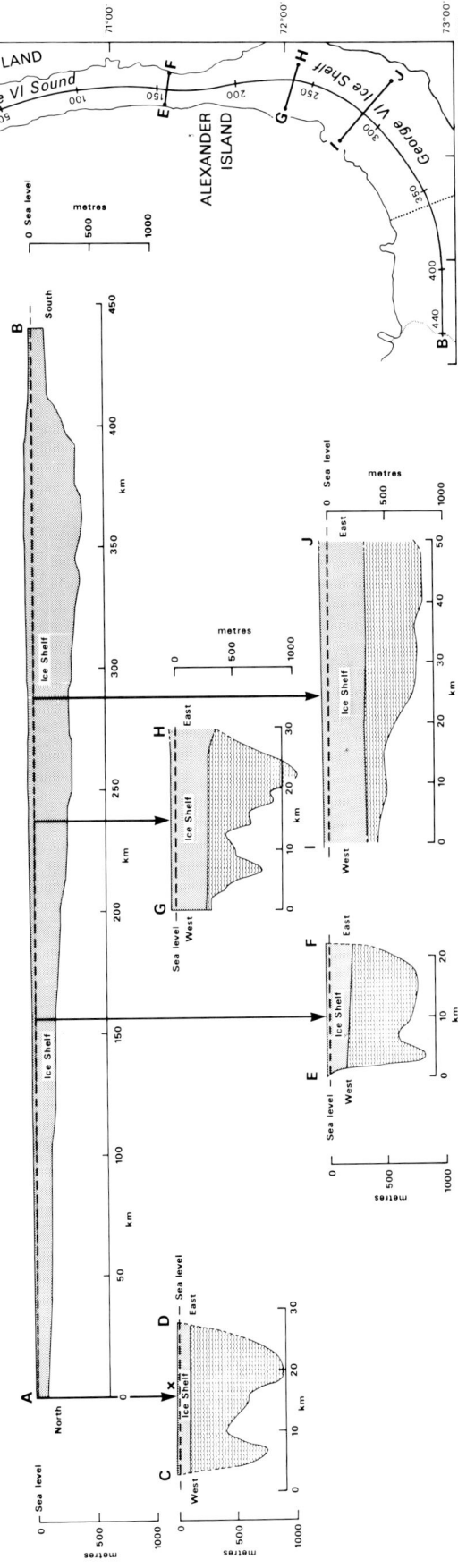

Fig. 2. Ice shelf thickness on a central line (AB) along George VI Ice Shelf and known bedrock topography. Bedrock soundings along the line CD were determined by plumb line [Loynes et al., 1984], along line GH by seismic shooting [Crabtree et al., in press] and also along lines EF and IJ by seismics [M. Maslanyj, personal communication, 1984]. The dotted line shows the area of deepest shelf ice across the sound. The 5-month record was measured by an Aanderaa RCM4 meter located as shown by a cross near the CD transect.

Fig. 3. Photograph taken in a typical rift in shelf ice 100 m thick enabling access to the sea both under the ice shelf and within its draught. The gentle rise from sea ice to shelf ice may just be made out in the background.

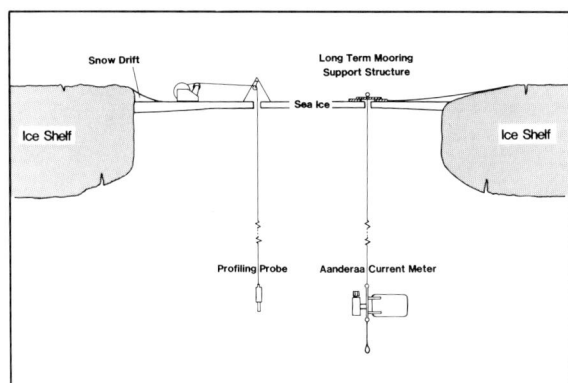

Fig. 4. Typical cross section of a rift shown with the arrangements used for long-term current meter observations and summer profiling.

and often imperceptible in poor weather. This local thinning is caused by preferential melting of the exposed corner at the ice shelf base. The "ice pump" mechanism of Lewis and Perkin [1983] can also produce a smooth transition from ice shelf to sea ice as in McMurdo Sound [Lewis and Perkin, this volume]. This behavior, however, relies on relieved supercooling and is unlikely to be effective in our situation where the large sensible heat flux reduces the presence of sufficiently cold water to a very thin boundary layer adjacent to the ice. A diagrammatic illustration of a typical rift cross section is shown in Figure 4 together with the general arrangement used for long-term current meter moorings and summer profiling.

Temperature and Salinity

Summer temperature-salinity (T/S) profiles have been taken along and up to 13 km back from the northern ice front of George VI Ice Shelf using water sampling bottles with mercury reversing thermometers and Aanderaa current meters left for short periods at several depths. Lennon et al. [1982] have discussed the sampling bottle results, which are tabulated in the appendix, and those from greater than 10 m depth are plotted on a T/S diagram in Figure 5. Very recent sampling at Carse Point and Hobbs Pool have shown the same T/S relationship in the water below the base of the ice shelf as at the northern ice front. Representative profiles of temperature and salinity against depth have been calculated from the data in Figure 5 and are shown in Figure 6.

The points in Figure 5 appear to confirm a single T/S dependence. Linear T/S relationships provide evidence for the mixing of two water masses. Although the points deviate from a linear mixing line by more than the uncertainty in the salinity or temperature determinations, this variability is common to all five sites and the difference between sites is random at any depth. Careful inspection of the data reveals that no single straight line fits the data from the surface to the seabed since there are two distinct breaks in slope. The first occurs at about 85 m depth and the second near 200 m. The ice shelf draught at the northern ice front is approximately 85 m so it seems reasonable to expect a change in the oceanographic regime at this depth. The second break at 200 m probably represents the transition from the mixing layer to the less stratified layer of advected Circumpolar Deep Water. Between 85 m and 200 m depth the gradient of the T/S line (dT/dS) is 2.41 ± 0.09°C per mil.

A theoretical characteristic T/S gradient for ice melting in seawater can be determined directly from conservation of energy and salt as Gade [1979] and Greisman [1979] have shown. From the analysis of Gade [1979] the resulting steady state gradient of the T/S relationship is

$$dT/dS = [L_f C_w^{-1} + (T-T_f) + (T_f-T_o) C_i C_w^{-1}] S^{-1} \qquad (1)$$

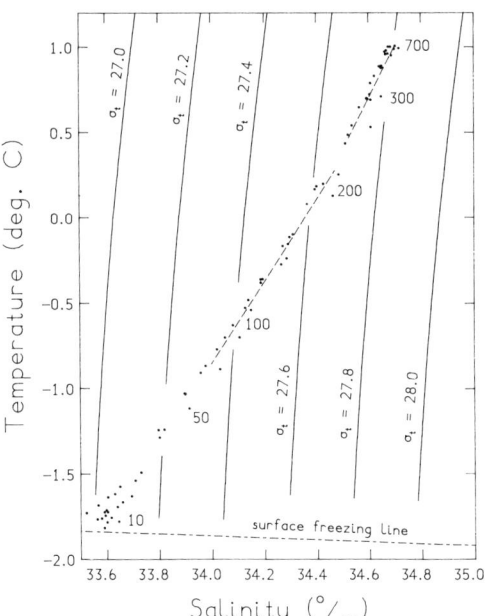

Fig. 5. Temperature-Salinity (T/S) relationship at the northern ice front from sampling bottles and reversing thermometers. Measurement depths are shown in meters below sea level. The dashed lines drawn through the data indicate linear least squares fits for the ranges 85-200 m and 250-700 m. The change in slope between the two lines is statistically significant.

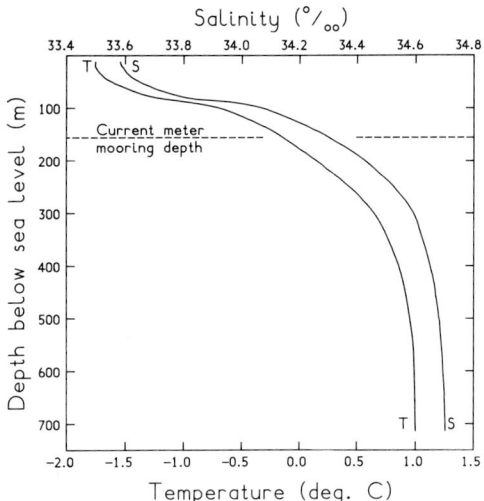

Fig. 6. Representative profiles of temperature and salinity from sampling bottle and reversing thermometer observations taken at five sites along the northern ice front. The 5-month current meter record taken in this area was obtained from an Aanderaa RCM4 moored at the depth indicated.

where T_f is the freezing temperature corresponding to the salinity at the level of the bottom of the ice shelf, L_f is the latent heat of fusion at T_f, T_o is the ice temperature before becoming warmed by the underlying sea, and C_i and C_w are respectively the specific heats of ice and water.

For the range of temperature and salinity observed in Figure 5, equation (1) predicts a theoretical value for dT/dS of 2.50 ± 0.05°C per mil, depending on the values chosen for L_f and T_o. There is excellent agreement between theory and observation suggesting that ice melting in seawater is the primary thermodynamic process. The northern part of George VI Sound is the first known example where there is clear confirmation of the theoretical gradient throughout the water column. Beneath the Ross Ice Shelf, for example, the heat which melts the ice comes from a warm core of water at intermediate depth which is modified not only by ice melt but also by mixing with shelf water of a different character [Jacobs et al., 1979]. As a result the T/S gradient is dependent on the melt rate and has a value of 0.65°C per mil. Even so, some profiles taken along the ice front of the Ross Ice Shelf have T/S gradients near 2.5°C per mil, but only over a limited depth range below the temperature minimum.

The 5-month RCM4 record began in the summer on February 11, 1980, and terminated prematurely on July 15, 1980, when the recording mechanism failed. The record provided an opportunity to examine the T/S relationship at the fixed depth of 156 m. Care had to be taken in the analysis since the fluctuations in temperature and conductivity were small and the sensors had poor resolution. The results were recorded digitally, resulting in only 6 different conductivity values and 16 different temperature values occurring during the record. A histogram of the number of occurrences of each temperature value was drawn up and smoothed to give a distribution curve. The six observed conductivities were then mapped on this distribution so that each conductivity value corresponded to the observed number of records of that value. This provided a temperature-conductivity data pair at each of the five crossover points between one conductivity value and the next. The gradient of the T/S diagram drawn from the five data pairs is 2.30 ± 0.05°C per mil. The measured T/S gradients of 2.41°C per mil from summer profiling between 85 and 200 m and 2.30°C per mil from the long-term mooring at 156 m agree well, confirming that the thermocline is controlled both in summer and winter by the same processes. The small discrepancy between the theoretical gradient of 2.50°C per mil and the observed gradients may be due to surface runoff which percolates through rifts in the ice shelf diluting the salinity without absorbing latent heat from the sea. Such a contribution from surface runoff is not inconsistent with previous estimates [Reynolds, 1981b].

Northern Circulation

Circumpolar Deep Water provides the only sustained southward flow in the Southern Ocean, replacing the cold water that spreads northward in the surface and bottom layers. Bottom water is not found on the continental shelf of Marguerite Bay, since Circumpolar Deep Water extends to the seafloor, where its maximum temperature and salinity are observed [U.S. National Oceanographic Data Center, 1974]. Tidal currents in the north of George VI Sound rarely exceed 0.1 m s^{-1} and non-tidal currents, with one exception, are even smaller. George VI Sound is ice covered, either by the ice shelf or by shore-fast sea ice, which discounts the possibility of wind-driven circulation although some low-frequency flow may be caused by local atmospheric pressure gradients.

The generally weak flow at the northern ice front is contrasted by a narrow outflow jet confined to the western side of the channel. The existence of the jet was revealed by profiles made with an electromagnetic current meter at sites 3 and 6 km from the western shore. At the 3 km site, measurements were taken during two consecutive summer seasons. Eighty hours of data were gathered with individual records varying between 20 min and 15 hours at 10 depths from 50 to 350 m [Lennon et

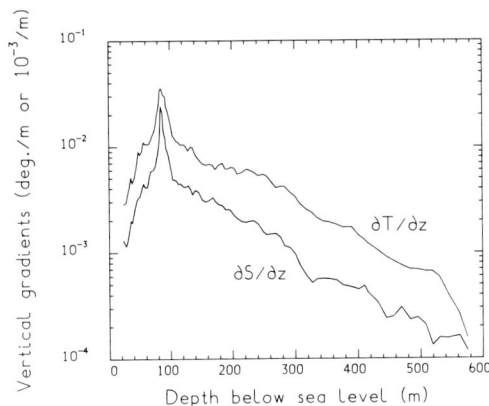

Fig. 7. Vertical temperature and salinity gradients for the representative profiles at the northern ice front shown in Figure 6, plotted logarithmically. Gradients less than 5×10^{-4} should be considered speculative.

al., 1982]. In addition a 7-day record from an Aanderaa RCM4 curent meter moored in the jet shows that the direction of the flow did not deviate from its mean direction by more than 30° at any time during the record. These records show that at the 3 km site a relatively strong flow (0.1-0.25 m s^{-1}) influences the upper 100 to 150 m of the water column beneath the ice shelf. The vertical structure of the jet has been constant during the two seasons of measurements despite changes in current strength. Further east at the 6-km site, the jet almost disppears. Its maximum speed is approximately half that of the 3-km site and the jet influences only a narrow depth range of 60 m directly below the ice shelf base [Lennon et al., 1982]. This suggests that the jet is a buoyant outflow collected to the west by Coriolis force. This hypothesis is supported by a calculation which shows that the horizontal extent of the jet is approximately equal to the Rossby radius of deformation, λ, which gives a length scale for motion controlled by Coriolis effects [Pond and Pickard, 1983]. The value of λ calculated for the western boundary current is 5 km. An approximate numerical integration over the jet's cross-section indicates that its volume transport is some 5×10^4 m^3 s^{-1}.

The weak nature of the currents at the northern ice front and the domination of the vertical temperature and salinity by melting suggest that thermohaline circulation may be the controlling process under the ice shelf. In this case a steady state temperature or salinity field should satisfy the vector eddy diffusivity relationship

$$\mathbf{V} \cdot \nabla F = \mathbf{K} \cdot \nabla^2 F \qquad (2)$$

where F is the property field, V the velocity field, and K the vector eddy diffusivity.

Left-handed orthogonal coordinates are chosen with the x axis parallel to the base of the ice shelf southwards along the axis of the sound and the z axis downward. Equation (2) then becomes

$$u \partial F/\partial x + v \partial F/\partial y + w \partial F/\partial z$$
$$= K_x \partial^2 F/\partial x^2 + K_y \partial^2 F/\partial y^2 + K_z \partial^2 F/\partial z^2 \qquad (3)$$

where u, v, and w are the current speeds in the x, y, and z directions. In a circulation driven by the melting of ice there should be horizontal gradients of temperature and salinity in addition to the pronounced vertical gradients observed in profiles. Because of the nature of the local terrain no true synoptic data set is available to establish the size of these horizontal gradients. Furthermore, profiles measured at the same geographical areas in different austral summers show large variations in the isotherm depths within the mixing layer under the ice shelf. Thus, although there appears to be a sufficiently high density of sites on the ice shelf and in Marguerite Bay to determine the local temperature field in the summer months, reprofiling at selected sites has convinced us that the oceanographic regime is too variable for measurements taken in different years to be used for that purpose. For this reason we have decided to concentrate on a model of thermohaline circulation which emphasizes vertical flow at the expense of horizontal flow. There is a precedent for this approach for the ice shelf regime. Gade [1979] has used the concept of vertical advection balancing vertical diffusion in his theoretical study which models temperature and salinity profiles beneath ice shelves. Additionally, our sparse synoptic profiles give us no reason to suspect that horizontal gradients are more significant than vertical ones. It may seem unwise to use such a simplified model to discuss data taken at and near ice fronts, but in our area of study all the data seem to indicate that, despite the nearby ice front, the sample sites share a common oceanographic regime.

For these reasons we simplify equation (3) to consider vertical motion only, which then becomes

$$w \partial F/\partial z = K_z \partial^2 F/\partial z^2 \qquad (4)$$

Figure 7 displays $\partial T/\partial z$ amd $\partial S/\partial z$ shown logarithmically against depth for the averaged profiles of Figure 6 which represent the oceanographic regime of the northern part of the ice shelf. Since T and S are essentially linearly related we need only examine the temperature field. If w/K_z is constant, there is a linear solution for equation (4).

$$\partial T/\partial z = A \exp(wz/K_z) \qquad (5)$$

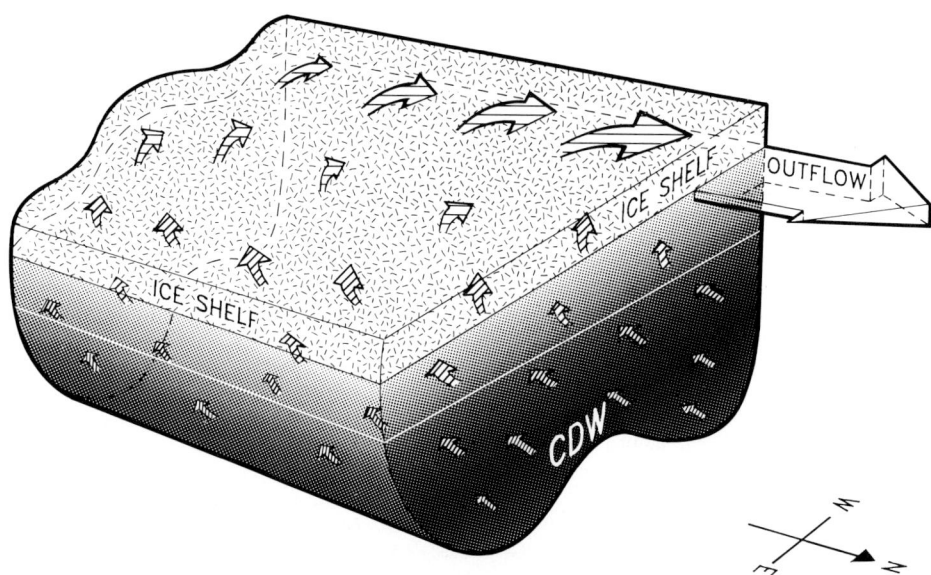

Fig. 8. Conjectured circulation at the northern George IV Sound under the ice shelf. Circumpolar Deep Water (CDW) flows southwards at depth under the ice. Sensible heat melts the ice shelf base causing a buoyant outflow which is gathered at the west by Coriolis force. The outflow, together with other stirring mechanisms, lifts more CDW into proximity of the ice shelf base, completing the cycle.

Within the draught of the ice shelf ln $(\partial T/\partial z)$ increases with depth, implying that the vertical velocity, w, is downward. This is feasible if melting occurs at the side ice walls of the rift, causing local upwelling as shown by Josberger [1979]. To preserve continuity this upwelling would have to be balanced by downwelling elsewhere and we may be witnessing this in the middle of the rift where the profiles were taken. Near the base of the ice shelf $\partial^2 T/\partial z^2 = 0$ and hence $w/K_z = 0$ indicating no vertical flow (a necessary boundary condition for a physically realistic model). Below the ice shelf base $ln(\partial T/\partial z)$ decreases with depth, implying that the flow is upwards. The vertical velocity cannot be quantitatively assessed without a value for K_z, the vertical eddy diffusivity, which is a property of the flow and difficult to evaluate.

The observations above lead to a model for the circulation at the north of George VI Sound: Circumpolar Deep Water is advected from Marguerite Bay southwards under the ice shelf at depths greater than 200 m and upwells supplying heat which melts the ice. A thermohaline convective layer occupies the water column below the ice shelf from approximately 85 m to 200 m depth which is associated with upwelling driven by the ice shelf melt. The water then flows away northwards with the ice melt in a surface jet concentrated to the west by Coriolis force. An illustrative representation of this flow field is shown in Figure 8. MacAyeal [this volume] has modeled what may be a similar circulation beneath the Ross Ice Shelf. The inflow of Circumpolar Deep Water proposed in our model should have a speed of 0.006 m s^{-1} if uniform in strength below 200 m. Such a low mean speed could only be detected by a long-term current meter record taken from below 200 m. Such an observation has not yet been attempted.

In principle, there is enough information to calculate the basal melt rate. The approximate velocity, temperature and salinity fields for the outflow are known. Integrating the product of salinity with speed over the cross section of the outflow gives the total salt flux. Similarly, integrating the product of temperature with speed gives a result proportional to the sensible-heat flux. Dividing these fluxes by the volume flux gives the mean salinity and temperature of the outflow. Since, in our model, the outflow is derived from Circumpolar Deep Water and ice melt, comparison of the outflow temperature and salinity with the values for Circumpolar Deep Water yields the proportion of the outflow supplied by ice melt. Both temperature and salinity comparisons indicate that 1% of the outflow is derived from ice melt. This implies that the outflow carries away some 16 km^3 yr^{-1} of ice melt from the ice shelf.

In order to convert this volume of ice into a basal melt rate it is necessary to find the area of the ice shelf which drains its ice melt to the north. One indication comes from oceanographic sampling within the confines of the ice shelf and capitalizes on changes in water characteristics as one travels from the

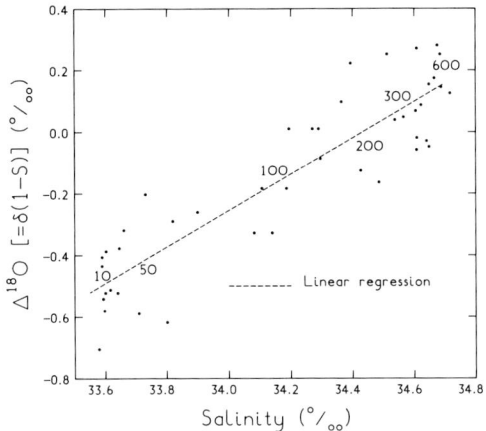

Fig. 9. Isotope-salinity relationship in the sea at the northern ice front with measurement depths shown in meters below sea level. The least squares linear fit to the data is shown as a dashed line. Adjusted δ values have been plotted following Paren and Potter [1984].

The outflow from under the ice shelf must balance the upwelling flux plus the ice melt, so independent knowledge of the upwelling velocity w also yields the area of the ice shelf affected by the northern circulation. Unfortunately it has only been possible to calculate an approximate value for the outflow flux and the thermohaline model only gives values for w/K_z (from Figure 7), rather than for w alone. The value of K_z is uncertain and should vary down the profile since it is determined by the stability of the water column and the speed of the flow. Previous estimates of K_z for an ice shelf regime vary from 1×10^{-4} m^2 s^{-1} [Jacobs et al., 1979] to 25×10^{-4} m^2 s^{-1} [Gade, 1979]. A more profitable approach may be to calculate the basal melt rate from other data and then infer a value for K_z. The calculations which follow suggest a most probable mean basal melt rate of some 2 m yr^{-1}, implying an average value of 8×10^{-4} m^2 s^{-1} for K_z which is not inconsistent with earlier estimates.

Oxygen Isotope and Salinity

Isotope and salinity data obtained at the northern ice front from water bottle sampling are tabulated in the appendix and those results from over 10 m depth are shown in Figure 9. The samples were analyzed by H. B. Clausen of the Geophysical Isotope Laboratory, University of Copenhagen, and the results have been discussed by Paren et al. [1983] and in more detail by Potter et al. [1984]. Between the depths of 10 m and 600 m oxygen isotope composition is linearly correlated with salinity. The linear relationship enables us to identify the oxygen isotope composition of the low- and high-salinity water sources which we know to be melting ice and Circumpolar Deep Water. Both oxygen isotope ratio and salinity are conservative properties in the mixing of two water types. The processes which alter the isotope composition of seawater are mixing, precipitation, evaporation, and, in polar regions, melting or freezing of ice. It is unfortunate that isotope-salinity variations caused by evaporation and precipitation can be somewhat similar to those determined by ice melting and freezing [Weiss et al., 1979, Figure 12]. For this reason the correct processes must be identified before making interpretations of such isotope data. Thermohaline arguments already presented show that the water properties under the ice at the sampling sites are almost wholly determined by melting. This eliminates basal freezing, sea ice formation, liquid precipitation, evaporation, and continental runoff as significant, because these processes would produce quite different T/S gradients. There remains only ice shelf melt and snow precipitation over the sea in Marguerite Bay to consider. Extensive current

southern ice front to the north. The deep water near the southern ice front has a maximum temperature of 1.06°C, compared to 1.00°C for deep water found in the same season at the northern ice front. The deep water salinities are also similar. Nearer the base of the ice shelf, temperature and salinity profiles from the two ice front regimes differ markedly. In the north, both temperature and salinity increase monotonically from the surface to the seafloor (Figures 5 and 6). In the south between Eklund Islands and the southern ice front, salinity increases with depth but a second temperature minimum is found at an intermediate depth. Recent sampling has shown that sites near the southern ice front share a common T/S profile dissimilar to those observed at Hobbs Pool and further north. At Hobbs Pool, under thicker ice (the ice shelf base is some 185 m below sea level), temperature profiles beneath the ice shelf base match those in the mixing layer at the northern ice front. Hobbs Pool, 160 km to the south of the northern ice front and 300 km distant from the southern ice front, clearly belongs to the northern regime and in common with both ice fronts has a seafloor temperature just above 1.00°C. These results constrain the southernmost limit of influence of the northern circulation to between Eklund Islands and Hobbs Pool. The natural barrier in this area is the inverted sill at the base of the ice shelf, where the ice thins towards both ice fronts. Using this divide as one extreme and Hobbs Pool as the other, the outflow at the north of George VI Sound is consistent with an average basal melt rate of between 1.1 and 3.6 m yr^{-1}. This implies that the upwelling rate lies between 110 and 360 m yr^{-1}.

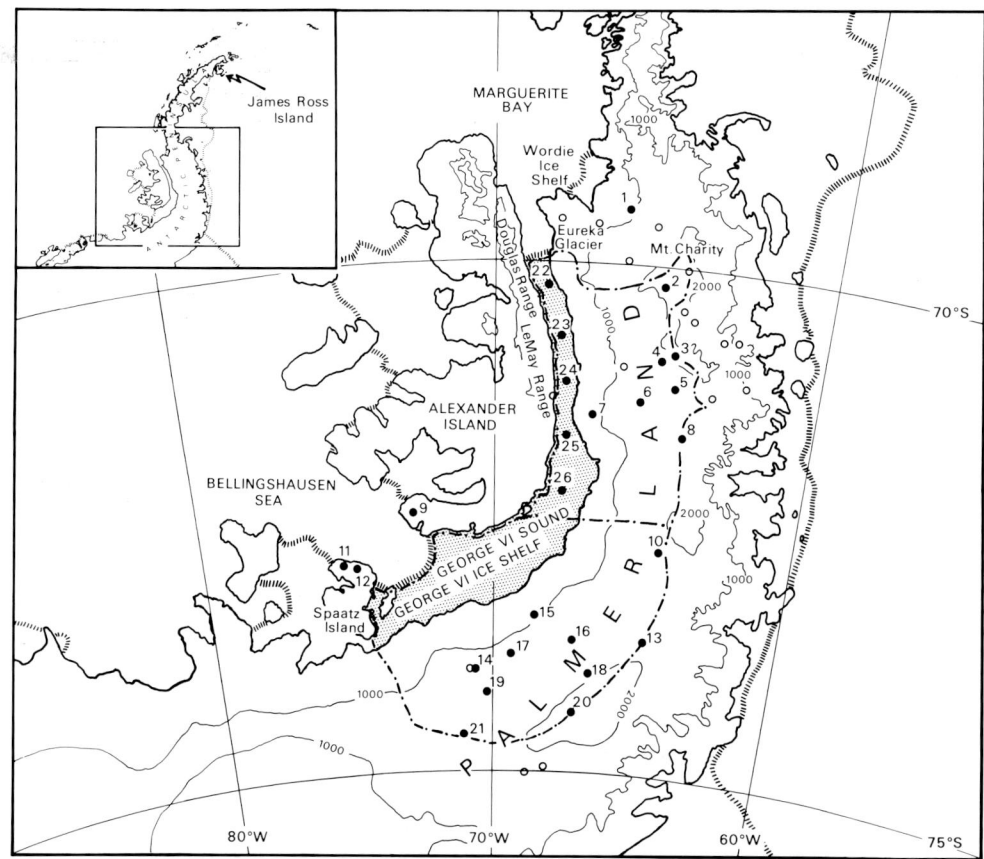

Fig. 10. George VI Ice Shelf (shaded) and its catchment (outlined by a dot-dashed line). Accumulation data sites are shown by solid circles numbered as in the text. Open circles represent sites of temperature data only. Elevation contours are shown at 1000-m intervals [Figure from Potter et al., 1984]. Reproduced by permission of the International Glaciological Society.

meter observations have found no net southward flow from Marguerite Bay toward the ice shelf in the zone of mixing. The local influence of melt water from the base of the ice shelf is therefore expected to be predominant.

If water of salinity S_M and isotope composition δ_M is formed as a result of the mixing of glacier ice of isotope composition δ_I with Circumpolar Deep Water of salinity S_{CDW} and isotope composition δ_{CDW}, then following Paren and Potter [1984] we see that

$$\Delta_M = \Delta_I + (S_M/S_{CDW})(\Delta_{CDW} - \Delta_I) \qquad (6)$$

where Δ is defined by $\Delta = \delta(1-S)$. There are 44 Δ-S data pairs plotted in Figure 9, taken from profiles at three sites extending from 10 m depth to the seafloor. A linear regression of the observations gives $\delta_I = -20.3°/oo$ (since $\Delta_I = \delta_I$) with a standard deviation of 1.5°/oo. The same regression indicates that the deep water of salinity 34.70°/oo and temperature 1.0°C has a δ_{CDW} value of +0.16 ± 0.04°/oo. This positive value for δ_{CDW} is unexpected when compared to other measurements in the Southern Ocean, summarized by Jacobs et al. [this volume]. All other deep Antarctic water masses have negative δ values with the Weddell Deep Water and Circumpolar Deep Water having values between -0.02 and -0.07°/oo. This discrepancy leads us to consider possible sources of error in our δ values. The most plausible are that some samples may have been partially evaporated or there may have been undetected biological action. An average of 2% evaporation of the samples would account for the discrepancy. Even so, providing such occurrences were random among the samples, this possibility does not affect the conclusions drawn from our value of δ_I with its inherent standard deviation of 1.5°/oo.

Mass and Isotope Conservation for George VI Ice Shelf

George VI Ice Shelf is derived from accumulation over the Pacific side of the Antarctic Peninsula, the ice shelf, and parts of Alex-

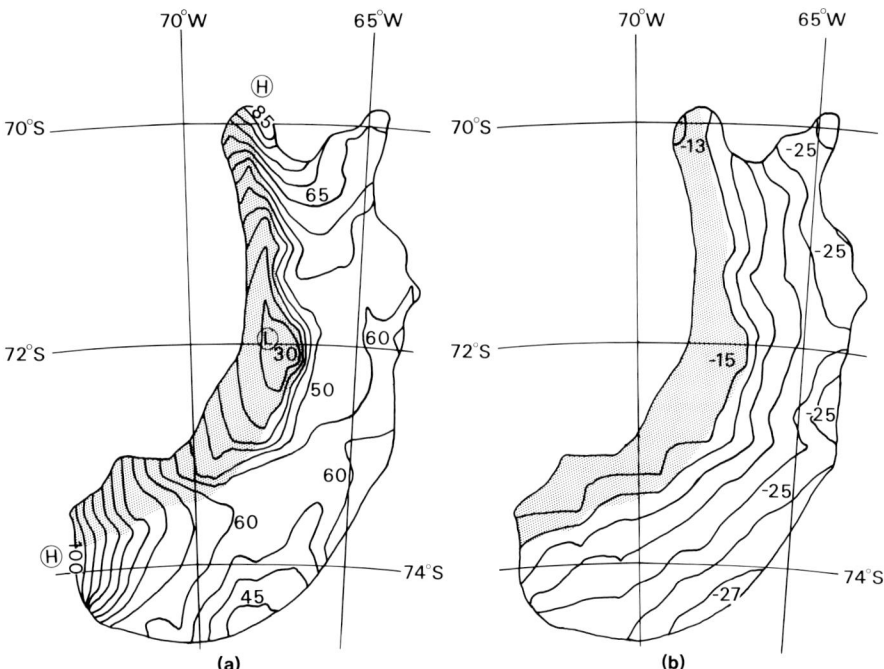

Fig. 11. The catchment area of George VI Ice Shelf shown in Figure 10 with (a) contours of modelled accumulation in centimeters per year of ice and (b) contours of the derived δ field in parts per thousand with respect to SMOW. The ice shelf is shown shaded. (Figure from Potter et al. [1984]. Reproduced by permission of the International Glaciological Society.)

ander Island. The major source of ice is the Antarctic Peninsula, whose ice streams almost reach the western margin of the ice shelf. In contrast, Alexander Island ice extends only a few kilometers into the ice shelf [Pearson and Rose, 1983].

The ice that accumulates within a catchment is formed from snow with a stable isotope composition which cannot be altered as the snow is compacted and recrystallizes into glacier ice. In a catchment which terminates in an ice shelf there must be a close association between the isotope composition of the present-day accumulation and that of old ice being removed by melting and iceberg calving. With both glaciological and oceanographic observations it is possible to investigate the transfer of mass and the ^{18}O isotope through a catchment, from deposition at the surface to melting from the bottom of an ice shelf. Because isotope ratios of precipitation are strongly dependent on changes in air temperature it is possible to examine whether any secular change may have occurred in a period comparable to the residence time of ice in the catchment.

To determine the mass and isotope balance of George VI Ice Shelf catchment it is necessary to know $\dot{a}$ and δ_a for the whole of the catchment, where $\dot{a}$ is the accumulation rate as a function of position and δ_a is the corresponding isotope ratio. The following calculation is a summary of that presented in Potter et al. [1984].

To show the range over which the isotope ratio varies for the ice terminating in the northern and southern parts of the ice shelf, the catchment was divided into northern and southern areas which were evaluated separately. Alexander Island represents only a small part of the catchment area and some of it lies in a precipitation shadow [Bishop and Walton, 1981]. For simplicity it has been neglected, and the income to the ice budget may be some 3-4% low as a result.

The two catchment sections are contained within the boundaries of the east coast of Alexander Island, the peninsula crest and assumed flowlines in the north and west. The catchment boundaries are shown in Figure 10 together with the sites at which the accumulation rate is known. The accumulation data for sites 1-14 were obtained from β radioactivity, density and isotope measurements on ice cores chiefly detailed by Peel and Clausen [1982]. Sites 15-21 were assessed from Behrendt's [1963, 1965] interpretation of Shimizu's [1964] stratigraphic studies on the Antarctic Peninsula traverse. Sites 22-26 were evaluated from stake measurements on George VI Ice Shelf [Bishop and Walton, 1981]. The data represent mean annual accumulations for dates varying from about 1930 for Behrendt's results to 1973 for Bishop and Walton's.

A smooth function of latitude, longitude, elevation, and their second order combinations was fitted to the accumulation data using a multiple regression to minimize the rms deviation between the function and data. Based on the known topography of the catchment, a contour plot of the accumulation function values for the catchment is shown in Figure 11a. The accumulation pattern reflects the path of maritime depressions moving in from the Bellingshausen Sea or on a southerly course from South America. These tracks lead to increased precipitation near the west and north coasts respectively. There is a marked accumulation shadow in the lee of the high mountains in the Douglas and LeMay ranges on Alexander Island. This reduces accumulation on part of the ice shelf to less than half that found elsewhere.

Oxygen isotope measurements are available from sites 1-14 within the catchment. A more complete set of values was derived from temperature measurements taking advantage of the close relationship that exists between temperature and the isotope ratio of precipitation, [Peel and Clausen, 1982]. There is some scatter of the available data about a smooth empirical relationship, but hardly more than is expected from the known interannual variability in the climate. Given the temperature of a site this enables us to assign an isotope ratio with only a small expected error (± 1.2 $^o/_{oo}$).

The temperature field of the catchment was determined by the multiple regression of temperatures against latitude and elevation from 53 sites in the Antarctic Peninsula [Potter et al., 1984] to give the following relationship:

$$T = -7.81 - 0.73 \times (\emptyset - 70 \text{ S}) - 6.38 \times 10^{-3} h \quad (7)$$

where $\emptyset$ is the latitude in degrees and h is the elevation in meters. This equation agrees very closely with the results of Martin and Peel [1978] and Reynolds [1978a] who addressed the same problem, and has a standard deviation of 1.0°C. The standard deviation is of the same size as interannual temperature variations over the period of the temperature measurements [Limbert, 1974, 1980]. The δ field, derived from equation (7) and the δ-T relationship, is shown as a contour map over the catchment in Figure 11b.

To evaluate the total accumulation and its isotope composition the values estimated by the empirical functions were integrated over the catchment. The results are shown in Table 1.

Equilibrium Mass Balance

In general, conservation of mass requires that

$$\int \dot{a} dx = \dot{A} = dM/dt + \Sigma \dot{M}_i \quad (8)$$

where $\dot{a}$ is the accumulation rate per unit area, $\dot{A}$ is the total accumulation rate, M is the total mass in the catchment, $\dot{M}_i$ are the mass loss rates and the integration is performed over the catchment area, x. The following calculation begins by assuming $dM/dt = 0$, the equilibrium condition, and evaluates the various $\dot{M}_i$.

Ablation occurs in summer at the low-lying margins of the ice sheet, and throughout the year at the bottom of the ice shelf. Basal melt is the major source of ablation; as little as 0.4 km^3 yr^{-1} of water percolates directly into the sea from summer surface melt on this ice shelf between 70°30'S and 72°S [Reynolds, 1981b]. Multiyear shore-fast sea ice up to 8 m thick in the north of George VI Sound dampens swell and tends to restrain the ice front. As a result, the ice front advances for several years before breaking back episodically. Comparison of Landsat satellite images shows that between 1974 and 1979 some 25 km^3 was lost in this way, probably late in the austral summer of 1976-1977. The long-time interval and irregularity of these surges makes it unreasonable to assess the rate of calving from changes in ice front position. Therefore the mass loss calculations for both ice fronts were performed using forward ice velocities integrated with ice front thickness and width to give an ice flux. For the north, Bishop and Walton [1981] gave surface ice velocities for a stake scheme some 40 km south of the ice front. These were extrapolated northwards following Paterson [1981, p. 174] to give an ice front velocity of 125 m yr^{-1} with a corresponding equilibrium ice loss of 0.3 km^3 yr^{-1}. The southern ice front is less well known and only a crude estimate of 300 to 400 m yr^{-1} ice front velocity could be made, leading to a speculative estimate of 3.5-4.6 km^3 yr^{-1} ice for the equilibrium calving loss. The only remaining significant mass loss in equation (8) is basal melting, which we evaluate by subtraction of the other terms. For the northern part of the ice shelf the equilibrium basal melt rate is 2.07 $\pm$ 0.23 m yr^{-1} and in the south the value is 2.17 $\pm$ 0.14 m yr^{-1} (Table 1).

If $dM/dt \neq 0$, then the ice shelf and catchment are not in equilibrium, and the predicted basal melt rates should be reinterpreted as the sum of basal melting and some change in thickness of either the ice shelf or catchment ice sheet. Reynolds [1983] calculated an approximate value for the ice flux across the grounding line into the north of George VI Sound by combining ice depths from radio echo sounding with ice velocities determined by theodolite survey of stake schemes. North of 72 30'S the grounding line flux has a value of 17.4 km^3 yr^{-1} ice. Based on the accumulation

TABLE 1. Accumulation Over Palmer Land and the Ice Shelf, Ice Losses Through Surface Drainage, Calving, and Calculated Equilibrium Basal Melt Rates for the Northern, Southern, and Entire Ice Shelf

	Ice Flux Into Ice Shelf	Ice Shelf Accumulation	Flux Into Ice Shelf + Ice Shelf Accumulation	Melt Water Drainage	Iceberg Calving	Equilibrium Basal Melt
Northern Part of Catchment						
$\bar{\delta}$, °/°°	-20.8 ± 0.5	-13.9 ± 1.0	-19.6 ± 0.5	-14.0 ± 1.0	-14.0 ± 1.0	-19.8 ± 0.5
$\dot{A}$, km^3 yr^{-1} ice	18.4 ± 2.1 $(17.4)^a$	3.7 ± 1.2	22.1 ± 2.4	0.4 ± 0.2	0.3 ± 0.1	21.3 ± 2.4 (2.07 ± 0.23 m yr^{-1})
Area, 10^3 km^2	31.7 ± 0.8	10.3 ± 0.1	42.0 ± 0.8			10.3 ± 0.1
Southern Part of Catchment						
$\bar{\delta}$, °/°°	-22.5 ± 0.4	-15.0 ± 0.9	-20.7 ± 0.4		-15.0 ± 1.0	-21.5 ± 0.5
$\dot{A}$, km^3 yr^{-1} ice	27.1 ± 1.8	8.4 ± 1.0	35.5 ± 2.0		4.0 ± 0.6	31.6 ± 2.1 (2.17 ± 0.14 m yr^{-1})
Area, 10^3 km^2	45.6 ± 1.2	14.6 ± 0.1	60.1 ± 1.2			14.6 ± 0.1
North and South of Catchment						
$\bar{\delta}$, °/°°	-21.8 ± 0.3	-14.7 ± 0.7	-20.3 ± 0.3	-14.0 ± 1.0	-14.9 ± 1.0	-20.8 ± 0.3
$\dot{A}$, km^3 yr^{-1} ice	45.6 ± 2.7	12.1 ± 1.6	57.6 ± 3.1	0.4 ± 0.2	4.3 ± 0.6	52.9 ± 3.2 (2.12 ± 0.13 m yr^{-1})
Area, 10^3 km^2	77.2 ± 1.5	24.9 ± 0.2	102.1 ± 1.5			24.9 ± 0.2

^{a}J.M. Reynolds [1983]
(Reproduced from Potter et al. [1984], the Journal of Glaciology by permission of the International Glaciological Society).

map of Figure 11a, the necessary steady state ice flux across the same length of grounding line is 18.4 km^3 yr^{-1}, some 6% larger than Reynolds' estimate. This discrepancy is small compared to the uncertainties in each calculation. One interpretation of this result is that the present-day accumulation balances the contemporary ice flow into the sound, confining any nonequilibrium behavior to the ice shelf. The equilibrium basalt melt rates can, in this case, be considered as the arithmetic sum of basal melt and ice shelf thickening. If the ice shelf is also in equilibrium a basal melt rate of 53 km^3 yr^{-1} is predicted. Kotlyakov et al. [1978] estimated the total loss for Antarctic ice shelves to be 320 km^3 yr^{-1}. If it is in equilibrium, George VI Ice Shelf would appear to be a major contributor to the ice melt budget of Antarctica, supplying one-sixth of the total ice shelf melt despite its small size.

Bishop and Walton [1981] predict a mean equilibrium melt rate of some 2 m yr^{-1}, which is consistent with the result produced here. If the mean melt rate is less than 2 m yr^{-1} there will be an overall thickening of the ice shelf. The oceanographic observations previously presented suggest that the mean melt rate must be at least 1 m yr^{-1} and probably no more than 2 m yr^{-1}, limiting any overall thickening to 1 m yr^{-1} and making thinning an unlikely possibility. Since the present rate of accumulation will support the grounding line mass flux it is expected that any thickening would continue for some time to come. The northern ice front, however, is retreating, presumably in response to an earlier dearth in the ice supply. Any net thickening should cause the ice fronts to advance again as this effect propagates along the ice shelf.

Equilibrium Isotope Balance

$$\int \delta_a \dot{a} dx = \delta_t \, dM/dt + \Sigma \delta_i \dot{M}_i \qquad (9)$$

where δ_a is the oxygen isotope ratio of the accumulation $\dot{a}$, δ_t is the mean isotope ratio of the thickening or thinning ice, and the δ_i are average isotope values for the mass losses $\dot{M}_i$. Once again, the following calculation begins by considering $dM/dt = 0$. Hence $\bar{\delta}$ is defined by

$$\bar{\delta} = \int \delta_a \, \dot{a} dx / \int \dot{a} dx = \int \delta_a \dot{a} dx / \dot{A} \qquad (10)$$

so that $\bar{\delta}$ represents the mean isotope ratio of the accumulation. Considering ice losses to be primarily calving (δ_c, $\dot{M}_c$), melting (δ_m, $\dot{M}_m$), and surface drainage (δ_d, $\dot{M}_d$) it follows that

$$\delta_m = (\bar{\delta}\dot{A} - \delta_c\dot{M}_c - \delta_d\dot{M}_d)/(\dot{A} - \dot{M}_c - \dot{M}_d) \qquad (11)$$

δ_m is an equilibrium estimate of the isotope ratio of melting ice derived solely from glaciological evidence. The ice emerging at each ice front is chiefly derived from local precipitation so δ_c has been given the appropriate value from Figure 11b.

Table 1 displays the isotope ratios calculated for different parts of the catchment from glaciological observations. A deduction from the data is that in equilibrium the δ value of ice melting in the sea is $-19.8 \pm 0.5°/_{oo}$ in the north of the catchment and $-21.5 \pm 0.5°/_{oo}$ in the south. If our circulation model is correct, the 0.4 km^3 yr^{-1} of surface melt ($\delta = -14°/_{oo}$) which drains into the sound will mix with the northern outflow. Even so, the expected isotope ratio in the sea is only reduced to $-19.7°/_{oo}$. This result is close to the value of $-20.3°/_{oo}$ derived from oceanographic measurements using the Δ-S relationship (equation (6)). We are led to conclude that the ice melting at the northern ice front is virtually all from the ice shelf with little contribution from sea ice or snow precipitated over the sea since their expected δ values of $+3°/_{oo}$ and $-13°/_{oo}$ are each so dissimilar.

Analysis of a deuterium record from an ice core spanning the years 1781-1981 from James Ross Island in the Antarctic Peninsula shows no persistent temperature trend for this 200-year period [Merlivat, presentation at IAMAP Symposium at IUGG, Hamburg, 1983]. We therefore anticipate that our accumulation δ values will be representative of the last century. The δ value deduced from the Δ-S relationship in the sea at the northern ice front is relevant to accumulation from the present time to a few thousand years ago now melting from the underside of the ice shelf. We might expect any secular warming to produce a discrepancy between the two δ values. The maximum difference between the oceanographically derived δ value and the δ values for the accumulation in the north and south of the catchment is $-0.6°/_{oo}$, insignificant compared to the expected error. From an estimate of the mutual standard deviation for the δ values there is an 80% confidence level that the melting ice was, on average, deposited in a climate no more than $2°/_{oo}$ "colder" than at present, corresponding to some 2.5°C. Alternatively, it has been suggested that the Antarctic Peninsula climate has cooled over the last few thousand years. Clapperton and Sugden [1982] have found barnacle shells in an ice shelf moraine on Alexander Island at the edge of George VI Ice Shelf. Radiocarbon dating indicated a minimum shell age of 6200 ± 60 years. Their implication is that at this time the climate must have been some 2-4°C warmer than at present. If this were so, and temperatures have warmed roughly linearly with time, we might expect to see some $+0.8°/_{oo}$ difference between our two δ

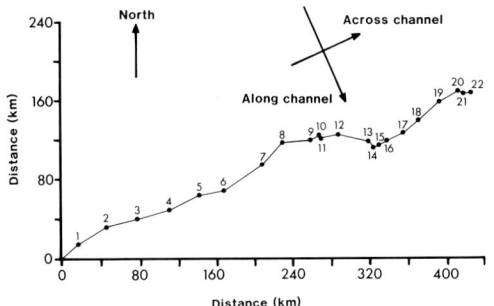

Fig. 12. The progressive vector diagram, annotated weekly, from February 11, 1980 to July 15, 1980, for the current meter record from the northern ice front. The mean flow is occasionally of very low amplitude but its direction is well aligned across the narrow channel. Mean flow near the channel margins has yet to be investigated.

values. There is an 80% certainty that our δ values do not come from distributions with such a separation.

We conclude that although there may be quite large variations in the temperature and accumulation rate over the western flank of the Antarctic Peninsula with periods ranging from decades to several centuries, there is no strong evidence for a secular change in the Antarctic Peninsula over the last few millennia.

Long-Term Flow and Tidal Interaction

The 5-month RCM4 record from 156 m has been analyzed and this, together with some preliminary results from two 13-month RCM4 records from the same site at 116 and 201 m, enables us to examine tidal effects and low-frequency flow. An analysis of the 5-month record is given in Loynes et al. [1984].

As expected from our summer observations, currents were weak to moderate at the mooring site. The mean speed was 5.2 cm s^{-1} and the maximum 16.3 cm s^{-1}. The Savonius rotor was stalled for 12% of the record. A progressive vector diagram of the current record is shown in Figure 12. The calculation was made from the original time series and annotated at weekly intervals. The average velocity of the mean flow was 3.4 cm s^{-1} toward 067°T (true north). Toward the middle of the data record the currents became weaker and more variable in direction. Before the end of week 9, on April 15, 1980, the average speed of the mean flow was 5.3 cm s^{-1}, whereas afterwards it reduced to an average of 2.4 cm s^{-1}. The direction of the flow was parallel to the nearby ice front and also parallel to the rift in the ice shelf at the mooring site. The mean flow is thus perpendicular to the long axis of George VI Sound. This unexpected behavior may be a local effect or part of a more general circulation. Possible causes will be examined later.

Spectral Analysis of Currents

The spectra of the along and across channel components are shown in Figure 13. Along channel there are distinct energy peaks in tidal species 1, 2, 3, and 4 (approximate periods of 24, 12, 8, and 6 hours, respectively). The semidiurnal tides are dominated by S_2. There is also significant energy at low frequencies (lower than 0.02 cph). Across channel there is distinctly less energy associated with the tidal frequencies whereas at low frequencies there is rather more energy as a result of the across channel flow.

The total kinetic energy of the current record, KE_T can be divided between kinetic energy of the mean flow, KE_M, and the fluctuating kinetic energy, KE_F. Per unit mass and for each orthogonal component,

$$KE_T = KE_M + KE_F$$

where

$$KE_M = 1/2 \ x^2$$

$$KE_F = 1/2 \ \sigma^2$$

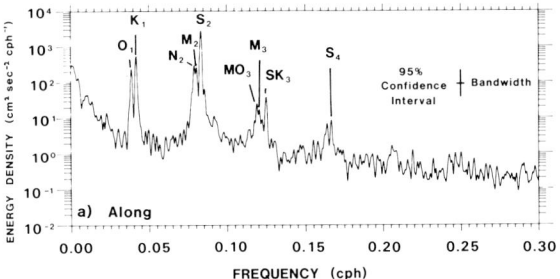

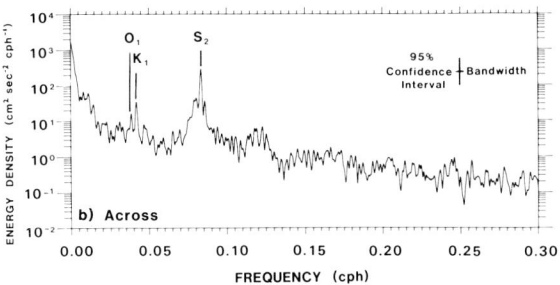

Fig. 13. Spectral current energy densities at the northern ice front (a) resolved along the channel and (b) resolved across the channel. Semidiurnal tides clearly dominate along the channel, although across the channel they are superseded by the low-frequency flows of 4-day period or more.

TABLE 2. Kinetic Energy Distribution at the North of George VI Sound

Direction	Mean Flow Energy cm² s⁻²	Fluctuating Flow Energy, cm² s⁻²				
		Low Frequency	Diurnal	Semi-diurnal	Ter-diurnal	Remaining Noise
Along the channel	0.00	1.66 (24%)	1.02 (15%)	3.95 (56%)	0.12 (2%)	0.25 (3%)
Across the channel	5.91	3.52 (82%)	0.11 (3%)	0.55 (13%)	0.04 (1%)	0.06 (1%)
Total along and across the channel	5.91	5.18 (46%)	1.13 (10%)	4.50 (40%)	0.16 (1%)	0.31 (3%)

Percentages show the contribution of each frequency band to the total fluctuating energy for that row.

TABLE 3. Tidal Signals at the North of George VI Sound

Constituent Symbol	Predicted Tidal Height[a] (Adjusted Tide Gauge Data)		Observed Tidal Current[b] Current Along the Channel		Observed Tidal Current[b] Current across the Channel		Observed Temperature Signal[b]	
	H, cm	G, deg.	V, cm s⁻¹	G, deg.	V, cm s⁻¹	G, deg.	T, x 10⁻³ °C	G, deg.
Q_1	5.8	81	0.26	334	<0.05		1.6	289
O_1	25.8	71	1.23	0	0.20	72	5.1	280
P_1	10.5	80	0.61	22	0.11	128	1.7	287
K_1	30.3	85	1.74	26	0.36	91	7.0	306
J_1	1.8	133	0.21	40	0.05	326	1.5	
$2N_2$	2.1	111	0.29	41	0.16	132	1.8	199
μ_2	2.4	116	0.29	51	0.09	60	<1.5	
N_2	8.2	144	1.08	64	0.34	106	<1.5	
M_2	15.2	259	1.00	168	0.23	275	<1.5	
T_2	1.7	62	0.55	62	0.29	230	<1.5	
S_2	23.9	67	3.87	20	1.25	125	9.4	321
K_2	6.7	68	1.07	12	0.29	125	4.4	339
MO_3			0.38	3	0.17	136	1.8	296
M_3	1.1-2.4[c]	353[c]	0.24	245	0.06	334	<1.5	
SO_3			0.11	104	0.08	325	<1.5	
MK_3			0.14	75	0.07	200	<1.5	
SK_3	1.3-2.8[c]	211[c]	0.39	93	0.13	237	1.8	292
SN_4			0.08	90	0.06	185	<1.5	
S_4			0.17	68	0.05	92	1.9	337

[a] Data from Cartwright [1980]. Phases have been adjusted to 70°01'S, 48°46'W.
[b] P_1, $2N_2$, μ_2, T_2, and K_2 may be unreliably resolved.
[c] Unadjusted values.

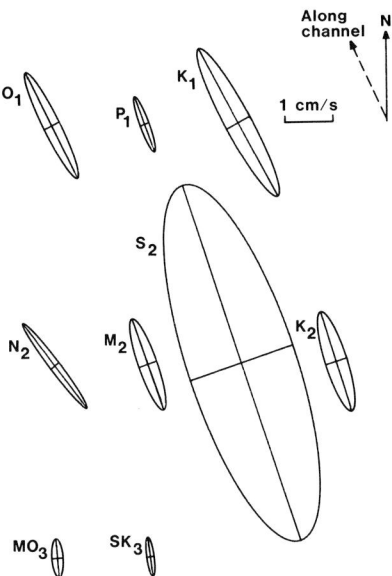

Fig. 14. Tidal current elipses for the major constituents. All the ellipses shown are described anticlockwise.

where x and σ^2 are the mean and variance of the current component.

The contribution to the fluctuating kinetic energy from different frequency bands is calculated by integration of the energy density over the band. Table 2 presents the values. Energy in the low-frequency and semidiurnal bands dominates the spectrum. Their combined energy comprises 86% of KE_F; 88% of the semidiurnal energy is associated with tidal flow along the channel. The greater part of the energy in the low-frequency band (68%) is across channel, which is the direction of the mean flow. The mean flow probably has variations with periodicity longer than the length of the record and it is suspected that the low-frequency currents and the mean flow share a common driving mechanism.

Harmonic Analysis of Currents

A spectral analysis gives information about energy at all frequencies but has poor resolution and confidence limits. To examine the tidal signal alone, a more precise calculation is to perform a harmonic analysis which evaluates energy at predetermined frequencies. A harmonic tidal analysis was applied to the 155-day current record. The accuracy of the analysis has been estimated from perturbation experiments on the data and from statistical theory [Godin, 1970]. Except for weaker partners of doublets or triplets, which have larger errors, expected errors are typically ±0.05 cm s^{-1} in amplitude and ±5° in phase. Phase errors escalate rapidly as the amplitude is reduced to near 0.05 cm s^{-1}. Constituents with nearly equal frequencies may be unreliably resolved if their beat frequency does not complete one cycle during the record. Results for the significant resolved constituents are given in Table 3. The amplitudes are given for along and across channel directions which are generally identical to the major and minor axes of the tidal ellipses. Tidal ellipses for the major constituents are shown in Figure 14. Table 3 also shows the major tidal height constituents predicted for the site. The prediction is derived from the analysis by Cartwright [1980] of tide gauge measurements at Barry Island (68°08'S, 67°05'W), and at lakes on the ice shelf which communicated with the sea, Ablation lake, and Hobbs Pool. The principal diurnal tidal currents southward along the channel have phases which precede tidal height by an average of 63°. The semidiurnal currents precede tidal heights by some 69° although the strongest constituents, S_2 precedes by only 47°. All the major constituents thus exhibit a strong standing wave nature as well as a southward progressive wave component.

A concise method of revealing the character of the tide is to calculate its complex admittance function, Z, relative to the equilibrium tide [Munk and Cartwright, 1966]. For tidal height in George VI Sound, Cartwright [1960] has published an exact analysis. The tidal currents admittance function has been calculated and $\ln|Z|$ is displayed in Figure 15 with the phase, θ, associated with the major axis of the tidal ellipse. Cartwright's results for tidal height at Ablation Lake are also shown for comparison.

The diurnal constituents show no unusual features. The values of $|Z|$ for both current and height remain roughly constant and the

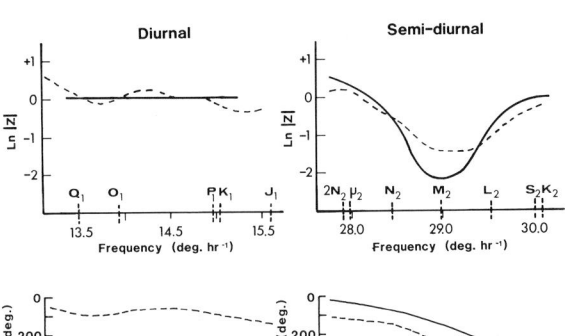

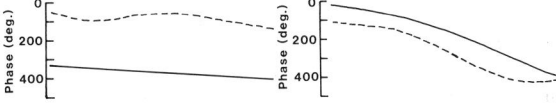

Fig. 15. Tidal current admittance amplitude and phase variation with frequency for the diurnal and semidiurnal bands at the northern ice front. Previous tidal height results for Ablation Lake from Cartwright [1980] are shown as dashed lines for comparison.

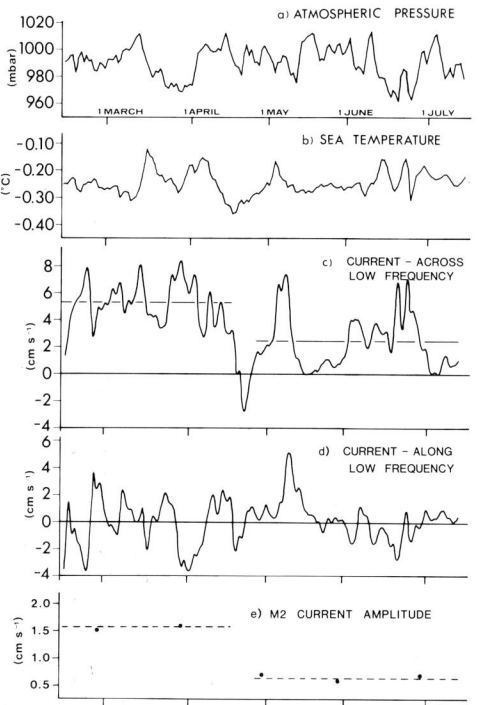

Fig. 16. Oceanographic and meteorological parameters and their variation from February 11, 1980 to July 15, 1980. A change in low frequency across channel currents and M_2 tidal current occurs in mid-April. Dahsed lines in Figure 16c and 16e show mean values before and after the transition.

phases increase linearly by less than 100° across the band. In contrast, the $|Z|$ curve for the semidiurnal currents dips in the region of M_2 and the phase swings by 360° across the band. The $M_2:S_2$ tidal current ratio is 0.26. A clear dip in the tidal height admittance is found near N_2 for the Argentine Islands (600 km north of George VI Sound) and near M_2 for Ablation Lake in George VI Sound, where the phase changes by around 360° with an $M_2:S_2$ ratio of 0.64. In the vicinity of an amphidrome, a change in phase of around 180° should occur across the amphidromic band [Cartwright, 1980] but tidal currents can remain very strong. The numerical model for the M_2 tide produced by Schwiderski [1979] shows no amphidromes in Drake Passage or near the west coast of the Antarctic Peninsula. The semidiurnal tidal heights and currents exhibit similar features which is uncharacteristic of an amphidrome and probably indicates a selective absorption of tidal energy. Cartwright [1980] discussed the considerable "age" of the tide on the west coast of the Antarctic Peninsula: 4 days at the Argentine Islands and 7 days at Ablation Lake. Webb [1973] had suggested that tidal "ages" over 2.5 days are associated with strong topographic tidal dissipation and local tidal resonance. The topography and ice covering of George VI Sound may provide just such an environment. Doake [1978] has discussed tidal dissipation in ice shelves and concludes that if the ice behavior is governed by parameters observed in creep tests then globally significant tidal dissipation occurs at ice shelf grounding lines.

The 155-day record was segmented into 5 x 29 day periods and a harmonic analysis performed on each period to obtain five independent estimates of the major tidal constituents. Except for M_2, the amplitudes remain virtually constant throughout the record. M_2, however, suffers an amplitude reduction of some 50% between the second and third periods, shown in Figure 16. This event coincides with the abatement of the across channel flow. The phase of M_2 remains steady but the sense of rotation for periods 3 and 5 changes to clockwise. Clockwise tidal rotation is unusual; every other constitutent rotates anticlockwise. It is relevant to enquire whether similar trends have been observed in tidal heights or currents elsewhere in Antarctic waters. D.E. Cartwright [personal communication, 1982] confirms that there are no significant seasonal variations in M_2 in the 9-year tidal height record from the Argentine Islands. However, Shesterikov [1959] at Mirnyy (66°30'S, 93°00'E) observed that the M_2 tidal height amplitude decreased as the sea ice formed, though no other constituent was affected. At Scott Base, near McMurdo Sound (77°53'S, 166°42'E), Gilmour et al. [1962] detected no seasonal changes in M_2 from several months of tidal height data. Foldvik and Kvinge [1974] observed a rapid reduction in the K_1 and O_1 current amplitudes in July/August in the deep water on the continental slope of the Weddell Sea. Foldvik and Kvinge explain the decay as a removal of a baroclinic resonance by advection of a different water type. There are no marked seasonal changes in temperature or salinity at 156 m at our mooring site. Even so, a change in the stratification at other depths or in Marguerite Bay could be tuned to the baroclinic modes of the M_2 tide and also cause changes in the across channel flow. The surface M_2 tide would not ordinarily be affected.

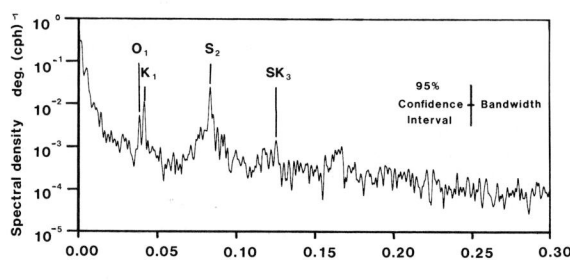

Fig. 17. Temperature spectrum for the 5-month record.

This mechanism may explain the concurrent changes in the across channel flow and M_2 current amplitude.

Excluding M_3, all species 3 and 4 tides are a result of nonlinear combinations of species 1 and 2. Figure 13 and Table 3 show that in this frequency range MO_3 and SK_3 have the largest amplitudes. Frictional effects can produce nonlinear coupling but MO_3 and SK_3 are second-order interactions, whereas frictional processes tend to generate third-order interactions such as M_6, $2MS_6$, S_6 which are not observed. Cartwright [1980] detected SK_3 in the tidal height records of George VI Sound and suggested that it may be caused either by an asymmetric response of the ice shelf to water level changes, or by the drag of tidal currents on the tide gauge wire. The latter option can now be discounted since the Aanderaa meter is not affected by viscous drag in the same way. The SK_3 amplitude thus appears to be a genuine response to the ice shelf-ocean system to the tide-generating forces.

Temperature Variations

Large fluctuations are present in the sea temperature record. In the 10 days following April 7, 1980, the temperature fell steadily by 0.27°C. This period ends with the coldest recorded temperature. No progressive cooling from summer to winter is evident. Over the 5-month period the temperature range was 0.35°C with a standard deviation of 0.049°C. A smoothed record of the temperature time series is shown in Figure 16.

A spectral analysis of the temperature record is shown in Figure 17. Significant energy exists at low frequencies between 0.00 and 0.03 cph, and there are distinct peaks corresponding to tidal species 1 and 2, with some indication of species 3 energy. Results of the harmonic analysis of the temperature signal are given in Table 3. With the exception of weaker partners of doublets or larger todal groups, typical errors are $\pm 1.5 \times 10^{-3}$ °C in amplitude and $\pm 8°$ in phase. Temperature variations at tidal frequencies are unlikely to be caused by an instrumental effect such as drag exerted on the instrument by tidal currents since the tides are so weak. A more plausible cause is a physical effect such as advection or a vertical movement of the instrument with respect to the isotherms. Diffusion is an unlikely cause of an oscillatory signal, and so we delete the diffusive terms from equation (3) to give the time-dependent temperature equation

$$\partial T/\partial t = -u\partial T/\partial x - v\partial T/\partial y - \omega\partial T/\partial z \quad (12)$$

We use the left-handed coordinate system with the x axis aligned parallel to the sound increasing southward and the y axis across the sound increasing eastward as before. If vertical movement of the instrument is ignored, equation (12) gives the recorded variation of temperature with time. We consider two possible driving mechanisms. First, in the limited regime of the rift, horizontal temperature gradients may be larger than on the scale considered for the general circulation and advection may provide the signal. The tidal currents convey water a distance of only some 270 m along channel and 90 m across channel, so the inferred horizontal temperature gradients would have only a local validity. Applying this interpretation to the strongest constituents implies approximate horizontal temperature gradients of $+3 \times 10^{-5}$ °C m^{-1} southward along the channel and $+4 \times 10^{-5}$ °C m^{-1} eastward across the channel. These gradients are rather large but may apply locally due to the topographic effects of the ice shelf rift below which the meter was moored.

Second, locally forced or freely propagating internal tides may drive the temperature signal. This leads to

$$\zeta_i = T_i/(\partial T/\partial z) \quad (13)$$

where ζ_i is the vertical displacement of the isotherm for constituent (i) and T_i is its temperature signal. Since $\partial T/\partial z$ is positive $(+7.7 \times 10^{-3}$ °C m$^{-1})$, T_i and ζ_i should be in phase. Table 3 lists the amplitudes and phases of the temperature signal and sea surface elevation. For the diurnal constituents internal wave motion would have to be locally driven, as the diurnal periods are outside the inertial subrange between the Brunt Väisälä period (22 min) and the inertial period (12.73 hours). For the principal diurnal constituents (O_1, P_1 and K_1), ζ_i is a factor of 2.6, 2.1, and 3.0 larger than the surface elevations, respectively, and phase lagged by $214 \pm 7°$. Turning to the principal semidiurnal constituents (N_2, M_2, S_2, and K_2), no tight grouping of enhancement factors emerge from a similar analysis. This is not surprising since the principal semidiurnals are all within the inertial subrange and so can generate freely propagating internal waves. With the temperature signal interpreted as a vertical displacement, the potential and vertical kinetic energy density at frequency ω can be calculated, and following Fofonoff [1969] it is possible to test whether the energy distribution is compatible with freely propagating internal wave motion

Potential Energy/Horizontal K.E. =

$$[N^2/(N^2-\omega^2)][(\omega^2-f^2)/(\omega^2+f^2)] \quad (14)$$

$$P.E./Total\ K.E. = [N^2(\omega^2 - f^2)]/$$
$$[(N^2-\omega^2)(\omega^2+f^2) + \omega^2(\omega^2-f^2)] \quad (15)$$

where N is the Brunt-Väisälä frequency and f is the inertial frequency.

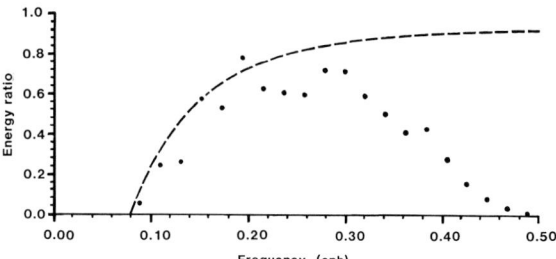

Fig. 18. Frequency dependence of the vertical potential to horizontal kinetic energy ratio. The theoretical line for freely propagating internal waves is shown dashed.

Unfortunately, spectral densities cannot be calculated for frequencies higher than 0.5 cph (since the data record has an hourly time base) yet the inertial subrange spans 0.079 to 2.73 cph. This restricts the comparison to less than 20% of the inertial subrange, where the two energy expressions have almost identical values. For this reason we test only the potential/horizontal kinetic energy. Figure 18 shows that the semidiurnal temperature signal is roughly consistent with internal wave motion but that above 0.25 cph the signal appears too small to be generated by internal waves. This falloff in energy is probably caused by the poor temperature resolution of the instrument which causes it to fail to respond to such small temperature oscillations, rather than a genuine physical effect. This leads us to conclude that freely propagating internal wave motion is a possible mechanism for the observed temperature signal at semidiurnal and higher frequencies. Phillips [1977] discusses internal waves and the possibility of internal wave energy residing in several modes at each frequency. Normally, most energy will reside in the lowest mode, in which case the entire thermocline heaves up and down in phase. If $(N^2-\omega^2) \geq 0$ over a considerable depth, a significant part of the internal wave energy may occur in higher modes with a corresponding variation in amplitude and phase with depth. Figure 19 shows the profile of Brunt Väisälä frequency at the northern ice front deduced from Figure 6. There is a sharp peak at about 80-90 m corresponding to the pycnocline at the base of the ice shelf. For semidiurnal frequencies $(N^2 - \omega^2) \geq 0$ over virtually all of the water column. If higher internal tidal modes are significant, a seasonal change in the vertical density profile could affect the modal structure of the internal tides and this process may be related to the attenuation of the M_2 tidal current discussed previously.

Low-Frequency Motions and Their Origins

The current spectra have shown there is a significant proportion of energy (46%) in the low-frequency band (Figure 13). Only a small part of this energy is coherent with the monthly and fortnightly tides; a major part of the flow is therefore due to other factors. Local wind stress is an unlikely cause, as all but a small fraction of the sea near by is ice covered throughout the year. Figure 16 shows time series plots of mean daily atmospheric pressure, sea temperature, and low-frequency current velocities. Beyond mid-April there is a good negative correlation between atmospheric pressure and across channel current speed. This is circumstantial evidence that the low-frequency circulation is, at least in part, atmospherically driven. To investigate this hypothesis, the low-frequency currents should ideally be compared with gradients of the local pressure field. Regrettably, lack of data prevents such calculations for the area between Marguerite Bay and Ronne Entrance for the 5-month record, but should be possible for part of the two 13-month records.

Figure 12 shows that the mean flow varies in magnitude but rarely in direction. The flow may be a local feature, in which case possible causes are the topography of the rift below which the instrument was moored or the presence of the ice front 6 km to the north; both are aligned parallel to the mean flow. Local interactions at the ice front may have a strong steering effect on the flow which could be particularly pronounced in the stratified water under the ice shelf. Alternatively, incoming tidal currents and internal waves may become rectified on reaching the partial barriers presented by the ice front and the rift.

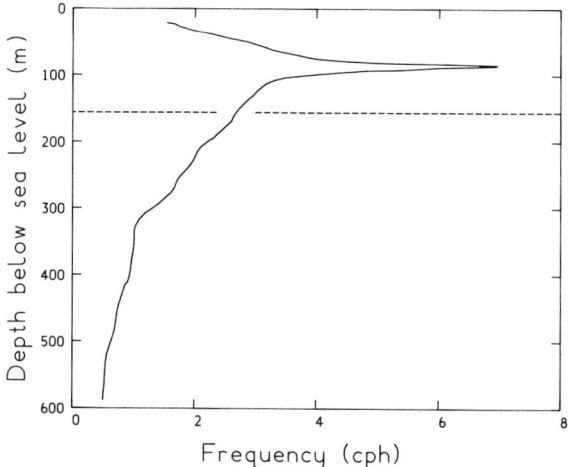

Figure 19. Profile of the Brunt Väisälä frequency using the profiles in Figure 6. As in Figure 6, the dashed line shows the mooring depth of the meter which provided the 5-month record. Below 110 m the curve has an approximately exponential decay with an e^{-1} decay depth of some 225 m.

Such a process is described by Robinson [1981], but the tidal currents are unlikely to be energetic enough to produce the observed mean flow. Preliminary results from the two 13-month records at the same site show that the across channel mean flow persists at least down to 201 m depth with only a few degrees change in heading. There is also a high correlation between variations in the mean flow at the depths of 116 and 201 m throughout the 13 months. This correlation implies that the driving mechanism affects the whole thermohaline convective layer (85 to 200 m depth) and could perhaps be generated by a general north-south slope of the isopycnals. The aim of present work is to drill through the ice shelf at sites between the northern ice front and the basal sill near 70°W to assess the evidence for such an estuarine type circulation.

TABLE A1. Temperature, Salinity, and (Where Available) Oxygen Isotope Ratio From Reversing Thermometers and Bottle Sampling for Five Sites Near the Northern Ice Front of George VI Ice Shelf

Depth, m (±1%)	Temperature, °C (±0.007)	Salinity, (±0.005)	δ^{18}, °/oo (±0.03)	σ_t
Site 1 (70°01.1'S, 68°46.1'W): Profile on December 30, 1979				
0.0	-1.656	32.821	-0.89	26.405
10.4	-1.816	33.595	-0.42	27.038
20.8	-1.783	33.605	-0.54	27.045
31.2	-1.756	33.621	-0.53	27.057
52.0	-1.538	33.714	-0.61	27.127
78.0	-1.287	33.805	-0.64	27.193
104.0	-0.698	34.110	-0.19	27.418
130.0	-0.360	34.189	-0.19	27.468
156.0	-0.154	34.292	+0.01	27.542
182.0	+0.081	34.366	+0.10	27.589
208.0	+0.254	34.487	-0.17	27.677
260.0	+0.540	34.538	+0.04	27.702
312.0	+0.709	34.649	-0.05	27.781
416.0	+0.885	34.641	-0.03	27.763
520.0	+0.952	34.688	+0.15	27.796
624.0	+0.994	34.717	+0.13	27.817
Site 2 (69°57.8'S, 68°39.7'W): Profile on January 16, 1980				
1.0	-0.389	18.240		14.593
10.4	-1.766	33.568		27.014
20.8	-1.729	33.527		26.980
31.2	-1.684	33.572		27.016
52.0	-1.573	33.655		27.080
78.0	-1.117	33.918		27.280
88.4	-0.868	33.981		27.322
104.0	-0.699	34.055		27.374
114.4	-0.526	34.131		27.428
130.0	-0.379	34.190		27.470
156.0	-0.238	34.288		27.542
208.0	+0.186	34.403		27.613
260.0	+0.485	34.523		27.693
312.0	+0.695	34.598		27.741
416.0	+0.888	34.650		27.770
520.0	+0.973	34.663		27.775
624.0	+0.980	34.669		27.779
Site 3 (69°59.5'S, 68°36.3'W): Profile on January 9, 1980				
1.0	+0.248	18.592	-7.96	14.878
10.4	-1.724	33.595	-0.45	27.036
20.8	-1.779	33.650	-0.39	26.081

TABLE A1. (continued)

Depth, m (±1%)	Temperature, °C (±0.007)	Salinity, (±0.005)	δ^{18}, °/oo (±0.03)	σ_t
\multicolumn{5}{c}{Site 3 (continued)}				
31.2	-1.723	33.607	-0.40	27.045
52.0	-1.666	33.665	-0.33	27.091
78.0	-1.241	33.823	-0.30	27.207
104.0	-0.627	34.085	-0.34	27.396
130.0	-0.359	34.197	+0.01	27.474
156.0	-0.166	34.273	+0.01	27.527
208.0	+0.168	34.396	+0.23	27.609
260.0	+0.435	34.513	+0.26	27.688
312.0	+0.646	34.566	+0.05	27.718
338.0	+0.690	34.609	+0.28	27.750
364.0	+0.787	34.608	-0.06	27.743
380.0	+0.829	34.623	+0.09	27.752

Site 4 (70°02.0'S, 69°04.7'W): Profile on January 14, 1980

Depth, m	Temperature	Salinity	δ^{18}	σ_t
1.0	-1.637	33.608		27.044
10.4	-1.631	33.700		27.118
20.8	-1.618	33.635		27.065
31.2	-1.243	33.801		27.189
52.0	-1.030	33.900		27.261
78.0	-0.909	33.962		27.307
88.4	-0.887	34.036		27.367
104.0	-0.771	34.024		27.352
114.4	-0.538	34.153		27.447
130.0	-0.273	34.267		27.527
156.0	-0.097	34.312		27.555
208.0	+0.130	34.466		27.667
312.0	+0.698	34.594		27.738
416.0	+0.881	34.651		27.771
520.0	+0.962	34.675		27.785
624.0	+0.989	34.699		27.803
728.0	+1.007	34.703		27.805

Site 5 (70°04.1'S, 68°40.1'W): Profile on February 14, 1980

Depth, m	Temperature	Salinity	δ^{18}	σ_t
1.0	-1.511	17.189	-8.52	13.723
10.4	-1.761	33.585	-0.73	27.028
20.8	-1.744	33.599	-0.56	27.039
31.2	-1.712	33.603	-0.60	27.042
52.0	-1.693	33.645	-0.54	27.076
78.0	-1.491	33.736	-0.21	27.144
104.0	-1.032	33.903	-0.27	27.264
130.0	-0.480	34.143	-0.34	27.436
156.0	-0.113	34.299	-0.09	27.545
208.0	+0.200	34.429	-0.13	27.633
260.0	+0.531	34.610	-0.02	27.760
312.0	+0.722	34.606	+0.07	27.745
416.0	+0.880	34.650	+0.16	27.771
520.0	+0.958	34.666	+0.18	27.778
624.0	+1.002	34.685	+0.26	27.790
728.0	+1.002	34.676	+0.29	27.783

Site numbers are those of the map in Figure 1b of Lennon et al. [1982]. Salinities were determined using a salinometer corrected for the 1978 Practical Salinity Scale after Lewis and Perkin [1981]. Isotope values were determined by H. B. Clausen of Copenhagen.

Acknowledgments. The authors would like to thank M. Pedley and J. Tighe for their fieldwork in 1983-1984 which has helped modify some aspects of this paper. We also thank J. Loynes and J. Gould of the Institute of Oceanographic Sciences, Wormley, England, for their constructive criticism of the manuscript and preliminary analysis of the 13-month mooring records. Finally, we are indebted to S. Jacobs and anonymous referees, who have clarified many aspects of Antarctic oceanography.

References

Behrendt, J.C., Seismic measurements on the ice sheet of the Antarctic Peninsula, J. Geophys. Res., 68, 5973-5990, 1963.

Behrendt, J.C., Densification of snow on the ice sheet of Ellsworth Land and the southern Antarctic Peninsula, J. Glaciol., 5(40), 451-460, 1965.

Bishop, J.F., and J.L.W. Walton, Bottom melting under George VI Ice Shelf, Antarctica, J. Glaciol., 27(97), 429-447, 1981.

British Antarctic Territory ice thickness map, Alexander Island, 1:500,000 series BAS 500 R Sheet 1, edition 1, Cambridge, 1983.

Cartwright, D.E., Analyses of British Antarctic Survey tidal records, Br. Antarct. Sur. Bull., 49, 167-179, 1980.

Clapperton, C.M., and D.E. Sugden, Late quaternary glacial history of George VI Sound area, West Antarctica, Quat. Res., 18, 243-267, 1982.

Crabtree, R.D., B.C. Storey, and C.S.M. Doake, The structural evolution of George VI Sound, Antarctic Peninsula, in Symposium on Geophysics of the Polar Regions: XVIII General Assembly of the International Union of Geodesy and Geophysics, in press, 1985.

Doake, C.S.M., Dissipation of tidal energy by Antarctic ice shelves, Nature, 275, 304-305, 1978.

Doake, C.S.M., State of balance of the ice sheet in the Antarctic Peninsula, Ann. Glaciol., 3, 77-82, 1982.

Fofonoff, N.P., Spectral characteristics of internal waves in the ocean, Deep Sea Res., 16, 58-71, 1969.

Foldvik, A., and T. Kvinge, Bottom currents in the Weddell Sea, Geophys. Inst. Rep., 37, 43 pp., Univ. of Bergen, 1974.

Gade, H.G., Melting of ice in seawater: A primitive model with application to the Antarctic ice shelf and icebergs, J. Phys. Oceanogr., 9, 189-198, 1979.

Gilmour, A.E., W.J.P. Macdonald, and F.G. Van der Hoeven, Winter measurements of sea currents in McMurdo Sound, N. Z. J. Geol. Geophys., 5, 778-789, 1962.

Godin, G., The resolution of tidal constituents, Int. Hydrogr. Rev., 133-143, 1970.

Greisman, P., On upwelling driven by the melt of ice shelves and tidewater glaciers, Deep Sea Res., 26, 1051-1065, 1979.

Jacobs, S.S., R.G. Fairbanks, and Y. Horibe, Origin and evolution of water masses near the Antarctic continental margin: Evidence from $H_2^{18}O/H_2^{16}O$ ratios in seawater, this volume.

Jacobs, S.S., A.L. Gordon, and J.L. Ardai, Jr., Circulation and melting beneath the Ross Ice Shelf, Science, 203, 439-443, 1979.

Josberger, E.G., Laminar and turbulent boundary layers adjacent to melting vertical ice walls in salt water, Ph.D. thesis, Univ of Washington, Seattle, 1979.

Kotlyakov, V.M., K.S. Losev, and I.A. Loseva, The ice budget of Antarctica, Polar Geogr., 2, 251-262, 1978.

Lennon, P.W., J. Loynes, J.G. Paren, and J.R. Potter, Oceanographic observations from George VI Ice Shelf, Antarctic Peninsula, Ann. Glaciol., 3, 178-183, 1982.

Lewis, E.L., and R.G. Perkin, The practical salinity scale 1978: Conversion of existing data, Deep Sea Res., 28A, 307-328, 1981.

Lewis, E.L., and R.G. Perkin, Supercooling and energy exchange near the Arctic Ocean surface, J. Geophys. Res., 88, 7681-7685, 1983.

Lewis, E.L., and R.G. Perkin, The winter oceanography of McMurdo Sound, Antarctica, this volume.

Limbert, D.W.S., Variations in the mean annual temperature for the Antarctic Peninsula, 1904-72, Polar Rec., 17, 303-306, 1974.

Limbert, D.W.S., Surface temperature changes in Antarctica, including the Antarctic Peninsula, 1944-1978, Br. Antarct. Sur. Ann. Rep., 1979-80, 25-28, 1980.

Loynes, J., J.R. Potter, J.G. Paren, Current temperature, and salinity beneath George VI Ice Shelf, Antarctica, Deep Sea Res., 9A, 1037-1055, 1984.

MacAyeal, D.R., Evolution of tidally triggered meltwater plumes below ice shelves, this volume.

Martin, P.J., and D.A. Peel, The spatial distribution of 10 m temperatures in the Antarctic Peninsula, J. Glaciol., 20(83), 311-317, 1978.

Munk, W.H., and D.E. Cartwright, Tidal spectroscopy and prediction, Proc. Ro. Soc. London, Ser. A, 259, 533-581, 1966.

Paren, J.G., and J.R. Potter, Isotopic tracers in polar seas and glacier ice, J. Geophys. Res., 89, 749-750, 1984.

Paren, J.G., J.R. Potter, and J. Loynes, Using oxygen isotope tracers to determine the ocean circulation under an ice shelf, International association of meteorology and atmosphere physics program and abstracts, IUGG 18th General Assembly, Hamburg, p. 32, 1983.

Paterson, W.S.B., The Physics of Glaciers, 2nd ed., Pergamon, New York, p. 174, 1981.

Pearson, M.R., and I.H. Rose, The dynamics of George VI Ice Shelf, Br. Antarct. Sur. Bull., 52, 205-220, 1983.

Peel, D.A., and H. Clausen, Oxygen-isotope and total beta-radioactivity measurements on 10 m ice cores from the Antarctic Peninsula, J. Glaciol., 28(98), 43-55, 1982.

Phillips, O.M., The Dynamics of the Upper Ocean, Cambridge Monogr. Mech. and Appl. Math., 2nd ed., Cambridge University Press, New York, pp. 209-211, 1977.

Pond, S., and G.L. Pickard, Introductory Dynamical Oceanography, 2nd ed., Pergamon, New York, p. 161, 1983.

Potter, J.R., J.G. Paren, and J. Loynes, Glaciological and oceanographic calculations of the mass balance and oxygen isotope ratio of a melting ice shelf, J. Glaciol., 30(105), 161-170, 1984.

Reynolds, J.M., The distribution of mean annual temperatures in the Antarctic Peninsula, Br. Antarct. Sur. Bull., 54, 123-133, 1981a.

Reynolds, J.M., Lakes on George VI Ice Shelf, Antarctica, Polar Rec., 20, 425-432, 1981b.

Reynolds, J.M., Geophysical studies of the ice of the Antarctic Peninsula, Ph.D. thesis, pp. 19-20, Council for Nat'l. Acad. Awards, Brit. Antarc. Surv., Cambridge, 1983.

Robinson, I.S., Tidal vorticity and residual circulation, Deep Sea Res., 28A, 195-212, 1981.

Schwiderski, E.W., Global ocean tides, II, The semi-diurnal principal lunar tide (M_2) atlas of tidal charts and maps, Rep. NSWC TR 79-414, Naval Surface Weapons Center, Dahlgren, Va., 1979.

Shesterikov, N.P., Sea-level variations in the Mirnyy area, Sov. Antarct. Exped., Inf. Bull., Engl. Transl., 2, 18-21, 1959.

Shimizu, H., Glaciological studies in West Antarctica, 1960-1962, in Antarctic Snow and Ice Studies, Antarct. Res. Ser., vol. 2, edited by M. Mellor, pp. 37-64, AGU, Washington, D.C., 1964.

U.S. National Oceanographic Data Center, Inventory of the Antarctic, USNODC, October 1974, National Oceanographic Data Center, Washington, D.C., 1974.

Vanney, J.R., and G.L. Johnson, GEBCO bathymetric sheet 5.18 (Circum-Antarctic), this volume.

Webb, D.J., On the age of the semi-diurnal tide, Deep Sea Res., 20A, 852-857, 1973.

Weiss, R.F., H.G. Ostlund, and H. Craig, Geochemical studies of the Weddell Sea, Deep Sea Res., 26A, 1093-1120, 1979.

(Received March 1, 1984; accepted July 20, 1984.)

ORIGIN AND EVOLUTION OF WATER MASSES NEAR THE ANTARCTIC CONTINENTAL MARGIN: EVIDENCE FROM $H_2^{18}O/H_2^{16}O$ RATIOS IN SEAWATER

Stanley S. Jacobs and Richard G. Fairbanks

Lamont-Doherty Geological Observatory of Columbia University, Palisades, New York 10964

Yoshio Horibe

Toyogaoka, Tama-shi, Tokyo 206, Japan

Abstract. We use measurements of the temperature, salinity and oxygen isotope content of seawater in the Ross Sea and beneath the Ross Ice Shelf to define water types and differentiate between melting, freezing, and mixing processes. The Ross Sea and Weddell Sea have remarkably similar temperature, salinity and $\delta^{18}O$ characteristics, and tongues of relatively warm and very cold water that traverse the continental shelves between the deep ocean and glacial ice. The inflows provide heat for melting of the land ice, confirmed by the low $\delta^{18}O$ content (-0.71°/oo) of seawater directly beneath the Ross Ice Shelf. Ventilation of the deep ocean at the Slope Front adjacent to the continental margin is most strongly influenced by Low Salinity Shelf Water. High Salinity Shelf Water resulting from sea ice freezing in shore leads and polynyas in the western Ross Sea may regulate the subsurface flow of warm water onto the continental shelf. We outline the water, ice, marine precipitation, heat, salt and $\delta^{18}O$ budgets for the circumpolar Antarctic continental shelf. With an iceberg calving rate of 1.3×10^3 km^3 yr^{-1}, net basal melting of 0.4 m yr^{-1} will occur under the floating portions of an equilibrium Antarctic ice sheet. This corresponds to a shelf water residence time of 6 years, effective sea ice production over the shelf of 1.9 m, and an apparent bottom water production rate of 13×10^6 m^3 s^{-1}. Ice Shelf Water formed beneath the glacial ice contributes about 20% of the shelf water component of bottom water. As much brine is required to balance glacial melting as to provide the salinity increase between waters transported on and off the continental shelf.

Introduction

Several decades ago it was commonly assumed that disintegration and melting of continental ice was the primary determinant of the polar characteristics of Antarctic surface water, shelf water and bottom water [e.g., Pettersson, 1904; David, 1914; Drygalski, 1928]. However, after the work of Brennecke (1921), sea surface processes were recognized as the dominant influence in making Antarctic waters cold and dense. The potential role of glacial ice was then ignored, for several good reasons. It was apparent that melting would increase water column stability, and the potential volume of glacial meltwater was shown to be much less than that of marine precipitation over the entire Southern Ocean area below 45°S. There were few data from the continental shelf regions and none from beneath the ice shelves. The low thermal conductivity of glacial ice posed severe limits to sea ice accretion, and the significance of the freezing temperature dependence upon pressure was not fully appreciated.

Recent oceanographic observations near the Antarctic continent and concern about the role of the high latitudes in climatic change have renewed interest in glacial melting. The ice shelves, glaciers, and icebergs that fringe the coastline float in water that is cold even by Antarctic standards, but commonly warmer than the in situ melting point. The extreme depletion of $H_2^{18}O$ in glacial ice relative to seawater makes the ratio $H_2^{18}O/H_2^{16}O$ a potentially useful tracer of glacial melting and of circulation and mixing on and near the continental shelf. However, the impact of glacial meltwater must be differentiated from that of local precipitation, also greatly depleted in $H_2^{18}O$, and from the fractionation that occurs during freezing. Weiss et al. (1979) found that shelf water in the Weddell Sea contained meltwater from the base of the Filchner Ice Shelf and linked the melting rate by mass balance calculations to the rate of bottom water formation. We attempt here to expand on their work, using temperature/salinity (Θ/S) and oxygen-18/salinity ($\delta^{18}O$/S) relationships to decipher the origin and evolution of each identifiable water type in the Ross Sea. Based upon comparisons with other Antarctic oxygen isotope data, these results are extended to the circumpolar continental shelf. Estimates are then made of salt flux, sea ice thickness, heat transport, and shelf water residence time, consistent with mass balance of the Antarctic ice sheet.

The $H_2^{18}O/H_2^{16}O$ Ratio ($\delta^{18}O$) in Polar Waters

Epstein and Mayeda [1953] postulated that isotopic fractionation led to precipitation depleted in $\delta^{18}O$ at high latitudes and elevations. Meltwater from the resulting glacial ice would spike surface waters, resulting in characteristic end points on world ocean $\delta^{18}O$-/S diagrams. Craig and Gordon [1965] estimated that freshwater discharged to Antarctic waters from the glaciated continent would represent only 2 to 8% of that added by marine precipitation and thus could not strongly influence seawater $\delta^{18}O$ content, except immediately adjacent to Antarctica. By applying parameters appropriate to the continental margin, we estimate below that glacial meltwater exceeds the impact of marine precipitation on the $\delta^{18}O$ content of shelf water.

There is relatively little runoff from the continental ice in the form of surface meltwater streams [Bull, 1971], so most of the glacial ice $\delta^{18}O$ signal in seawater must derive from subsurface melting. Another, probably less significant, way that discharge from the ice sheet enters the shelf waters is via shore leads and small polynyas around the ice sheet perimeter [e.g. Priestley, 1914; Cavalieri and Martin, this volume; Kurtz and Bromwich, this volume; Zwally et al., this volume]. Winds will move snow off the ice sheet to these open water areas [Kotlyakov et al., 1977]. That displaced precipitation may be incorporated into the forming sea ice, be directly entrained into the water column, or enter it later when the sea ice melts. Also, meltwater that exists regionally at the ice sheet base [Robin et al., 1970] may coalesce into streams and flow into the sea.

Redfield and Friedman [1969] investigated the relative influence of meteoric water, meltwater, and brine on polar and deep ocean waters. They noted that in some confined Arctic straits and fjords, about three-quarters of the apparent subsurface freshwater component is derived from glacial ice. They combined meteoric water, snow and glacial ice to arrive at an average deuterium value equivalent to $-32.5°/oo$ in $\delta^{18}O$. In the Ross Sea the $\delta^{18}O$ content of precipitation is about $20°/oo$ higher (more positive) than that of glacial meltwater. Both of those sources appear to have a greater impact than surface freezing on the $\delta^{18}O$ content of shelf water.

Methods

Water samples and associated hydrographic data were collected as part of the Ross Ice Shelf Project from December 1976 through February 1979. The complete data set and standard procedures were reported by Jacobs and Haines [1982]. Approximately half the $\delta^{18}O$ water samples used in this study were processed at Lamont-Doherty Geological Observatory. These samples were equilibrated with CO_2 gas [Epstein and Mayeda, 1953] that was analyzed on a VG Micromass 903 triple collector mass spectrometer. The sample processing time was significantly reduced through elimination of the sample freezing/thawing steps used in the Epstein and Mayeda method by simultaneously pumping out the atmosphere of air from six samples [Roether, 1970; Fairbanks, 1982]. One sample of Lamont-Doherty's North Atlantic Deep Water standard water is processed under identical conditions with each set of five water samples [Fairbanks, 1982]. The precision is $\pm 0.03 °/oo$, based on the standard deviation of the mean.

The remaining water samples were processed at the Ocean Research Institute, University of Tokyo, to a precision of $\pm 0.02 °/oo$, based upon the average difference between paired analyses [Horibe et al., 1973]. No discrepancies are apparent between the results from the Horibe and Fairbanks laboratories. The relative precision (analytical precision divided by the magnitude of natural variations) of $^{18}O/^{16}O$ and D/H ratio measurements is comparable. Little additional information is gained by measuring both hydrogen and oxygen isotopes in shelf water samples.

The $H_2^{18}O/H_2^{16}O$ ratio is expressed as the fractional difference between the ratio in the sample to the ratio in standard mean ocean water (SMOW) [Craig, 1961], according to the following notation:

$$\delta^{18}O \; (°/oo) = \frac{^{18}O/^{16}O_{Sample} - ^{18}O/^{16}O_{SMOW}}{^{18}O/^{16}O_{SMOW}} \times 1000$$

Comparisons Between Southern Ocean $\delta^{18}O$ Data Sets

Figure 1 shows the locations of our $\delta^{18}O$ measurements in the Ross Sea. About 70% of the Ross Sea water samples analyzed for $\delta^{18}O$ content were taken near temperature or salinity extrema, near the sea surface, or within well-defined water types. These samples were averaged for each water type, which was then assigned to one of five general categories (Table 1). Three types were subdivided to show the different characteristics in water near or underneath the ice shelf. With few exceptions, deep and bottom waters exist only north of the continental shelf break. The High Salinity Shelf Water and Low Salinity Shelf Water, together with 'warm' inflows from

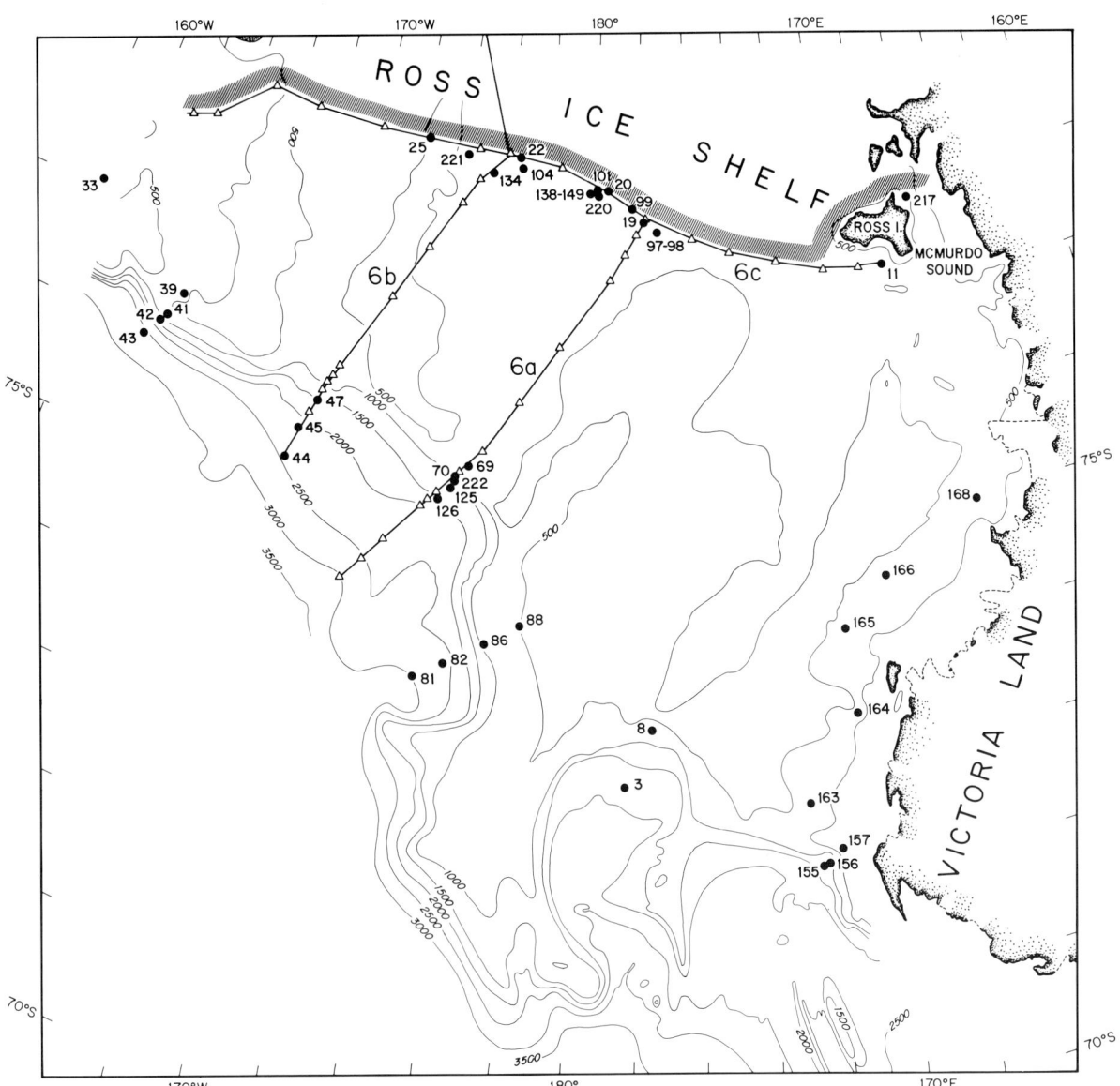

Fig. 1. Numbered Ross Sea stations provided the oxygen-18 ($\delta^{18}O$) measurements used in this study. Sections show locations of vertical temperature and salinity profiles in Figures 6a, 6b, and 6c. Data were taken December 1976 to January 1977, January to February 1978, and February 1979 [Jacobs and Haines, 1982]. Bottom topography after Heezen and Bentley [1972].

the deep ocean, account for about 75% of the water on the continental shelf [Carmack, 1977].

The Ross Sea data and Weiss et al. [1979] observations combine summer measurements from a large continental shelf region and the adjacent deep ocean. The corresponding water types occupy similar locations on $\delta^{18}O/S$ and Θ/S diagrams (Figures 2 and 3), suggesting that the same processes influence salinity and $\delta^{18}O$ content in these large marginal Antarctic seas. We tentatively extrapolate this analogy to the remaining continental shelf region, although there is evidence [cf. Potter and Paren, this volume] that the Amundsen-Bellingshausen sector may be different in several respects.

Average $\delta^{18}O/S$ values for the Ross Sea and elsewhere in the Southern Ocean appear in Figure 4. The Craig and Gordon [1965], GEOSECS, and SOUTHERN CROSS deep ocean data range over less than 0.3°/oo in $\delta^{18}O$, while the Ross Sea and Weiss et al. [1979] Weddell Sea measurements span nearly 0.7°/oo. With an analytical

TABLE 1. Properties of Ross Sea $\delta^{18}O$ Samples Used in This Study

Water Mass	Abbreviation (Samples)	Depth, m Average	1σ	Potential Temperature, θ Average	1σ	Salinity Average	1σ	$\delta^{18}O$, °/∞ Average	1σ
Deep Water									
Circumpolar Deep Water	CDW (12)	622	306	+1.17	0.25	34.704	0.024	-0.07	0.05
Surface waters									
Antarctic Surface Water	AASW (17)	7	11	-0.96	0.57	34.117	0.179	-0.31	0.08
	$AASW_G$ (13)	19	25	-1.17	0.33	34.242	0.125	-0.50	0.08
Temperature Minimum	TMIN (11)	87	46	-1.64	0.18	34.296	0.049	-0.44	0.15
Shelf Waters									
High Salinity Shelf Water	HSSW (11)	556	258	-1.91	0.02	34.838	0.054	-0.42	0.05
	$HSSW_B$ (8)	572	21	-1.86	0.01	34.823	0.008	-0.50	0.09
Low Salinity Shelf Water	LSSW (8)	258	155	-1.59	0.21	34.526	0.038	-0.45	0.04
Warm Core	WMCO (7)	312	129	-0.84	0.33	34.539	0.019	-0.34	0.09
Ice Shelf Waters									
Deep Ice Shelf Water	DISW (12)	382	84	-2.03	0.08	34.678	0.038	-0.54	0.06
Shallow Ice Shelf Water	SISW (3)	119	23	-2.04	0.04	34.363	0.016	-0.58	0.04
	$SISW_B$ (10)	400	41	-2.14	0.03	34.417	0.027	-0.71	0.06
Bottom Water									
Low Salinity Bottom Water	LSBW (16)	1583	736	-0.16	0.33	34.650	0.034	-0.26	0.06
High Salinity Bottom Water	HSBW (3)	2087	435	+0.34	0.10	34.710	0.007	-0.17	0.03

resolution of 0.02 to 0.03°/∞, $\delta^{18}O$ thus has the dynamic range to be a good water mass tracer on the Antarctic continental shelf.

The Craig and Gordon [1965] measurements on surface, deep, and bottom waters are about 0.15°/∞ lower in $\delta^{18}O$ than the more recent data (see also Broecker, et al., 1985). This may result from different calibration standards or from the limited number or location of samples in the earlier study. The Craig and Gordon [1965] bottom water samples were from two Eltanin Cruise 12 stations in the northwest Weddell Sea, where particularly fresh and cold water was observed on the slope. They reported an $\delta^{18}O$ content of -0.45°/∞ for both surface and bottom water there, which suggested that some bottom water contained very little entrained deep water or had access to source water very low in $\delta^{18}O$.

The Redfield and Friedman [1969] deep water point in Figure 4 was calculated from the deuterium mean in their Table 2. An associated Circumpolar Deep Water temperature of only 0.75°C makes it hard to account for the relatively high $\delta^{18}O$ value. The Potter and Paren [this volume] deep water value in Figure 4 is the average of all measurements with salinity ⩾34.6 in their Figure 9. Its $\delta^{18}O$ content is 0.15 to 0.2 °/∞ higher than other recent CDW measurements. If indicative of a systematic offset, the surface water value calculated from their Figure 9 salinities of 34.2 would decrease to around -0.55°/∞ in $\delta^{18}O$. Their measurements were made on water samples taken through a rift in the 85-m draft George VI Ice Shelf, where surface samples may be less sig-

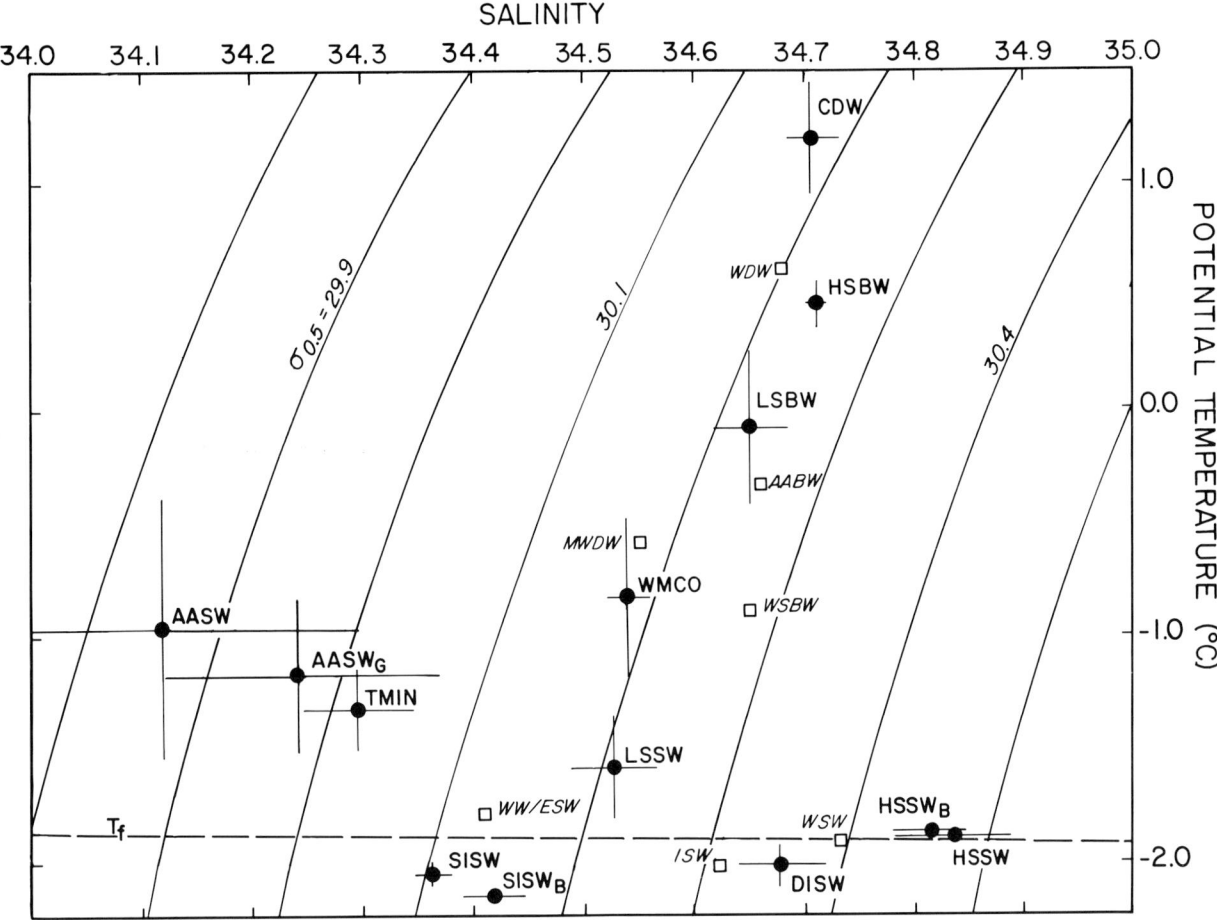

Fig. 2. Potential temperature/salinity (Θ/S) diagram of water types in the Ross Sea (solid circles with 1σ error bars) and Weddell Sea (open squares). The corresponding abbreviations for the Ross Sea appear in Table 1 and are discussed in the text. Isopycnal surfaces are referenced to 500 dbar, and T_f is the surface freezing temperature (UNESCO, 1978). The Weddell Sea data are from Foster and Carmack [1976], Carmack and Foster [1977], and Gammelsrød and Slotsvik [1981]. WDW, Weddell or Warm Deep Water; MWDW, Modified Warm Deep Water; WW, Winter Water; ESW, Eastern Shelf Water; ISW, Ice Shelf Water; WSW, Western Shelf Water; WSBW, Weddell Sea Bottom Water; AABW, Antarctic Bottom Water.

nificant, regionally, than the 85-m salinity of 33.91 and $\delta^{18}O$ of -0.31‰ (J. Potter, personal communication, 1984). In either case, adjusted $\delta^{18}O$ values would bring those observations into the range of waters that are associated with melting glacial ice in the Ross Sea.

Deep ocean $\delta^{18}O$ profiles in the Ross and Weddell Seas are similar, showing a slight subsurface $\delta^{18}O$ minimum and a prominent subsurface maximum in the deep or 'oldest' water at several hundred meters depth (Figure 5a). On the continental shelf, most of the water column has an $\delta^{18}O$ content well below -0.3‰ (Figures 5b-5d), in sharp contrast to the deep ocean situation. The near-surface pattern is reversed on the shelf, with an $\delta^{18}O$ maximum derived from the offshore deep water in the upper 300 m. Adjacent to the ice shelf, this subsurface maximum is overridden by a thicker, low-$\delta^{18}O$ surface layer. Measurements on samples obtained from beneath the ice shelf at J-9 (82°22.5'S, 168°37.5'W) display considerable scatter, with a tendency toward bimodal values in the upper layer (350-450 m). This may be an artifact of the sampling procedure, carried out over a 20 hour period and occasionally complicated by freezing problems [Jacobs and Haines, 1982]. Glacial ice near the base of the Ross Ice Shelf at J-9 has an $\delta^{18}O$ range from -40 to -43‰ [Grootes and Stuiver, 1983]. If that variability is typical of melting regions at the ice shelf base, it could account for some of the scatter, par-

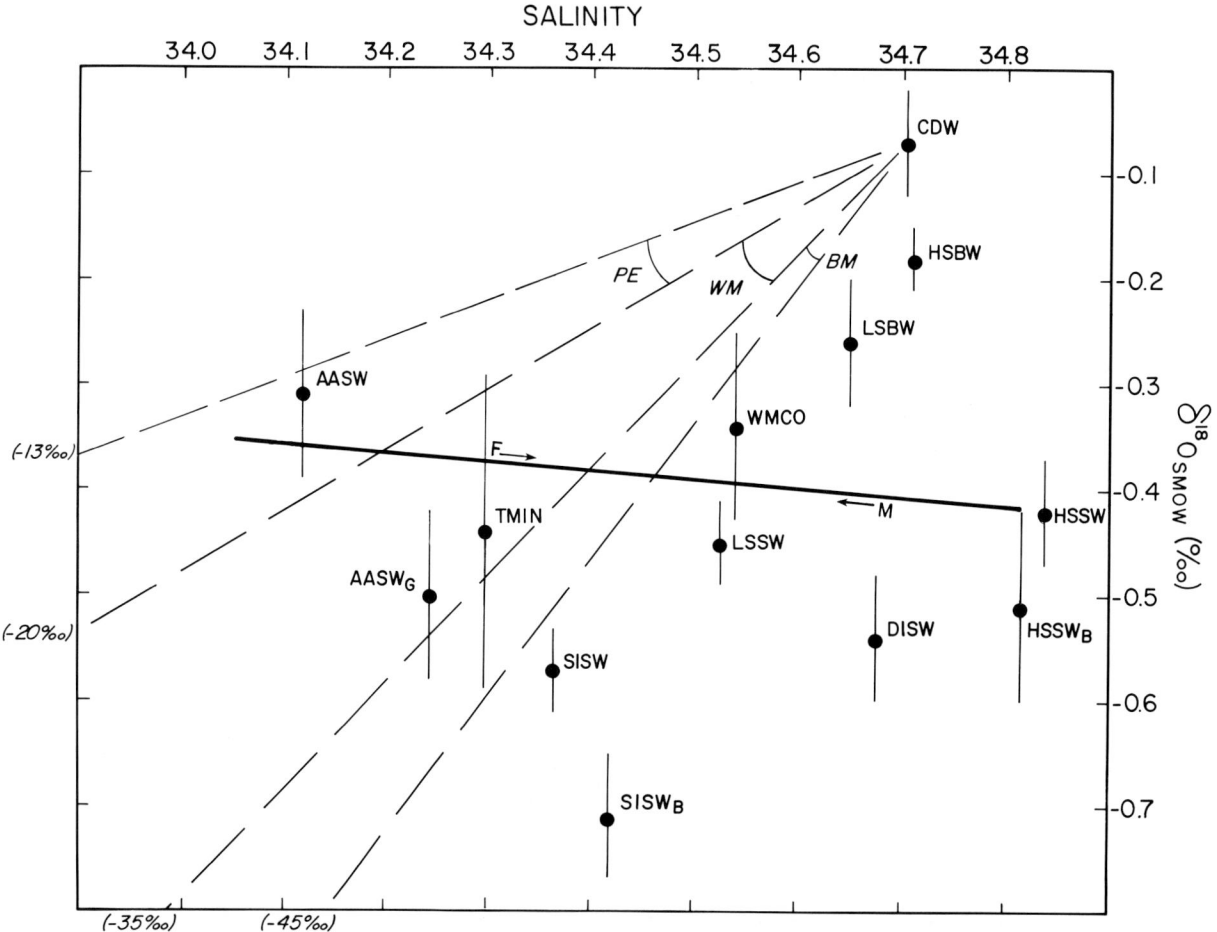

Fig. 3. Oxygen-18/salinity ($\delta^{18}O/S$) diagram for Ross Sea water types. Abbreviations and error bars as in Figure 2 and Table 1. The dashed lines define envelopes encompassing the probable $\delta^{18}O$ content of marine precipitation (PE) in the Ross Sea and wall meltwater (WM) or basal meltwater (BM) from the Ross Ice Shelf. The solid line indicates the $\delta^{18}O$ fractionation for sea ice freezing (F) or melting (M) at the sea surface or beneath the ice shelf.

ticularly if tidal oscillations interleave the products of basal melting and freezing.

The Circumpolar Deep Water Source

Circumpolar Deep Water (CDW) is continually renewed by North Atlantic Deep Water where the two merge in the South Atlantic to form the midsection of the Antarctic Circumpolar Current. CDW shoals gradually as it approaches Antarctica within that current, reaching its minimum depth beneath the Antarctic Divergence and its southernmost extent near the continental margin. At that point its zonal component of flow has generally reversed from east to west, and its distinguishing characteristics are a relatively high temperature and salinity (Figures 6a and 6b), high silicate, and low oxygen concentration. As the only external-origin water mass near the continental shelf, all other water types there must be derived from it by mixing and cooling and by the input of precipitation, meltwater, or brine.

Warm Deep Water [Fofonoff, 1956; Foster and Carmack, 1976] and Weddell Deep Water [Gordon, 1982] refer to the water mass that circulates at intermediate depths within the Weddell Gyre that occupies the western and central Weddell-Enderby Basin. Weddell Deep Water (WDW) is lower in salinity and temperature than its CDW parent (Figure 2), which is cooled and freshened during its circuit of Antarctica. WDW is also exposed along its lengthy southern and western flanks to fresh and cold continental margin waters as it circulates within the Weddell Gyre [Deacon, 1979]. The protective arm of the Antarctic Peninsula potentially allows a long residence time within this gyre, while promoting more severe atmospheric conditions at the sea surface [Schwerdtfeger, 1974]. In

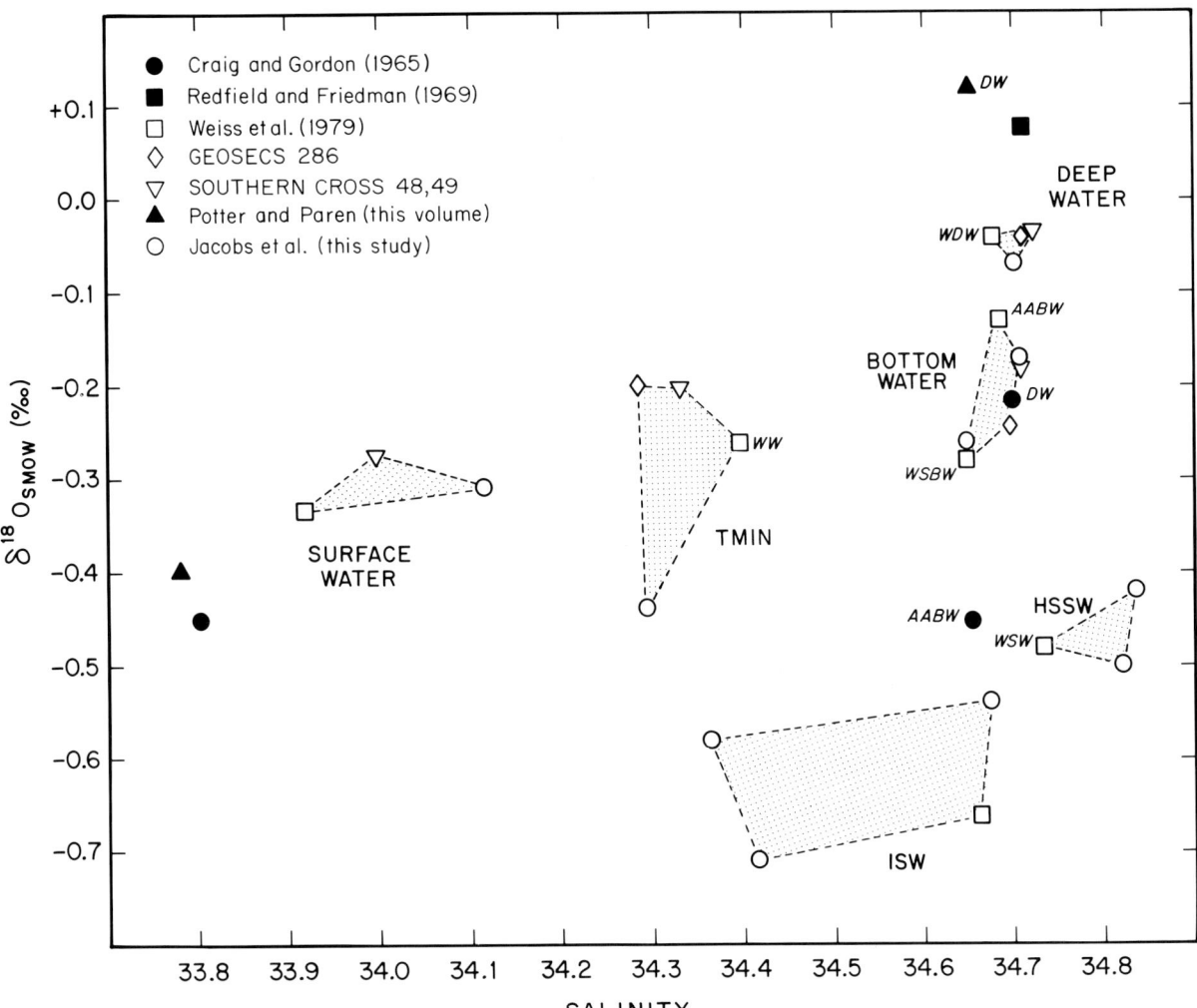

Fig. 4. Oxygen-18/salinity ($\delta^{18}O/S$) diagram combining several Southern Ocean data sets. The Craig and Gordon [1965] surface and bottom water averages are from the Weddell Sea; their deep water point is from the Pacific-Antarctic sector. The Weiss et al. [1979] observations were in the Weddell Sea. GEOSECS station 286 (66°05'S, 173°40'E) and SOUTHERN CROSS stations 48 (68°00.1S, 170°06.9'W) and 49 (69°29.4'S, 169°57.3'W) are averaged over several depths, after conversion from deuterium measurements [Horibe and Ogura, 1968] using the relationship $\delta D = 8.3 \delta^{18}O - 0.8$ [Weiss et al., 1979]. The Potter and Paren [this volume] data were taken in a rift through the George VI Ice Shelf. Abbreviations are identified in the text, Table 1 and Figure 2.

addition, Weddell Deep Water overlies and mixes with the coldest and freshest bottom water in the Antarctic [Carmack and Foster, 1975b]. Combined with upwelling in the gyre, these factors reduce water column stability and facilitate deep ventilation from the sea surface [Gordon, 1978]. WDW does not appear to have a lower $\delta^{18}O$ content than CDW (Figures 3 and 4), perhaps because our Ross Sea CDW samples were taken close to the continental shelf (Figure 1). GEOSECS Station 293 at 52°40'S, 178°05'W has CDW $\delta^{18}O$ values above +0.1°/oo.

The Surface Waters

Antarctic Surface Water

We use the term Antarctic Surface Water (AASW) here to describe only the mixed layer at the summer sea surface. North of the continental margin this feature is usually 50 to 100-m deep and marked in the vertical by a relatively low salinity and high temperature [Jacobs and Haines, 1982]. Over the continental shelf most surface salinities range from

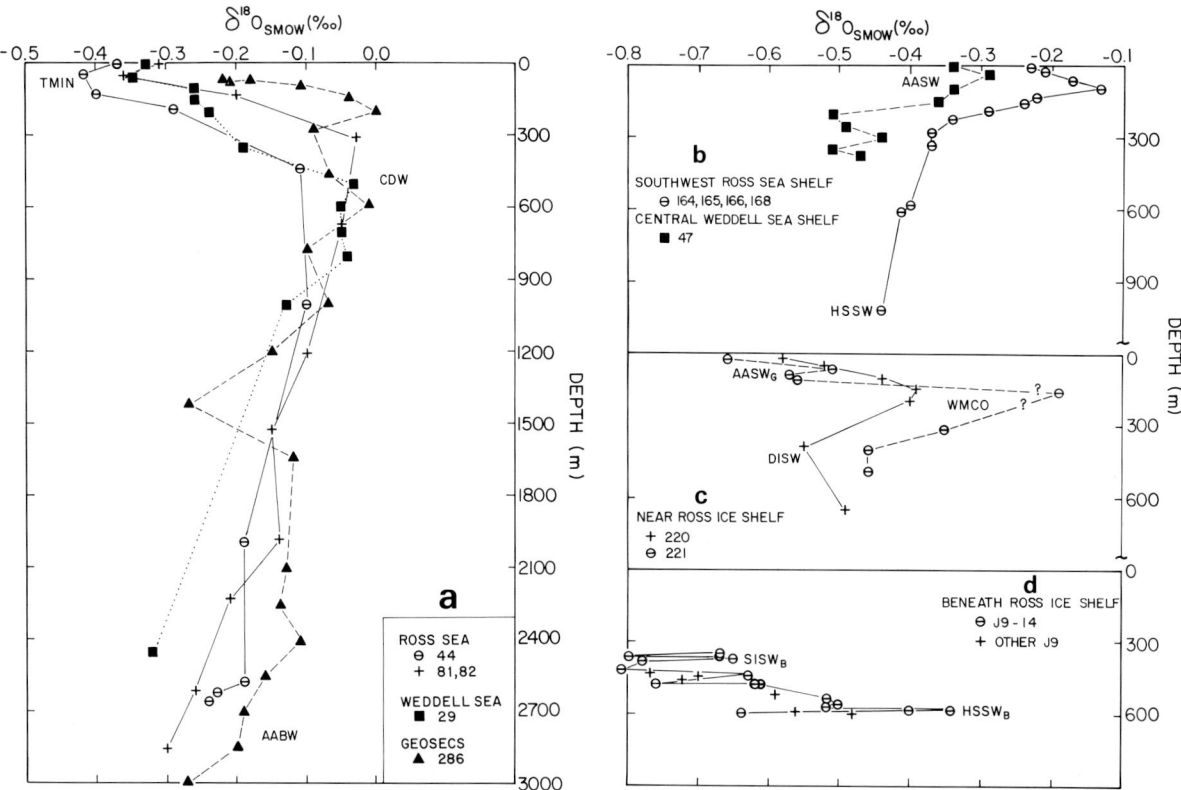

Fig. 5. (a) Profiles of $\delta^{18}O$ versus depth north of the Antarctic continental shelf. The Ross Sea and Weddell Sea stations [Jacobs and Haines, 1982; Weiss et al., 1979] are on the continental rise or slope. GEOSECS station 286 is 700 km north of the Ross Sea continental shelf. (b)-(d) Profiles of $\delta^{18}O$ versus depth on the Ross Sea and Weddell Sea continental shelves. Stations 220 and 221 are near the Ross Ice Shelf and slightly east of sections 6a and 6b, respectively. The J-9 measurements were made on water samples retrieved from a hole through the ice shelf at the southern end of section 6b. Abbreviations as in previous figures.

34.1 to 34.5, and surface temperatures uncommonly exceed 0°C except in the southwest Ross Sea (Figures 6a-6c). AASW presumably envelops and renews the underlying temperature minimum during winter, as on the left-hand side of Figure 6a. Local melting, heating, and vertical mixing cause the AASW to be the most variable water type in its summer Θ/S properties (Table 1). The Ross Sea AASW $\delta^{18}O$ samples were mostly taken within a few meters of the sea surface, with the deepest subsurface sample from 33 m.

Antarctic Surface Water evolves from CDW that has upwelled and been modified by interactions with the atmosphere, ice, and temperature minimum. On a $\delta^{18}O/S$ diagram that includes only the surface and deep waters (Figure 7a), AASW lies within an envelope (PE) originating at the CDW source and projected to a precipitation-evaporation range of -13 to -20°/oo in $\delta^{18}O$. The heavier of these end points (-13 °/oo) is the average of our measurements on a snow sample from the fast ice in Robertson Bay (71°25'S, 170°E) and a Leningradskaya Station (69°30'S, 159°23'E) sample reported by Morgan [1982]. The lighter value (-20°/oo) is representative of measurements reported for snow samples taken on the northern edge of the Ross Ice Shelf [Clausen et al., 1979]. Given the known variability in $\delta^{18}O$ content of precipitation near the coastline [Bromwich and Weaver, 1983], these values furnish an approximate range for the Ross Sea.

The $H_2^{18}O$ enrichment of sea ice relative to the water from which it forms ($\delta^{18}O$ = 1.00270) is about 2.7°/oo [Craig and Horn, 1968]. This makes the slope of the freezing/melting line on a salinity/$\delta^{18}O$ diagram about 0.08°/oo in $\delta^{18}O$ for each part per thousand salinity. Winter freezing of the sea surface will thus increase surface water salinity and slightly decrease its $\delta^{18}O$ content, moving AASW along the fractionation line in the direction designated by the 'F' arrow in Figure 7a. Summer melting of sea ice will reverse the AASW shift along the same line, as shown by the 'M' arrow

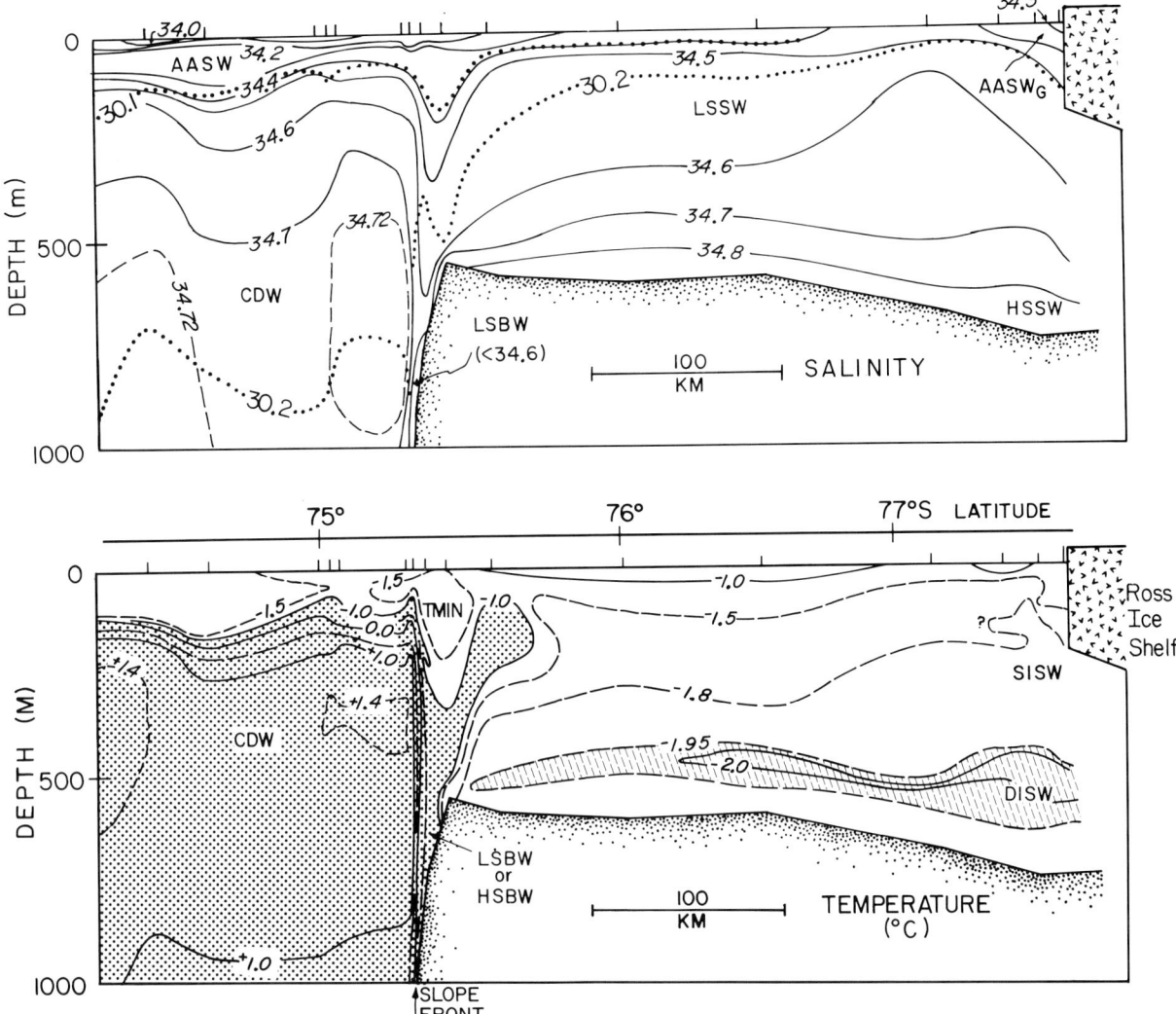

Fig. 6a. Salinity and temperature versus depth along transect 6a in Figure 1. Data from December 1976, STD stations of Jacobs and Haines [1982]. Vertical ticks indicate station locations. Dotted isopycnal surfaces in the salinity section are referenced to 500 dbar. At these continental shelf depths, potential temperatures would be about 0.03°C colder than the in situ temperatures shown.

in Figure 7a. This reversal shows that AASW can also be produced on the continental shelf, from a mixture of shelf water (see below), and sea ice meltwater (Figure 7b).

It does not appear from Figure 7a that most AASW can be strongly influenced by melt-driven upwelling near icebergs, as postulated for the Weddell Sea [Neshyba, 1977]. If that were the case, AASW would fall in an envelope bounded by the freshwater end points of Antarctic glacial ice, i.e., between about -20 and -45°/oo in $\delta^{18}O$. Surface water observations less than 10 km from the Ross Ice Shelf do show a meltwater influence, with an average $\delta^{18}O$ content 0.2°/oo lower than the remaining AASW (Figure 7a). These $AASW_G$ samples also average about 0.15 more saline than AASW, but are slightly lower in $\delta^{18}O$ and salinity than the temperature minimum (Figure 3). This precludes $AASW_G$ formation from a mixture of the AASW and TMIN (Figure 2), and surface freezing alone cannot account for the $AASW_G$ $\delta^{18}O/S$ property, given the slight fractionation line slope (Figure 7a). $AASW_G$ falls within a meltwater envelope (WM) defined by CDW and the -20 to -35°/oo $\delta^{18}O$ content of glacial ice exposed, top to bottom, at the north wall (barrier) of the ice shelf. We have not found unaltered CDW in direct contact with the Ross Ice Shelf, but it does appear near the seafloor under the shelf ice in George VI Sound [Potter and Paren, this volume].

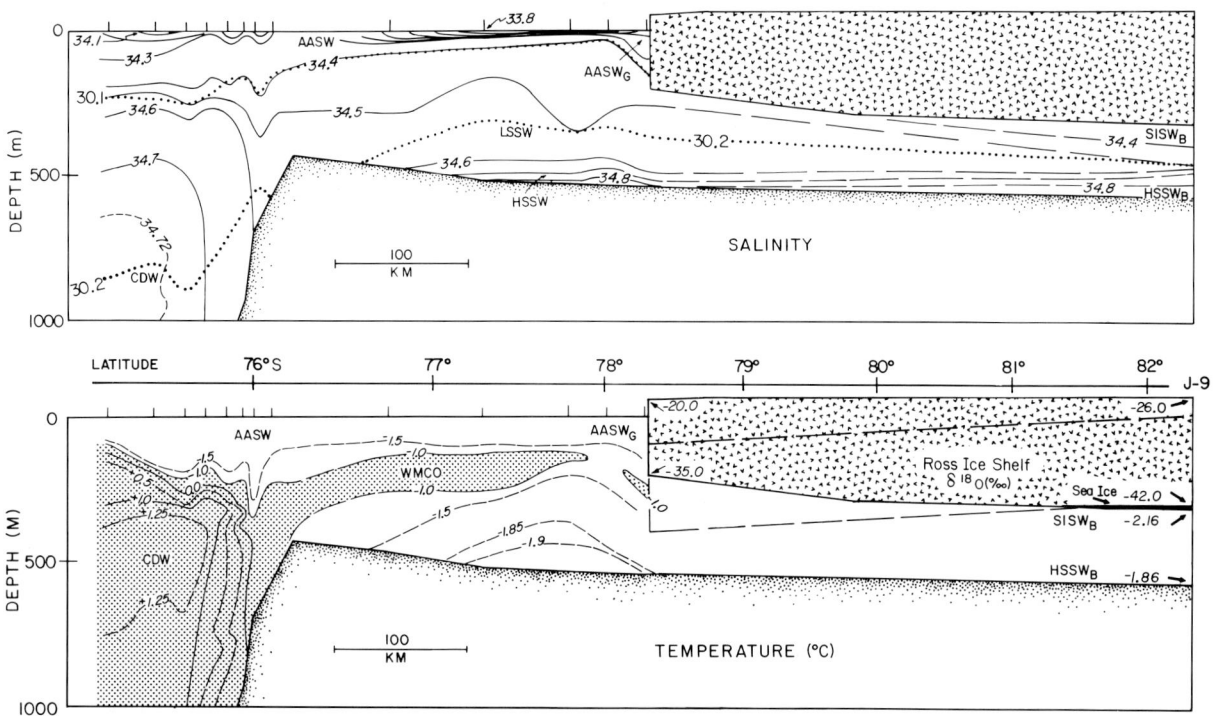

Fig. 6b. Salinity and temperature versus depth along transect 6b in Figure 1, to the position of the hole through the ice shelf at J-9 (82°22.5'S, 168°37.5'W). Representative $\delta^{18}O$ values are shown for the glacial ice, with surface accumulation and basal melting and freezing schematically illustrated. See caption for Figure 6a.

Summer surface water salinities reach 34.4 at some locations along the barrier and could mix with meltwater (dotted line in Figure 7a) to form $AASW_G$. The narrow band of ocean along the Antarctic ice shelves is a highly dynamic location year-round. Byrd [1930] noted that "often during winter after a stiff blow from the south, the darkness of a water sky indicated that the Ross Sea was open." The offshore winds frequently open shore leads between the sea ice and glacial ice [Naval Polar Oceanography Center, 1974-1984; Kurtz and Bromwich, this volume]. The rapidly moving, cold, dry air will locally increase evaporation and sea ice formation [Gill, 1973; Zwally et al., this volume] the effects of which penetrate deep into the water column [Pillsbury and Jacobs, this volume]. The resulting salinity (density) increase in the strong coastal current will be countered to some extent by meltwater from icebergs, from the adjacent barrier, and from snow blown off the continent.

Temperature Minimum

The Temperature Minimum (TMIN) lies in a halocline at the base of the AASW, south of the Polar Front. It is best developed north of the continental margin [Gordon and Molinelli, 1982], deepest and coldest over the upper continental slope (Figures 6a and 6b) and often difficult to differentiate from other water types on the continental shelf. The TMIN is sometimes referred to as Winter Water (WW) since it results from autumn-winter cooling of AASW to the surface freezing temperature (T_f in Figure 2), with the subsequent addition of brine during sea ice formation. That designation is less useful on the continental shelf, where water types formed at the surface in winter can extend to depths exceeding 1000 m and yet not reach temperatures as low as other water formed at shallower depths there year-round. The TMIN summer temperature is usually above T_f because it has mixed with the warmer CDW below and AASW above.

The $\delta^{18}O$/salinity diagram (Figure 7a) suggests that factors other than surface cooling and freezing also influence formation of the temperature minimum. The TMIN is about 0.2 higher in salinity than AASW but is depleted in $\delta^{18}O$ more than can be accounted for by surface freezing. Is this lower $\delta^{18}O$ increment from marine precipitation or glacial meltwater? Precipitation (P - E) in the Ross Sea is about 30 cm yr^{-1} [Baumgartner and Reichel, 1975]. However, some fraction of the snowfall will presumably be deposited on the pack ice and be transported to the north before it

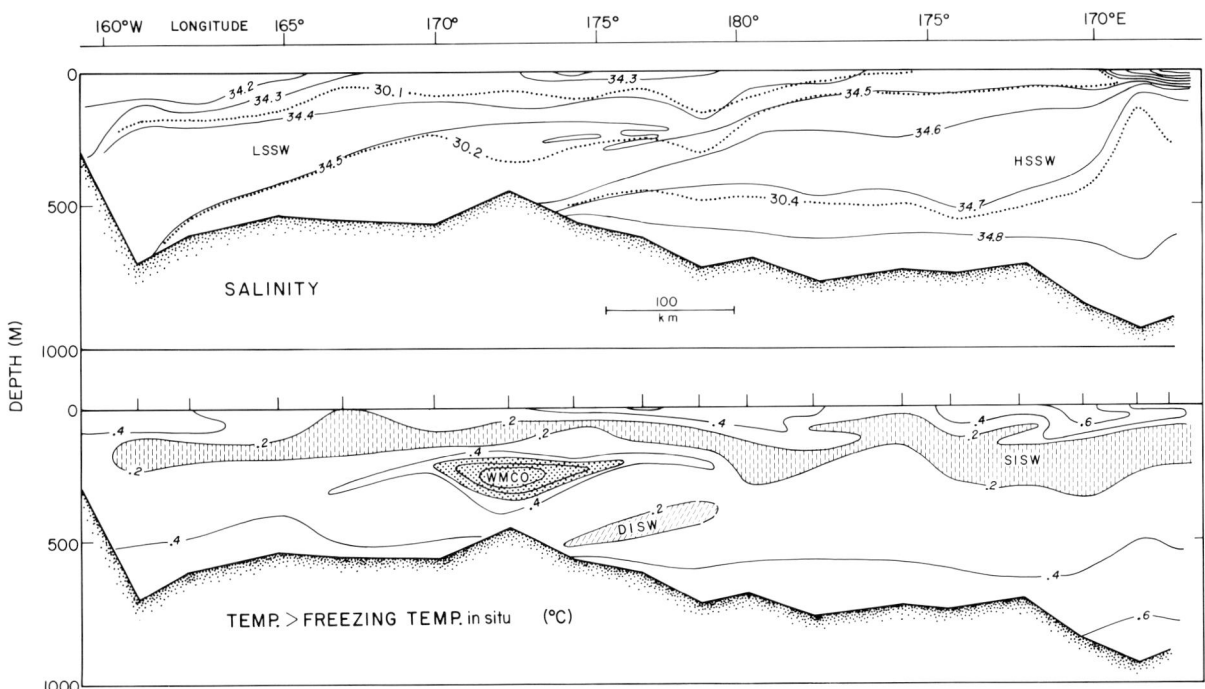

Fig. 6c. Vertical sections of salinity and temperature above the freezing point (UNESCO, 1978) constructed from stations within 3 km of the Ross Ice Shelf Barrier in Figure 1. See caption for Figure 6a. Shallow Ice Shelf Water (SISW) has been defined (Table 1) from water samples taken along the eastern side of the barrier, but may extend into the upper High Salinity Shelf Water (HSSW) as depicted here.

melts. As a first-order estimate, we assume that the presence and northward movement of pack ice will halve net precipitation into the continental shelf surface water. A total of 15 cm yr^{-1} of precipitation at a mean $\delta^{18}O$ content of -16.5°/oo (midpoint in the -13 and -20°/oo range defined above) distributed over a 90-m surface layer (TMIN depth in Table 1) would lower its $\delta^{18}O$ content by 0.03°/oo. Combined surface freezing and precipitation could thus lower the AASW $\delta^{18}O$ content by about 0.05°/oo, considerably less than the 0.13°/oo difference between TMIN and AASW.

The TMIN position near AASW$_G$ in Figure 7a suggests that glacial meltwater may be a factor in development of the temperature minimum, as postulated earlier [Pettersson, 1904; Drygalski, 1928]. A slightly higher salinity for the temperature minimum, such as that shown by the Weddell Sea WW in Figure 2, would allow it to mix isopycnally with Shallow Ice Shelf Water (SISW, see below). As the voluminous TMIN at the continental margin (Figure 6a) increases in salinity from deep winter convection, its $\delta^{18}O$ content will be decreased by accumulated glacial meltwater [Jacobs et al., 1979a].

In the Ross Sea the TMIN is lower in $\delta^{18}O$ content than the AASW, while the corresponding Weddell Sea TMIN (WW) is slightly higher. The Ross Sea TMIN is highly variable (Figure 3), and several of the Weddell Sea values in Figure 4 are from a single station. However, if the two data sets are representative, the $\delta^{18}O$ difference might result from lower stability and greater mixing between the TMIN and deep water in the Weddell Sea [Martinson et al., 1981].

The Shelf Waters

High Salinity Shelf Water

High Salinity Shelf Water (HSSW) is the densest water type in the Antarctic oceans. It is nearly isothermal at the sea surface freezing point, with stability maintained by salinity that increases with depth. Its low temperature and high salinity (Table 1; Figure 6a) clearly differentiate HSSW from any deep water that intrudes onto the shelf. Carmack [1977] defined HSSW as shelf water above 34.6, corresponding to a density threshold necessary for it to play a role in bottom water formation. Countryman and Gsell [1966] and Jacobs et al. [1970] referred to the Ross Sea variant as Ross Sea Shelf Water, while several authors have used the term Western Shelf Water (WSW) for the Weddell Sea version (Figure 4).

A variety of High Salinity Shelf Water,

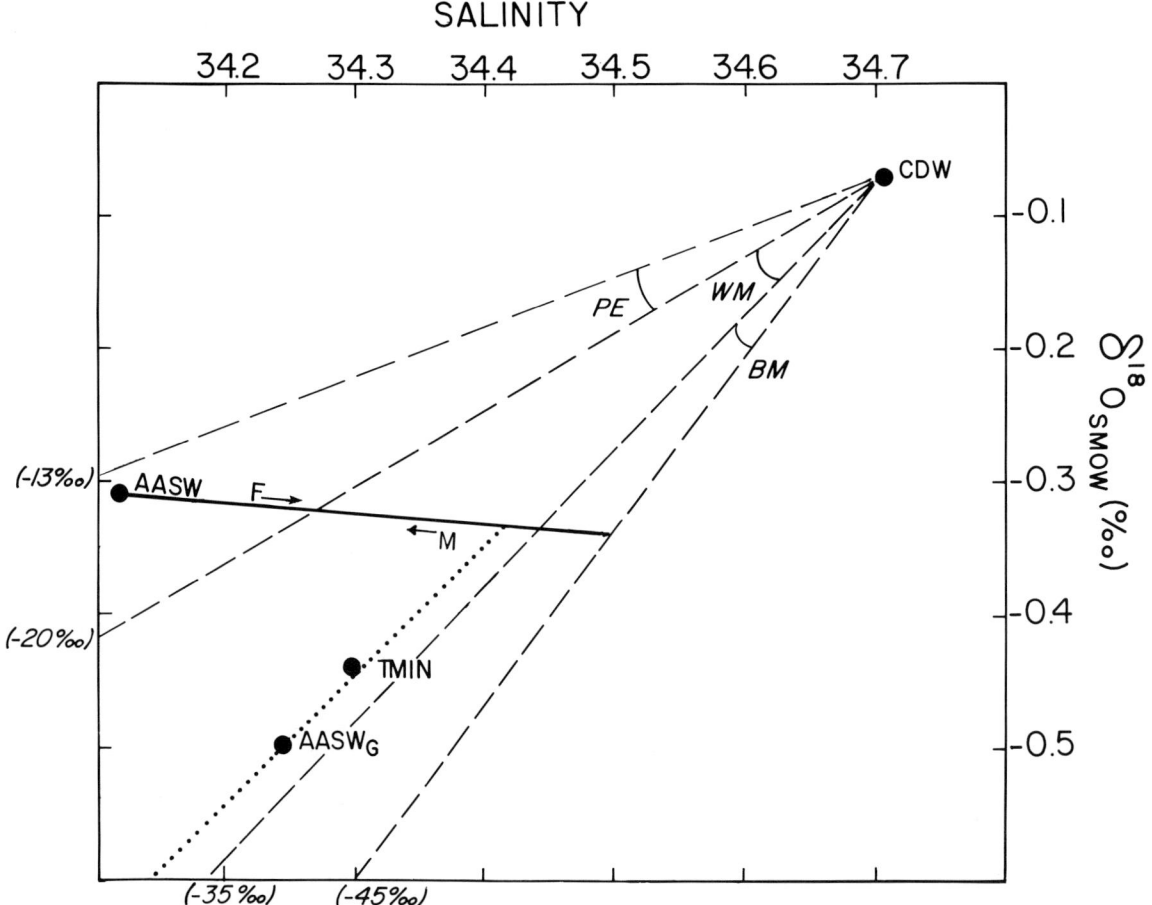

Fig. 7a. Oxygen-18/salinity diagram illustrating the evolution of surface waters (AASW, TMIN and $AASW_G$) from deep water (CDW) in the Ross Sea. The dotted line shows how the temperature minimum (TMIN) and surface water near glacial ice ($AASW_G$) could be generated by combinations of precipitation, freezing and glacial melting.

$HSSW_B$, was observed in a 50 to 75-m boundary layer at the seafloor under the Ross Ice Shelf at J-9 (Figure 6b). It is marginally warmer and fresher than HSSW and is lower in $\delta^{18}O$ content by about $0.1^o/oo$ (Figures 2 and 3; Table 1). Its $\delta^{18}O$ and salinity suggest that $HSSW_B$ simply results from glacial ice melting into HSSW, but it occurs to the right of that meltwater envelope in Figure 7c. Historical HSSW salinities have been observed above 34.9 [Jacobs et al., 1970], and Barrett [1975] reported 35.0 in an ice shelf crevasse. These saltier components could have mixed with glacial meltwater to influence $HSSW_B$, but its sea surface freezing point temperature also indicates a primary origin north of the ice shelf.

Gill [1973] and Killworth [1974] proposed that HSSW is formed by surface freezing along the ice shelf, accompanied by deep overturning, westward flow and upwelling. Mean surface flow along the barrier is known to be westward in summer, and some subsurface currents also have westerly components [Jacobs and Haines, 1982; Pillsbury and Jacobs, this volume]. However, HSSW has a higher $\delta^{18}O$ content than most other water types along the ice front ($AASW_G$, SISW, DISW, and LSSW in Figure 3), whereas the input of brine from surface freezing would lower their $\delta^{18}O$ values.

On a Θ/S diagram, HSSW occurs near the high-salinity end of the surface freezing line, T_f (Figure 2). Brine from sea surface freezing, combined with some glacial meltwater and precipitation, could produce the observed HSSW characteristics. This is schematically illustrated in Figure 7b, with the freshwater components contributing most of the $\delta^{18}O$ change. While the HSSW salinity increases with depth, its $\delta^{18}O$ content decreases (Figure 5b). This characteristic could result from onshore intrusions of higher-$\delta^{18}O$ water that initially override and then are transformed into HSSW. Preferential accumulation of glacial meltwater with depth would be consistent

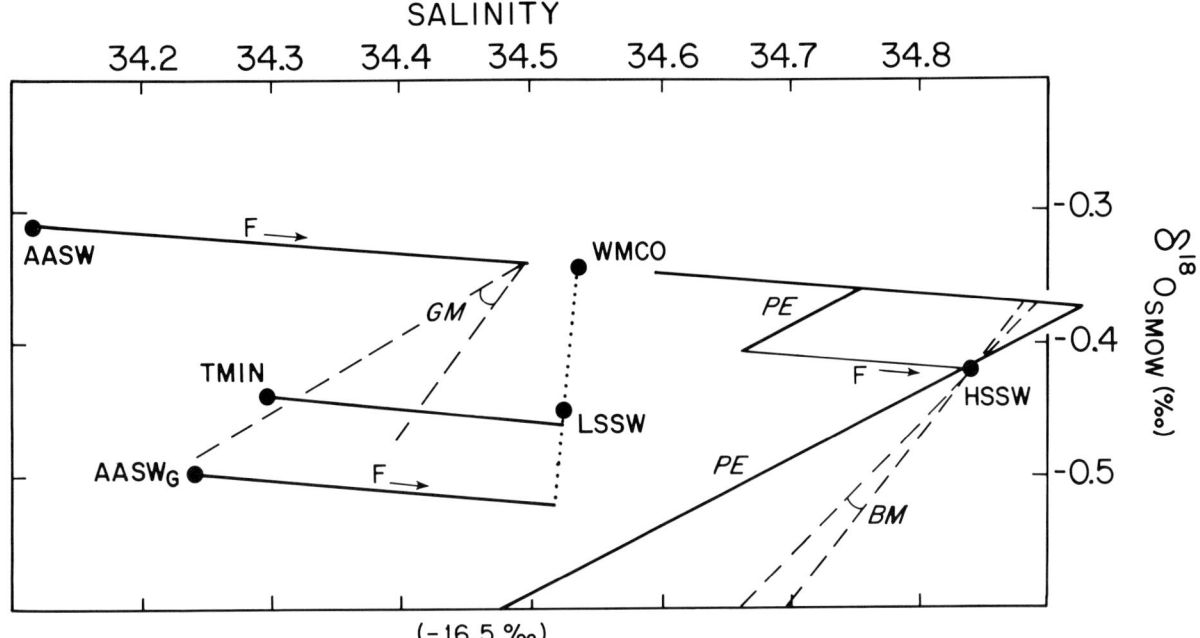

Fig. 7b. Oxygen-18/salinity diagram illustrating the formation of shelf waters (HSSW and LSSW) from surface waters (AASW, TMIN and AASW$_G$) in the Ross Sea. See caption for Figure 3. Some of the freezing (F), precipitation (PE), glacial melting (GM and BM) and mixing (dotted line) paths are more likely than others. Freezing AASW cannot produce WMCO, which has a relatively high temperature.

with the pressure effect on the freezing point, but the near-isothermal sea surface freezing temperature of HSSW shows that it has yet to be exposed to the glacial ice. The chlorofluoromethane distribution in HSSW [Trumbore et al., 1984] and the composite vertical profile in Figure 5b suggest that dense plumes containing brine and continental precipitation may descend through the water column or along the sloping seafloor at some locations.

A large amount of HSSW in the Ross Sea probably forms by sea surface freezing along the Victoria Land coast, particularly in the Terra Nova Bay polynya at 74°50'S, 164°30'E. Szekielda [1974] may have been the first to consider the oceanographic significance of that polynya, suggesting it was kept open by vertical heat diffusion from upwelling and/or volcanic activity. The summer hydrographic data do not suggest the presence of submarine hot spots. Intrusions of warm water have been observed well north of Terra Nova Bay during summer (see Figure 3 in Dunbar et al. [this volume]) and as far south as the Ross Ice Shelf during winter [Pillsbury and Jacobs, this volume]. However, the probable high density of water in the southwestern Ross Sea during winter may preclude the subsurface lateral import of much sensible heat from the CDW. In any case, Kurtz and Bromwich [this volume] have shown that the polynya can be maintained by the combination of katabatic winds off the continent and the blocking effect of the Drygalski Ice Tongue. They estimate the formation and continual removal of up to 65 m yr^{-1} of sea ice in Terra Nova Bay. Only 3 m of ice would be needed to increase the salinity of a 50-m, 33.1 salinity summer surface layer to a level where brine could be pumped directly into the HSSW.

The weekly Antarctic ice charts [Naval Polar Oceanography Center, 1974-1984] show that leads along the Ross Ice Shelf tend to be concentrated at its western end. In combination with Victoria Land polynyas, that could help account for the observed west to east salinity gradient in the Ross Sea and lessen the need for upwelling HSSW on the western continental shelf [Killworth, 1974]. We assume that highly saline creeks inland along the Victoria Land coast [Keys, 1979] and any brine pockets trapped under the ice sheet [Hughes, 1972] will contribute an insignificant amount of salt, if any, to the HSSW.

Ross Sea HSSW appears to be saltier (denser) than its counterpart in the Weddell Sea (Figure 2). This could result from lower-salinity source water to the Weddell Sea shelf (WDW versus CDW), or the ratio of polynyas to shelf area may be less there than in the Ross Sea. Cavalieri and Martin [this volume] have demonstrated a good correlation along the Wilkes Land Coast between more saline shelf

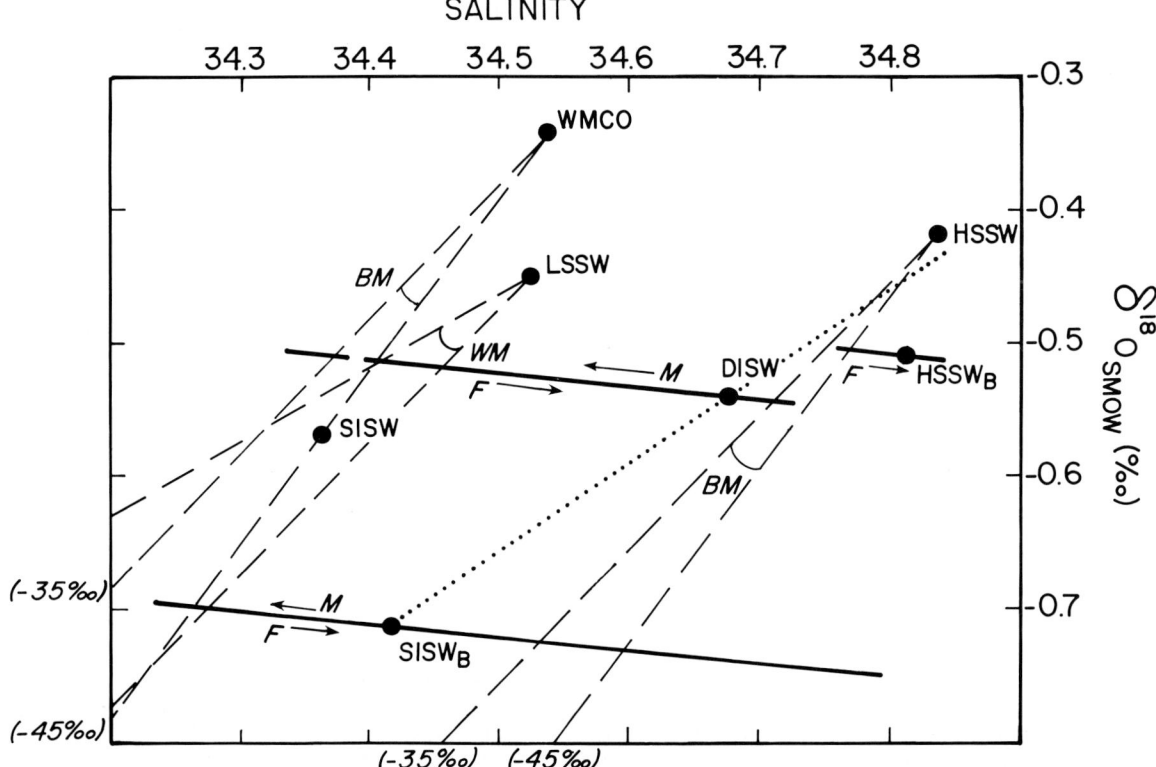

Fig. 7c. Oxygen-18/salinity diagram illustrating the production of ice shelf waters (SISW, SISW$_B$, and DISW) from shelf waters (WMCO, LSSW, HSSW and HSSW$_B$) in the Ross Sea. See captions for earlier figures.

water and large polynyas with high ice production rates. In addition, the continental shelf is largely ice-free at the end of summer in the western Ross Sea but not in the western Weddell Sea [Naval Polar Oceanography Center, 1974-1984]. Presumably, somewhat more brine will be released to the underlying water column if the winter period does not begin with the sea surface blanketed by an insulating layer of pack ice. Temporal variability of 0.05 to 0.10 in salinity has been observed over the past two decades in HSSW near Ross Island [Jacobs, 1985].

Low Salinity Shelf Water

Low Salinity Shelf Water (LSSW) is the most widespread water type on the Antarctic shelf [Carmack, 1977], occupying the eastern sides of the large embayments and much of the remaining continental shelf [Gordon, 1974a, Figure 5]. The Weddell Sea version has also been referred to as Eastern Shelf Water (ESW) [Carmack and Foster, 1977]. LSSW is slightly higher in temperature but considerably lower in salinity (density) than HSSW (Figures 6b and 6c). On a Θ/S diagram (Figure 2), LSSW occurs midway between the surface waters and HSSW. The seasonal cycle of surface water conversion to HSSW and back to surface water could also produce some intermediate type like LSSW, but the geographical separation of these two shelf waters complicates that scenario.

An $\delta^{18}O$ diagram identifies the temperature minimum (TMIN) as the simplest immediate precursor to LSSW (Figure 7b). The TMIN is deepest along the continental margin (Figures 6a and 6b), and its salinity will increase as it moves to the west [Gill, 1973; Solomon, 1983], particularly during the winter salting cycle. Dynamic topography and vertical sections [Carmack and Foster, 1975a] suggest a bifurcation of this slope current where the continental shelf widens. The portion that takes a southerly bearing onto the shelf would renew the LSSW. In the Weddell Sea, for example, the Temperature Minimum (WW) and Low Salinity Shelf Water (ESW) have similar Θ/S characteristics [Carmack and Foster, 1977], with the former occupying the oceanic domain and the latter occupying the shelf domain [Carmack, 1977, Figure 3].

LSSW could also be formed or further modified by the freezing of surface water in combination with glacial melting, as illustrated by the GM envelope between the AASW and TMIN freezing lines in Figure 7b. In addition, AASW$_G$ may freeze and overturn to a depth where

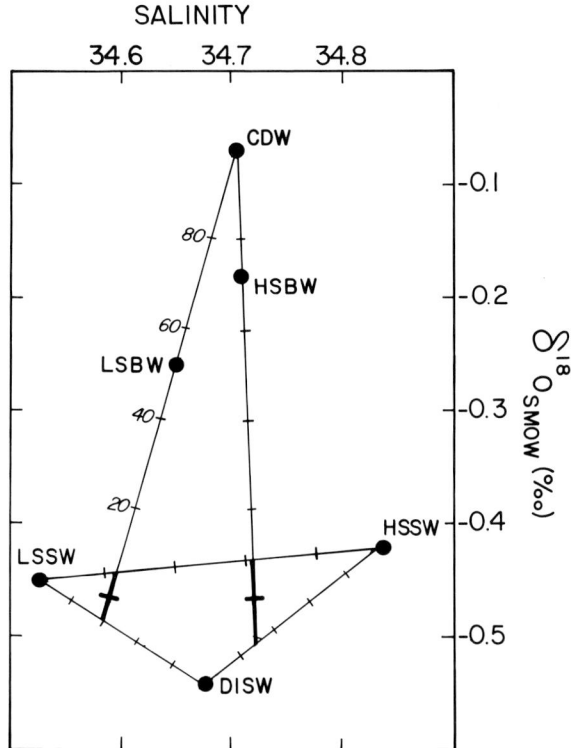

Fig. 7d. Oxygen-18/salinity diagram illustrating possible composition of bottom waters (LSBW and HSBW) from deep water (CDW) and shelf waters (LSSW, DISW, and HSSW) in the Ross Sea. Ticks denote the percentage of each water type, as labeled on the CDW-LSBW line.

it can mix with "warm" water (WMCO; see next section) that has intruded onto the shelf, i.e. via the dotted line through LSSW to WMCO in Figure 7b. All of these processes suggest fairly rapid renewal and some degree of glacial melting. Since LSSW is warmer and more depleted than HSSW in $\delta^{18}O$ content, there may be a greater melting potential in the eastern Ross Sea. Greater topographic relief beneath the eastern side of the Ross Ice Shelf [Bentley et al., 1974] will expose the eastern basal ice to a large vertical extent of water at temperatures well above the in situ melting point. Some mass balance and radio echo-sounding data have indicated that basal melting may be greatest just east of Ross Island and west of Roosevelt Island [Crary et al., 1962; Neal, 1979]. Those regions flank the locus of the warmest subsurface water we have observed along the barrier (Figure 6c).

Warm Core

On some continental shelf transects (Figure 6b), cores of water warmer than that at the sea surface (WMCO) extend at intermediate depths from the deep ocean southward to the glacial ice. The existence of these features has been known for some time, and they have commonly been referred to as Modified Circumpolar Deep Water or Warm Deep Water [Countryman and Gsell, 1966; Jacobs et al., 1970; Foster and Carmack, 1976]. The Ross Sea Warm Core (WMCO) can be defined as an intermediate-depth feature with a fairly narrow salinity range centered near 34.54 (Table 1), and a temperature near -1.0°C where it reaches the ice shelf. It also appears to upwell into the near-surface layers, at least where the isopycnals shoal in the western Ross Sea along the ice shelf (Figure 6c). WMCO influence is reflected in the shallow $\delta^{18}O$ maxima on stations 164 and 221 in Figures 5b and 5c. At a salinity of 34.436, the station 221 value was not included in the WMCO average, but it is associated with relatively low dissolved oxygen, probably derived from north of the continental shelf.

For the Weddell Sea variant (MWDW), Foster and Carmack [1976] postulated formation from an equal mixture of WDW and WW seaward of the continental shelf break, and its subsequent intrusion onto the shelf. To produce the WMCO Θ/S properties observed in the Ross Sea (Figure 2), a mixture of CDW and TMIN would require subsequent cooling of about 1°C or a salinity increase of 0.15. Those thermohaline alterations may well occur during a crossing of the Slope Front, but a cross-isopycnal mixture of CDW and TMIN may not be generated offshore at a rate adequate to sustain the WMCO. It may be more significant that the CDW, LSSW, and SISW share a common density horizon (Figure 2), allowing isopycnal mixing between these deep and shelf waters to give rise to the WMCO. Such a mode of formation would be consistent with the WMCO $\delta^{18}O$/S properties (Figure 3).

The warm intrusions onto the shelf appear to be continuous and may persist throughout the year. It seems unlikely that the 300-km-long feature in Figure 6b could have been set up during the brief period between the end of winter and those December observations. Year-long temperature measurements near the Ross Ice Shelf do show that the WMCO shifts, cools, or is periodically obliterated there by deep overturning during winter [Pillsbury and Jacobs, this volume]. Repeated summer transects along the Ross Ice Shelf demonstrate considerable variability in WMCO shape and position. Similar intrusions at other locations around the continent display large temperature differences [Jacobs, 1985]. Nearly unaltered CDW (∼1°C) invades the shelf along the Amundsen-Bellingshausen coastline [Potter and Paren, this volume]. An influx on the narrow continental shelf north of the Mertz Glacier Tongue [Jacobs and Haines, 1982, station 188] is relatively high in temperature and salinity

(+0.5°C and 34.66). The Weddell Sea intrusion is cooler than the one in the Ross Sea, consistent with its cooler Weddell Deep Water source [Gordon, 1983].

Ice Shelf Waters

There have been several references to Ice Shelf Water or its equivalent in the literature [e.g., Sverdrup, 1940; Lusquinos, 1963; Jacobs et al., 1970; Carmack and Foster, 1975a; Jacobs et al., 1979a; Weiss et al., 1979; Smith et al., 1984; Foldvik et al., this volume]. Identification has been based upon location near subsurface glacial ice and temperatures below the sea surface freezing point. Analyses of $\delta^{18}O$/S characteristics now show two subsurface variants in the Ross Sea, with quite different circulation patterns. They are here labeled Shallow Ice Shelf Water (SISW) and Deep Ice Shelf Water (DISW).

Shallow Ice Shelf Water

Shallow Ice Shelf Water is located at 50- to 250-m depths along the Ross Ice Shelf (Figure 6c) and can sometimes be traced for short distances north of the ice (Figure 6a). Its salinity is intermediate between that of surface and shelf waters, but it is lower in temperature and $\delta^{18}O$ content (Figures 2 and 3). SISW falls in the LSSW-wall melting (WM) envelope (Figure 7c) and so probably forms from melting along the barrier or other ice walls [Jacobs et al., 1981] or beneath the seaward edges of the ice where active tidal recirculation is likely [MacAyeal, this volume (a)]. At some locations, SISW is shallower than the lower northern edge of the Ross Ice Shelf, which implies upwelling if it has emerged from beneath the ice. Gammelsrød and Slotsvik [1981] reported upwelling of as much as 160 m over a 4-hour period near the Ronne Ice Shelf. A few observations very near the ice front [e.g. Jacobs and Haines, 1982, station 135] show this water can be supercooled with respect to the in situ pressure.

Foldvik and Kvinge [1974] proposed that water emerging from beneath an ice shelf at its in situ freezing point would become supercooled by upward displacement, allowing ice crystals to form in the ascending water. The STD and CTD instruments we used were not well suited to verify the presence of ice crystals, but upwelling off the ice front is probable, given the density field. The low-salinity surface layer wedged along the ice (Figure 6b) results in a north to south slope of isopycnals, so that water emerging from beneath the ice would shoal simply to retain its density horizon. In addition, winds at the time of station 135 cited above were offshore (150°) at 20 knots (10.3 m/s). Lewis and Perkin [this volume] find evidence for Ice Shelf Water upwelling at the end of winter in McMurdo Sound.

Oceanographic measurements beneath the Ross Ice Shelf at J-9 revealed a 75- to 100-m boundary layer next to the ice, with Θ/S characteristics similar to Shallow Ice Shelf Water (Figures 2 and 6b). Its $\delta^{18}O$ content averaged -0.71°/oo, with one -0.81°/oo sample the lowest yet reported for Antarctic seawater (Figures 3 and 7c). Because of its $\delta^{18}O$/S property we have defined this water type as a subcategory of Shallow Ice Shelf Water ($SISW_B$). $SISW_B$ is depleted in $\delta^{18}O$, apparently from basal melting of the Ross Ice Shelf, but it occupies an intermediate position between basal meltwater (BM) envelopes in Figure 7c and thus cannot be generated entirely by glacial melting.

The probable source for $SISW_B$ is the WMCO which flows year-round into the sub-ice shelf cavity at temperatures well above the pressure melting point [Pillsbury and Jacobs, this volume]. As it moves beneath the ice, this current will encounter and melt layers of glacial ice that are successively lower in $\delta^{18}O$ content, from -35°/oo near the ice front to about -42°/oo in the midshelf region [Dansgaard et al., 1977; Grootes and Stuiver, 1983]. This $\delta^{18}O$ gradient in the basal ice occurs because of increasing surface accumulation and basal melting as the ice shelf moves north, schematically illustrated in Figure 6b. Eventually the inflow will reach and melt sea ice that has frozen onto the ice shelf base since it moved off the continent, causing the boundary layer to evolve in the "M" direction along the fractionation line in Figure 7c. The basal melting will produce a fresher mixture than the observed $SISW_B$. When all sensible heat has been removed from the water, sea ice will begin to form at the ice-water interface, and $SISW_B$ properties will move in the "F" direction along the fractionation line toward higher salinity and lower $\delta^{18}O$. We sampled $SISW_B$ at some point after it had passed into the basal freezing region, as evidenced by a 6 m basal layer of sea ice at J-9, with ice crystals projecting into the water column [Zotikov et al., 1980]. This analysis neglects mixing between $SISW_B$ and the underling $HSSW_B$ and heat conduction into the ice [Jacobs et al., 1979b]. It also assumes that the initially warm inflow would remain in contact with the basal ice between the barrier and J-9. MacAyeal [this volume (b)] presents a model whereby the WMCO would turn north after contact with the ice to emerge as a plume of SISW.

Gilmour [1979] proposed that the upper boundary layer at J-9 evolved from HSSW melting the glacial ice. HSSW has access to the ice shelf base along at least its western side, and there is probably a net flow southward through McMurdo Sound [Gilmour, 1975; Heath, 1977]. While HSSW is near the sea surface freezing temperature ($\simeq$ -1.9°C), it is 0.2° to 0.3°C above the pressure melting point at the depth of the ice shelf base. As above, HSSW could first melt glacial ice and then sea ice to attain the $\delta^{18}O/S$ characteristic of $SISW_B$. However, the amount of glacial melting necessary to convert HSSW into $SISW_B$ may be prohibitively high [Jacobs et al., 1979b], and if $SISW_B$ is widespread beneath the shelf, it could insulate the ice from the HSSW. The distribution of water types in Figure 7c suggests that HSSW is a more likely source for Deep Ice Shelf Water.

Deep Ice Shelf Water

Deep Ice Shelf Water (DISW) is the mode generally noted in the literature, because of its great lateral extent, extreme temperature, and association with the major ice shelves (Figure 6a). It has also been observed between the Mertz and Ninnis Glacier Tongues (station 197 of Jacobs and Haines [1982]) and near the Drygalski Ice Tongue [S. Jacobs et al., unpublished data, 1984].

From the properties of Ice Shelf Water and High Salinity Shelf Water (WSW) in the Weddell Sea, Weiss et al. [1979] found that glacial ice exposed to 34.77 seawater beneath the Filchner Ice Shelf would have an $\delta^{18}O$ content of -54 ± 9°/oo. Basal ice could acquire an $\delta^{18}O$ content that low at the highest elevations in East Antarctica [Morgan, 1982]. However, Foldvik et al. [this volume] show higher WSW salinities (above 34.8) in the southwest Weddell Sea, so that ice at the base of the Ronne and Filchner Ice Shelves need not average as low as -54°/oo in $\delta^{18}O$. In the Ross Sea, a similar HSSW-ISW projection (Figure 7c) intersects the ordinate around -26°/oo, rather high in $\delta^{18}O$ for glacial ice likely to be exposed at the Ross Ice Shelf base. Given the large extrapolations, allowance must be made for inaccuracies due to the small number of samples, but other processes besides glacial melting may again be involved.

Since DISW is offset to the left of the HSSW-BM envelope in Figure 7c, its simplest origin would seem to involve HSSW sequentially melting glacial ice and sea ice off the ice shelf base. The WMCO and LSSW could also be sources of DISW, after evolution through a Shallow Ice Shelf Water stage and considerable sea ice deposition on the ice shelf base. In addition, some DISW could result from vertical mixing or double diffusion where cold fresh $SISW_B$ overlies warmer, saltier $HSSW_B$ (Figures 2, 6b, and 7c). Tidal or cross-frontal mixing of these components may occur beneath the ice shelf and lead to an emergent plume of DISW [MacAyeal, this volume (a,b)].

The Bottom Waters

Gill [1973], Foster and Carmack [1976] and others have used Θ/S diagrams to model the formation of bottom water in the Weddell Sea. In an oversimplification of their analyses, Warm Deep Water (WDW) mixes with the overlying Winter Water (WW) to produce a Modified Warm Deep Water (MWDW). That mixture intrudes onto the shelf, combines with Western Shelf Water and flows off the shelf to form Weddell Sea Bottom Water (WSBW). Subsequent vertical mixing between WSBW and WDW results in the classical Antarctic Bottom Water (AABW). The Θ/S values associated with these water types are shown in Figure 3, along with the corresponding Ross Sea averages. The Weddell Sea models were idealized, of necessity, but have several shortcomings and limited applicability to the rest of the continental margin. For example, MWDW was assumed to form by cross-isopycnal mixing over the deep ocean and to exhibit properties midway between those of WDW and WW. However, its onshelf mode far exceeds that offshelf [Carmack, 1977], and its measured temperature [Foldvik et al., this volume] is much cooler than predicted. Once on the shelf, mixing between MWDW and WSW was supposed to produce the WSBW, although that type is only observed off the shelf. Large-scale vertical mixing would then result in world class AABW.

More recent observations suggest a somewhat different picture. First, the intermediate to deep levels of the Antarctic basins can be ventilated by water of intermediate density, either near the continental slope [Jacobs and Georgi, 1977; Carmack and Killworth, 1978] or over the deep ocean [Gordon, 1978; Killworth, 1983]. This is significant for world ocean abyssal waters, which derive their characteristics primarily from deep water and not from bottom water in the Antarctic [Mantyla and Reid, 1983]. Second, it is now apparent that Ice Shelf Water plays a role in setting bottom water properties [Weiss et al., 1979; Jacobs et al., 1979a]. Foldvik et al. [this volume] have measured essentially undiluted ISW flowing down the Weddell Sea continental slope at a rate of 10^6 m^3 s^{-1}. The ISW in Figure 2 (from Station 15 in Gammelsrod and Slotsvik [1981]) lies on the projection of a line from WDW through AABW and WSBW. The simplest way to produce bottom water, then, is to mix deep water and ice shelf water. Third, the Low Salinity (Eastern) Shelf Water is essentially the TMIN (Winter Water) of the continental shelf region. However, it occupies isopycnal surfaces in common with deep water at the Slope Front (Figure 6b) and is thus better situated than the TMIN to influence deep and bottom water properties. One way to illustrate these concepts is to evaluate the $\delta^{18}O/S$ characteristics of the major water masses near the continental shelf break, after differentiating incipient bottom water by its salinity, which may be either higher or lower than that of the adjacent deep water.

Low Salinity Bottom Water

Most Antarctic Bottom Water in the world ocean has a lower salinity than the overlying deep water, consistent with the relative abundance of Low Salinity Bottom Water (LSBW) along the circumpolar continental shelf. LSBW displays the lowest temperature, salinity, dissolved oxygen and $\delta^{18}O$ content as WSBW in the Weddell Sea. This occurs because the Weddell Deep Water that contributes to it is fresher and cooler than deep water elsewhere along the margin. It does not necessarily follow that a greater volume of bottom water is generated in the Weddell Sea, although proportionally more is probably collected and recirculated there by the very long continental margin and blocking arm of the Antarctic Peninsula.

In the Ross Sea, LSBW is intermediate in Θ/S and $\delta^{18}O/S$ properties between CDW and LSSW (Figures 2 and 3) but slightly displaced toward the high salinity components. Figure 7d illustrates possible mixtures of CDW with various percentages of the three shelf waters that are found near the continental shelf break. LSBW is shown to be composed of 49-55% CDW, 40-27% LSSW, 0-18% DISW, and 11-0% HSSW. The range of uncertainty lies in that part of the CDW-LSBW projection that crosses the shelf water triangle (LSSW-HSSW-DISW). In any case, LSBW contains about equal portions of deep and shelf water and has properties that are more strongly influenced by LSSW than by HSSW.

Low Salinity Shelf Water has never received credit for its potentially important role in deep water modification and bottom water formation. Since the work of Brennecke [1921], there has been a focus on the scenario of adding brine to surface water to increase its density sufficiently for it to mix with deep water to form bottom water. This has perhaps led to a preoccupation with Western Shelf Water (WSW) and Winter Water (WW) and to a concern with bottom water formation rather than deep water ventilation. However, there is more LSSW than HSSW, even with the HSSW threshold at a salinity of 34.6, and LSSW vastly exceeds the WW in volume [Carmack, 1977]. Ross Sea LSSW is slightly denser than the CDW, whereas the Weddell Sea ESW appears to be slightly less dense than the WDW (Figure 2). For nonlinearities in the seawater equation of state to be an important factor in bottom water production, ESW salinity must exceed about 34.5 to combine with Weddell Deep Water [Fofonoff, 1956]. That it does this may be evidenced by a volumetric Θ/S maximum centered slightly above 34.5 in Figure 4 of Carmack and Foster [1977]. Indeed, if Weddell Sea Bottom Water (WSBW) is more than a simple mixture of deep water and ice shelf water, then it must include a significant ESW component to balance any WSW it contains (Figure 2).

High Salinity Bottom Water

High Salinity Bottom Water (HSBW), earlier referred to as Ross Sea Bottom Water [Jacobs et al., 1970; Gordon and Tchernia, 1972], appears on the continental slopes north and west of the Ross Sea. Only three HSBW samples were available for this study, so the $\delta^{18}O/S$ averages in Table 1 may not fully represent the HSSW component. Nonetheless, we can estimate from Figure 7d that this HSBW is composed of 69-75% CDW, 12-0% LSSW, 0-18% DISW, and 19-7% HSSW. The Deep Ice Shelf Water also contributes up to 18% of this bottom water type, even though DISW reaches the continental margin at restricted locations and accounts for a relatively small proportion of the shelf waters [Carmack, 1977]. The northward movement of DISW from the ice shelf base to the continental shelf break (Figure 6a) suggests that melting glacial ice exerts a strong influence on the $\delta^{18}O$ content of bottom water.

HSBW is not widespread, even north of the Ross Sea where a large volume of High Salinity Shelf Water extends 400 m above any sill that might block flow to the north. HSSW is denser than CDW (Figure 2) and might be expected to spill unaltered off the continental shelf and flow down and along the slope. Late winter measurements in the northwest Ross Sea may yet show HSSW on the slope there. However, the pressure gradient across the Slope Front must be balanced by Coriolis forces on the along-front currents, so that most HSSW recirculates on the shelf. Some HSSW is altered in the process of melting glacial ice, and some will evolve back into summer surface water on the western shelf. Also, the production rate of HSSW with its greater salt requirement may be lower or more variable than that of LSSW.

Preliminary Budget Considerations

Estimates of volumes and renewal rates for individual water masses require a good sample grid and time series measurements of currents and other parameters. However, by grouping the shelf waters together and considering $\delta^{18}O$ flux through the system, first-order intercomparisons can be made of mass, heat, salt, and ice transports. The Ross Sea data closely resemble Weddell Sea observations and can be combined with Carmack's [1977] volumetric calculations to estimate budgets for the circumpolar continental shelf that are consistent with mass balance of the ice sheet.

We will make the assumption that shelf water is renewed from equal portions of deep and surface water. The seaward sides of Figures 6a and 6b show that the water columns above shelf break depth are made up of one-third to one-half surface water. Since the Slope Front is not well developed above 100-200 m, surface water may be less constrained than deep water from moving onto the shelf. Charts of dynamic topography indicate surface flow toward the ice shelves in the eastern and central sectors of the large embayments and away from the ice in the western sectors [Klepikov and Grigoryev, 1966; Carmack and Foster, 1975a].

Warren [1981] noted the apparent anomaly between onshelf transport of surface water and simultaneous offshelf transport of sea ice, required by the Gill [1973] model to provide brine for the shelf water. Pack ice motion may not be closely coupled to that of the thick surface layer, however, and it is now evident that substantial volumes of deep water or its derivatives intrude onto the shelf at intermediate depths (Figure 6b). This water is about 0.3 higher in salinity than AASW, necessitating less sea ice formation to maintain a salt balance on the shelf. Another consequence of this lateral subsurface influx is that more heat is transported onto the continental shelf, in some locations along density surfaces that have access to the ice shelf base.

Mass Balance of the Antarctic Ice Sheet

Most glaciological studies show a slight positive mass balance for the Antarctic ice sheet, but reliable data exist only for the accumulation side of that budget. An oxygen isotope profile through the ice sheet at Byrd Station (80°S, 121°W) indicated a negative mass budget in that basin, which drains into the Ross Ice Shelf [Johnson et al., 1972]. Global sea level trends may give weak evidence for a slight negative budget, but most of the sea level rise over the past century has been attributed to thermal expansion of seawater and receding temperate glaciers [Gornitz et al., 1982; Meier, 1984]. Robin [1979] indicated that basal melting beneath the thicker parts of ice shelves is much greater than is generally appreciated and could be sufficient to bring the mass balance of the Antarctic ice sheet into approximate equilibrium. For an equilibrium ice sheet of area 14.1×10^6 km^2 and an average accumulation (P - E) of 14.1 cm yr^{-1} [Baumgartner and Reichel, 1975], the volume of glacial meltwater released to the Southern Ocean would be about 2.0×10^3 km^3 yr^{-1}.

That portion of the ice sheet that melts into the shelf waters can be divided into basal melting under ice shelves, wall melting along ice features that rim the coastline, and some portion of iceberg melting. Estimates for these terms vary widely due to the different data sets, methods, and assumptions employed and because of real spatial and temporal variability. Bull [1971] summarized several mass balance calculations, giving a range that would correspond to 4-50 cm yr^{-1} of basal melting. Indirect determinations of melting near the ice shelf fronts range as follows: 60-224 cm yr^{-1} under the Ross Ice Shelf [Crary, 1964; Crary et al., 1962], 100 cm yr^{-1} beneath the Brunt Ice Shelf [Thomas, 1973], 120-360 cm yr^{-1} under the George VI Ice Shelf [Potter and Paren, this volume], 320 cm yr^{-1} under the Ronne Ice Shelf [Robin et al., 1983], 400 cm beneath the Amery Ice Shelf [Morgan and Budd, 1978], and 960 cm yr^{-1} under the Filchner Ice Shelf [Behrendt, 1968]. Relatively high melting rates beneath the northern edges of the ice shelves are believed to result from tidal pumping. Thomas and MacAyeal [1982] calculate a decrease from 70 cm to 0 ± 10 cm of melting 100 km south of the Ross Ice Shelf Barrier. There is also some evidence for basal melting deep within the sub-ice shelf cavities [Robin et al., 1983; MacAyeal, 1984].

Basal freezing of 2-4 cm yr^{-1} may occur over significant areas beneath the larger ice shelves [Zotikov et al., 1980], as latent heat is conducted into the ice from seawater at the in situ freezing point. For basal freezing rates much higher than that [Morgan, 1972], the seawater either must have access to meltwater from the base of the continental ice sheet [Kotlyakov et al., 1977], access to larger heat sinks, such as in basal crevasses [Barrett, 1975; Clough, 1975], or it must deposit ice as it upwells beneath the ice shelf. In the latter case, the depression in freezing point with increasing pressure [Gordon, 1974b; Foldvik and Kvinge, 1974; Robin, 1979; MacAyeal, this volume (a); Lewis and Perkin, this volume] could result in sea ice deposition at upwelling locations and basal ice removal where the seawater moved to deeper levels. This ice redistribution could change the impact of submarine rises on grounding and flow of an ice shelf.

During summer the vertical walls of ice shelves and glaciers are exposed to relatively warm water, and year-round the basal areas will be near water that is above the in situ melting temperature. Glaciers and ice shelves occupy 45% of the Antarctic coastline [Grosvald et al., 1971], so the exposed area of the vertical walls will be about 1.5×10^4 km^2, two orders of magnitude less than the basal area. If wall melting exceeds basal melting by a factor of 10 [Neshyba and Josberger, 1980], it will then contribute about 10% as much meltwater. From data in Morgan and Budd [1978], the production rate of icebergs from the Antarctic continent is about 1.3×10^3 km^3 yr^{-1}, of which one-third are observed in waters south of the Antarctic Divergence. These southern icebergs will melt more slowly than those in warmer water to the north, but some are grounded and others track the continental margin for long periods of time [e.g., Tchernia and Jeannin, 1983]. We postulate that one-sixth of the annual iceberg discharge, about 0.2×10^3 km^3 yr^{-1}, will melt directly into the shelf waters or into waters that are subsequently transported onto the shelf. That meltwater would amount to 8 cm yr^{-1} if distributed over the entire continental shelf. Significantly lower and higher estimates of annual iceberg calving may be found in the literature [e.g., M. Mellor in Gow, 1965; Orheim, 1985].

From the $\delta^{18}O$ content of shelf water relative to its source we can then obtain a figure for basal melting that is consistent with mass balance of the ice sheet. In that case, 0.7×10^3 km^3 yr^{-1} of in situ melting and 0.2×10^3 km^3 yr^{-1} of iceberg melting would be input to the shelf water. This corresponds to 36 cm yr^{-1} of glacial meltwater (GM) which would be composed of 8 cm of iceberg meltwater (IM), 25 cm of ice shelf basal meltwater (BM), and 3 cm of ice shelf wall meltwater (WM). Converted to equivalent melting of the attached ice features, that corresponds to 42 cm of net basal melting and 420 cm of wall melting. These values are strongly dependent upon the rates and locations of iceberg calving and attrition and are not sufficiently accurate to warrant corrections for the different densities of ice and seawater. Our ice shelf bottom melting rate is about twice that estimated by Kotlyakov et al. [1977].

The $\delta^{18}O$ Flux

Surface and deep water moves onto the shelf with an average $\delta^{18}O$ content of -0.22°/oo, assuming it is made up of 50% CDW, 25% AASW, and 25% TMIN (Table 1). On the shelf, its $\delta^{18}O$ content will be lowered by marine precipitation, glacial meltwater, and the net freezing of sea ice. From the TMIN section above, 15 cm yr^{-1} of -16.5°/oo precipitation will lower the $\delta^{18}O$ content of shelf water by 0.005°/oo yr^{-1}. A decrease of 0.025°/oo yr^{-1} will result from the 36 cm yr^{-1} of -36°/oo glacial meltwater. The decrease in shelf water $\delta^{18}O$ from sea ice freezing will be only 0.002°/oo yr^{-1}, based upon the salt flux considerations below and the freezing line slope on an $\delta^{18}O$ diagram (Figure 3). The average $\delta^{18}O$ content of shelf water can be estimated as -0.42°/oo from Table 1 and a recipe modified from Carmack [1977]: shelf water = 10% AASW + 5% AASW$_G$ + 20% WMCO + 35% LSSW + 20% HSSW + 5% DISW + 5% SISW. The net change in $\delta^{18}O$ between source and shelf water is then -0.20°/oo, but over what period of time?

The relationships between shelf water $\delta^{18}O$ content and residence time are illustrated in Figure 8. Combining effective surface freezing and precipitation (F+P) with different amounts of glacial melting (GM) gives rise to a family of curves originating at the source water $\delta^{18}O$ value of -0.22°/oo and crossing the average shelf water $\delta^{18}O$ content of -0.42°/oo at different residence times. Assuming that all water exported from the shelf combines with an equal portion of deep water to produce bottom water, then the formation rates would be as shown. If there were no glacial meltwater (F+P+0GM), the shelf water residence time would be nearly 30 years, and present day bottom water formation would be less than 3×10^6 m^3 s^{-1}. From a heat budget and from radiocarbon data, Gordon [1975] and Stuiver et al. [1983] have calculated Southern Ocean bottom water formation rates of 38 and 41×10^6 m^3 s^{-1}. Estimates of bottom water formation along the continental margin range from 5 to 20×10^6 m^3 s^{-1} [Killworth, 1973; Carmack, 1977]. The 5 to 40×10^6 m^3 s^{-1} range accentuated by the heavy horizontal bar in Figure 8 is equivalent to the input of 8 to 134 cm yr^{-1} of glacial meltwater to the continental shelf. With 36 cm yr^{-1} of glacial meltwater corresponding to the ice sheet mass balance, shelf water residence time would then be 6.2 years, its exchange rate with the deep ocean 6.5×10^6 m^3 s^{-1}, and bottom water production would be 12.9×10^6 m^3 s^{-1}. For comparison, Gill [1973] estimated a 3.5- to 7-year shelf water residence time in the Weddell Sea.

Heat Transport

Heat transport onto the circumpolar continental shelf and its subsequent dissipation can be estimated from the previous data and arguments. Proportioning the source water as above and using Table 1 averages, the inflow will have an average temperature of -0.11°C. With the export of water from the shelf as in the LSBW case, its temperature would be -1.70°C. The mass exchange above would then correspond to a net southward heat transport of approximately 4.3×10^{13} W. This is roughly equivalent to an average flux of 0.4 W cm^{-2} across a 500-m-deep shelf break at 70°S. Geothermal heat flow, at an average rate of 5.0×10^{-6} W cm^{-2} over the continental shelf, will supply only 0.01×10^{13} W.

To produce 36 cm of meltwater over the continental shelf area at 335 J cm^{-3} for the phase change and to account for heat flux into the glacial ice would require about 1.0×10^{13} W. To melt 95 cm of sea ice (see next section) would require 2.5×10^{13} W, leaving a net heat flux to the atmosphere of around 3.2 W m^{-2}. Vertical heat flux to the sea ice and atmosphere would then be 14.8 W m^{-2}, within Allison's [1981] range of 10-15 W m^{-2} during the ice growth season at Mawson Station, also on the continental shelf. The heat transport

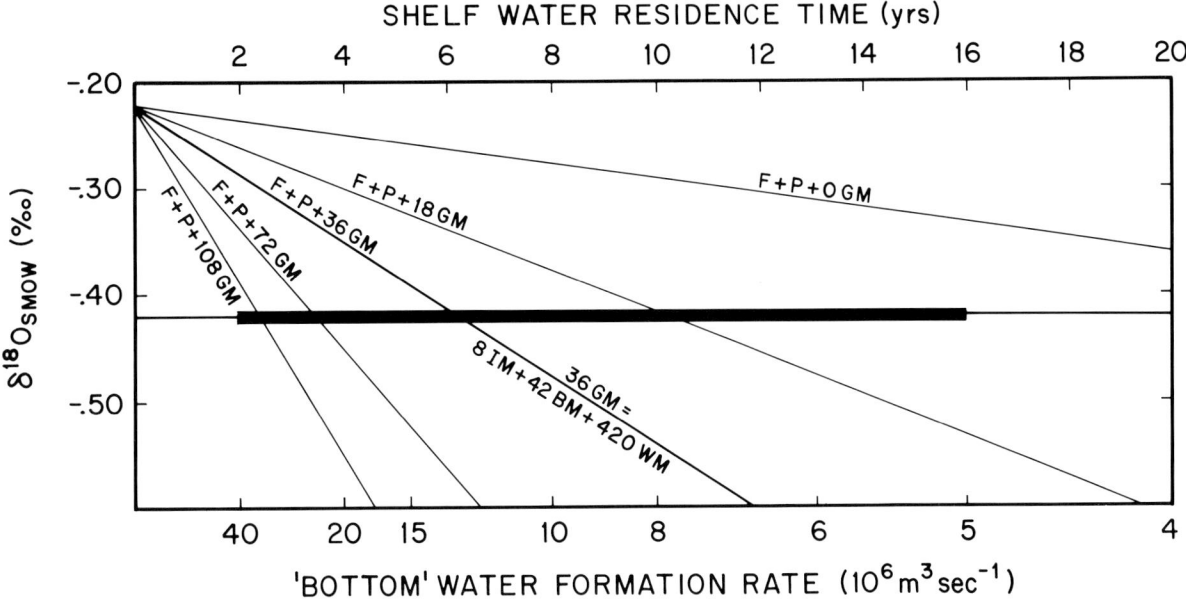

Fig. 8. Relationship between residence time and average $\delta^{18}O$ content of shelf water for fixed rates of sea ice freezing (F = 95 cm) and precipitation (P = 15 cm) and different amounts of glacial meltwater (GM = 0 to 108 cm yr^{-1}). Seawater moving onto the continental shelf is assumed to be half surface water and half deep water (25% AASW, 25% TMIN and 50% CDW), at an average $\delta^{18}O$ content of -0.22°/oo. Seawater leaving the shelf is assumed to be LSBW (70% LSSW, 20% DISW, and 10% HSSW), at an average $\delta^{18}O$ content of -0.42°/oo. The solid bar indicates a range of commonly accepted bottom water formation rates, assuming equal mixtures of shelf and deep water. For the ice sheet mass balance example (GM = 36), 8 cm of iceberg melt (IM), and 28 cm of ice shelf melt are added to the shelf water each year; that ice shelf melt corresponds to 42 cm yr^{-1} of basal melting (BM) and 420 cm yr^{-1} of wall melting (WM).

associated with water mass exchange thus appears sufficient to balance these preliminary budgets.

Gordon [1981] calculated an order of magnitude larger ocean to atmosphere heat flux (31 W m^{-2}) over the deep ocean between 60°S and 70°S. However, there is a proportionately larger temperature gradient between subsurface and surface waters over the deep Antarctic ocean basins than exists over the continental shelf. Pillsbury and Jacobs [this volume] estimate a heat flux beneath the Ross Ice Shelf that would represent about 40% of the oceanic heat transported across the Ross Sea continental shelf break, assuming a circumpolar average.

Salt Flux and Sea Ice Thickness

Some fraction of sea ice is exported from the continental shelf, resulting in net deposition of brine to the underlying shelf waters. From the water mass characteristics above, seawater is imported to the shelf at an average salinity of 34.455 and is exported at 34.589. The brine it acquires in a 6.2-year residence time is equivalent to the formation of 37 cm of sea ice with a residual salinity of 4.1, from AASW at 34.1. Salt must also be drained from 58 cm of sea ice to balance the 15 cm of net precipitation (P - E) and 36 cm of glacial meltwater. These figures total 95 cm but do not include sea ice that forms and melts on the shelf. In the absence of quantitative information, we will continue with the working assumption that half the ice formed on the shelf melts there. Sea ice formation of 1.9 m yr^{-1} is then needed to balance sea ice melting, precipitation, glacial melting, and the water mass salt flux.

Most reported sea ice thicknesses on the continental shelf have been based upon measurements in fast ice near Antarctic stations. These 1-year growths commonly fall in the 1.5 - 2.5 m range [Allison, 1981; U.S. Navy, unpublished data 1971-1977]. The pack ice observed from ships working on the continental shelf is often thinner than that, perhaps due to the onset of summer melting and the ships' avoidance of thick ice. Net ice formation over the shelf may be considerably higher due to ridging and intense freezing in coastal leads and polynyas. The latter process may be particularly effective where offshore winds continually move the newly formed ice away from the coastline [Gill, 1973; Killworth, 1974; Cavalieri and Martin, this volume; Kurtz and Bromwich, this volume]. Some of this northward flux may be balanced by recirculation on the shelf, or by ice that moves onto the shelf at the eastern sides of the large embayments. Ackley [1979] has argued that ridging could increase effective ice thickness over the shelf to 3.7 m, and that there is relatively little ablation at the top or bottom. Satellite data interpreted by the Naval Polar Oceanography Center [1974-1984] show that south to north internal opening of the Ross Sea in early summer can span 5° of latitude in a month. While that could be accomplished by a 20 cm s^{-1} wind-driven northward transport, some melting at the bottom and fringes of floes is suggested by their scalloped and honeycombed surfaces.

Zwally et al. [this volume] have calculated that freezing in leads and polynyas may be equivalent to the formation of 4.6 to 7.8 m of ice over the entire Antarctic continental shelf. That is several times the observed ice thicknesses, implying a high rate of ice export and more brine left on the shelf than is accounted for in the budgets above. We return to this point in the discussion below.

Discussion

The $H_2^{18}O/H_2^{16}O$ ratio is a useful tracer for glacial melting, precipitation, sea ice freezing, and water mass mixing near the Antarctic continental margin. Realistic assumptions as to source and residence time of water on the continental shelf lead to first-order estimates for bottom water formation, sea ice thickness, heat, and salt flux. For example, water on the Ross Sea continental shelf is 0.20°/oo lower in $\delta^{18}O$ content than its probable source water. With the addition of 15 cm yr^{-1} of precipitation (P - E) and 36 cm yr^{-1} of glacial meltwater, its 6.2-year residence time corresponds to a shelf water export of 6.5 x 10^6 m^3 s^{-1} and "bottom" water formation of 12.9 x 10^6 m^3 s^{-1}. Southward transport of heat onto the circumpolar continental shelf is then 4.3 x 10^{13} W, of which about 25% goes into warming and melting glacial ice, and 60% into melting sea ice. If half of the sea ice that forms each year on the continental shelf eventually melts north of the shelf break, it will leave behind the brine from an average 95 cm of ice, or about 2.8 g cm^{-2}. Approximately 60% of that salt would balance the glacial ice meltwater and 40% would balance the salt differential between water imported to and exported from the shelf. These numbers are not uniquely determined, but they are internally consistent and within the bounds of earlier estimates.

Since waters on the Ross and Weddell continental shelves display similar Θ/S and $\delta^{18}O/S$ characteristics, we have taken the Ross Sea data as typical of the circumpolar continental shelf. In fact, these summer observations were selected from available observations to portray specific water masses and may not otherwise be representative. There are significant thermohaline and other differences between vertical sections at various locations across the continental margin and between reoccupied stations at the same location. For example, Figure 6b shows a late December WMCO intrusion onto the continental shelf along a transect where this feature is best developed in the Ross Sea. If the WMCO temperature were substituted for CDW temperature in the heat transport calculation, then inflow to the Ross Sea continental shelf would be considerably cooler than -0.11°C. Further, the Ross Ice Shelf occupies about 30% of the total Antarctic ice shelf area, while the continental shelf to its north encompasses only about 15% of the circumpolar continental shelf [Vanney and Johnson, this volume]. With these altered regional parameters, about 25 cm yr^{-1} would be melted off the Ross Ice Shelf, rather than the 42 cm yr^{-1} circumpolar average calculated above. This lower value may be nearer to the actual melting rate for the major ice shelves [e.g., Jacobs et al., 1979b; Thomas and MacAyeal, 1982], with greater melting experienced by ice that borders the more northerly coastline. Even the large ice shelves are not likely to experience the same basal melt rates. Intermediate-depth intrusions onto the continental shelf are warmer in the Ross Sea than in the Weddell Sea (compare Figure 6b here with Figure 9 of Foldvik et al. [this volume]). Whether the relatively thinner Ross Ice Shelf or the internal opening of the Ross Sea pack ice in spring are due in part to this warmer inflow remains to be determined.

Zwally et al. [this volume] have estimated that effective pack ice formation on the shelf could range from 4.6 to 7.8 m yr^{-1}. The higher number is close to the production required by the Killworth [1974] model to maintain the observed shelf water density structure. However, their range is also similar to values calculated by Cavalieri and Martin [this volume] for ice production rates in wind-driven polynyas along the Wilkes Land Coast during 3 austral winter months, where a maximum might be expected, and more than twice that modeled for the southwest Weddell Sea [Hibler, 1984]. Effective ice formation thicknesses calculated by Zwalley et al. [this volume] are 2 to 4 times those observed on summer expeditions, which implies a high sea ice export rate. Gill [1973] also used a high export rate, but he would have left the brine from only 1.7 m of ice on the shelf. Our budgets require the brine from only 95 cm of ice to balance precipitation, glacial ice melting and water mass exchange. With an export rate of 75%, how can the salt from another 2.5 to 4.9 m of ice be accounted for? Doubling effective precipitation, residual sea ice salinity, and the input of glacial meltwater would require 84 cm more ice. Such an increase in glacial melting would greatly decrease the number of icebergs north of the continental shelf, or be at the expense of mass balance of the ice sheet. Altering the proportions of source and sink waters, e.g., to 2 parts surface water and 1 part deep water, and to equal parts LSBW and HSBW, would together necessitate 77 cm more ice. However, those fractions are inconsistent with the relative water mass volumes observed on the shelf and in the deep ocean [Carmack, 1977]. Decreasing shelf water residence time by, say, a factor of 2, would require only one-third of a meter more sea ice to balance the water mass exchange. It would seriously perturb the $\delta^{18}O$ budget, however, with significant implications for the glacial melting rate (Figure 8).

A short shelf water residence time implies a high bottom water formation rate, or the export of shelf water before it attains deep or bottom water densities. By this latter process, 100 m of surface water leaving the shelf with its salinity raised from 34.2 to 34.5 could account for another 1 m of sea ice. However, exported surface water will presumably be replaced not only by more 34.2 surface water, but also by subsurface water of higher salinity. It is thus difficult to account for the salt from an average sea ice formation on the shelf as high as 7.8 m yr^{-1} which may indicate that an average ice formation rate of 0.17 m/day [Zwally et al., this volume] is high for Antarctic leads and polynyas.

The region over the continental slope (Figures 6a and 6b) is a sink or source for several Antarctic water masses. Its Slope Front is a well-developed but highly variable feature that influences mass and property transfers between the shelf and deep ocean and appears to be important to biological activity [Ainley and Jacobs, 1981]. Although the continental shelves are generally believed to be the primary location of surface water buoyancy loss in the Antarctic at the present time, it is possible that deep water ventilation may be less coupled to this region than to local overturning associated with the deep TMIN at the Slope Front. Weyl [1968] postulated that the shallow polar seas would not have existed during glacial periods, and there is evidence that the Antarctic ice sheet did extend at least to the continental shelf break during previous glacial advances [Kellogg et al., 1979]. Bottom water "activity" may have been greater or lesser during glacial epochs than it is today [Lohmann, 1978]. If activity and volume are directly related, and shelf water does play a major role in bottom water formation, then more shelf ice might well be correlated with less bottom water. Perhaps ventilation of the deep ocean is concentrated along the Slope Front or in "chimneys" north of the continental margin during glacial periods. It is conceivable that ice shelves might at times extend over much of the continental shelves without being grounded there. If deep water continued to provide sufficient heat to the cavities beneath that glacial ice, the production of Ice Shelf Water and its input to the deep ocean would seem likely.

The demonstrated links between the deep sea and Antarctic glacial ice raise several questions related to climate change. What will be the response of the Antarctic continental shelf circulation to projected global increases in atmospheric temperature? Salt flux on the continental shelf influences the density field, which appears to restrict warm water access to the central sector of the ice shelf base (Figure 6c). If a warmer atmosphere results in less sea ice production, will that shift the density field so that more "warm" water then reaches the ice shelves? We have estimated that oceanic heat now accounts for about one-third of the wastage of the Antarctic ice sheet, via direct melting of its attached, floating ice forms. If ocean temperatures increase considerably, will this melting accelerate more rapidly than feedbacks such as snow accumulation on the continent? If not, will the ice sheet mass balance become negative and contribute to the global rise of sea level?

Acknowledgments. Financial support for this work was provided by the National Science Foundation, grants DPP-79-18674 and DPP-81-19863 (S.S.J.) and OCE-82-07998 (R.G.F.). R. Weiss, P. Killworth, P. Grootes, A. Gordon, and others made helpful comments on the manuscript. Word processing and drafting was done by E. Free, J. Wright, C. Anderson, J. Hubschman, S. Porta and B. Batchelder. Lamont-Doherty Geological Observatory contribution 3748.

References

Ackley, S.F., Mass-balance aspects of the Weddell Sea pack ice, J. Glaciol., 24(90), 391-405, 1979.

Ainley, D.G., and S.S. Jacobs, Sea-bird affinities for ocean and ice boundaries in the Antarctic, Deep Sea Res., 28A(10), 1173-1185, 1981.

Allison, I., Antarctic sea ice growth and oceanic heat flux, IAHS AISH Publ., 131, 161-170, 1981.

Barrett, P.J., Seawater near the head of the Ross Ice Shelf, Nature, 256(5516), 390-392, 1975.

Baumgartner, A., and E. Reichel, The World Water Balance: Mean Annual Global, Continental and Maritime Precipitation, Evaporation and Run-Off, 179 pp., Elsevier, New York, 1975.

Behrendt, J.C., The structure of the Filchner Ice Shelf and its relation to bottom melting, IAHS AISH Publ., 86, 488-496, 1968.

Bentley, C.R., J. W. Clough, and J.D. Robertson, RISP geophysical work, Antarctic Journal of the United States, 9(4), 157-159, 1974.

Brennecke, W., Die ozeanographischen Arbeiten der Deutschen Antarktischen Expedition 1911-12, Ark. Dtsch. Sewarte, 39(1), 214 pp., 1921.

Broecker, W.S., T. Takahashi, and T. Takahashi, Sources and flow patterns of deep ocean waters as deduced from potential temperature, salinity and initial phosphate concentration, J. Geophys. Res., in press, 1985.

Bromwich, D.H., and C.J. Weaver, Latitudinal displacement from main moisture source controls δ^{18} of snow in coastal Antarctica, Nature, 301, 145-147, 1983.

Bull, C., Snow accumulation in Antarctica, in Research in the Antarctic, Publ. 93, edited by L. Quam, pp. 367-421, AAAS, Washington, D.C., 1971.

Byrd, R.E., Little America, 422 pp., Putnam, 1930.

Carmack, E.C., Water characteristics of the Southern Ocean south of the Polar Front, in A Voyage of Discovery, edited by M. Angel, pp. 15-42, G. Deacon 70th Anniversary Volume, Supplement to Deep Sea Res., Pergamon, New York, 1977.

Carmack, E.C., and T.D. Foster, Circulation and distribution of oceanographic properties near the Filchner Ice Shelf, Deep Sea Res., 22(2), 77-90, 1975a.

Carmack, E.C., and T.D. Foster, On the flow of water out of the Weddell Sea, Deep Sea Res., 22, 711-724, 1975b.

Carmack, E.C., and T.D. Foster, Water masses and circulation in the Weddell Sea, in Polar Oceans, edited by M. Dunbar, pp. 151-165, Proceedings of the Polar Oceans Conference, Montreal, May 1974, Arctic Institute of North America, Calgary, 1977.

Carmack, E.C., and P.D. Killworth, Formation and interleaving of abyssal water masses off Wilkes Land, Antarctica, Deep Sea Res., 25, 357-370, 1978.

Cavalieri, D.J., and S. Martin, A passive microwave study of polynyas along the Antarctic Wilkes Land Coast, this volume.

Clausen, H.B., W. Dansgaard, J.O. Nielsen, and J.W. Clough, Surface accumulation on Ross Ice Shelf, Antarct. J. U.S., 14(5), 68-72, 1979.

Clough, J.W., Bottom crevasses in the Ross Ice Shelf, J. Glaciol., 15(73), 457-458, 1975.

Countryman, K.A., and W.L. Gsell, Operations Deep Freeze 63 and 64; Summer oceanographic features of the Ross Sea, U.S. Naval Oceanogra. Off., Tech. Rep. 190, 193 pp., Suitland, Md., 1966.

Craig, H., Standard for reporting concentrations of deuterium and oxygen-18 in natural waters, Science, 133, 1833-1834, 1961.

Craig, H., and L.I. Gordon, Deuterium and oxygen 18 variations in the ocean and marine atmosphere, in Proceedings of a Symposium on Marine Chemistry, Occ. Publ. 3-1965, edited by D. Schink and J. Corliss, pp. 277-374, University of Rhode Island, 1965.

Craig, H., and B. Hom, Relationships of deuterium, oxygen 18, and chlorinity in the formation of sea ice, EOS Trans. AGU, 49, 216-217, 1968.

Crary, A.P., Melting at the ice-water interface, Little America Station, J. Glaciol., 5(37), 129-130, 1964.

Crary, A.P., E.S. Robinson, H.F. Bennett, and W.W. Boyd, Glaciological regime of the Ross Ice Shelf, J. Geophys. Res., 67(7), 2791-2807, 1962.

Dansgaard, W., S.J. Johnsen, H.B. Clausen, C. U. Hammer, and C.C. Langway, Jr., Stable isotope profile through the Ross Ice Shelf at Little America, V, Antarctica, IAHS AISH, Publ. 118, 322-325, 1977.

David, T.W.E., Antarctica and some of its problems, Geogr. J., 43(6), 605-630, 1914.

Deacon, G.E.R., The Weddell Gyre, Deep Sea Res., 26A, 981-995, 1979.

Drygalski, E. Von, The oceanographical problems of the Antarctic, Spec. Publ. 7, pp. 269-283, translated from German, American Geographical Society, New York, 1928.

Dunbar, R.B., J.B. Anderson, E.W. Domack, and S.S. Jacobs, Oceanographic influences on sedimentation along the Antarctic continental shelf, this volume.

Epstein, S., and T. Mayeda, Variation of O^{18} content of waters from natural sources, Geochm. Cosmochim. Acta, 4, 213, 1953.

Fairbanks, R.G., The origin of continental shelf and slope water in the New York Bight and Gulf of Maine: Evidence from $H_2^{18}O/H_2^{16}O$ ratio measurements, J. Geophys. Res., 87(C8), 5796-5808, 1982.

Fofonoff, N.P., Some properties of sea water influencing the formation of Antarctic bottom water, Deep Sea Res., 4, 32-66, 1956.

Foldvik, A., and T. Kvinge, Conditional stability of sea water at the freezing point, Deep Sea Res., 21(3), 169-174, 1974.

Foldvik, A., T. Gammelsrød, and T. Torresen, Circulation and water masses on the southern Weddell Sea shelf, this volume.

Foster, T.D., and E.C. Carmack, Frontal zone mixing and Antarctic Bottom Water formation in the southern Weddell Sea, Deep Sea Res., 23, 301-317, 1976.

Gammelsrød, T., and N. Slotsvik, Hydrographic and current measurements in the southern Weddell Sea 1979/80, Polarforschung, 51(1), 101-111, 1981.

Gill, A.E., Circulation and bottom water production in the Weddell Sea, Deep Sea Res., 20(2), 111-140, 1973.

Gilmour, A.E., McMurdo Sound hydrological observations, 1972-73, N. Z. J. Mar. Freshwater Res., 9(1), 75-95, 1975.
Gilmour, A.E., Ross Ice Shelf sea temperature, Science, 203(4379), 438-439, 1979.
Gordon, A.L., Varieties and variability of Antarctic Bottom Water, Colloq. Int. C. N. R. S., 215, 33-47, 1974a.
Gordon, A.L., RISP Oceanographic Investigations, in Ross Ice Shelf Project (RISP) Science Plan, Univ. of Nebraska, Lincoln, 41-58, 1974b.
Gordon, A.L., General ocean circulation, in Numerical Models of Ocean Circulation, pp. 39-43, National Academy of Science, Washington, D.C., 1975.
Gordon, A.L., Deep Antarctic convection west of the Maud Rise, J. Phys. Oceanogr., 8(4), 600-612, 1978.
Gordon, A.L., Seasonality of Southern Ocean sea ice, J. Geophys. Res., 86(C5), 4193-4197, 1981.
Gordon, A.L., Weddell Deep Water variability, J. Marine. Res., 40 Suppl., 199-217, 1982.
Gordon, A.L., Comments about the ocean role in the Antarctic glacial ice balance, Proc. CO_2 Research Conference: CO_2, Science and Concensus, pp. 75-96, U.S. Department of Energy, Washington, D.C., 1983.
Gordon, A.L., and E.J. Molinelli, Thermohaline and chemical distributions and the Atlas data set, in Southern Ocean Atlas, 11 pp. and 233 plates, Columbia University Press, New York, 1982.
Gordon, A.L., and P. Tchernia, Waters of the continental margin off Adelie Coast, Antarctica, in Antarctic Oceanology II: The Australian-New Zealand Sector, Antarct. Res. Ser., vol. 19, edited by D.E. Hayes, pp. 59-69, AGU, Washington, D.C., 1972.
Gornitz, V., S. Lebedeff, and J. Hansen, Global sea level trend in the past century, Science, 215, 1611-1614, 1982.
Grootes, P.M., and M. Stuiver, Ross Ice Shelf oxygen isotope profile at J-9, Antarct. J. U.S., 18(5), 107-109, 1983.
Grosvald, M.G., V.M. Kotlyakov, and P.A. Shumskii, Ice shelves (shelf glaciers): Definition, terminology and classification, in Antarctica Commission Reports, edited by V.A. Bugaev, translated from Russian by Israel Program for Scientific Translation, Jerusalem, 1971.
Heath, R.A., Circulation across the ice shelf edge in McMurdo Sound, Antarctica, in Polar Oceans, edited by M.J. Dunbar, pp. 129-149, Arctic Institute of North America, Calgary, 129-149, 1977.
Heezen, B.C., and C. Bentley, Plate 1, Antarctic Map Folio Series, no. 16, American Geographical Society, New York, 1972.
Hibler, W.D. III, The role of sea ice dynamics in modeling CO_2 increases, in Climate Processes and Climate Sensitivity, edited by J. Hansen and T. Takahashi, Geophys. Monogr. Ser., vol. 29, pp. 238-253, AGU, Washington, D.C., 1984.
Horibe, Y., and N. Ogura, Deuterium content as a parameter of water mass in the ocean, J. Geophys. Res., 73(4), 1239-1249, 1968.
Horibe, Y., K. Shigehara, and Y. Takakuwa, Isotope separation factor of carbon dioxide-water system and isotopic composition of atmospheric oxygen, J. Geophys. Res., 78(15), 2625-2629, 1973.
Hughes, T., Is the West Antarctic Ice Sheet disintegrating?, Ice Streamline Cooperative Antarctic Project, ISCAP Bull. 1, 77 pp., Institute of Polar Studies, Ohio State Univ., Columbus, 1972.
Jacobs, S.S., Oceanographic evidence for land ice/ocean interactions in the Southern Ocean, in Glaciers, ice sheets and sea level: Effect of a CO_2-induced climatic change, Report of a Workshop, Seattle, Sept. 13-15, 1984, National Academy Press., Wash., D.C., 1985.
Jacobs, S.S., and D.T. Georgi, Observations on the southwest Indian/Antarctic Ocean, in A Voyage of Discovery, edited by M. Angel, pp. 43-84, G. Deacon 70th Anniversary Volume, Supplement to Deep Sea Res., Pergamon, New York, 1977.
Jacobs, S.S., and W.E. Haines, Ross Ice Shelf Project, Oceanographic Stations, 1976-1979, Tech. Rep. 82-1, Lamont-Doherty Geological Observatory, Palisades, N.Y., 1982.
Jacobs, S.S., A.F. Amos, and P.M. Bruchhausen, Ross Sea oceanography and Antarctic Bottom Water formation, Deep Sea Res., 17, 935-962, 1970.
Jacobs, S.S., A.L. Gordon, and A.F. Amos, Effect of glacial ice melting on Antarctic Surface Water, Nature, 277(5696), 469-471, 1979a.
Jacobs, S.S., A.L. Gordon, and J.L. Ardai, Circulation and melting beneath the Ross Ice Shelf, Science, 203, 439-442, 1979b.
Jacobs, S.S., H.E. Huppert, G. Holdsworth, and D.J. Drewry, Thermohaline steps induced by melting of the Erebus Glacier Tongue, J. Geophys. Res., 86(C7), 6547-6555, 1981.
Johnson, S.J., W. Dansgaard, H.B. Clausen, and C.C. Langway, Jr., Oxygen isotope profiles through the Antarctic and Greenland ice sheets, Nature, 235, 429-434, 1972.
Kellogg, T.B., R.S. Truesdale, and L.F. Osterman, Late Quaternary extent of the West Antarctic Ice Sheet: New evidence from Ross Sea cores, Geology, 7(5), 249-253, 1979.
Keys, J.R., Saline discharge at the snout of Taylor Glacier, Antarctica, N.Z.J. Antarct. Res., 2(1), 20-21, 1979.
Killworth, P.D., A two-dimensional model for the formation of Antarctic Bottom Water, Deep Sea Res., 20, 941-971, 1973.
Killworth, P.D., A baroclinic model of motions on Antarctic continental shelves, Deep Sea Res., 21(10), 815-838, 1974.

Killworth, P.D., Deep convection in the world ocean, Rev. Geophys. Space Phys., 21(1), 1-26, 1983.

Klepikov, V.V., and Yu.A. Grigoryev, Water circulation in the Ross Sea, (in Russian), Inf. Byull. Sov. Antarkt. Eksped., no. 56, 1966. (Engl. Transl., Information Bulletin of the Soviet Antarctic Expedition, 6(1), 52-54, New York, 1966.)

Kotlyakov, V.M., K.S. Losev and I.A. Loseva, The ice budget of Antarctica, (in Russian), Izvestiya Akademii Nauk SSSR, Seriya Geograficheskaya, 1, pp. 5-15, 1977. (Engl. Transl., Polar Geogr., 2, 251-262, 1978).

Kurtz, D.D., and D.H. Bromwich, A recurring, atmospherically-forced polynya in Terra Nova Bay, this volume.

Lewis, E.L., and R.G. Perkin, The oceanography of McMurdo Sound, this volume.

Lohmann, G.P., Response of the deep sea to ice ages, Oceanus, 21(4), 58-64, 1978.

Lusquinos, A., Extreme temperatures in the Weddell Sea, Arbok Univ. Bergen Naturvitensk. Mat. Ser., 23, 19pp., 1963.

MacAyeal, D.R., Thermohaline circulation below the Ross Ice Shelf: A consequence of tidally induced vertical mixing and basal melting, J. Geophys. Res., 89(C1), 597-606, 1984.

MacAyeal, D.R., Tidal rectification below the Ross Ice Shelf, Antarctica, this volume (a).

MacAyeal, D.R., Evolution of tidally triggered meltwater plumes below ice shelves, this volume (b).

Mantyla, A.W., and J.L. Reid, Abyssal characteristics of the World Ocean waters, Deep Sea Res., 30(8A), 805-833, 1983.

Martinson, D.G., P.D. Killworth, and A.L. Gordon, A convective model of the Weddell Polynya, J. Phys. Oceanog., 11(4), 466-488, 1981.

Meier, M.F., Contribution of small glaciers to global sea level, Science, 226, 1418-1421 1984.

Morgan, V.I., Oxygen isotope evidence for bottom freezing on the Amery Ice Shelf, Nature, 238, 393-394, 1972.

Morgan, V.I., Antarctic Ice Sheet surface oxygen isotope values, J. Glaciol., 28(99), 315-323, 1982.

Morgan, V.I., and W.F. Budd, Distribution, movement and melt rates of Antarctic icebergs, in Proceedings of the 1st International Conference on Iceberg Utilization, edited by A. Husseiny, pp. 220-228, Pergamon, New York, 1978.

Naval Polar Oceanography Center, Weekly Antarctic Ice Charts, Navy-NOAA Joint Ice Center, Suitland, Md. 1974-1984.

Neal, C.S., The dynamics of the Ross Ice Shelf revealed by radio echo-sounding, J. Glaciol., 24(90), 295-307, 1979.

Neshyba, S., Upwelling by icebergs, Nature, 267(5611), 507-508, 1977.

Neshyba, S., and E.G. Josberger, On the estimation of Antarctic iceberg melt rate, J. Phys. Oceanogr., 10(10), 1681-1685, 1980.

Orheim, O., Iceberg discharge and the mass balance of Antarctica, in Glaciers, ice sheets, and sea level: Effect of a CO_2-induced climatic change, Report of a workshop, Seattle, 13-15 Sept., 1984, National Academy Press, Wash. D.C., 1985.

Pettersson, O., On the influence of ice-melting upon oceanic circulation, Geogr. J., 24, 285-333, 1904.

Pillsbury, R.D., and S.S. Jacobs, Preliminary observations from long-term current meter moorings near the Ross Ice Shelf, this volume.

Potter, J.R., and J.G. Paren, Interaction between ice shelf and ocean in George VI Sound, Antarctica, this volume.

Priestly, R.A., Work and adventures of the northern party of Captain Scott's Antarctic Expedition, 1910-13, Geogr. J., 43, 1-14, 1914.

Redfield, A.C., and I. Friedman, The effect of meteoric water, meltwater and brine on the composition of polar sea water and of the deep waters of the ocean, Deep Sea Res., 16, 197-214, 1969.

Robin, G. DeQ., Formation, flow and disintegration of ice shelves, J. Glaciol., 24(90), 259-271, 1979.

Robin, G. deQ., C.W.M. Swithinbank, and B.M.E. Smith, Radio echo exploration of the Antarctic ice sheet, IAHS AISH Publ., 68, 97-128, 1970.

Robin, G. DeQ., C.S.M. Doake, H. Kohnen, R.D. Crabtree, S.R. Jordan, and D. Moller, Regime of the Filchner-Ronne Ice Shelves, Antarctica, Nature, 302(5909), 582-586, 1983.

Roether, W., Water-CO_2 exchange set-up for the routine ^{18}oxygen assay of natural waters, Int. J. Appl. Radiation Isotopes, 21, 379-387, 1970.

Schwerdtfeger, W., Antarctic Peninsula and the temperature regime of the Weddell Sea, Antarctic J. U.S., 9(5), 213-214, 1974.

Smith, N.R., D. Zhaoquian, K.R. Knowles, and S. Wright, Water masses and circulation in the region of Prydz Bay, Antarctica, Deep Sea Res., 31(9), 1121-1147, 1984.

Solomon, H., Vertical mixed layer convection in the Weddell Sea, Atmos. Ocean, 21(2), 187-206, 1983.

Stuiver, M., P.D. Quay, and H.G. Ostlund, Abyssal water carbon-14 distribution and the age of the world oceans, Science, 219(4586), 849-851, 1983.

Sverdrup, H.U., Hydrology, Section 2, Discussion, Rep. B.A.N.Z. Antarc. Res. Exped. 1921-1931, Ser. A, 3, Oceanography, Part 2, Section 2, 88-126, 1940.

Szekielda, K.H., The hot spot in the Ross Sea: Upwelling during wintertime, Tethys, 6, 105-110, 1974.

Tchernia, P., and P.F. Jeannin, Quelques Aspects de la Circulation Oceanique Antarctique Reveles par l'Observation de la Derive d'Icebergs (1972-1983), 92 pp., Centre d'Etudes Spatiales Expeditions Polaires Francaises (CNRS), Museum d'Histoire Naturelle de Paris, 92 pp., 1983.

Thomas, R.H., Dynamics of the Brunt Ice Shelf, Coats Land, Antarctica, Br. Antarct. Sur., Sci. Rep., 79, 45 pp., 1973.

Thomas, R.H., and D.R. MacAyeal, Derived characteristics of the Ross Ice Shelf, Antarctica, J. Glaciol., 28(100), 397-412, 1982.

Trumbore, S., S.S. Jacobs, and W.M. Smethie, Jr., Distribution of chlorofluoromethanes (F-11 and F-12) in the Ross Sea, EOS, 65 (45), 915, abstract, 1984.

UNESCO, Freezing point of sea water, Eighth report of the Joint Panel of Oceanographic Tables and Standards, Appendix 6, UNESCO Tech. Pap. Mar. Sci., 28, 29-35, 1978.

Warren, B.A., Deep circulation of the world ocean, chap. 1, in Evolution of Physical Oceanography, edited by B. Warren and C. Wunsch, 623 pp., MIT Press, Cambridge, Ma., 1981.

Weiss, R.F., H.G. Ostlund, and H. Craig, Geochemical studies of the Weddell Sea, Deep Sea Res., 26(10A), 1093-1120, 1979.

Weyl, P., The role of the oceans in climate change: A theory of the ice ages, Meteorol. Monogr., 8, 37-62, 1968.

Zotikov, I.A., V.S. Zagorodnov, and J.V. Raikovsky, Core drilling through the Ross Ice Shelf (Antartica) confirmed basal freezing, Science, 207, 1463-1465, 1980.

Zwally, H.J., J.C. Comiso, and A.L. Gordon, Antarctic offshore leads and polynyas and oceanographic effects, this volume.

(Received January 10, 1984; accepted November 30, 1984).

PRELIMINARY OBSERVATIONS FROM LONG-TERM CURRENT METER MOORINGS NEAR THE ROSS ICE SHELF, ANTARCTICA

R. Dale Pillsbury

College of Oceanography, Oregon State University, Corvallis, Oregon 97331

Stanley S. Jacobs

Lamont-Doherty Geological Observatory of Columbia University
Palisades, New York 10964

Abstract. We present an overview of current and temperature measurements at 200-m to 500-m depths near the Ross Ice Shelf from late January to mid-August 1978 and from February 1983 through January 1984. These observations are interpreted in relation to the thermohaline stratification along the ice shelf in January 1984. Nine instruments were moored for one year between 172°W and 176°W, where historical summer data have suggested a persistent, relatively warm flow from the Circumpolar Deep Water southward across the continental shelf and into the sub-ice shelf cavity. Current directions were remarkably constant through 1983 with mean annual southward or westward components from 5 to 9 cm s^{-1}. Maximum current speeds exceeded 40 cm s^{-1}. Velocity spectra showed significantly higher energy levels during the winter period of sea ice formation along the ice shelf. Marked temperature decreases in winter can be correlated with the presence of sea ice cover above the mooring sites. Temperatures ranged from a March minimum of -2.19°C in Ice Shelf Water to a July maximum of -0.14°C during a midwinter period of warm intrusions. Mean annual temperatures between -1.41° and -1.52°C, 0.5° to 0.75°C above the in situ freezing point, were obtained from six instruments that spanned the 20-km^2 warm core. Preliminary transport estimates indicate that the ocean supplies sufficient heat to melt about 150 km^3 yr^{-1} of ice off the base of the Ross Ice Shelf. The basal melting rate is subject to considerable uncertainty, in part due to the recirculation of sensible heat back into the open Ross Sea.

Introduction

Austral summer oceanographic stations in the Ross Sea have repeatedly shown subsurface regions of relatively warm water along the Ross Ice Shelf Barrier [Crary, 1961; Jacobs et al., 1970]. These features apparently arise from the Circumpolar Deep Water north of the continental shelf and have been postulated to flow into the sub-ice shelf cavity, where they could induce considerable basal melting [Jacobs et al., 1979a]. At other locations along the barrier, subsurface regions with very cold temperatures (below the sea surface freezing point) have been ascribed to warming and melting of the shelf ice by seawater that has cir-

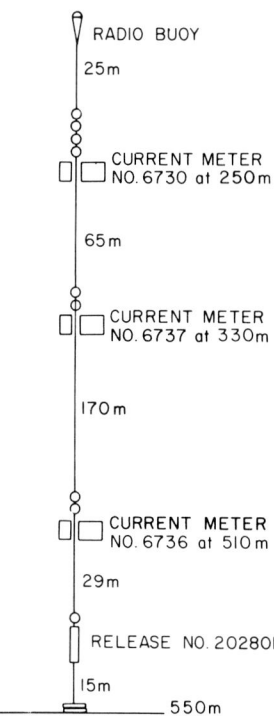

Fig. 1. Design for mooring P with top to bottom placement of relocation beacon, buoyant glass floats, current meters, release device, and weights. Somewhat different actual water depth and final array configuration resulted in set depths of 253 m, 327 m, and 508 m for the three instruments.

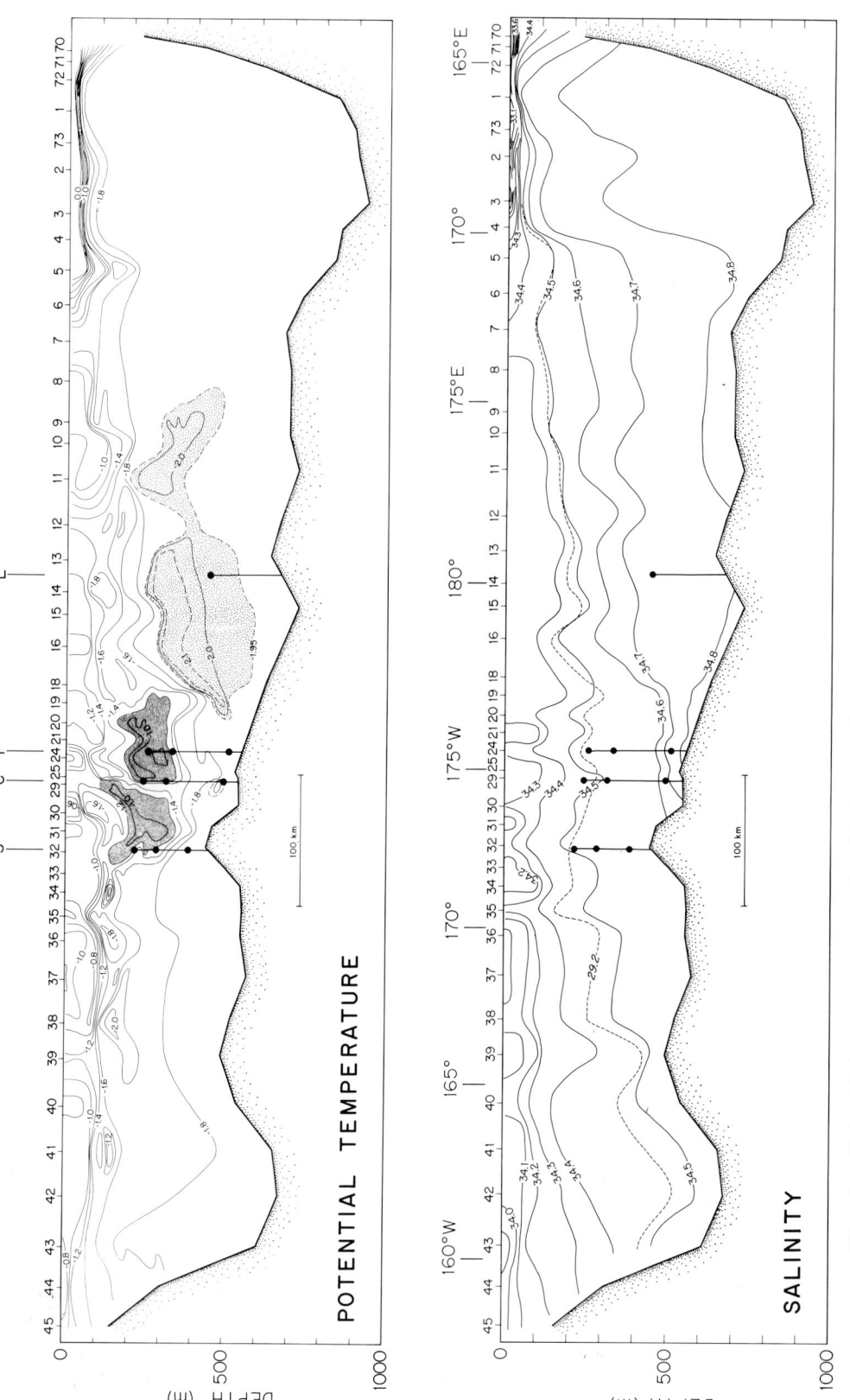

Fig. 2. Potential temperature and salinity sections constructed from oceanographic stations along the northern edge of the Ross Ice Shelf and Ross Island (stations 45 through 73) and across McMurdo Sound (stations 73 through 70). The data are taken from conductivity/temperature/depth casts made from the U.S. Coast Guard icebreaker Polar Sea between January 22 and February 2, 1984. The shaded regions depict "warm core" temperatures above $-1.2°C$ (>200 m) and Ice Shelf Water temperatures colder than $-1.95°C$. The 29.2 isopycnal, referenced to 500 dbar, closely parallels the 34.5 isohaline. At sites S, C, P, and L the current meter positions are illustrated in relation to the thermohaline structure.

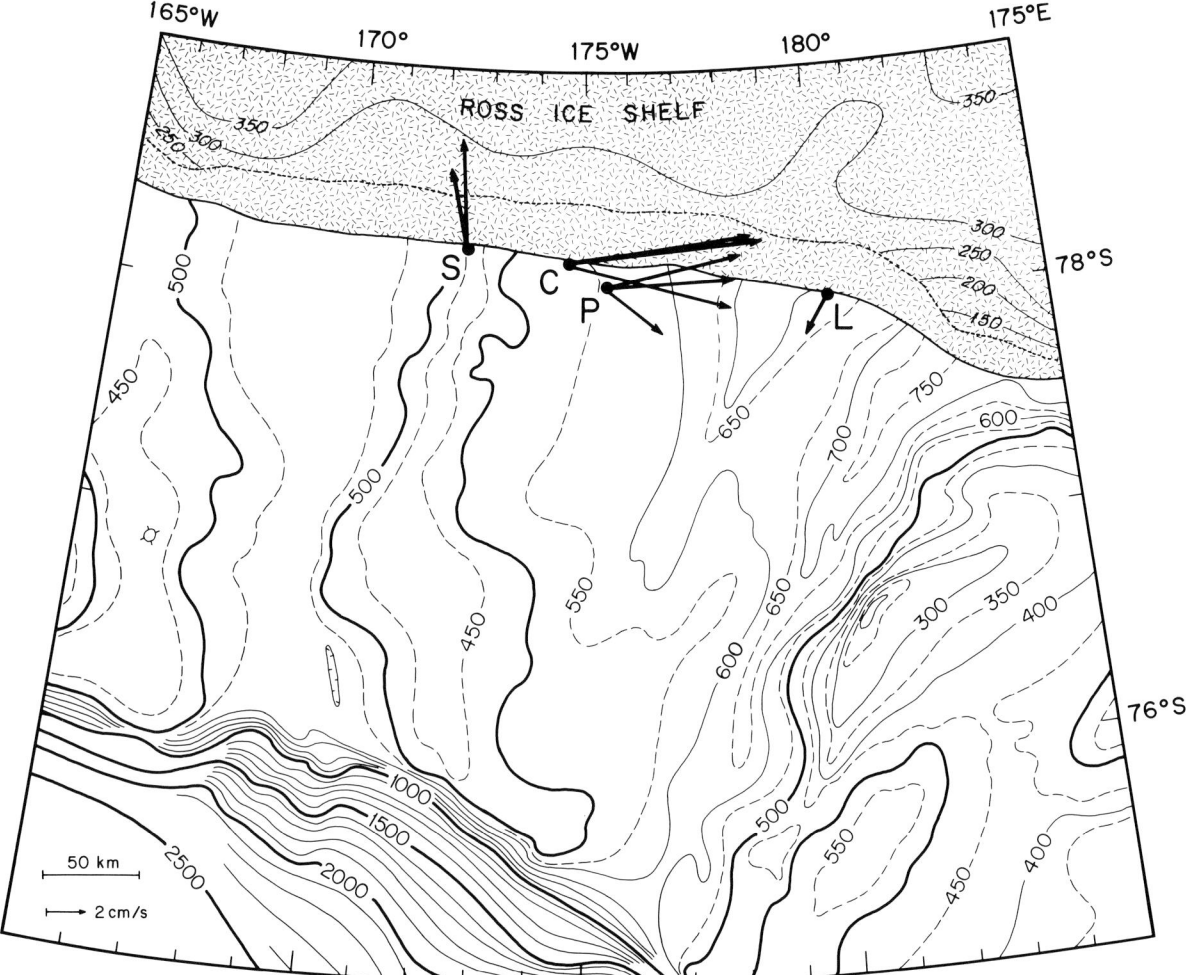

Fig. 3. Current meter mooring sites relative to the Ross Ice Shelf Barrier and the submarine topography of the Ross Sea continental shelf (Hayes and Davey [1975]). Mooring coordinates were site S: 78°13.6'S, 172°29.4'W; site C: 78°11'S, 174°39'W; site P: 78°05.5'S, 175°30'W; site L: 78°01'S, 179°46'E. Ice shelf contours represent ice thickness and terminate at a dashed line showing the barrier position in 1962, well south of the position we charted from the U.S. Coast Guard icebreaker Glacier in February 1983 [Jacobs, 1985]. Arrows represent mean current vectors over the 354-day (sites S, C, and P) and 204-day (site L) recording periods.

culated in the large cul-de-sac beneath [Jacobs et al., 1979b; MacAyeal, this volume (a)]. Analogous "warm cores" and "Ice Shelf Waters" have been observed in the Weddell Sea and elsewhere on the Antarctic continental shelf [Carmack and Foster, 1975; Gammelsrød and Slotsvik, 1981; Foldvik et al., this volume (a); Jacobs et al., this volume]. In summer, the "warm" intrusions vary in temperature by more than 2°C from one location to another, with the highest observed temperatures in the Bellingshausen Sea [Jacobs, 1985; Potter and Paren, this volume]. The configuration of the Antarctic continental shelf and ice shelves in relation to the circumpolar deep ocean is shown on the chart accompanying Vanney and Johnson [this volume].

Most past oceanographic measurements have been limited to the summer season, so there has been no direct information regarding the year-round persistence of these warm and cold features. It was also not known whether the direction and magnitude of flow could accomplish a significant heat transport into the cavity beneath the ice shelf. Initial attempts in 1978 to obtain yearlong current, temperature, and salinity data were only partially successful, resulting in one 7-month record in the Ice Shelf Water [Jacobs and Haines, 1982]. A larger field experiment was

TABLE 1. Current Speeds, Current Vectors, and Temperatures for Annual and Quarterly Periods at the Mooring Sites in Figures 2 and 3

Site-Depth Quarter	Start	Duration, days	Current Speed, cm s^{-1} Mean	Current Speed, cm s^{-1} Maximum	Current Vector, cm s^{-1} u	Current Vector, cm s^{-1} v	Temperature, °C Minimum	Temperature, °C Mean	Temperature, °C Maximum
S-211	Feb. 5, 1983	354.8	7.5	28.4	0.4	-5.0	-2.07	-1.50	-0.15
1	Feb. 5, 1983	88.7	7.2	27.4	0.4	-3.8	-2.07	-1.40	-0.97
2	May 5, 1983	88.7	8.3	28.4	0.6	-7.2	-1.87	-1.67	-0.15
3	Aug. 2, 1983	88.7	8.4	27.8	-0.3	-6.4	-1.88	-1.63	-0.44
4	Oct. 20, 1983	88.7	6.0	19.0	1.0	-2.7	-1.86	-1.31	-0.56
S-283	Feb. 5, 1983	354.8	...	...	...	...	-2.04	-1.52	-0.47
1	Feb. 5, 1983	88.7	...	...	...	...	-2.04	-1.27	-0.94
2	May 5, 1983	88.7	...	...	...	...	-1.87	-1.55	-0.47
3	Aug. 2, 1983	88.7	...	...	...	...	-1.88	-1.69	-0.71
4	Oct. 29, 1983	88.7	...	...	...	...	-1.87	-1.56	-0.85
S-383	Feb. 5, 1983	354.8	7.3	32.6	0.9	-3.6	-2.04	-1.75	-0.75
1	Feb. 5, 1983	88.7	7.3	31.1	1.6	-2.4	-2.04	-1.66	-0.84
2	May 5, 1983	88.7	7.1	29.7	0.1	-4.5	-1.87	-1.72	-0.75
3	Aug. 2, 1983	88.7	8.4	32.6	0.2	-5.3	-1.88	-1.82	-1.37
4	Oct. 29, 1983	88.7	6.4	21.2	1.7	-2.3	-1.88	-1.81	-1.44
C-237	Feb. 5, 1983	354.7	10.5	41.0	-8.8	-1.4	-2.04	-1.41	-0.14
1	Feb. 5, 1983	88.7	11.8	38.6	-10.0	-2.0	-2.04	-1.32	-0.84
2	May 5, 1983	88.7	10.8	35.2	-8.8	-1.0	-1.85	-1.59	-0.14
3	Aug. 1, 1983	88.7	11.2	41.0	-9.4	-1.2	-1.86	-1.65	-0.51
4	Oct. 29, 1983	88.7	8.2	26.9	-7.1	-1.2	-1.83	-1.09	-0.46
C-310	Feb. 5, 1983	353.6	10.6	40.3	-8.2	1.8	-1.87	-1.42	-0.29
1	Feb. 5, 1983	88.4	11.2	34.4	-9.2	1.6	-1.87	-1.24	-0.73
2	May 4, 1983	88.4	11.1	34.6	-8.4	2.4	-1.85	-1.35	-0.29
3	Aug. 1, 1983	88.4	11.5	40.3	-8.8	2.3	-1.86	-1.66	-0.63
4	Oct. 28, 1983	88.4	8.5	23.9	-6.5	1.0	-1.85	-1.42	-0.75
C-492	Feb. 5, 1983	354.7	9.1	34.5	-7.6	-1.2	-2.19	-1.76	-0.85
1	Feb. 5, 1983	88.7	9.1	31.6	-7.6	-1.3	-2.19	-1.77	-0.85
2	May 5, 1983	88.7	9.7	31.6	-8.2	-1.1	-1.84	-1.67	-0.90
3	Aug. 1, 1983	88.7	10.4	34.5	-9.3	-1.1	-2.05	-1.76	-1.45
4	Oct. 29, 1983	88.7	7.4	21.1	-5.3	-1.4	-2.08	-1.83	-1.53
P-253	Feb. 5, 1983	350.1	9.4	41.6	-5.9	-0.4	-1.86	-1.43	-0.25
1	Feb. 5, 1983	87.5	10.4	35.7	-8.5	-1.7	-1.86	-1.17	-0.74
2	May 3, 1983	87.5	8.7	30.2	-4.8	-0.4	-1.83	-1.57	-0.25
3	July 30, 1983	87.6	10.2	41.6	-5.2	-0.3	-1.86	-1.73	-0.95
4	Oct. 25, 1983	87.5	8.5	33.6	-5.2	0.6	-1.86	-1.27	-0.50
P-327	Feb. 5, 1983	354.1	8.9	41.7	-6.2	-1.4	-1.93	-1.50	-0.57
1	Feb. 5, 1983	88.5	9.9	37.0	-8.6	-2.3	-1.93	-1.26	-0.63
2	May 4, 1983	88.5	8.1	26.8	-5.3	-1.5	-1.84	-1.39	-0.57
3	Aug. 1, 1983	88.5	9.6	41.7	-5.5	-1.2	-1.87	-1.76	-1.10
4	Oct. 28, 1983	88.5	8.1	34.7	-5.6	-0.7	-1.88	-1.60	-0.86
P-508	Feb. 5, 1983	354.1	8.7	33.9	-2.6	2.0	-2.15	-1.82	-1.19
1	Feb. 5, 1983	88.5	8.4	28.3	-4.5	3.6	-2.12	-1.85	-1.25
2	May 4, 1983	88.5	8.0	31.7	-3.8	1.6	-1.99	-1.72	-1.19
3	Aug. 1, 1983	88.5	10.3	31.0	-1.4	1.4	-2.15	-1.82	-1.48
4	Oct. 28, 1983	88.5	8.2	33.9	-0.8	1.4	-2.11	-1.87	-1.57
L-444	Jan. 26, 1978	204.0	6.2	29.6	1.0	1.5	-2.09	-1.93	-1.81
1	Jan. 26, 1978	102.0	4.4	12.8	0.7	0.9	-2.09	-1.94	-1.81
2	May 8, 1978	102.0	8.1	29.6	1.3	2.1	-2.09	-1.92	-1.81

Current components are positive to the north and east.
Mooring coordinates were site S: 78°13.6'S, 172°29.4'W; site C: 78°11'S, 174°39'W; site P: 78°05.5'S, 175°30'W; site L: 78°01'S, 179°46'E.

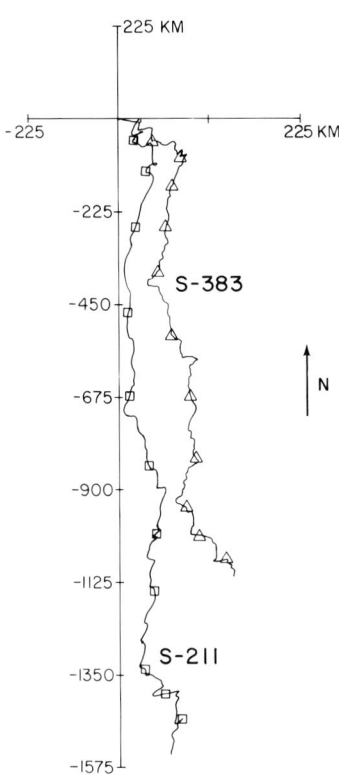

Fig. 4. Progressive vector diagram for the 211-m and 383-m depths at site S. The records span 354.8 days, starting at 1702 UT on February 5, 1983. Symbols mark the beginning of each month.

initiated in early February 1983, with the deployment of nine current meters on three moorings in the Ross Sea warm core. These instruments were recovered in late January 1984, at which time a larger array, 22 current meters on nine moorings, was set along the full length of the barrier [Jacobs et al., 1984]. In this paper we present an overview of results from the three moorings recovered in early 1984 and from the 7-month 1978 record. We relate the current meter data to the summer hydrography and seasonal ice cover and make some preliminary estimates of heat transport. Conductivity data from the current meters are not discussed here. Slow changes in the cell constant, apparently due to absorption of seawater, resulted in errors for which possible corrections are being evaluated.

Experimental Design

Design constraints on the 1983-1984 array of current meter moorings were considerable. Oceanographic stations along the barrier have shown the warm and cold features to be 100-300 km wide and 100-300 m thick. The cross-sectional area beneath the barrier, including McMurdo Sound, exceeds 300 km^2. Resources available for the 1983 field program limited the moorings to three, each with three current meters (Figure 1). We decided to concentrate initially upon the warm core and to locate all moorings near the ice shelf. The instrument sites, here designated as S, C, and P, are shown in Figure 2 relative to the temperature and salinity fields determined from early 1984 conductivity/temperature/depth (CTD) stations. Also shown in Figure 2 is the single 1978 mooring, designated as L. Most CTD stations were taken within 3 km of the Ross Ice Shelf. Moorings S, C, and L were set about 5 km and mooring P about 10 km north of the barrier. The Ross Ice Shelf moves northward at about 1 km yr^{-1} along this sector [Bentley, 1985], a rate that apparently has not been balanced by equivalent calving of icebergs over the past two decades [Jacobs, 1985]. A relatively strong surface current [U.S. Naval Oceanographic Office, 1960] has been observed to set westward along the barrier during the summer.

The 1983 moorings were placed near a rise in the seafloor, one of several rises that trend north-south across the Ross Sea continental shelf (Figure 3). Oceanographic station data taken during several summer seasons have always shown one or more warm cores above or slightly to the east of this feature, which is the shallowest point on an east-west section along the barrier (Figure 2). The rise also appears to limit the eastward extension of dense High Salinity Shelf Water (HSSW) [Jacobs et al., this volume], represented in Figure 2 by the water with salinity greater than 34.7. The vertical positions of the upper instruments on these moorings were determined by requirements that they be in or near the warm core but below the depth of probable iceberg keels. The middle meter on each mooring was situated 70-75 m below the shallow one for redundancy in the event of meter failure and for measurements of vertical coherence and modal structure. The deepest levels were chosen to be outside the probable warm core and near the eastern boundary of HSSW, which is renewed by sea surface freezing during winter in the western Ross Sea [Jacobs et al., this volume]. The single instrument at site L was intended to be within the Ice Shelf Water outflow. The current meter positions in Figure 2 do not coincide with the maximum and minimum temperatures measured on summer 1984 oceanographic stations, but repeated CTD and expendable bathythermograph (XBT) casts along the barrier [Jacobs and Haines, 1982] have revealed considerable variability over periods of a few days. Current meter depths are derived from pressure sensors, accurate to about 10 m, where C-310, for example, indicates data from the instrument that was 310 m below the sea surface at mooring C.

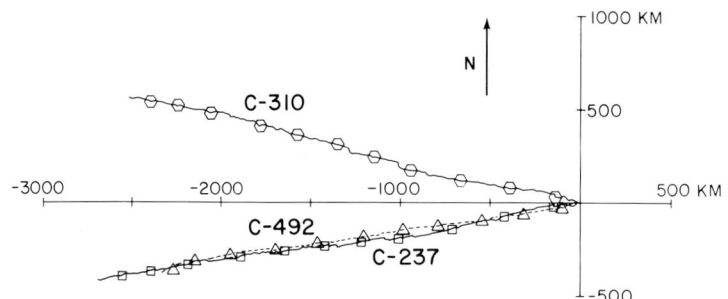

Fig. 5. Progressive vector diagram for the 237-m, 310-m, and 492-m depths at site C. The records span 354.7 days, starting at 1103 UT on February 5, 1983. Symbols mark the beginning of each month.

Current Meter Data Return

Aanderaa current meters were used; they recorded current speed, direction, temperature, and pressure each hour. The single instrument at site L yielded 204 days of data from January 26, 1978, through August 18, 1978, at which time the clock battery apparently failed. All instruments on moorings S, C, and P operated for the full installation period of about 355 days, beginning on February 5, 1983. The middle instrument on mooring S (S-283) showed very low speeds relative to the other current meters, a single value of compass direction, and apparently normal temperatures. Our conclusion is that this instrument was incorrectly installed or was fouled by the mooring line. Calibration techniques, statistics, and descriptive plots for each record appear in work by Pillsbury et al. [1974, 1985]. For the 1983 deployment the temperature sensors were calibrated over a range of about -1.3° to +5.6°C, with the data from -1.3° to +2.0°C used for data reduction. A total of 30 data points were taken over this restricted calibration range, giving standard deviations for the nine instruments of 0.0007° to 0.0011°C. An error analysis of the calibration system used at Oregon State University indicates that the temperature data over the range of -2.0° to +5.5°C is accurate to ±0.008°C, which is the resolution of the Aanderaa temperature sensor. The minimum recording threshhold was 1.7 cm s^{-1}. The current meter data are summarized in Table 1 and in several representative figures. Record length vector means are plotted in Figure 3.

Currents

Mean flows at the four mooring sites (Figure 3) were directed southward toward the ice shelf cavity (site S), westward more or less parallel to it (sites C and P), or northward away from it (site L). The flow was remarkably uniform in direction at each of the instrument locations on moorings S, C, and P (Table 1; Figures 4-6). Significant meandering occurred only at the westernmost instrument (L-444), which operated within the Ice Shelf Water and during a different year (Figure 7). The flows at sites P and S appear to

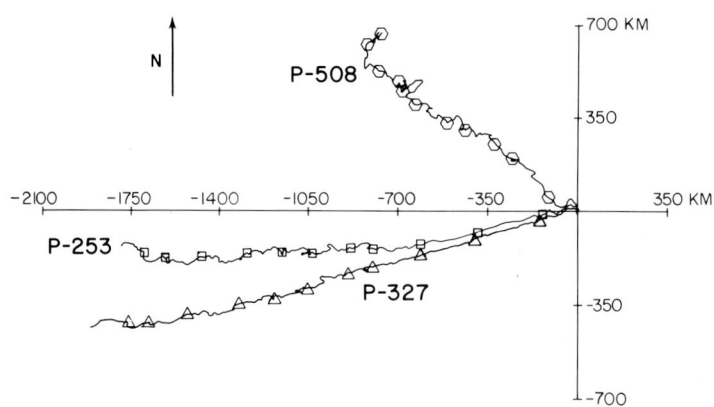

Fig. 6. Progressive vector diagram for the 253-m, 327-m, and 508-m depths at mooring P. The records span 350.1 days at 253 m and 354.1 days at 327 m and 508 m, starting at 0601 UT on February 5, 1983. Symbols mark the beginning of each month.

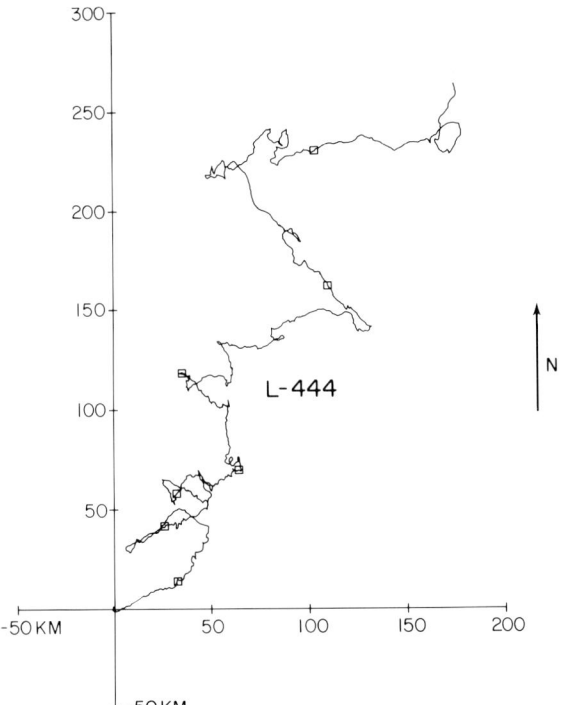

Fig. 7. Progressive vector diagram for the 444-m depth at site L. The record spans 204.0 days, starting at 1045 UT on January 26, 1978. Symbols mark the beginning of each month.

be vertically coherent over 100-200 m of the water column (Figures 4 and 6; Table 1). The records from mooring C (Figure 5) appear somewhat anomalous in that the top and bottom instruments recorded almost exactly the same direction, while the middle instrument differed by 22°, more closely resembling the current direction at P-508. However, the flow directions at C-492 and P-508 are consistent with the deep circulation that would be expected from the density field and the minor topographic relief west of the primary seafloor rise at stations 31 and 32 (Figure 2).

The full quadrant change in mean current direction between sites S and C is striking, particularly in combination with the directional constancy at each location. The summer temperature field along the central barrier (Figure 2) does appear to show two separate warm cores, a bifurcation that might be related to the seafloor rise near S. This submarine bank continues beneath the ice shelf, where basal topography of the ice may also exert its influence on the flow regime. In addition, salinity (density) increases along the barrier to the west of these moorings and may increase in a similar fashion to the southwest beneath the ice. If High Salinity Shelf Water fills most of the cavity beneath the western half of the ice shelf, lower-density inflows would be diverted around it to the south or west.

Whether or not the warm core always divides near the barrier, it is clear that southward flow is strongest near its eastern side (site S) and that westward flow predominates in its central region (Figures 2 and 3). The change in slope of the 29.2 isopycnal near station 19 in Figure 2 suggests that the westward flow at sites C and P may turn to the south or upwell on the western side of the warm core. The vertical undulations of isohalines and isopycnal in Figure 2 bear some resemblance to the seafloor topography and suggest the possibility of alternating flow in and out of the sub-ice shelf cavity. However, this may in part

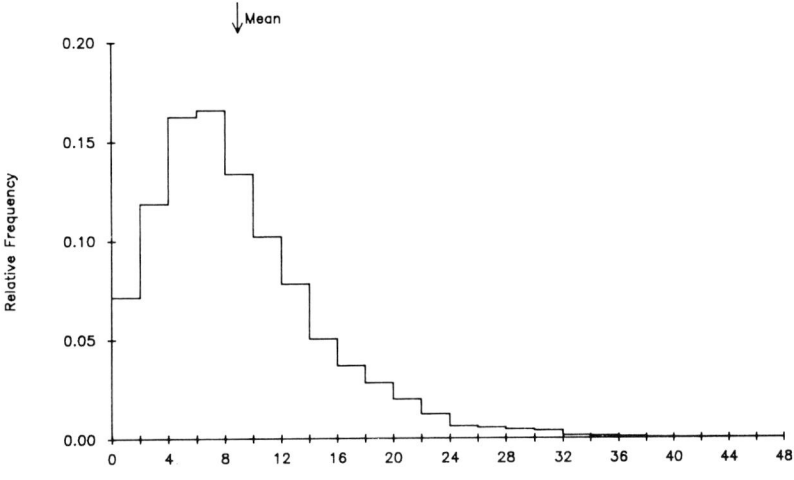

Fig. 8. Relative frequency of current speeds, from the 327-m depth at site P.

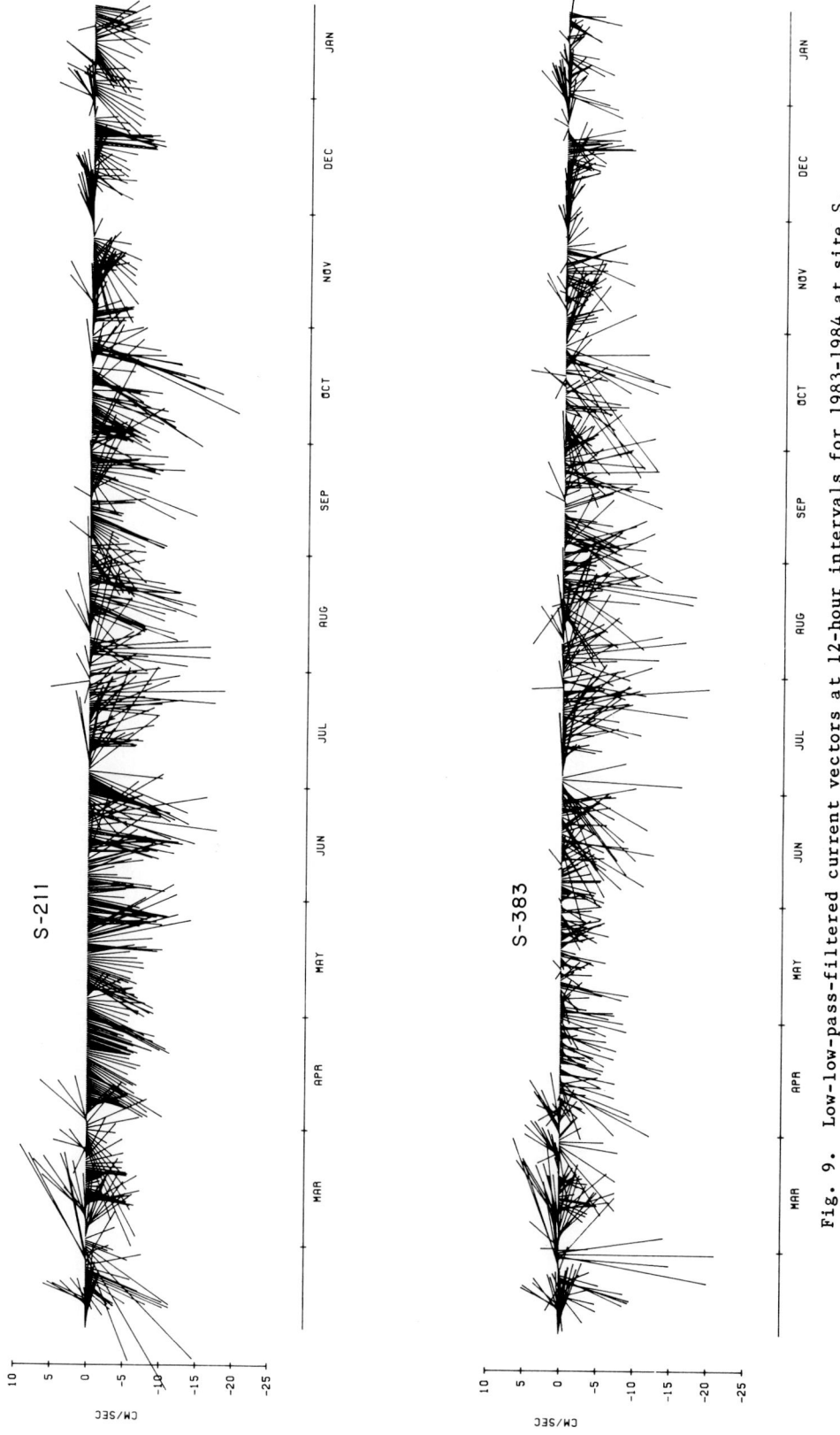

Fig. 9. Low-low-pass-filtered current vectors at 12-hour intervals for 1983-1984 at site S.

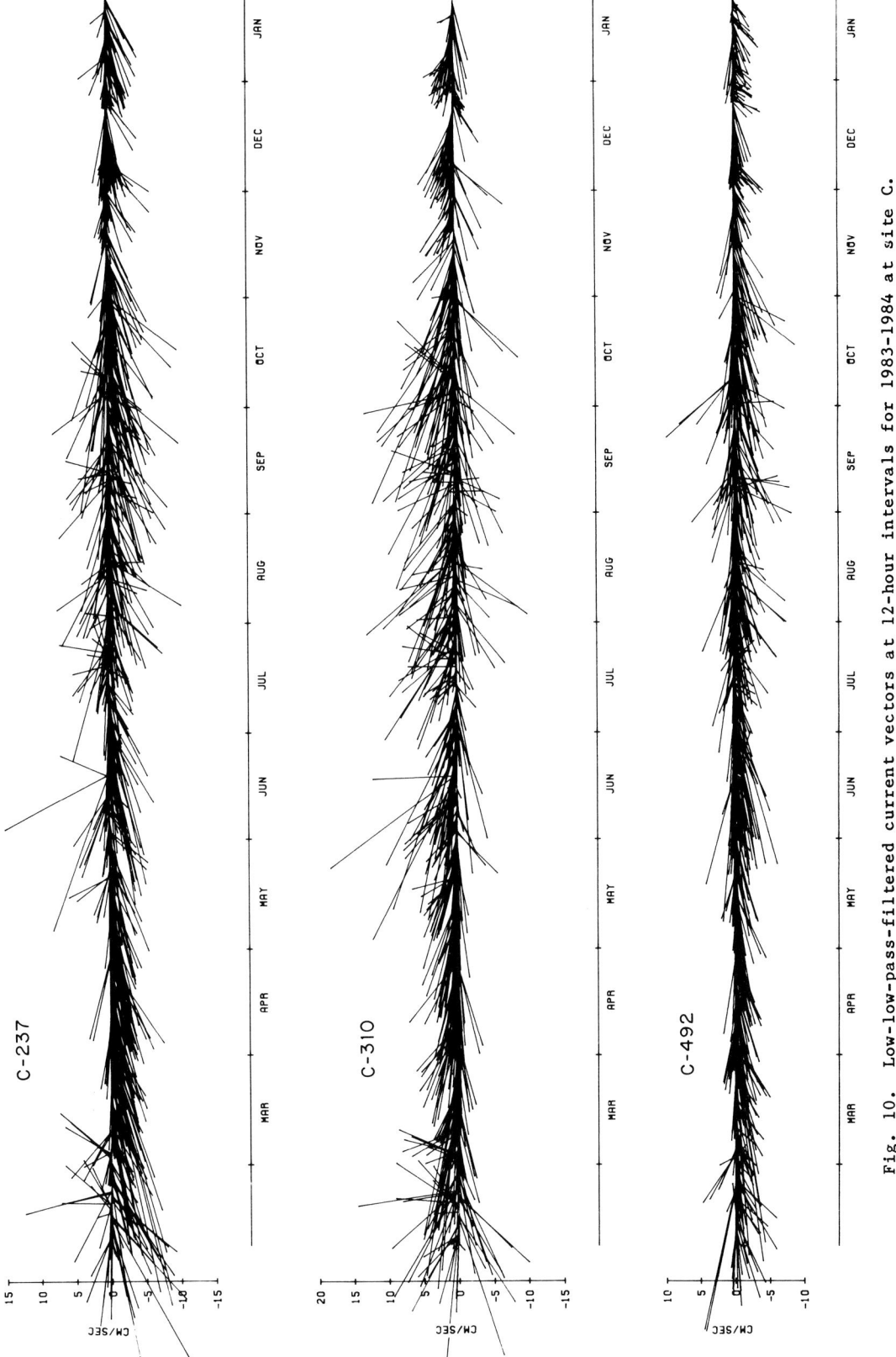

Fig. 10. Low-low-pass-filtered current vectors at 12-hour intervals for 1983-1984 at site C.

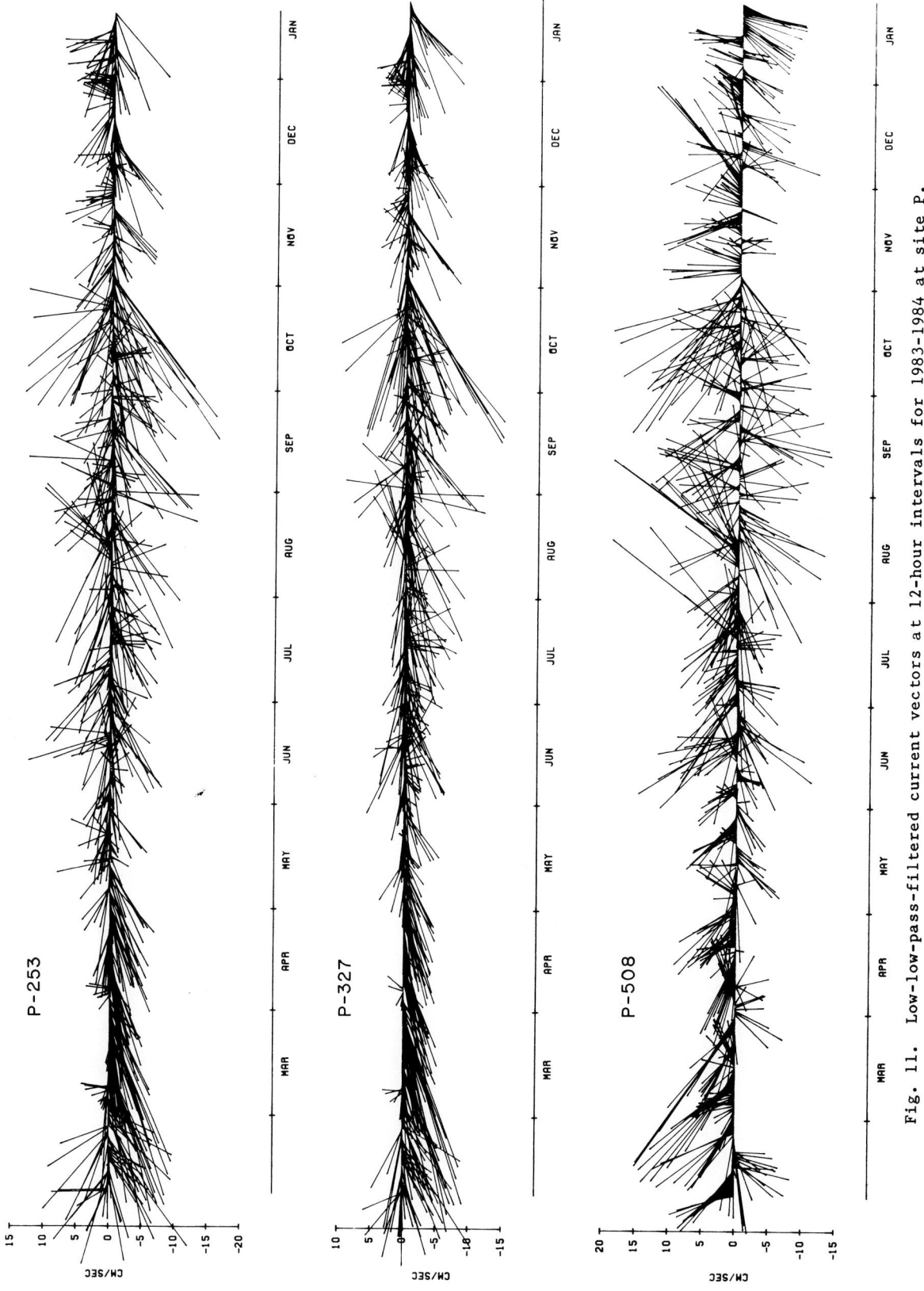

Fig. 11. Low-low-pass-filtered current vectors at 12-hour intervals for 1983-1984 at site P.

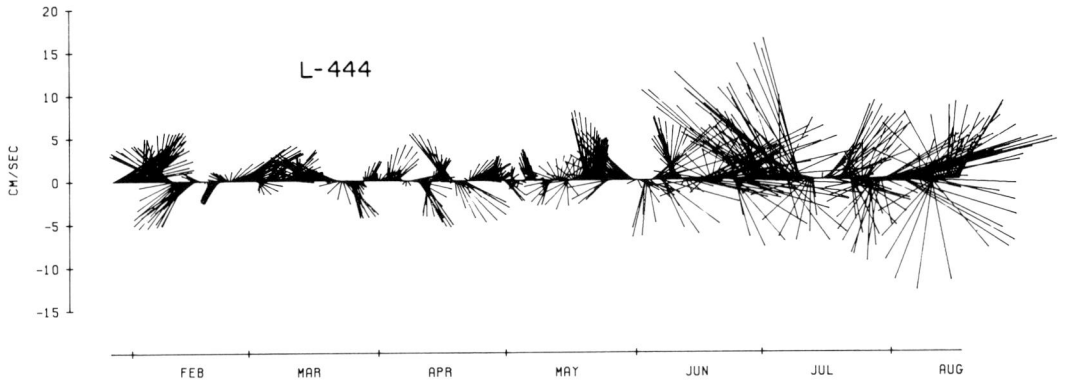

Fig. 12. Low-low-pass-filtered current vectors at 12-hour intervals for 1978 at site L.

be an artifact of CTD station distance from the shelf ice, which varied from 0.2 to 20 km depending upon current meter operations, wind direction, and other factors beyond our control. Summer isohalines commonly rise toward the sea surface in the few tens of kilometers north of the ice shelves [Sverdrup, 1953; Jacobs et al., this volume], which may account for some of the isohaline undulations in Figure 2, such as the 34.5 shoaling at stations 16 and 35.

Current velocities at all sites were higher during winter than during summer. Current speeds averaged over the full records were between 6.2 and 10.6 cm s^{-1} (Table 1). Quarterly means ranged from 4.4 to 11.8 cm s^{-1}, with the lowest values generally recorded during the November through January period. Maximum speeds ranged from 28.4 cm s^{-1} (S-211) to 41.7 cm s^{-1} (P-327). Four of the speed maxima were above 40 cm s^{-1} and occurred on the upper two instruments at sites C and P during the late winter (August through Octo-

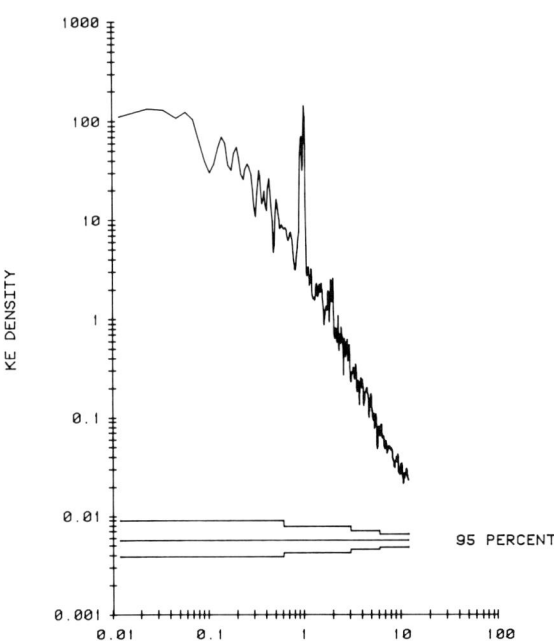

Fig. 13. Kinetic energy spectra (unfiltered; both components) for velocity at the 310-m depth at site C. The yearlong time series was divided into untapered segments of about 2000 hours, and a spectrum was calculated from each segment. The segment spectra were averaged to produce a composite spectrum, then smoothed by multiple passes through a Hanning filter. Smoothing increased to the right in order to preserve legibility and achieve a satisfactory balance between frequency resolution and peak significance.

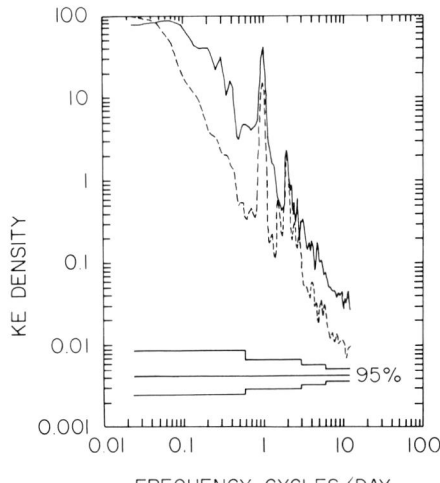

Fig. 14. Kinetic energy spectra for velocity at the 444-m depth at site L. See caption for Figure 13. The dashed line is for the period January 26 to May 8, 1978, and the solid line for the period May 8 to August 18, 1978.

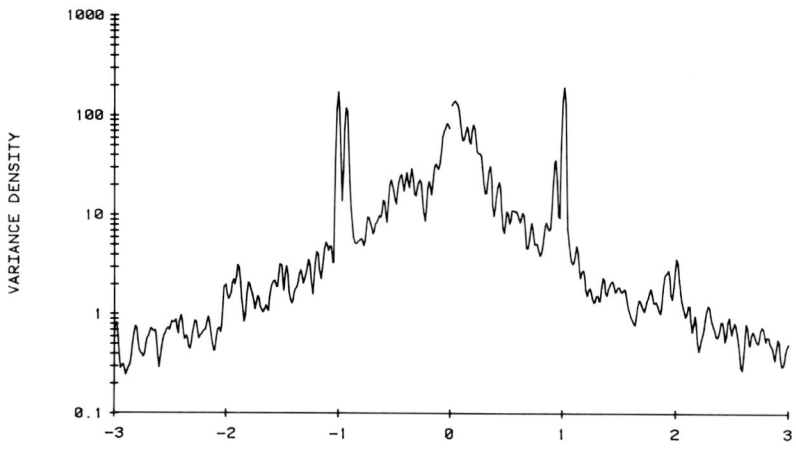

Fig. 15. Rotary spectrum for the 310-m depth at site C. See also Figure 13.

ber) quarter. There was only a slight decrease in speed with depth on the multi-instrument moorings. The current speed distributions were similar at all instruments and are illustrated by the P-327 record in Figure 8. The relative frequency maximum ranged from the 2-4 cm s^{-1} band on L-444 to the 8-10 cm s^{-1} band on C-237.

Stick diagrams of 12 hour mean current vectors (Figures 9-11) illustrate the seasonality and shorter-term variability in the mean flow. Current vectors during late winter (August through October) were much larger than during the November through January summer period (Table 1), suggesting that the circulation is primarily thermohaline and not wind-driven.

The predominantly southward flow at mooring S was rarely interrupted during the winter period (early April through November), but northeasterly components were frequently recorded from December through March (Figure 9). The predominantly westward flow at site C showed high directional coherence from mid-March through mid-May and again in November and December (Figure 10). Eastward components were measured for less than 4% of the year at C-492, and these flow reversals usually occurred simultaneously at the two shallower levels. Somewhat more variable directions were recorded at mooring P, with noticeable differences between the shallow and deep instruments (Figure 11). Coherent westward flow appeared

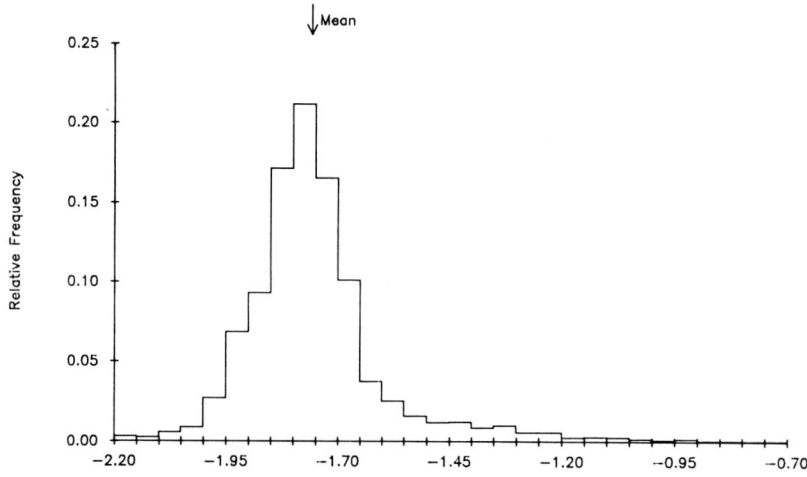

Fig. 16. Relative frequency of temperatures for 8513 observations at the 492-m depth at site C from February 5, 1983, to January 26, 1984.

simultaneously at P-253 and P-327 and on mooring C. The P-508 and L-444 records are punctuated by abrupt shifts in current direction every several days, interspersed by periods of more gradually evolving directions (Figures 11 and 12). The time between these direction changes appears to be shorter in late winter. These instruments were well within the High Salinity Shelf Water or Ice Shelf Water, apparently a different flow regime than waters lying above and to the east. In addition, P-508 may have registered some vertical movement of the pycnocline that defines the upper boundary of HSSW at its eastern extent (Figure 2).

Spectra and Tides

The velocity power spectrum for the data collected at C-310 is shown in Figure 13. Within the 95% confidence limits there are no differences between this spectrum and any of the others, either on the same mooring or on other moorings. There is a slight indication that mesoscale variability increases to the west.

The inertial period at the latitude of the moorings is 12.23 hours, which is nearly equal to the principal lunar period of 12.42 hours. However, even the combination of inertial and principal lunar spectral peaks in Figure 13 does not equal the magnitude of the luni-solar diurnal (23.93 hours) or the principal lunar diurnal (25.82 hours) periods. Both of these peaks are by far the most energetic in the spectrum. There is also a large amount of energy in the band from 10 to 100 days.

Significant differences exist between kinetic energy densities during the summer and winter periods. In Figure 14 the January 26 to May 8, 1978, and May 8 to August 18, 1978, intervals are compared for site L-444. Energy levels were higher during midwinter except at the inertial and principal lunar semidiurnal periods. The diurnal forcing potential exceeds the semidiurnal by a factor of 5.5 at 78°S, which accounts for the extreme energy difference between these currents (see also Nowlin et al. [1982]). It is less apparent why the seasonal semidiurnal peaks are the same height in Figure 14. Foldvik et al. [this volume (b)] report a winter breakdown of the diurnal tidal current component on a mooring near the Weddell Sea continental shelf break. Wind influence is probably slight at 444 m, so the increased energy level from May through July suggests enhanced thermohaline activity during winter. Sea ice begins to form in the Ross Sea in early March [Fleet Weather Facility, 1979], but there was a lag until late May before stronger lateral currents were recorded at L-444.

The rotary spectrum shown in Figure 15 has its frequency axis chosen to emphasize the tidal and inertial bands (see, for example,

O'Brien and Pillsbury [1974]). From this presentation it is clear that the principal diurnal and semidiurnal tides will have nearly linear hodographs. In Figure 15 there is little or no energy in the inertial band, which is on the positive side of the spectrum in the southern hemisphere. There is more inertial energy in the spectra from the deep record on P-508, but still very little compared with other records obtained, e.g., on the Oregon and California continental shelves. The source of the inertial energy on those temperate, shallower continental shelves is due to the passage of storms. The low level of inertial energy in the Ross Sea records may be due in part to the ice cover during winter when the more intense storms may occur. Site P-508 is within the High Salinity Shelf Water, which shoals over a wide area in the western Ross Sea (Figure 2) and is probably renewed at its deeper levels during winter convection [Jacobs et al., this volume].

There is high vertical coherence at low and tidal frequencies on each of the three moorings, but no significant horizontal coherence except at the tidal frequencies. Moorings C and P are in phase at the diurnal band, while the phase from C to S indicates eastward propagation at about 100 cm s^{-1}.

Temperature

Relative frequency diagrams for the temperatures measured during 1983 are unimodal (Figure 16) for the three deepest instruments and bimodal (Figure 17) at shallower levels. The bimodal distributions could result from seasonal changes associated with deep overturning during the formation of sea ice, or from synoptic weather patterns that shift the warm core location. The unimodal distribution illustrated by the C-492 record in Figure 16 shows that only 10% of the temperatures are colder than -1.9°C, indicative of High Salinity Shelf Water or Ice Shelf Water. The warm mode at P-253 had temperatures between -1.0 and -1.3°C, characteristic of the warm core (Figure 17). The colder mode at P-253 and more than half of the C-492 temperatures are above -1.85°C, rather warm for sea surface freezing at the salinities likely to occur along the barrier in winter. A surface freezing temperature of -1.85°C corresponds to a salinity of 33.85 [UNESCO, 1978], which is low even for summer conditions along the ice shelf (Figure 2). The L-444 temperature frequency distribution is bimodal in Figure 18 but would be unimodal if plotted at the same scale as P-253. Its temperatures are entirely within a narrow band (-1.8°C to -2.1°C), characteristic of High Salinity Shelf Water and Ice Shelf Water (ISW). L-444 extends for only 204 days, but its -1.81°C temperature maximum (Table 1) is more than 0.5°C colder than at all other

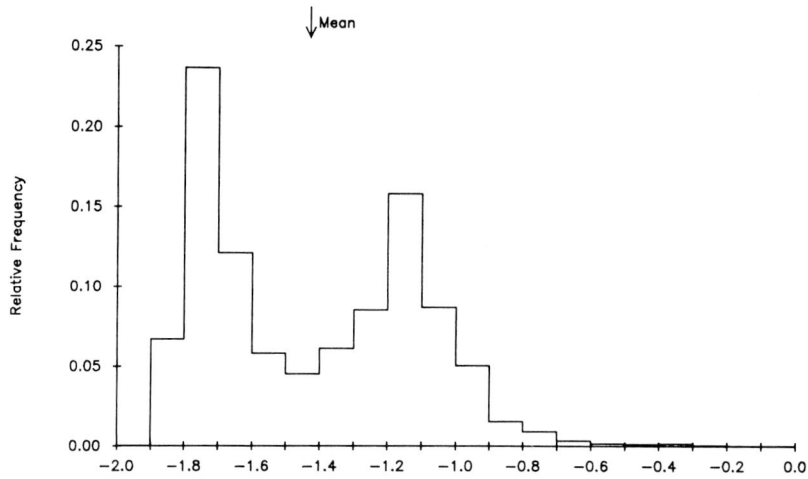

Fig. 17. Relative frequency of temperatures for 8498 observations at the 253-m depth at site P from February 5, 1983, to January 25, 1984.

locations during the same period. This suggests that warm intrusions do not reach this far west at this depth along the barrier. These data do not reveal whether the warm core and ISW are simply shifted during winter or are mixed with surface water, as brine is deposited from sea ice that is formed and driven away from the barrier [Gill, 1973; Zwally et al., this volume].

The yearlong temperature records at sites S, C, and P (Figures 19-21) show strong seasonal patterns, particularly at the upper two levels on each mooring. The shallowest levels, from 211 to 253 m, are characterized by the usual presence of the warm core for about half of the year, from early or middle November through some time in April. The winter to summer transition is abrupt, with water that for months has rarely been warmer than -1.5°C replaced over a few days time by water that, for the next several months, is rarely colder than -1.5°C. This rapid termination of the

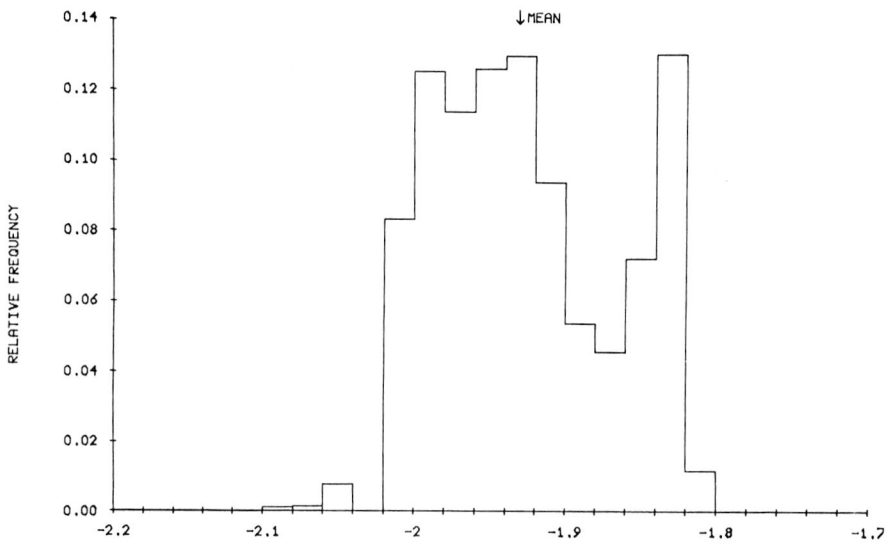

Fig. 18. Relative frequency of temperatures for 4896 observations at the 444-m depth at site L from January 26 to August 18, 1978.

winter regime would seem to indicate that the warm core has been displaced only a short distance during the period of sea ice formation along the barrier. The summer to winter transition is more gradual, extending over 1 or 2 months in the March-May period.

Establishment of the summer temperature regime at intermediate depths (283-327 m) lags that at the shallower levels (211-252 m). This may occur because the warm inflows must override a thicker column of denser shelf water in early summer. After sea ice production ceases, the denser shelf waters are not being renewed, but are being eroded at the top by warmed and freshened surface water, while continuing to lose volume to the deep ocean at the continental shelf break. Once the summer mode has been set up, it persists longer at the deeper levels, presumably because it takes longer for the water column to be cooled to these depths by the vertical mixing associated with sea ice formation.

In spite of the relatively close spacing of these moorings, there are significant differences in the dates at which temperatures indicative of surface freezing events penetrate to the shallowest instruments (211-253 m). For example, the first occurrence of temperatures colder than -1.8°C is in late April on moorings S and C but not until early June on mooring P. This may indicate that the overturning associated with surface freezing occurs over fairly limited regions or that the warm core inhibits sea ice formation by enhanced vertical heat flux at some locations. The observed winter temperatures need not have originated directly above the instruments, of course, but may result from advection after deep mixing elsewhere. In any case, the coldest events all reach minimum temperatures around -1.85°C, as indicated by the dashed baselines in Figures 19-21. This temperature is characteristic of Low Salinity Shelf Water in this sector [Jacobs et al., this volume]. There is some tendency for the baseline to drift toward slightly lower temperatures as the winter season progresses, which would be consistent with increasing salinity of the surface water during that time interval. However, the baseline temperatures are several hundredths of a degree warmer than would be expected for sea surface freezing at the observed summer salinities (Figure 2) and probable higher winter salinities. A mixed layer 0.035°C above the surface freezing point beneath late winter pack ice in the Weddell Sea was ascribed by Gordon and Huber [1984] to the upward flux of heat from deep water. In this case the temperature differential presumably results from entrainment of the warmer subsurface water during overturning events.

The coldest winter temperature at the three shallowest levels was -1.89°C, recorded at S-211 during September 1983. If that temperature represents surface freezing conditions, it would be equivalent to a salinity of 34.5. That salinity is about 0.2 higher than the summer surface salinity near site S but is still well below the 34.7 threshold for High Salinity Shelf Water (Figure 2). In addition, the quarterly temperature means at sites S, C, and P at no time reach the values below -1.9°C that are characteristic of High Salinity Shelf Water. It would thus appear that if High Salinity Shelf Water is generated along the Ross Ice Shelf Barrier as suggested by Gill [1973] and Killworth [1974], it is not formed along the central sector covered by these moorings (see also Jacobs et al. [this volume]). This does not preclude the possibility that continued salinization of the westward currents at sites C and P, if those flows remain north of the barrier, could generate shelf water along the western side of the ice shelf. Persistent winter leads along that sector [Zwally et al., this volume] could augment that process.

The winter record is punctuated by numerous warm events, particularly during July. These warm events appear at all three sites but are more prominent at site S (Figure 19). The warmest temperature on all of the records was measured during this midwinter period at C-237 (-0.14°C). The highest temperature at S-211 (-0.15°C) was reached after a day of ESE flow at speeds around 14 cm s^{-1}. Several of the temperature maxima extend to the deep (383 m) level at site S but not to the deep (492 m and 508 m) levels at sites C and P. These July events were usually warmer at the intermediate depths on moorings C and P than at the shallow depths. It is not obvious why the warmest waters occur along the Ross Ice Shelf in midwinter. Possibly, the rate of sea ice formation in the Ross Sea was most intense during the preceding several weeks, with deep overturning on the shelf increasing the rate of water mass exchange between shelf and deep ocean regimes. Warmer water drawn onto the shelf during that period would take about 2 months to reach the barrier traveling at 5 cm s^{-1} (current vector v at S-211 in Table 1). In addition, the warm core may be diverted or kept north of the barrier during an early winter period of dense water formation there.

The yearlong temperature records are coherent in the vertical direction particularly during the winter warm events at shallow and intermediate levels (Figures 19-21). The July warm events do not appear to be coherent in the horizontal direction. At the deep levels (383-508 m) the warmest temperatures occur during the summer to winter transition period, perhaps from downward mixing of the warmer water above. For example, warm peaks in late April and early May on P-508 occur simultaneously with cold events at shallower levels on the same mooring. In some instances, warm

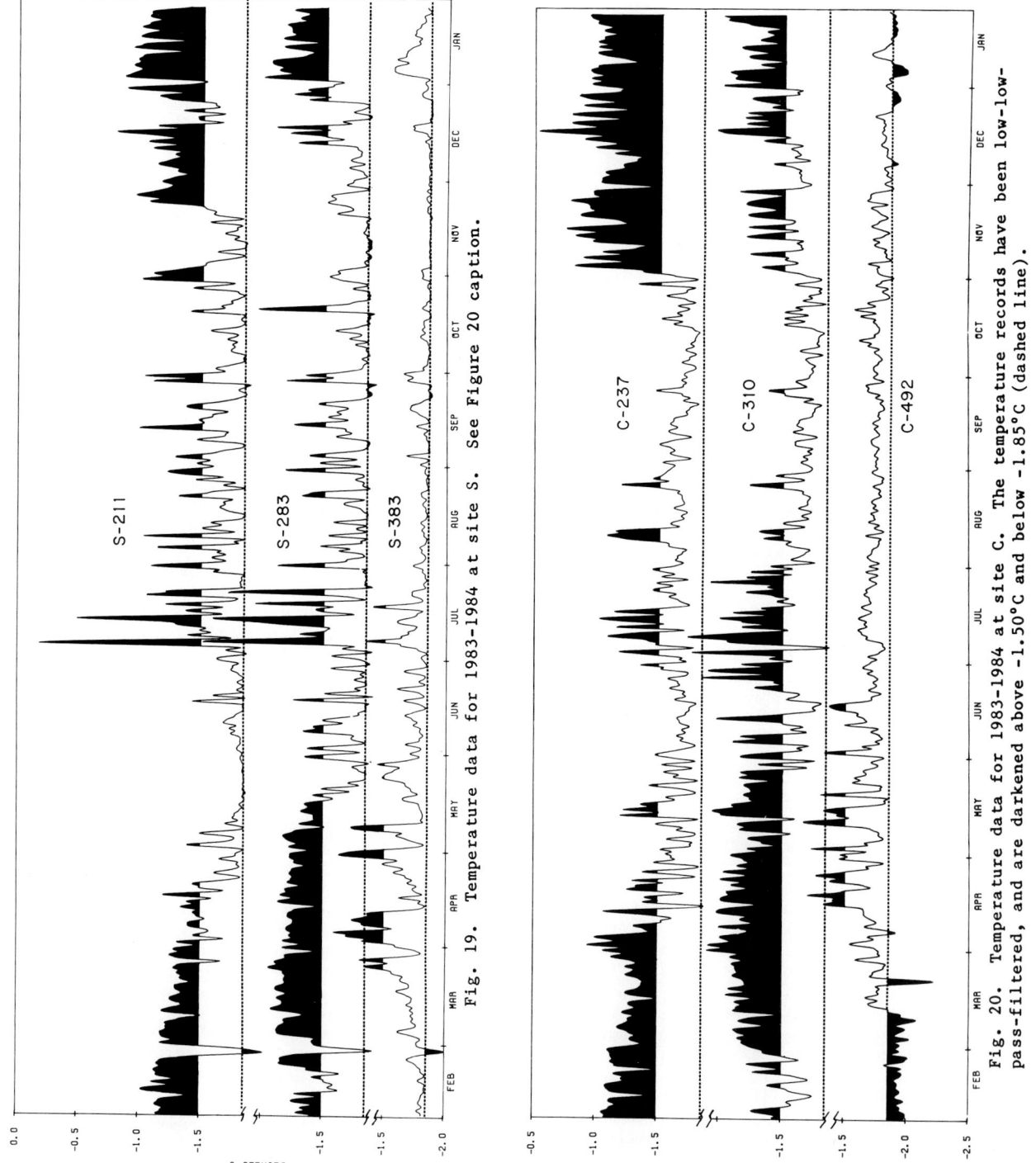

Fig. 19. Temperature data for 1983-1984 at site S. See Figure 20 caption.

Fig. 20. Temperature data for 1983-1984 at site C. The temperature records have been low-low-pass-filtered, and are darkened above -1.50°C and below -1.85°C (dashed line).

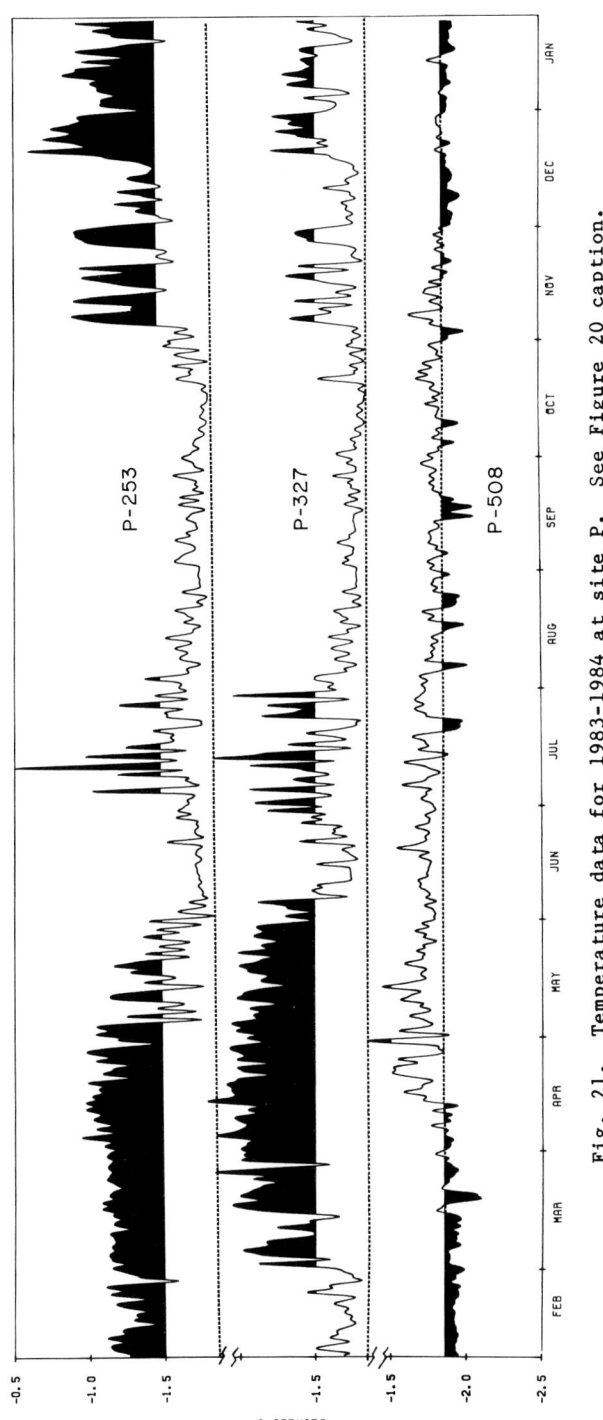

Fig. 21. Temperature data for 1983-1984 at site P. See Figure 20 caption.

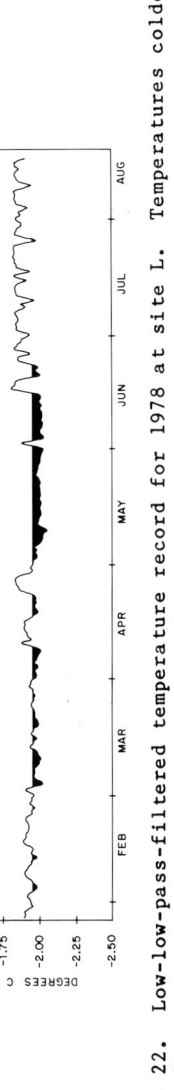

Fig. 22. Low-low-pass-filtered temperature record for 1978 at site L. Temperatures colder than -1.95°C are darkened.

temperatures appear at the deep levels before intermediate-level warm temperatures have been eroded.

An unusual and very cold episode in late February at site S (Figure 19) shows Ice Shelf Water extending throughout the water column (211-383 m). Prior to this event the currents at S-211 had tracked nearly ENE for about 42 hours at an average speed near 15 cm s^{-1}, sufficient to have advected water from beneath the ice shelf. Many near-bottom temperature minima at site S approach the -1.85° baseline coincidently with minima at shallower depths. At sites C and P most of the deep temperatures colder than -1.85°C occur during summer and begin to predominate at about the same time that warmer waters are becoming reestablished at shallower depths. The very low temperatures imply a sub-ice shelf origin and may signify an increase in the size or eastward shift in the position of the ISW outflow during summer (Figure 2).

The mean temperature at L-444 (Table 1) is at least 0.1°C colder than that observed at any of the other locations during the same February-August interval, consistent with its position in or near the Ice Shelf Water outflow. The coldest temperatures, mostly well below the sea surface freezing point, were observed from March through mid-June (Figure 22), a period during which the other deep instruments recorded their highest temperatures. The generally warmer temperatures during July (average -1.86°C) are above the summer range for this location and are also warmer than would be expected from local surface freezing. The northeast drift at L-444 (Figure 7), shows that much of the water passing that site must come from beneath the ice shelf, and the above-freezing temperatures indicate that considerable excess heat is being recirculated back into the open Ross Sea.

Sea Ice

Several correlations are apparent between these temperature and current velocity records and the presence or absence of sea ice above the moorings. In 1983, sea ice first appeared at sites S, C, and P between March 4 and 10 [Naval Polar Oceanography Center, 1985], at about the time that 211-m temperatures began to decrease at site S and intermediate depth temperatures began to increase at sites C and P (Figures 19-21). An intermittent lead appeared along the eastern Ross Ice Shelf (to 170°W) during April. At about the same time, major temperature decreases or fluctuations occurred at the shallowest and deepest levels on moorings S and C, but no appreciable activity was evident at intermediate depths. A lead opened west of 175°W along the barrier in early November and extended to 165°W by November 17; the barrier was generally ice free by December 8, 1983. This typical spring opening of the Ross Sea pack ice along the barrier coincided roughly with the return of the warm core, first at sites P and C, then later at site S. In 1978, sea ice appeared above mooring L in the March 2-6 interval [Fleet Weather Facility, 1979], at about the time the coldest ISW temperatures began to dominate the record (Figure 22). Full ice cover was attained by March 13-16, and this condition prevailed along the barrier through August when that mooring ceased operating.

Currents are stronger when sea ice is present and presumably being produced above the mooring sites. The shallow current records tend to show less variability in direction for the 2-3 months beginning in early March (sites C and P) or mid-April (site S) when sea ice may be forming most rapidly. All show a general decrease in velocity beginning in early November (Figures 9-11), at about the time that persistent leads indicate that seasonal production of sea ice along the barrier is coming to an end.

The apparent correlations noted above between temperature, current, and sea ice fields may result from synoptic-scale forcing rather than local events. In addition, the Antarctic ice charts [Naval Polar Oceanography Center, 1985] do not show narrow leads that may occur along the barrier as sea ice is formed and driven offshore.

Heat Transport

From this limited data set, we will make a preliminary estimate of heat transport into the cavity beneath the ice shelf. The currents and temperatures at S-211 and S-283 (Table 1; Figures 4 and 19) show that water at a mean annual temperature of -1.5° flows steadily under the barrier all year at this location. There is a smaller southerly component at three of the four upper levels on mooring C and P, at temperatures from -1.4° to -1.5°C. The narrow range of these annual temperature means and the consistent current directions indicate a remarkable stability in the supply of source water for this flow under and along the ice. Jacobs et al. [this volume] believe that the warm inflow may be the primary source for a shallow variety of Ice Shelf Water that upwells along the barrier (e.g., at stations 14, 36, and 38 in Figure 2). They identify High Salinity Shelf Water, generated at the winter sea surface in the open Ross Sea, as the most probable source of the deep Ice Shelf Water shaded in Figure 2.

For the purposes of this discussion we will assume that the warm core can be divided into two parts, as it appears in Figure 2, and has appeared on some earlier transects along the barrier [Jacobs and Haines, 1982]. Taking the 7.5-km^2 eastern half salinity to be 34.5 (Figure 2) and its mean annual temperature and velocity to be -1.5°C and -5.0 cm s^{-1} (S-211 in Table 1), then 8.6×10^{11} W of sensible heat are transported into the sub-ice shelf cavity, relative to a 211-m freezing temperature of -2.05°C [UNESCO, 1978]. That heat transport would be sufficient to warm by 10°C and then melt 73 km^3 yr^{-1} of ice off the ice shelf base, equivalent to 14 cm s^{-1} if averaged over the 540,000-km^2 basal area of the ice shelf [Zumberge and Swithinbank, 1962]. The heat transport is about double that estimated by Jacobs et al. [1979b], who assumed a larger region of inflow but a significantly lower mean current velocity. The heat transport is not as sensitive to the variable velocity during the year as might be expected. For example, if quarterly averages at S-211 (Table 1) are substituted for the yearly average, the heat flux drops to 7.7×10^{11} W. The less energetic circulation during summer is nearly balanced by the higher mean temperature during that time. The southward flow at S-211 continues during winter, but water that overturns to this depth after being cooled to the sea surface freezing point will still move beneath the ice shelf at about 0.15°C above its in situ freezing temperature. That temperature differential can account for about one quarter of the heat transport and demonstrates the importance to year-round basal melting of the freezing point depression with pressure.

The remaining cross section of the cavity below the barrier is about 320 km^2, allowing for Ross Island and outflow from the warm core. We will assume a southward flow of 1.4 cm s^{-1} over half this cross section and an equal northward component over the other half, based upon the yearly average of v current vectors at sites C, P, and L in Table 1. The temperature data in Table 1 are not representative of the entire barrier cross section (Figure 3), so we will assume that this larger inflow is cooled by about 0.1°C while beneath the shelf ice. In that case the heat given up beneath the ice would be 9.4×10^{11} W, approximately the same as that calculated for the warm core. The combined heat sources could warm and melt 152 km^3 of glacial ice, equivalent to about 28 cm s^{-1} averaged over the ice shelf base. Jacobs et al. [this volume] estimate an average melt rate equivalent to 42 cm s^{-1} for all the Antarctic ice shelves, but the Ross Ice Shelf is relatively protected in its southerly location, where a lower-than-average melt rate could be expected.

It should be noted that oceanic heat transfer to the cavity beneath the Ross Ice Shelf is not sufficiently rapid to prevent bottom freezing, which probably occurs at a few centimeters per year, over considerable portions of the base [Zotikov et al., 1980]. Potter and Paren [this volume] have calculated a melt rate of 2.1 m yr^{-1} for the George VI Ice Shelf, the underside of which is exposed to considerably warmer inflows than is the Ross Ice Shelf. However, Sugden and Clapperton [1981] infer little bottom melting in the recent past beneath part of that ice shelf, from evidence in horizontal moraines along its western margin.

It is difficult to assign error bars to the heat transport estimate. Measurements at relatively few points have been taken to represent mean conditions over large areas. The jetlike eastern warm core represents only 5% of the cross section allocated to inflow, yet appears to carry about 50% of the heat flux. The western half of the warm core in Figure 2 has been ignored but could add ±50% to the estimated heat transport, depending upon whether it flows under the ice or turns back into the open Ross Sea. Tidal circulation of a warm flow along the barrier could account for higher melt rates under the northern edge of the ice shelf [Thomas and MacAyeal, 1982; MacAyeal, this volume (b)]. The thinnest part of the shelf ice is found directly west of 180° in Figure 3 [Bentley et al., 1979].

Steady, westward along-barrier flows like those at sites C and P may shoal along an isopycnal surface north of the barrier. Limited regions of warm water near the sea surface are often observed along the western barrier, for example, at stations 8 and 9 in Figure 2 and in McMurdo Sound (station 213 of Jacobs et al. [1981]). If these warm features persist during winter, they may contribute to the presence of leads [Zwally et al., this volume] and enhance the sea to air heat flux.

It is also apparent that a considerable portion of the sensible heat transported beneath the ice shelf must be recirculated to the open ocean and not utilized for melting and warming the glacial ice. Regions where summer temperatures approach the in situ freezing point make up a relatively small portion of the subbarrier cross section (Figure 2), and current velocities in the only persistent, well-defined outflow of Ice Shelf Water do not appear to be large (L-444, Table 1). Annual and quarterly mean temperatures at all instrument locations are well above the in situ freezing points, which will range from about -2.05°C at S-211 to about -2.30°C at P-508. Outflows above the freezing point may still represent net southward heat flux, due to mixing and entrainment beneath the ice. Not all water that circulates beneath the ice shelf will gain access to the ice shelf base, particularly where there is a well-developed boundary layer beneath the ice [Gill, 1973; Jacobs et al., 1979b]. However, observations and modeling [Potter and Paren, this volume; MacAyeal, this volume (a)] show the potential for upwelling and accumulation of meltwater into jetlike outflows.

Gordon [1981] and Hastenrath [1982] calculated that oceanic heat transport across 60°S is 54 x 10^{13} to 65 x 10^{13} W, and Jacobs et al. [this volume] estimate that 4.3 x 10^{13} W cross the circumpolar continental shelf break. The Ross Ice Shelf spans about 36° of longitude, so the transport of 18 x 10^{11} W beneath its barrier corresponds to about 3% and 40% of the heat crossing proportional parts of 60°S latitude and the continental shelf break.

The eastern half of the warm core transport will be 0.4 x 10^6 m^3 s^{-1} if it flows beneath the ice shelf at an average velocity of 5 cm s^{-1}. Adding the slower drift of 1.4 cm s^{-1} over the remaining cavity opening results in a considerably larger transport of 2.6 x 10^6 m^3 s^{-1}. If that exchange rate is typical for all Antarctic ice shelves, then the volume of shelf water that flows beneath the floating land ice is about the same as that which participates in bottom water formation [Jacobs et al., this volume].

Conclusions

1. A persistent subsurface thermal feature (warm core) along the central Ross Ice Shelf Barrier displayed mean annual southward flow of 5 cm s^{-1} near 172°W and mean annual westward flow of 6-9 cm s^{-1} near 175°W. Current directions were remarkably uniform during the entire recording interval, from February 1983 through January 1984. Annual temperatures averaged approximately -1.4 to -1.5°C.

2. Seasonal cooling at 200- to 500-m depths along the barrier can be attributed to deep convection associated with sea ice formation. Water mass modification by this process is not sufficiently intense to produce High Salinity Shelf Water between 172°W and 176°W.

3. The highest temperatures (near -0.15°C) were recorded during a series of midwinter (July 1983) intrusions that may have resulted from enhanced exchange between the shelf and deep ocean regimes during an early winter period of intense sea ice formation.

4. Currents were vertically coherent over at least 100-200 m of the water column and attained maximum speeds over 40 cm s^{-1}. A predominantly thermohaline circulation near the barrier is indicated by significantly higher spectral energies during the winter period.

5. Half the 15-km^2 warm core flowing under the barrier year-round at an average temperature of -1.5°C and velocity of 5 cm s^{-1} would transport 9 x 10^{11} W of sensible heat, sufficient to warm and melt about 75 km^3 yr^{-1} off the ice shelf base. Comparable heat transports may be accomplished by the other half of the warm core and also by slower drifts over the remaining cross section into the sub-ice shelf cavity. That would result in net melting of 28 ± 14 cm yr^{-1}, averaged over the ice shelf base. Large uncertainties are involved in these preliminary estimates, in part because of the unknown recirculation of sensible heat back into the open Ross Sea.

6. A single current meter moored within a core of very cold water near the 180° meridian displayed meandering flow to the northeast from February through August 1978. However, the coldest temperature (-2.19°C, or about 0.3°C below the sea surface freezing point) was recorded at another location in March 1983. These Ice Shelf Waters result from warming and melting of the base of the ice shelf.

7. From these Ross Sea data, it appears that mass transport beneath all the Antarctic ice shelves may be volumetrically similar to the shelf water component of bottom water that is formed near the Antarctic continental margin.

Acknowledgments. Financial and logistic support for this work was provided by the National Science Foundation, Division of Polar Programs, grants DPP-81-20677 (R.D.P.) and DPP-81-19863 (S.S.J.), with additional support from the U.S. Department of Energy through Interagency Agreement DE-AI01-84ER60201. J. Simpkins, D. Root, J. Ardai Jr., B. Huber, J. Zerener, and several marine science technicians on the U.S. Coast Guard icebreakers Glacier and Polar Sea assisted with the fieldwork. Word processing and drafting was by B. Hautau, D. Criscione, and B. Batchelder. K. Bryan, A. Gordon, and D. MacAyeal made helpful comments on the manuscript. Lamont-Doherty Geological Observatory contribution 3826.

References

Bentley, C.R., Glaciological evidence: The Ross Sea Sector, in Glaciers, Ice Sheets and Sea Level: Effect of a CO_2-Induced Climatic Change, National Academy Press, Washington, D.C., 1985.

Bentley, C.R., J.W. Clough, K.C. Jezek, and S. Shabtaie, Ice-thickness patterns and the dynamics of the Ross Ice Shelf, Antarctica, J. Glaciol., 24(90), 287-294, 1979.

Carmack, E.C., and T.D. Foster, Circulation and distribution of oceanographic properties near the Filchner Ice Shelf, Deep Sea Res., 22(2), 77-90, 1975.

Crary, A.P., Glaciological studies at Little America Station, Antarctica, 1957 and 1958, IGY Glaciological Report Series, IGY World Data Center A, Glaciology, 5, pp. 111-163, Am. Geogr. Soc., New York, 1961.

Fleet Weather Facility, Antarctic ice charts, 1977-1978, Suitland, Md., 1979.

Foldvik, A., T. Gammelsrød, and T. Tørresen, Circulation and water masses on the southern Weddell Sea shelf, this volume (a).

Foldvik, A., T. Kvinge, and T. Tørresen,

Bottom currents near the continental shelf beak in the Weddell Sea, this volume (b).

Gammelsrød, T., and N. Slotsvik, Hydrographic and current measurements in the southern Weddell Sea 1979/80, Polarforschung, 51(1), 101-111, 1981.

Gill, A.E., Circulation and bottom water production in the Weddell Sea, Deep Sea Res., 20(2), 111-140, 1973.

Gordon, A.L., Seasonality of Southern Ocean sea ice, J. Geophys. Res., 86, 4193-4197, 1981.

Gordon, A.L., and B.A. Huber, Thermohaline stratification below the Southern Ocean sea ice, J. Geophys. Res., 89(C1), 641-648, 1984.

Hastenrath, S., On meridional heat transport in the world ocean, J. Phys. Oceanogr., 12, 922-927, 1982.

Hayes, D.E., and F.J. Davey, Geophysical study of the Ross Sea, Antarctica, Initial Rep. Deep Sea Drill. Proj., 28, 887-907, 1975.

Jacobs, S.S., Oceanographic evidence for land ice/ocean interactions in the Southern Ocean, in Glaciers, Ice Sheets and Sea Level: Effect of a CO_2-Induced Climatic Change, National Academy Press, Washington, D.C., 1985.

Jacobs, S.S., and W.E. Haines, Ross Ice Shelf Project, Oceanographic Data, 1976-1979, Tech. Rep. 82-1, Lamont-Doherty Geological Observatory, Palisades, N.Y., 1982.

Jacobs, S.S., A.F. Amos, and P.M. Bruchhausen, Ross Sea oceanography and Antarctic Bottom Water formation, Deep Sea Res., 17, 935-962, 1970.

Jacobs, S.S., A.L. Gordon, and A.F. Amos, Effect of glacial ice melting on Antarctic surface water, Nature, 277(5696), 469-471, 1979a.

Jacobs, S.S., A.L. Gordon, and J.L. Ardai Jr., Circulation and melting beneath the Ross Ice Shelf, Science, 203, 439-442, 1979b.

Jacobs, S.S., H.E. Huppert, G. Holdsworth, and D.J. Drewry, Thermohaline steps induced by melting of the Erebus Glacier Tongue, J. Geophys. Res., 86(C7), 6547-6555, 1981.

Jacobs, S.S., W.M. Smethie, Jr., R.D. Pillsbury, and D.R. MacAyeal, Ross Sea oceanography, 1984, Antarct. J. U.S., 19(5), 72-73, 1984.

Jacobs, S., R.G. Fairbanks, and Y. Horibe, Origin and evolution of water masses near the Antarctic continental margin: Evidence from $H_2^{18}O/H_2^{16}O$ ratios in seawater, this volume.

Killworth, P.D., A baroclinic model of motions on Antarctic continental shelves, Deep Sea Res., 21(10), 815-838, 1974.

MacAyeal, D.R., Evolution of tidally triggered meltwater plumes below ice shelves, this volume (a).

MacAyeal, D.R., Tidal rectification below the Ross Ice Shelf, Antarctica, this volume (b).

Naval Polar Oceanography Center, Antarctic ice charts, 1983-1984, Suitland, Md., 1985.

Nowlin, W.D., J.S. Bottero, and R.D. Pillsbury, Observations of the principal tidal currents at Drake Passage, J. Geophys. Res., 87(C7), 5752-5770, 1982.

O'Brien, J.J., and R.D. Pillsbury, Rotary spectra in a sea breeze regime, J. Appl. Meteorol., 13, 820-825, 1974.

Pillsbury, R.D., J.S. Bottero, R.E. Still, and W.E. Gilbert, A compilation of observations from moored current meters, vol. 6, Oregon continental shelf, April-October 1972, Data Rep. 57, Ref. 74-2, School of Oceanography, Oregon State Univ., Corvallis, 1974.

Pillsbury, R.D., J. Simpkins, D. Root, and J. Bottero, A compilation of observations from moored current meters, vol. 16, Currents, temperature, conductivity and pressure in the Ross Sea, February 1983 - January 1984, data report, Oregon State Univ., Corvallis, in press, 1985.

Potter, J.R., and J.G. Paren, Interaction between ice shelf and ocean in George VI Sound, Antarctica, this volume.

Sugden, D.E., and C.M. Clapperton, An ice-shelf moraine, George VI Sound, Antarctica, Ann. Glaciol., 2, 135-141, 1981.

Sverdrup, H.U., The currents off the coast of Queen Maud Land, Saertr. Nor. Geogr. Tidsskr., 14 239-249, 1953.

Thomas, R.H., and D.R. MacAyeal, Derived characteristics of the Ross Ice Shelf, Antarctica, J. Glaciol., 28(100), 397-412, 1982.

UNESCO, Freezing point of seawater, Appendix 6, Eighth report of the joint panel on oceanographic tables and standards, UNESCO Tech. Pap. Mar. Sci., 28, 29-35, 1978.

U.S. Naval Oceanographic Office, Sailing Directions for Antarctica, H.O. Publ. 27, 2nd ed., revised to 1970, 433 pp., U.S. Government Printing Office, Washington, D.C., 1960.

Vanney, J.R., and G.L. Johnson, GEBCO bathymetric sheet 5.18 (Circum-Antarctic), this volume.

Zotikov, I.A., V.S. Zagorodnov and J.V. Raikovsky, Core drilling through the Ross Ice Shelf (Antarctica) confirmed basal freezing, Science, 207, 1463-1465, 1980.

Zumberge, J.H., and C. Swithinbank, The dynamics of ice shelves, in Antarctic Research, Geophys. Monogr. Ser., vol. 7, edited by H. Wexler et al., pp. 197-208, AGU, Washington, D.C., 1962.

Zwally, H.J., J.C. Comiso, and A.L. Gordon, Antarctic offshore leads and polynyas and oceanographic effects, this volume.

(Received January 15, 1985;
accepted May 24, 1985.)

TIDAL RECTIFICATION BELOW THE ROSS ICE SHELF, ANTARCTICA

D. R. MacAyeal

Department of the Geophysical Sciences, University of Chicago, Chicago, Illinois 60637

Abstract. Numerical tidal simulation of the Ross Sea shows that periodic tidal currents drive steady barotropic circulations having magnitudes of the order of 0.01 m/s along the sides of several topographic bumps and ridges formed by the combined relief of the seabed and ice shelf base. The sensible heat transport implied by this flow is estimated to induce 0.5 ± 0.25 m/yr basal melting over approximately 5.0×10^4 km^2 of the ice shelf area closest to the ice front. As a means of flushing the entire sub-ice cavity, tidal rectification is too weak and too spatially sporadic to account for geochemically derived renewal rates.

Introduction

Geochemical and hydrographic observations taken through bore holes and seaward of the ice front indicate active ventilation of the cavity below the floating Ross Ice Shelf [Michel et al., 1979; Jacobs et al., this volume]. Although water masses influenced by glacial meltwater are observed in the open Ross Sea, the pattern and strength of oceanic heat flux and basal melting within the expansive cavity is presently unknown. Consideration of the thick, integrated ice cover suggests that the cause of sub-ice shelf circulation must operate in an environment devoid of atmospheric contact. Ocean tides constitute one of many plausible causes for such flow because they are not wind driven and can exist in an ice-covered basin [Williams and Robinson, 1981].

Observations of the gravity field at the surface of the Ross Ice Shelf indicate vertical motions ranging to more than two meters in amplitude at frequencies corresponding to the dominant diurnal and semidiurnal ocean tidal constituents [Williams and Robinson, 1980]. The ocean tides are thus clearly active in the sub-ice shelf cavity. Due to the large horizontal expanse of the cavity, however, only a relatively small water volume can cycle across the ice front during tidal ebb and flow. The tides must, therefore, interact with the time independent components of circulation, or with the thermohaline stratification, to produce a widespread effect.

These two interactions have been examined using a numerical tidal model applied to the Ross Sea basin [MacAyeal, 1983; 1984b]. Turbulent mixing driven by tidal currents is shown to destroy thermohaline stratification in remote parts of the cavity where the water column is pinched between the ice shelf and the seabed [MacAyeal, 1984a]. This stratification would normally suppress heat transfer and basal melting; so zones of vigorous tidal mixing can, in principle, cause basal melting and trigger a large-scale buoyancy driven circulation [MacAyeal, 1984a].

Another possible mechanism by which the tides may ventilate the cavity is called tidal rectification and refers to the generation of time-independent currents by periodic tidal flows across depth contours or past coastal headlands [Zimmerman, 1981]. This mechanism forms the focus of this paper. Depth contours below an ice shelf are controlled by both the seabed elevation and ice shelf draft. As an individual water column is pumped across these contours, relative vorticity, or apparent spin, is imparted by the conservation of angular momentum associated with the earth's rotation [Pedlosky, 1979, p.37]. Preferential dissipation of anticyclonic vorticity by bottom friction in shallow depths drives a time-independent anticyclonic flow about the regions of shallow depth. This preferential dissipation therefore constitutes a transfer of momentum from the purely periodic components to the time-independent components of the flow.

The efficiency of this transfer increases when the characteristic length scale of depth variation matches the range of horizontal tidal movement of water columns [Zimmerman, 1980; Robinson, 1981]. Maps of the observed depth, the time-averaged tidal current magnitude, and the maximum range of tidally driven particle displacement are presented in Figures 1 through 3, respectively, to identify potential rectification sites. The maps of current magnitude are derived from the numerical simulation presented in MacAyeal [1984b]. A scan for regions having both steep depth change and strong tidal current magnitude indicates that two sites are likely to exhibit strong tidal rectification: (1) the ice front segment northwest of Roosevelt Island, and (2) the seabed ridge extending northwest of the

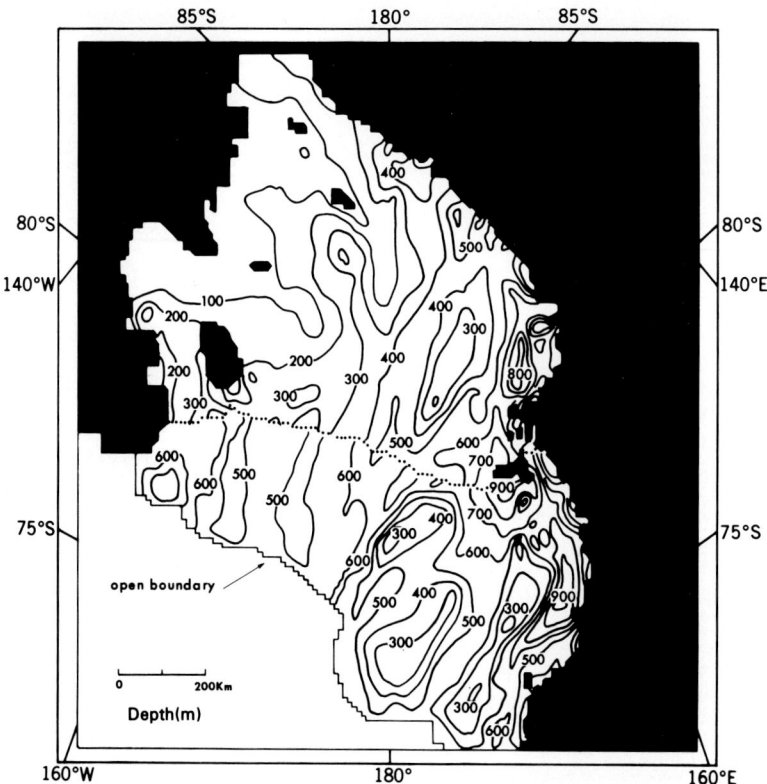

Fig. 1. The numerical domain, bounded by the mountainous coasts, the ice shelf grounding line, and the open boundary. Depth in the open portion of the Ross Sea was obtained from Hayes and Davey (1974). South of the ice front, indicated by the dotted line transecting the basin from the east to the west, the water depth is defined as the thickness of the seawater layer alone and is obtained from Greischar and Bentley (1980).

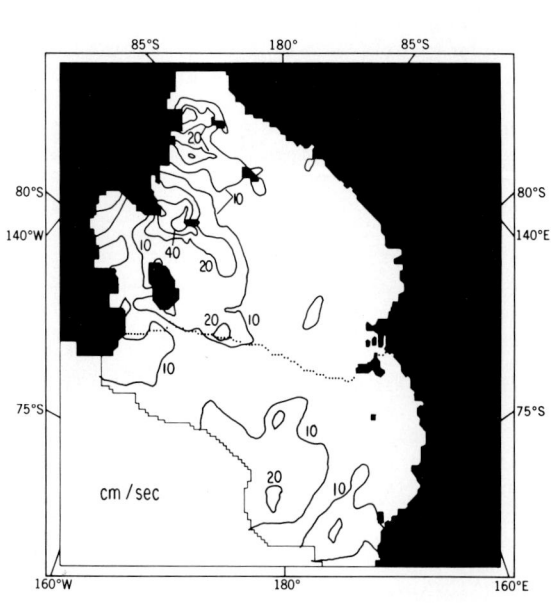

Fig. 2. The time-averaged tidal current velocity magnitude given by the simulation.

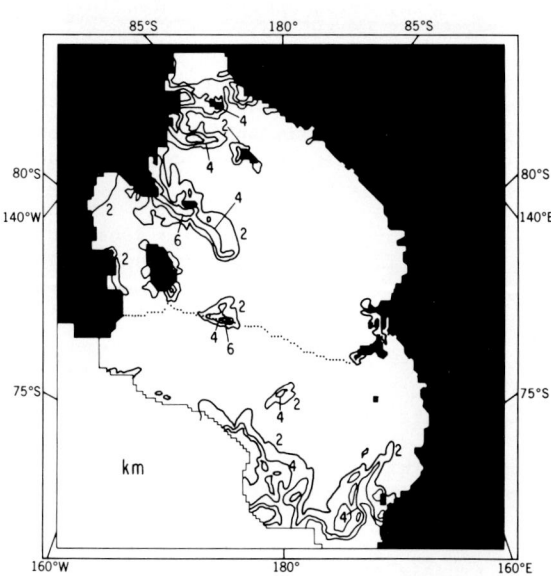

Fig. 3. The simulated time-averaged range of tidally driven horizontal displacement of individual water columns during one spring to neap tidal period (the time average of $|\underline{x} - \underline{x}_o|$ where $\underline{x}_o$ is the time-averaged position and $\underline{x}$ is the instantaneous position of a water column).

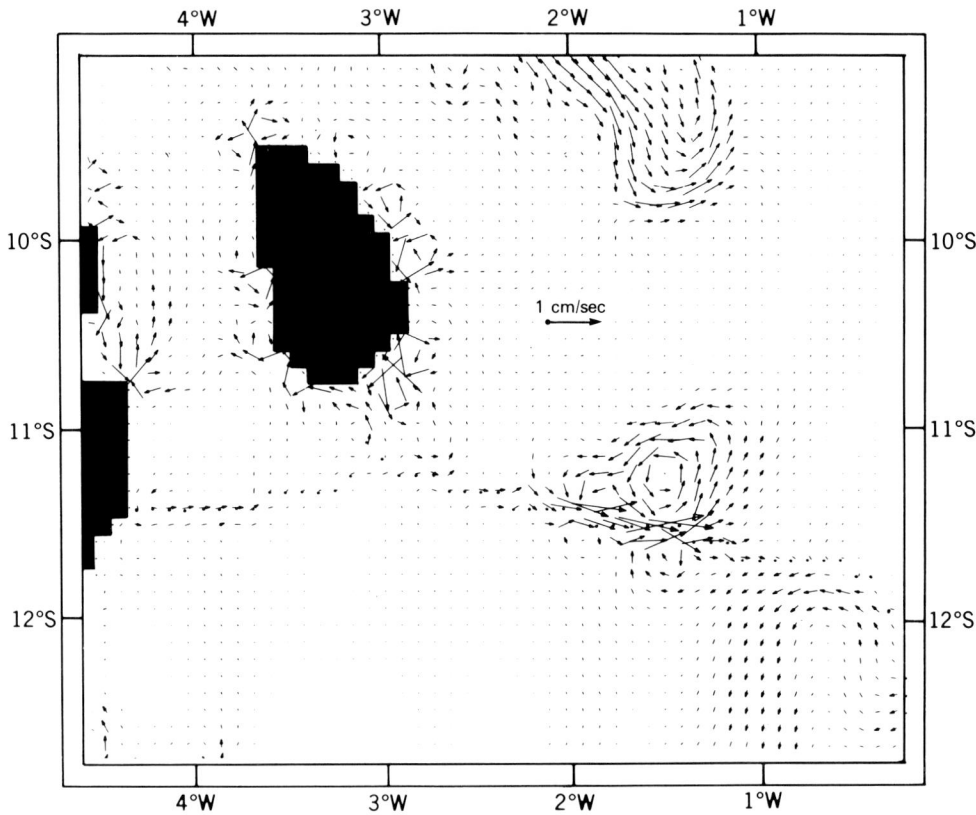

EULERIAN MEAN VELOCITY

Fig. 4. Tidal rectification produces two prominent anticyclonic eddies correlated with shallow depth in the vicinity of Roosevelt Island (the larger island). One eddy (in the center of the diagram) lies just south of the ice front and is correlated with a topographic bump visible in Figure 1. The diameter of this eddy is approximately 100 km. Its maximum flow is approximately 0.015 m/s and occurs along the ice front. The other prominent eddy (in the upper part of the diagram) correlates with a shallow seabed ridge extending northwest of the Steershead Ice Rise (the smaller island) also shown in Figure 1. There are numerous eddies of smaller scale produced along the coast of Roosevelt Island. The most prominent example consists of two eddies having opposite rotation that straddle the southern tip of Roosevelt Island. The rectangular coordinates along the edges of this map and on several subsequent figures are "grid coordinates", adopted here to be consistent with maps from which field data were digitized (e.g., Figure 1 of Bentley et al. [1979]). In the grid-coordinate system, the North Pole is rotated to the intersection of the equator and prime meridian. All directions, latitudes and longitudes explicitly referred to in the text are taken with respect to the geographical coordinate system. All map projections presented here are polar stereographic and are referenced to the geographic south pole.

Steershead Ice Rise (Steershead Crevasses) located at 81°S 160°W [Thomas et al., in press].

In addition to being favorable for rectification, the region northwest of Roosevelt Island is where the warmest subsurface water, [the "warm core" (WMCO), of Jacobs et al., this volume] is observed along the ice front. Possible rectification at this location may thus contribute substantially to the heat flux into the cavity.

To investigate the possibility of tidal rectification at the ice front site northwest of Roosevelt Island, the Ross Sea tides (O_1, K_1, P_1, M_2, S_2, N_2) were simulated using the model described in MacAyeal [1984b]. This simulation was integrated over a sufficiently long time period to allow an accurate assess-

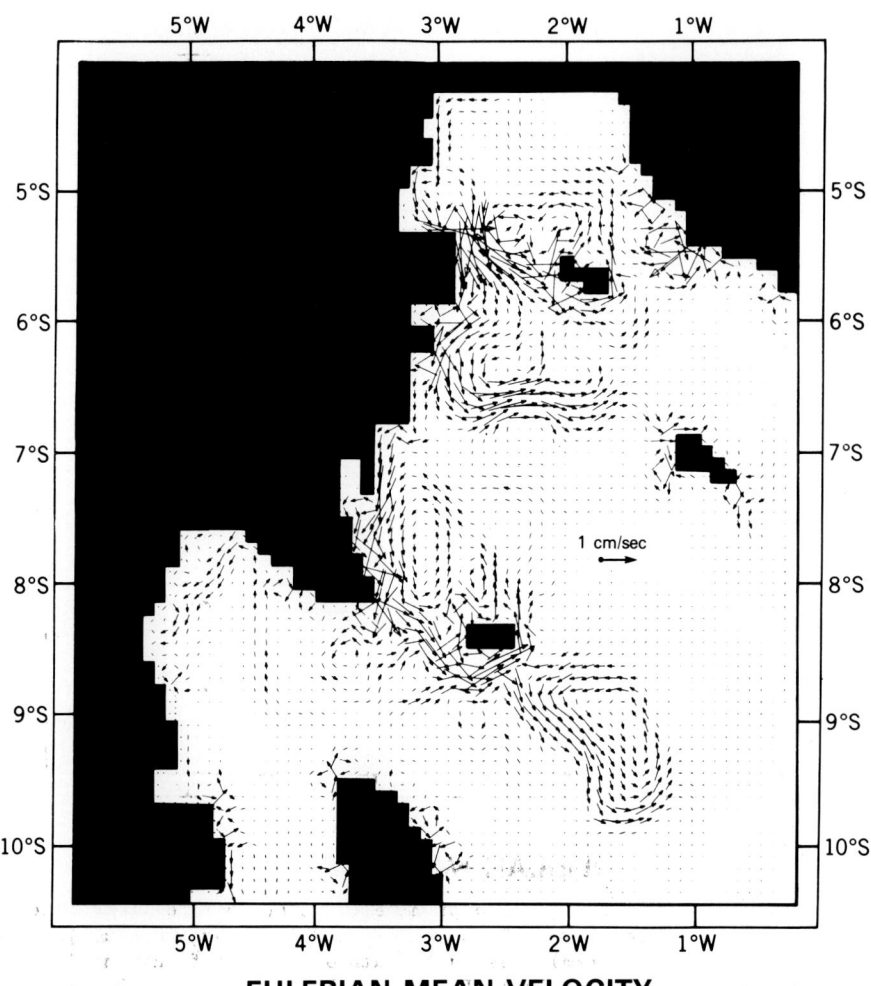

EULERIAN MEAN VELOCITY

Fig. 5. Tidal rectification, producing a strong anticyclonic circulation about the seabed ridge extending northwest of the Steershead Ice Rise.

ment of time-averaged quantities [MacAyeal, 1983]. Although the area of interest comprises a 300 x 300 km section of the Ross Sea, the numerical domain (Figure 1), consisted of the entire Ross Sea to facilitate model forcing and to isolate open boundary generated noise. Model forcing was accomplished by specifying the sea surface motions along the open boundary shown in Figure 1 using data supplied by Schwiderski's [1980] global-ocean tidal simulation. Verification of the simulated tidal fields was accomplished by comparison with the tidal amplitudes and phases observed at 10 ice shelf stations by Williams and Robinson [1980] and MacAyeal [1984b].

Governing Equations

The equations describing momentum and mass continuity are [Nihoul, 1975, p. 51]

$$\partial \underline{u}/\partial t = -\underline{u} \cdot \nabla \underline{u} - g\nabla \eta - f\underline{e}_z \times \underline{u} + \underline{\tau}/D \quad (1)$$

and

$$\partial \eta / \partial t = -\nabla \cdot (D\underline{u}) \quad (2)$$

where $\underline{u}$ is the depth-averaged velocity, η is the deviation of the sea surface or the ice shelf surface from the level of rest, $D=H+\eta$, where H is the undisturbed depth of the water layer, $f = -1.42 \times 10^{-4} s^{-1}$ is the Coriolis parameter at 78°S, $\underline{e}_z$ is the unit vector normal to the geoid, t is time, $g = 9.81 \text{ m s}^{-1}$ is the gravitational acceleration, and $\underline{\tau}$ is the frictional drag given by

$$\underline{\tau} = -k|\underline{u}|\underline{u} + D\nu\nabla^2\underline{u} \quad (3)$$

The first term on the right-hand side of equation (3) represents friction at the seabed and, where applicable, at the ice shelf base.

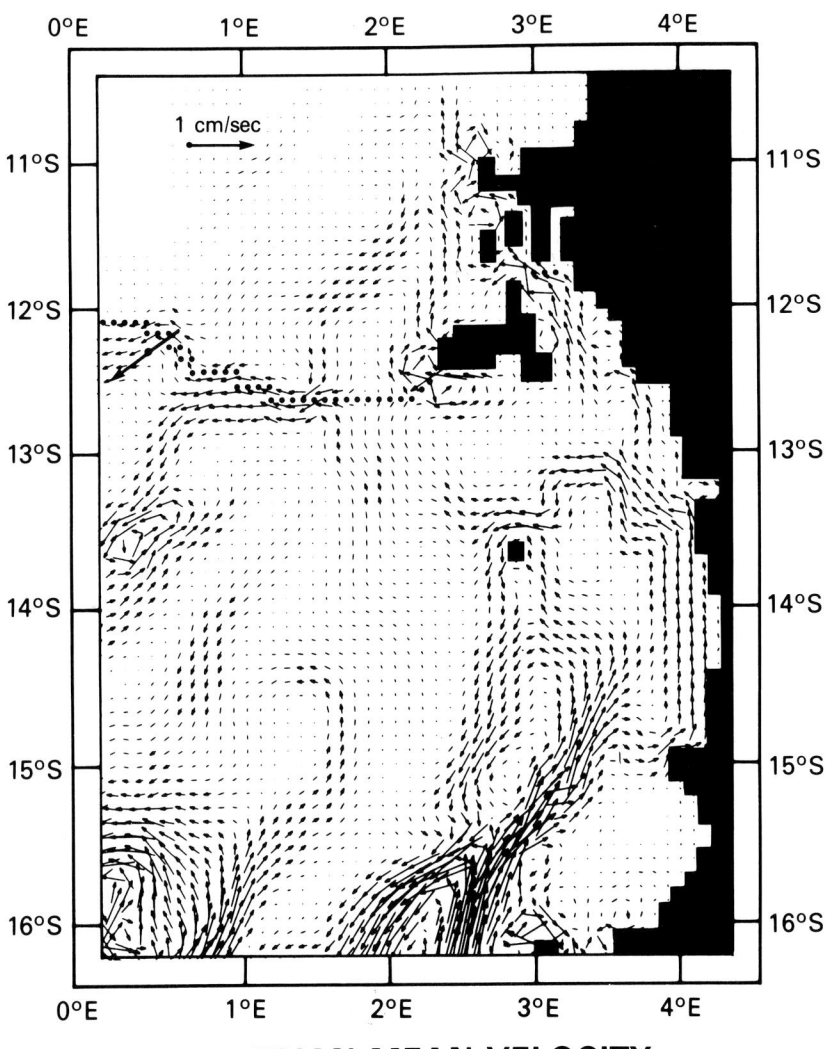

EULERIAN MEAN VELOCITY

Fig. 6. Strong simulated tidal rectification occurring along the sides of Pennell Bank in the northwestern portion of the Ross Sea. This flow penetrates through McMurdo Sound west of Ross Island and into the sub-ice shelf cavity largely because of a saddle in the seabed topography that divides the flow around Pennell Bank. This saddle can be seen in Figure 1 between the small island north of Ross Island and the Victoria Land coast.

This constitutive relation depends on the square of the velocity. The value of the non-dimensional parameter k is 2.5×10^{-3} in the open part of the Ross Sea and 5.0×10^{-3} below the ice shelf [Ramming and Kowalik, 1980, p.17]. The second term on the right-hand side of equation (3) represents viscous drag, included to suppress noise in the numerical simulation. The eddy viscosity ν equals 1.5×10^2 m²s. At coastal boundaries, equation (1) is replaced by a specification of the zero normal flow and a computationally efficient slip condition [MacAyeal, 1983].

Tidal rectification occurs when momentum is transferred from periodic tidal currents to the steady circulation [Huthnance, 1973]. This transfer is best illustrated by partitioning the momentum budget into a purely periodic part and a time-independent part. To accomplish this partitioning, the time-averaging operator is defined:

$$<\cdot> = \frac{1}{T} \int_{t_o}^{t_o+T} \cdot \, dt \quad (4)$$

where t_o is arbitrary and T is a time period of sufficient duration to allow partitioning of the variables into periodic and steady components

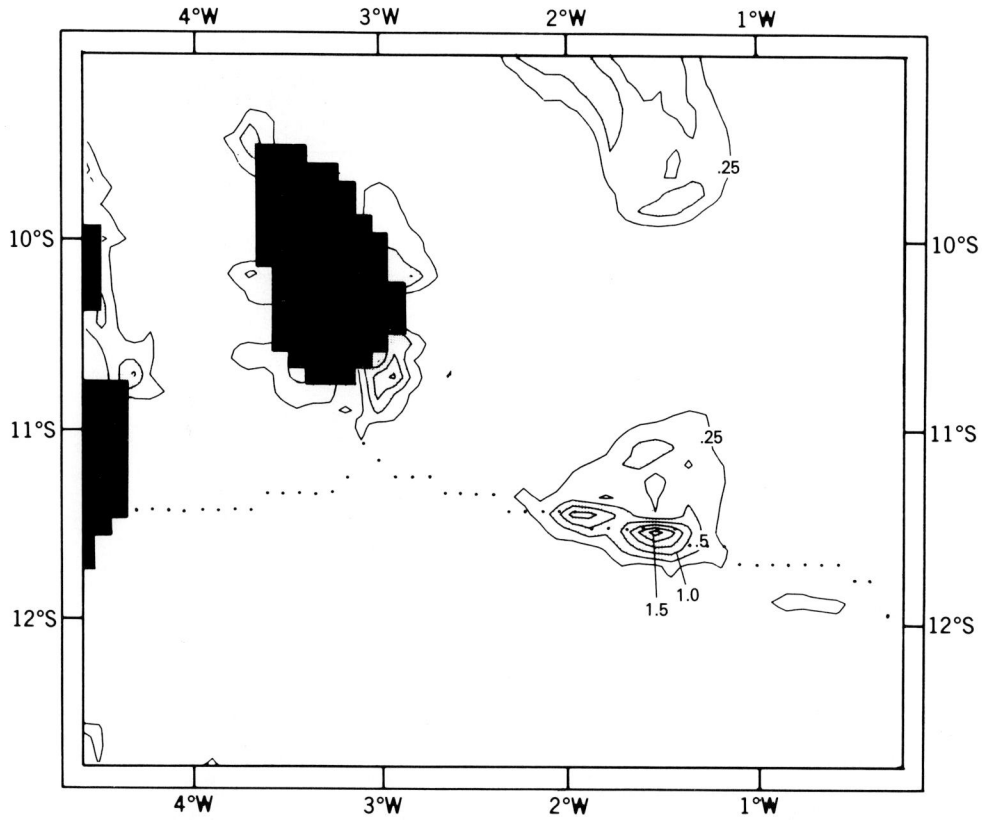

EULERIAN MEAN VELOCITY MAGNITUDE (cm/sec)

Fig. 7. The magnitude of the Eulerian mean flow near Roosevelt Island, displayed in vector form by Figure 4.

$$\underline{u} = \langle \underline{u} \rangle + \underline{u}' \quad (5)$$

$$\eta = \langle \eta \rangle + \eta' \quad (6)$$

with

$$\langle \underline{u}' \rangle = \langle \eta' \rangle = \langle \partial \underline{u}/\partial t \rangle = \langle \partial \eta/\partial t \rangle = 0 \quad (7)$$

The time-average momentum and mass budgets are obtained by applying the time-averaging operator to equations (1) and (2)

$$g\nabla\langle\eta\rangle + f\underline{e}_z \times \langle\underline{u}\rangle = -\langle\underline{u}'\cdot\nabla\underline{u}'\rangle - \langle\underline{u}\rangle\cdot\nabla\langle\underline{u}\rangle - \langle\underline{\tau}/D\rangle \quad (8)$$

and

$$\nabla \cdot (\langle D \rangle \langle \underline{u} \rangle) = -\nabla \cdot \langle \eta' \underline{u}' \rangle \quad (9)$$

The terms $-\langle\underline{u}'\cdot\nabla\underline{u}'\rangle$ and $-\nabla\cdot\langle\eta'\underline{u}'\rangle$ on the right-hand sides of equations (8) and (9) represent tidal forcing of the steady circulation.

Model Results

As expected from theory, strongest tidal rectification occurred in the simulation at sites along the ice front northwest of Roosevelt Island and near the Steershead Ice Rise. Strong rectification also occurred along the sides of Pennell Bank, but is separated from the sub-ice shelf regime. The values of u at these three rectification sites are displayed in Figures 4, 5 and 6; and the magnitude of u near Roosevelt Island is contoured in Figure 7. As expected, the rectified flow runs parallel to depth contours and is anticyclonic about shallows. In the southeastern corner (Figure 4), the depth contours are blocked by the coast, so flow running along depth contours there is forced to converge into an intensified boundary layer [MacAyeal, 1983].

Small eddies commonly occur in pairs of opposite rotation adjacent to various coastal headlands throughout the basin. Notable examples are seen in Figures 4, 5 and 6 off the southern tip of Roosevelt Island, the eastern tip of Ross Island, Minna Bluff, and along the ragged coastline of the Transantarctic Mountains. These vortex pairs are barely resolved by the numerical grid, so are not accurately reproduced in this simulation.

Verification of the simulated tidal rectification in the Ross Sea is best accomplished

TABLE 1. Comparison of Simulated Currents to Analytic Expressions

Position	Simulation	Robinson [1981]	Huthnance [1973]	Loder [1980]
Eulerian Mean Flow				
Ice front	1.5 cm/s	5.0 cm/s		9.0 cm/s
Seabed ridge	0.5 cm/s	0.3 cm/s	0.4 cm/s	0.4 cm/s
Lagrangian Mean Flow				
Ice front	1.0 cm/s			5.0 cm/s
Seabed ridge	0.25 cm/s			0.15 cm/s

by direct observations of the time-averaged and depth-averaged currents. Pillsbury and Jacobs [this volume] report preliminary analyses of records obtained from a current meter array deployed along the ice front northwest of Roosevelt Island. The data of Pillsbury and Jacobs [this volume] may be compared with Figures 4 and 7 to assess the match between observation and simulation.

The records at current meter sites "P" and "C" [Pillsbury and Jacobs, this volume] display steady westward flow along the ice front in a direction consistent with the simulation. The magnitude of the observed currents is approximately ten times larger than the simulated flow, however, and suggests that the simulation may produce insufficient flow strength in spite of correct flow orientation. The records at current meter site "S" [Pillsbury and Jacobs, this volume, Figure 3], indicate a southerly current at that location, which is east of the location where the observed and simulated flows are both westward. The flow direction at "S" is also consistent with the simulation, which shows a cyclonic eddy centered just north of the sub-ice cavity, with southerly flow diverted to the west as the ice front is approached. This suggests that data from a current meter mooring located closer to the ice front that "S" may be needed to verify the simulated flow. More current meter data from the field program will soon become available [S. Jacobs, personal communication, 1985] so it is premature to accept or reject the existence of tidal rectification at the ice front site northwest of Roosevelt Island. In any case, the results presented here are useful in that they embrace one of many possible mechanisms for sub-ice-shelf flow.

Verification of the simulated rectification elsewhere in the Ross Sea is hampered by the lack of current meter observations. A single current meter moored at a mid depth near the junction of the ice front and 180°W longitude provides a 7-month record of currents at hourly intervals [Jacobs and Haines, 1982]. The time-average velocity derived from this record is shown in Figure 6 by the large vector near grid coordinates 0°E, 12°S, and does not agree with the simulated depth-independent flow, driven by tidal rectification. This observation does not provide a way to distinguish between the depth-independent flow and the shear flow that may accompany thermohaline circulation. The disagreement is not, therefore, significant. Other current meter observations are available from locations in McMurdo Sound and at the J9 bore hole [Heath, 1977; Jacobs and Haines, 1982]. The additional observations do not, however, provide records of sufficient duration to differentiate the tidal currents from the steady currents.

A possible verification of the anticyclonic flow around Pennell Bank (Figure 6) is provided by the mapped areal extent of dense High Salinity Shelf Water (HSSW) given by Jacobs et al. [1970]. This areal extent is also shown in Figure 5 of MacAyeal [1984a], and reveals an inverted "S" pattern consistent with anticyclonic advection about the shallow bank. Anticyclonic advection may be caused by other processes as well, so the advection pattern provides more of a consistency check than an actual verification of the model result.

In addition to comparison between simulation and observations, comparison between simulation and analytic treatments of tidal rectification is also useful for model verification. The most appropriate comparison applicable to the rectification sites near Roosevelt Island is with analytic treatments developed by Huthnance [1973] and Loder [1980]. They solved for the circulation driven along an infinitely long seabed bank connecting two half-plane oceans of constant depth. This solution is compared in Table 1 to the simulated circulation along seabed topography extending to the northwest of the

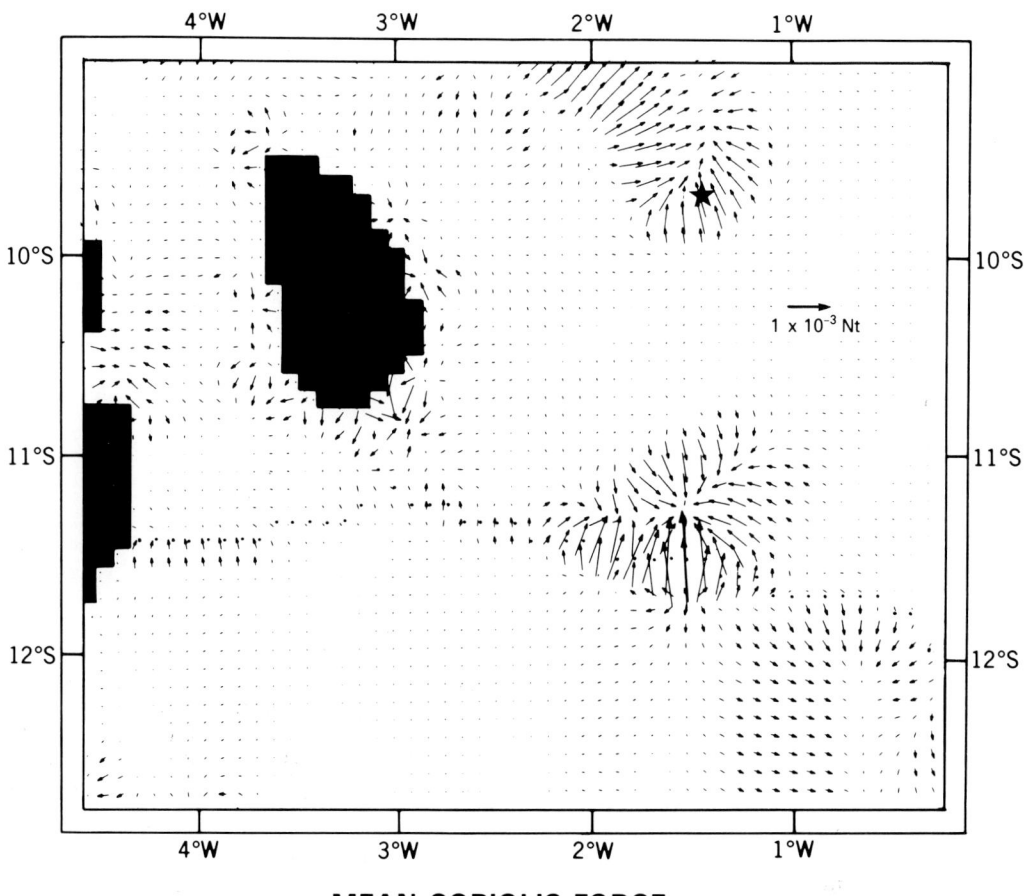

MEAN CORIOLIS FORCE

Fig. 8. Vectors representing the simulated time-averaged Coriolis force (per cubic meter of the water column), pointing left of the time-averaged flow and toward the centers of the two strong anticyclonic eddies shown in Figure 4. The star near 2°W, 10°S indicates where the momentum balance shown in Figure 12 is taken.

Steershead Ice Rise and along the ice front. An additional comparison is made between the simulated flow in these two areas and the theoretical estimates provided by Robinson [1981].

Loder [1980] extended his analytic treatment of rectification to include steplike topography such as that occurring at an ice front. His solution gives approximately 0.09 m/s of flow along the ice front, whereas the simulation gives only 0.015 m/s (Table 1). Poor agreement between model and theory in this instance is attributed to (1) insufficient spatial resolution provided by the model, and (2) the artificial horizontal momentum diffusion ($N = 1.5 \times 10^2$ m^2/s) used by the model to suppress grid-scale noise. Robinson [1981] also considered the theory of rectification along steplike topography, and showed that the current should monotonically decay away from the ice front with a length scale equal to the horizontal range of water-parcel movement. Applying Robinson's analysis to the simulated range of horizontal movement shown in Figure 3, the length scale of an ice front current should be approximately 6 km. This scale falls well below the 10 km grid resolution used by the model [MacAyeal, 1983]. Artificial momentum diffusion used in the model will thus suppress this current. The simulated length scale of off-ice-front decay is sufficiently wide, however, that the simulated transport integrated perpendicular to the ice front may be fortuitously in accord with the transport produced by an existing narrower, stronger jet. Preliminary results from a current meter array deployed within 10 km of the ice front show westward flow at rates in excess of 0.06 m/s at some locations [Pillsbury and Jacobs, this volume]. These observations are in better agreement with theory than the numerical simulations presented here, and suggest that the model results tend to underestimate the flow.

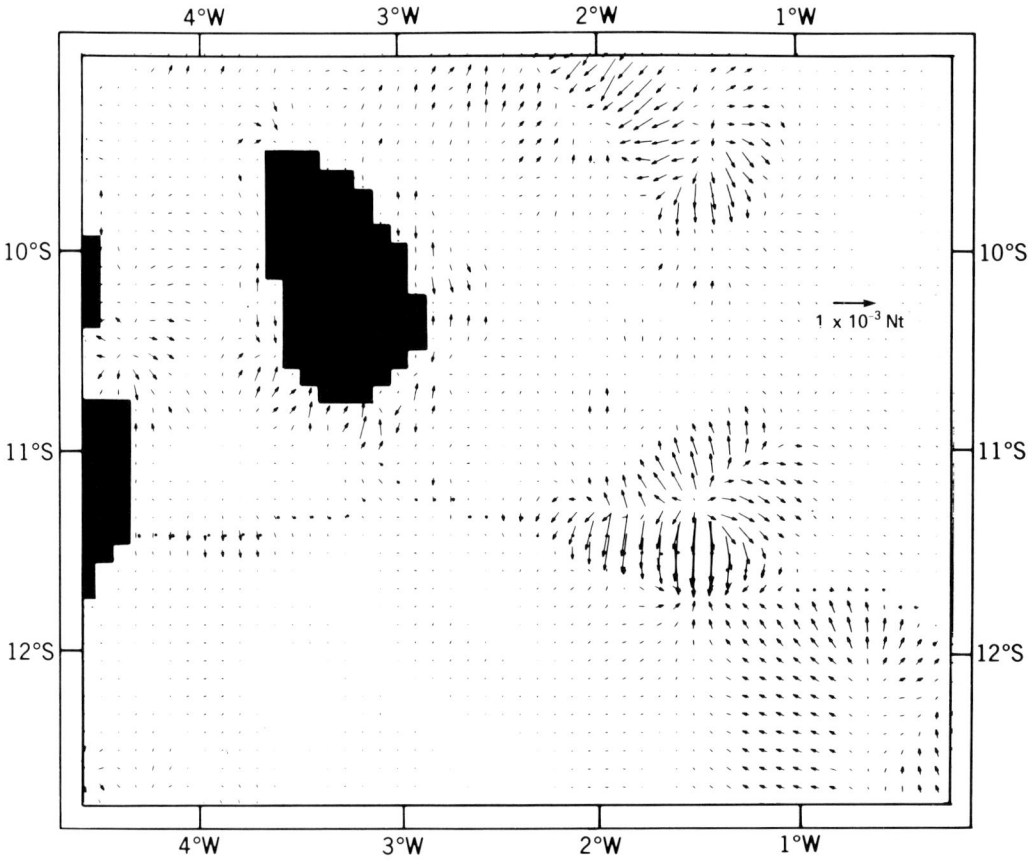

MEAN PRESSURE GRADIENT FORCE

Fig. 9. Vectors representing the time-averaged hydrostatic pressure-gradient force, pointing in a direction approximately opposite to that of the time-averaged Coriolis force shown in Figure 8.

Time-Averaged Momentum Budget

The time-averaged momentum budget expressed by equation (8) was reconstructed from model data and partitioned into the following four terms: (1) the time-averaged Coriolis force $<-\underline{e}_z \times \rho f \underline{u}>$, (2) the time-averaged hydrostatic pressure gradient force $<-\rho g \nabla \eta>$, (3) the time-averaged friction $<-\rho \underline{\tau}/D>$, and (4) the time-averaged momentum advection

$$<\rho \underline{u} \cdot \nabla \underline{u}> = <\rho \nabla |\underline{u}|^2/2 + \underline{e}_z \times \rho \zeta \underline{u}>$$

where ζ is the relative vorticity and ρ is the seawater density. The numerical model employed a finite-difference form of the depth-integrated momentum balance equations, so the momentum terms listed above were first determined in their flux form, and then divided by the time-averaged depth $<D>$ to obtain their values per cubic meter of the water column.

The momentum budget was balanced throughout most of the model domain except in the shallow region between the Steershead Ice Rise (81°S, 160°S) and Crary Ice Rise (83°S, 170°W) and the Siple Coast. The unbalanced force, caused mainly by excessive friction, was equal to the time-average momentum tendency $<\partial u/\partial t>$ computed from the model output. This error may be attributable to: (1) insufficient "spin up" after model initialization (5 days), (2) insufficient periods used to determine time-averaged quantities (13.97 days), or (3) over-simplification of the complex open boundary conditions [MacAyeal, 1983]. The momentum budget along the ice front and along the ridge northwest of the Steershead Ice Rise was balanced to the extent that any unbalanced force was less than 10% of any individual term in equation (8).

The momentum budget along the eastern section of the ice front and above the seabed ridge extending northwest of the Steershead Ice Rise is presented in Figures 8-11. The time-averaged hydrostatic pressure gradient force and the time-averaged Coriolis force

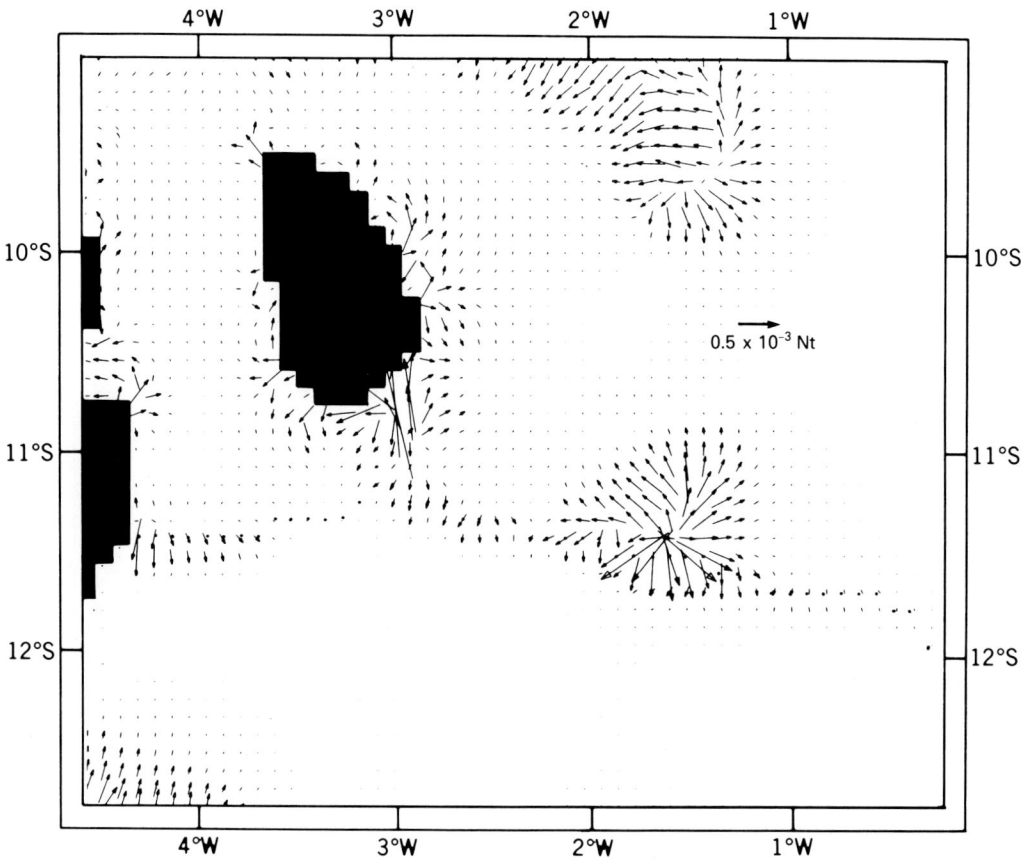

MEAN REYNOLDS STRESS DIVERGENCE

Fig. 10. The divergence of the time-averaged momentum flux (N/m^3), driving tidal rectification. This force is directed away from the centers of the anticyclonic flow near Roosevelt Island. The time-averaged dynamic pressure-gradient force, produced by stronger tidal currents over these shallow regions, combines with the time-averaged hydrostatic pressure gradient force to balance the time-averaged Coriolis force.

were largest but did not completely cancel, as would be required for geostrophic balance. The time-averaged Coriolis force tended to be approximately 30% larger than the time-averaged pressure gradient force, indicating that the time-averaged momentum advection balances both the excess time-averaged Coriolis force and the time-averaged friction force. Figure 12 displays the balance of forces at a point along the seabed ridge extending northwest of the Steershead Ice Rise (labeled with a star in Figure 4).

As suggested by Loder [1980], the Coriolis force on the time-averaged flow is balanced by the time-averaged gradient of the effective free surface elevation defined here as $<\eta> = <|\underline{u}'|^2/2g>$, where the time-averaged dynamic pressure, $<|\underline{u}'|^2/2>$, makes an important contribution. Figure 13 displays the time-averaged effective free surface elevation. The contours of Figure 13 may be regarded as streamlines of the time-averaged flow because of geostrophic balance.

Time-Averaged Vorticity Budget

Vorticity transport by tidal currents provides a simple way to explain tidal rectification [Zimmerman, 1980, 1981; Robinson, 1981]. Relative vorticity is imparted to the tidal currents by flow across isobaths or by bottom friction torque. Continuous removal of this relative vorticity by bottom friction forces the advection of relative vorticity by the tidal currents to be asymmetric over the tidal cycle. Convergence of this rectified vorticity flux replenishes the vorticity field of the residual circulation against frictional losses.

Zimmerman [1981] and Robinson [1981] de-

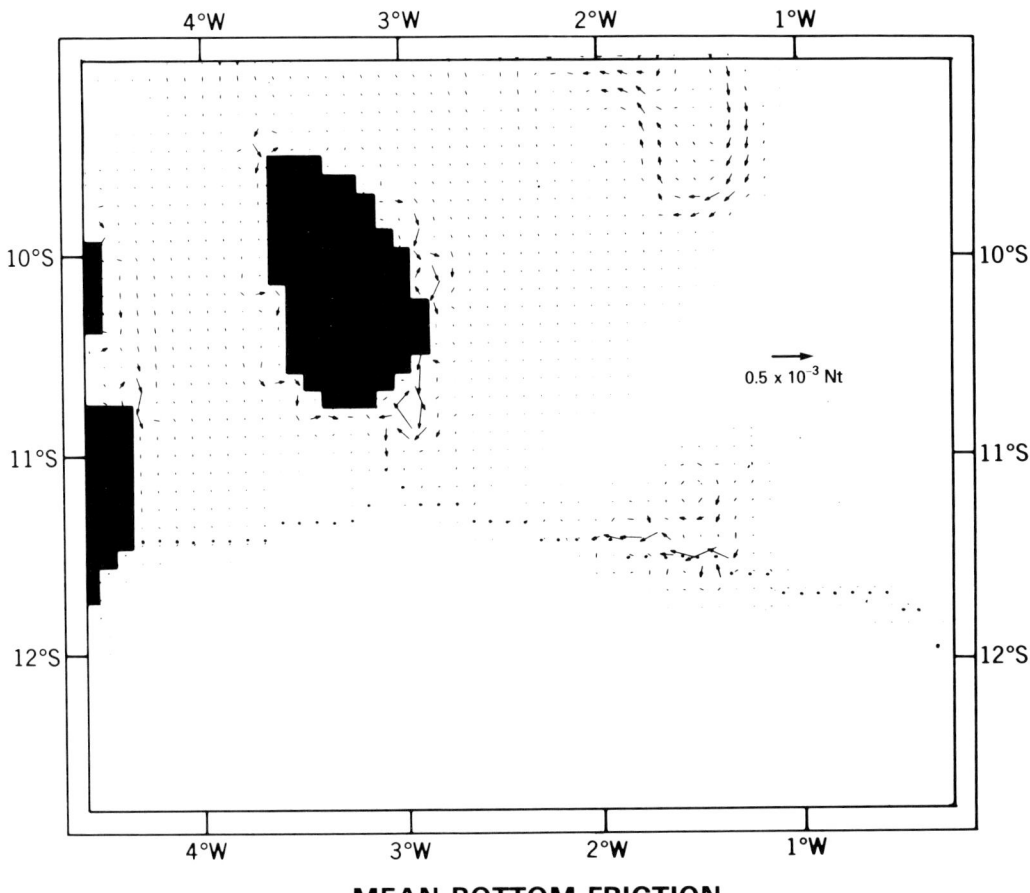

Fig. 11. The time-averaged bottom friction (N/m³) measured at a fixed geographic location, directed opposite to the time-averaged circulation. These vectors represent the opposite of the time-averaged force (per cubic meter of the water column above) transmitted to the sea bed.

vised a simple physical argument to explain how tidal currents set up vorticity fluxes, and their argument is repeated here. Consider the idealized situation depicted in Figure 14, adapted from Zimmerman [1981]. The relative vorticity of a water column displaced across the seabed topography is a function of its position with respect to the depth contours. The water column oscillating between points A and B has positive relative vorticity at B and negative relative vorticity at A, as dictated by the stretching of planetary vorticity filaments. In the absence of friction, the relative vorticity of the parcel is assumed to be zero at positions 0 and 0' midway between A an B. With bottom friction, potential vorticity is lost at point B and gained at point A. As the column passes position 0' on its way toward deep water, it will have slightly negative relative vorticity. Similarly, as the column passes position 0 on its way toward shallow water, it will have slightly positive relative vorticity. By these oscillatory motions, the water column considered in this example transports positive vorticity into shallow water. The entire ensemble of water columns surrounding the topography will behave similarly, leading to an accumulation of positive vorticity in the shallow water. This vorticity accumulation drives an anticyclonic circulation along the depth contours. Also shown in Figure 14 are other mechanisms by which frictional torque imparts relative vorticity. These mechanisms are not, however, important in the present context.

The time-averaged vorticity budget is constructed from the momentum budget by applying the curl operator [Robinson, 1981]

$$0 = -\nabla \cdot <\underline{u}'\zeta'> - \nabla \cdot (f + <\zeta>)<\underline{u}> = \nabla \times <\underline{\tau}/D> \quad (10)$$

where

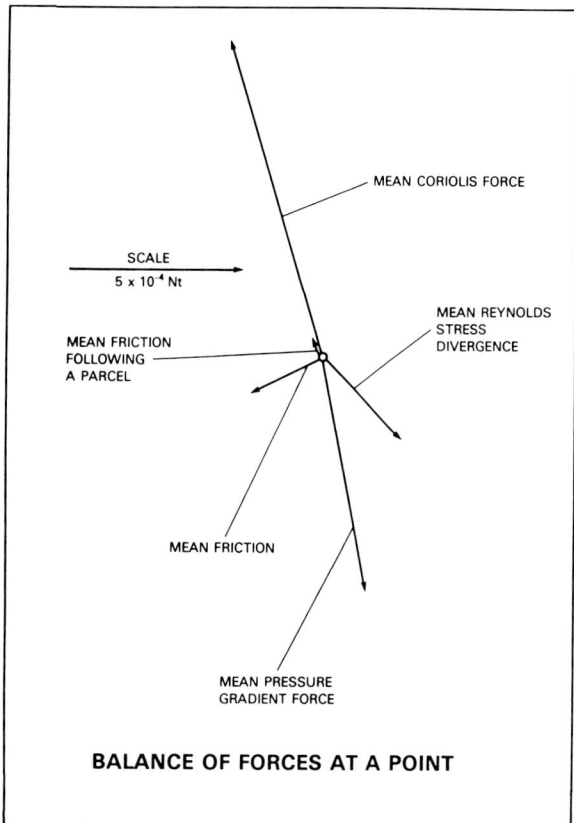

Fig. 12. The balance of time-averaged forces at the location indicated by a star in Figure 8, expanded for illustration.

$$\nabla \times \langle \underline{\tau}/D \rangle = -\nabla \times \langle (k|\underline{u}|/D)'\underline{u}' \rangle - \langle k|\underline{u}|/D \rangle \langle \zeta \rangle$$
$$- \langle \underline{u} \rangle \times \nabla \langle k|\underline{u}|/D \rangle \quad (11)$$

and $\zeta = \zeta' + \langle \zeta \rangle = (\partial v/\partial x - \partial u/\partial y)$ is the vertical component of the relative vorticity. In equation (11), the term representing the torque provided by eddy viscosity has been dropped. This simplifies the discussion and is justified by the small size of this term in the simulation. Equation (10) states that the convergence of the sum of the eddy vorticity flux, $\langle u'\zeta' \rangle$, and the flux of time-independent vorticity, $(f + \langle \zeta \rangle)\langle \underline{u} \rangle$, must balance frictional torque, $\langle \nabla \times (\underline{\tau}/\overline{D}) \rangle$.

Vorticity fluxes induced by topographic relief may be conveniently examined by constructing the circulation integral [Zimmerman, 1981]. Integrating equation (10) over the area enclosed by an isobath, assumed to be closed, and applying the divergence and Stokes theorems [Schey, 1973, p.44 and p.92], the following integral constraint is obtained

$$\int \langle k|\underline{u}|/D \rangle \langle \underline{u} \rangle \cdot d\underline{l} = -\int (f + \langle \zeta \rangle)\langle \underline{u} \rangle \cdot \underline{n} \, dl$$
$$-\int \langle \underline{u}'\zeta' \rangle \cdot \underline{n} \, d\underline{l} - \iint \nabla \times \langle (k|\underline{u}|/D)'\underline{u}' \rangle \, da \quad (12)$$

where $d\underline{l}$ is a vector element of path length along the depth contour, $\underline{n}$ is the unit vector pointing out of the enclosed region, and da is an element of the area enclosed by the contour. The left-hand side of equation (12) expresses the net drag on the circulation. This drag is balanced by a vorticity flux into the area enclosed by the isobath. The first term on the right-hand side of equation (12) is typically smaller than the second term because the time-averaged mass continuity equation implies that $\int \langle \underline{u} \rangle \cdot \underline{n} \, d\underline{l}$ is very small [Zimmerman, 1981]. The second and third terms on the right-hand side of equation (12) express the cross isobathic component of the eddy-vorticity flux and the net bottom friction torque imparted to the tidal currents within the area enclosed by the isobath.

The time-averaged vorticity budget reconstructed from the model data displays the workings of these rectification mechanisms along the eastern section of the ice front and the seabed ridge extending to the northwest of the Steershead Ice Rise. Figures 15 and 16 present diagrams of the first and last terms on the right-hand side of equation (10). To differentiate the important features from the background, the highest positive and lowest negative results of these terms are indicated by patterned shadings rather than by contours. The balance of terms was not exact because the finite-difference schemes used for vorticity analysis were different from those applied in the model, and because the time average of the vorticity tendency was not precisely zero. This latter effect caused the divergence of the flux of time-average vorticity by the steady circulation, $\nabla \cdot (f + \langle \zeta \rangle)\langle \underline{u} \rangle$, to exhibit anomalous unbalanced patterns similar to those expected of the instantaneous tidal fields. The time-average vorticity flux divergence, $\nabla \cdot \langle \underline{u}'\zeta' \rangle$, and the time-average vorticity dissipation, $\nabla \times \langle \underline{\tau}/D \rangle$, approximately canceled, however, consistent with the principal balance discussed above.

Figures 15 and 16 indicate that the mechanism of rectification described by the water column traveling between points A and B in Figure 14 applies along the flanks of the seabed ridge and along the ice front. Convergence and divergence of the vorticity flux is confined to parallel strips aligned with the depth contours: convergence occurs above the shallow areas, and divergence occurs above the deep areas. Convergence is predominantly balanced by frictional torque, implying that the circulation follows depth contours and the vorticity field associated with this time-

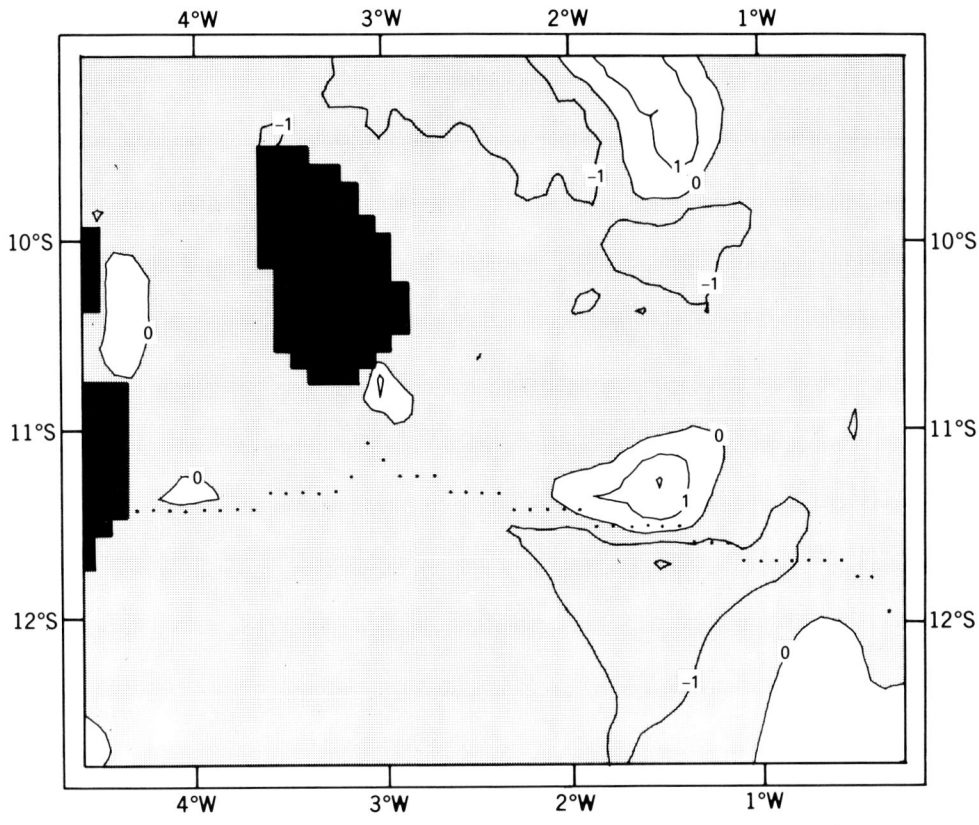

EFFECTIVE MEAN FREE SURFACE ELEVATION (10^{-3}m)

Fig. 13. The effective free surface elevation defined as the sum of the time-averaged hydrostatic pressure and the time-averaged dynamic pressure, with both divided by g.

averaged circulation correlates with topography. Although qualitative agreement between simulation and theory is excellent, a better quantitative display of the time-averaged vorticity budget would require significantly higher spatial resolution, and a longer period of simulation.

Ventilation of the Cavity Under the Ice Shelf

Large-scale transport of heat, salt, and passive tracers below the ice shelf is accomplished by water-parcel advection. The kinematics of individual water-parcel motions is called the Lagrangian description of motion [Longuet-Higgins, 1969] and has been precisely defined in the context of tidal rectification by Zimmerman [1979]. Following Zimmerman's [1979] analysis, the position $\underline{y}$ of a water column initially at position $\underline{x}$ (when t=0) is

$$\underline{y}(\underline{x},t) = \underline{x} + \int_0^t \underline{u}(\underline{y}(\underline{x},s),s)\, ds = \underline{x} + \underline{\xi} \quad (13)$$

where $\underline{u}(\underline{y},t)$ is the velocity field at the current position of the water column, and s is a dummy variable of integration. The second term on the right-hand side of equation (13) expresses the displacement of the water column since t=0. Equation (13) is generally a nonlinear integral equation that must be solved for $\underline{y}(\underline{x},t)$. In practice, however, approximations can be made to achieve an accurate expression for $\underline{y}(\underline{x},t)$ that is linear.

Longuet-Higgins [1969] and Zimmerman [1979] approximate $\underline{u}(\underline{y}(\underline{x},t))$ in equation (13) with a Taylor series

$$\underline{u}_1(\underline{y}(\underline{x},t),t) = \underline{u}(\underline{x},t) + \underline{\xi}\cdot\nabla\underline{u}(\underline{x},t) + \ldots \quad (14)$$

where the subscript 1 denotes the approximate Lagrangian velocity. A further approximation is made by substituting $\underline{u}(\underline{x},t)$ for $\underline{u}(\underline{y},t)$ in the expression for $\underline{\xi}$

$$\underline{\xi}(\underline{x},t) = \int_0^t \underline{u}(\underline{x},s)\, ds \quad (15)$$

The accuracy of this approximation, as discussed by Zimmerman [1979], depends upon the

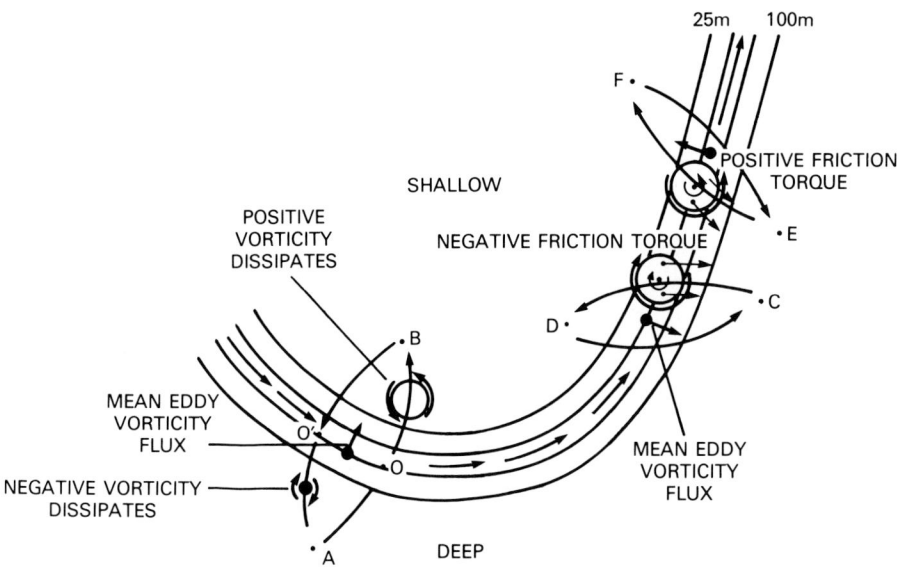

Fig. 14. Schematic diagram displaying the mechanism by which tidal currents drive steady circulation about shallow topography. This diagram is adapted from that conceived by Zimmerman (1981).

size of $\underline{\xi}$ relative to the spatial scale over which the velocity field varies.

The Lagrangian mean circulation, $<\underline{u}_1(\underline{y}(\underline{x},t),t)>$, is obtained by averaging $\underline{u}_1$ over an appropriately long time interval, chosen to filter out the tidal fields. The term $<\underline{\xi}\cdot\nabla\underline{u}(\underline{x},t)>$ is known as the "Stokes drift" [Longuet-Higgins, 1969] and accounts for spatial gradients of the tidal flow that alter the trajectories of water columns. The term $<\underline{u}(\underline{x},t)>$, previously referred to as the steady circulation, is also known as the "Eulerian mean velocity."

The intuitive distinction between the Eulerian mean and the Lagrangian mean flow is that the first can be obtained easily from fixed current meter data by averaging, and the second can be determined only by watching the trajectories of floats or other such passive tracers. Because the Lagrangian mean flow causes heat and mass transport, the Lagrangian mean flow rather than the Eulerian mean flow is of primary interest in determining the impact of the tidally driven flow on the sub-ice shelf environment.

Figures 17, 18 and 19 display the Stokes drift, the Lagrangian mean flow and its magnitude in the region near Roosevelt Island. The Stokes drift is directed opposite to the Eulerian mean flow shown in Figure 4, and is approximately two-thirds as strong. The strongest Stokes drift occurs along the ice front where the tidal current ellipses display a discontinuous change in magnitude and polarization [see MacAyeal, 1984b]. The Lagrangian mean flow along this same section of the ice front displays a double vortex with one eddy on either side of the ice front. In other portions of the model domain, except through McMurdo Sound, the relationship between the Eulerian mean and Lagrangian mean flows was similar [MacAyeal, 1983]. The Stokes drift was negligible through McMurdo Sound, and the Lagrangian mean flow was nearly the same as the Eulerian mean flow. This result suggests that currents observed in McMurdo Sound [Heath, 1977] do not require significant correction for the Stokes drift. As an independent test of the simulated Lagrangian mean flow, the flow near Roosevelt Island is compared in Table 1 with Loder's [1980] analytic expressions for the Lagrangian mean flow along the sides of an idealized seabed slope and an ice front.

Flushing of the cavity below the ice shelf is demonstrated by plotting trajectories of imaginary tracer particles advected by the simulated Lagrangian mean flow. Figures 20-22 display the tracer streaklines emitted from the ice front and from a position just to the west of the Steershead Ice Rise after 3 years, 7.5 years, and 15 years of advection. Penetration into the sub-ice shelf region is evident near Ross Island and the Bay of Whales, and along the western flank of the seabed ridge intersecting the ice front to the northwest of Roosevelt Island. A broad section of outflow occurs along the western half of the ice front.

It is evident from comparing the tracer trajectories to the depth contours (Figure 1) that Lagrangian mean movement follows isobaths. Cross ice front transport is thus somewhat inefficient because of the steplike

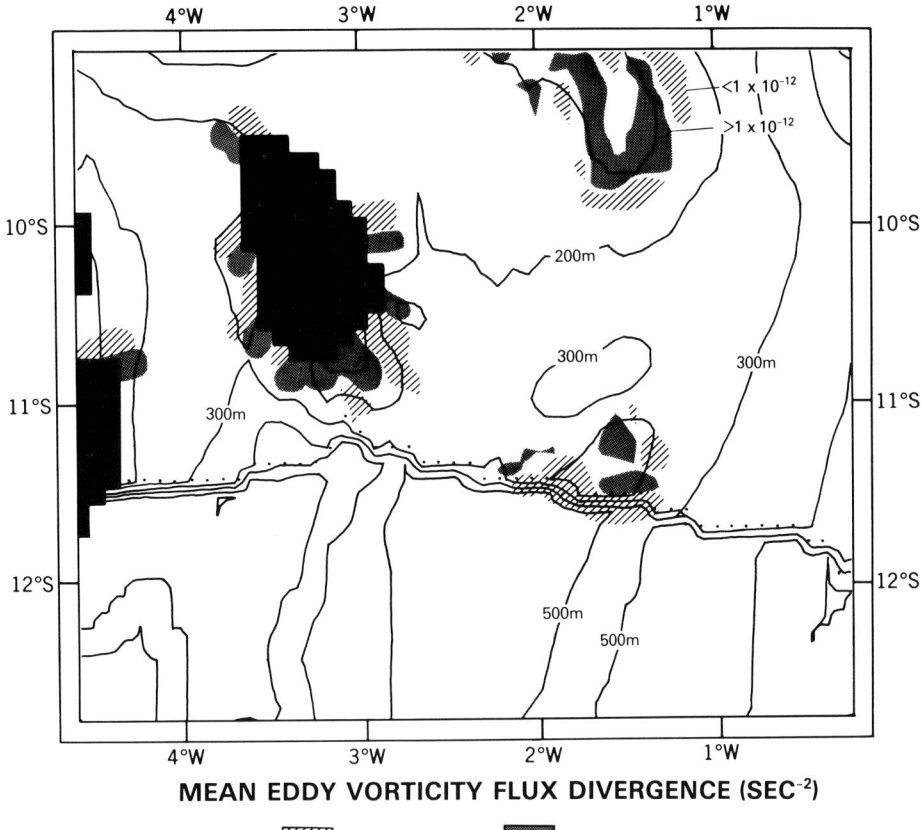

Fig. 15. The regions of strongest time-averaged vorticity flux divergence $\langle \nabla \cdot (u'\xi') \rangle$, indicated by shading superimposed on the depth contours in the vicinity of Roosevelt Island. For the anticyclonic flows near the ice front and along the seabed ridge northwest of Steershead Ice Rise, the time-averaged vorticity flux converges in shallow water and diverges in neighboring deep water. This time-averaged vorticity flux is balanced by the time-averaged vorticity dissipation presented in Figure 16 and is consistent with the tidal rectification mechanism displayed in Figure 14.

depth profile. Nevertheless, trajectories that do penetrate into the sub-ice shelf region generally originate along broad sections of the ice front and cross the ice front through narrow "windows". Once across the ice front, trajectories tend to turn back toward the ice front because few isobaths originating at the ice front extend very far to the south.

After 15 years of tracer advection, maximum penetration below the ice shelf is approximately 150 km. Trajectories attaining the deepest penetration follow the 300 m depth contour extending perpendicular to the ice front at the tidal rectification site northwest of Roosevelt Island. Trajectories originating in McMurdo Sound follow depth contours leading past Minna Bluff, but eventually head back toward the ice front on the eastern side of Ross Island. The large seabed ridge southeast of Minna Bluff prevents further penetration.

Trajectories originating near the Steershead Ice Rise also follow depth contours, but extend farther than the trajectories originating along the ice front. A high degree of dispersion is evident in this region; several trajectories originating close to one another head off in widely different directions. This is a common characteristic of Lagrangian mean transport in the vicinity of coasts and islands, and suggests that the effective horizontal eddy diffusivity in this region is high [P.Rhines, unpublished manuscript, Woods Hole Oceanographic Institution, 1978].

The flushing time required to completely renew the sub-ice shelf water mass, estimated on the basis of the simulated tracer advection patterns, is longer than 100 years. Tritium and carbon-14 concentrations observed below the ice shelf at the J9 bore hole suggest that the average turnover time is approximately 6-25 years and that the upper part of the water

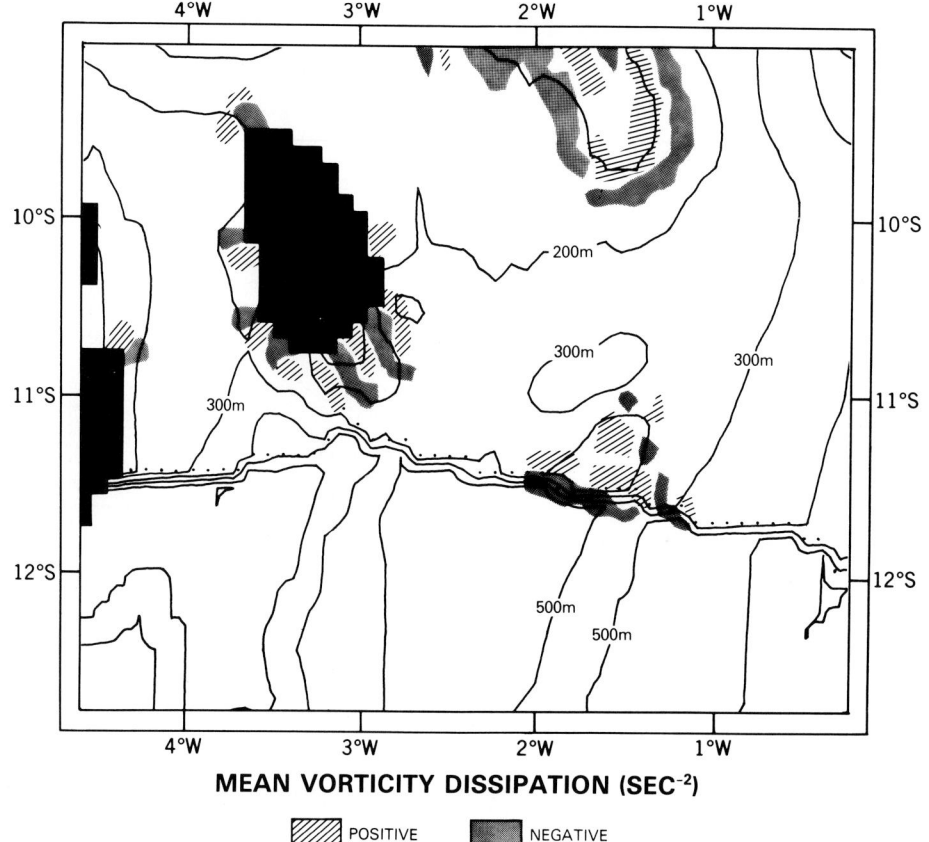

Fig. 16. The time-averaged vorticity dissipation rate $<\nabla \times \tau/D>$ balanced by the time-averaged vorticity flux divergence displayed in Figure 15.

column is renewed more quickly than the lower part [Michel, et al., 1979]. Biota such as fish and scavenging amphipods have been found below the ice shelf at J9 and near Minna Bluff [Bruchhausen et al., 1979; Lipps et al., 1979]. The presence of fish near Minna Bluff is consistent with the fast tracer transport from McMurdo Sound demonstrated in Figures 20 through 22. Both the biological and geochemical evidence suggest, however, that the deeper reaches of the sub-ice shelf region are flushed predominantly by other forms of ocean circulation.

Tracer transport from the ice front toward the continental slope is also relevant to questions concerning the production of Antarctic bottom water and oceanic processes at the junction of the Ross Sea and the circumpolar ocean. Tracer trajectories originating at the ice front reach the continental slope after approximately 15 years, and tend to follow the seabed trough transecting the open Ross Sea to the east of the Pennell Bank.

Hydrographic observations from the open part of the Ross Sea support the general patterns of simulated tracer transport, even though baroclinic flow may be dominant there. Figures presented by Jacobs et al. [this volume] show the observed summer temperature along three vertical sections. Subsurface waters with temperature in excess of -1°C generally originate north of the continental slope. One section perpendicular to the ice front shows a warm core (WMCO) extending from the continental slope toward an ice front position northwest of Roosevelt Island, where greatest sub-ice shelf inflow is indicated by the simulation. Cold water, with a temperature below -2°C, can be seen in another section extending away from the ice front and along the seabed trough east of the Pennell Bank. This region of outflow is also consistent with the simulated transport. The consistency between the areas of inflow and outflow indicated by the simulation and by the observed hydrographic structure suggests that the tidally driven flow may select the sites of warm water inflow and cold water outflow that would otherwise be associated with thermohaline flow.

Repeated hydrographic observations along the ice front in the vicinity of the warm core displayed in Figure 6c of Jacobs et al. [this volume] have indicated considerable temporal

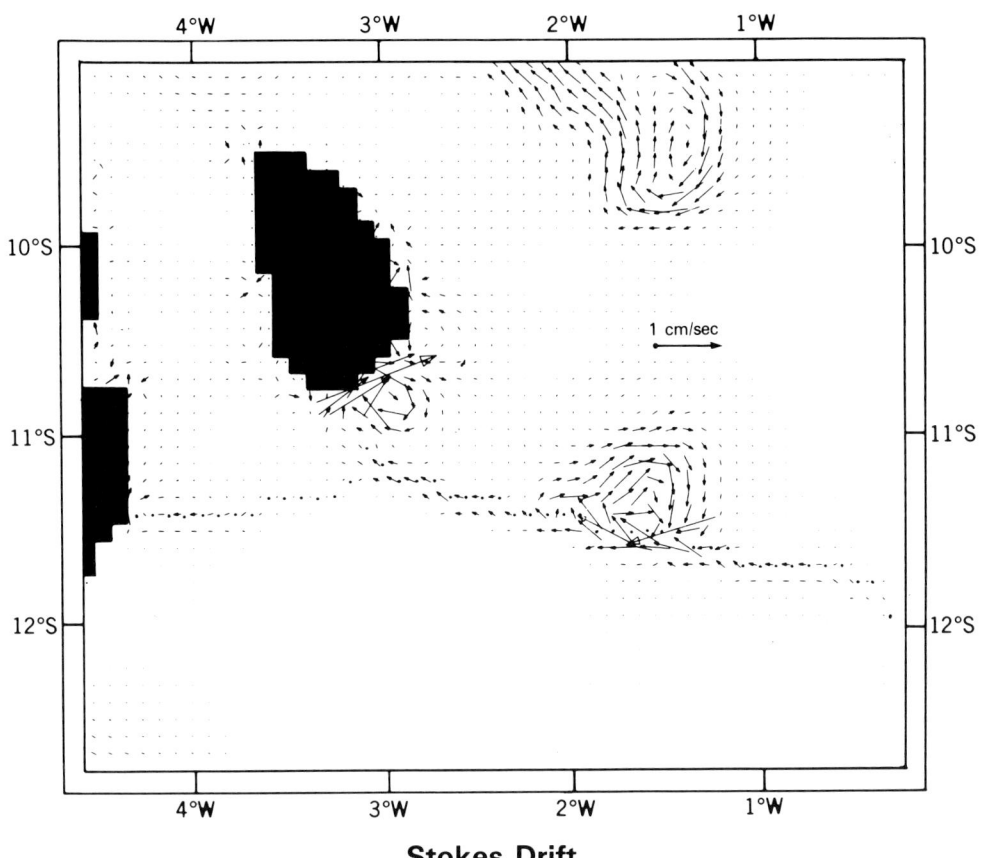

Fig. 17. The Stokes drift, cyclonic about shallow topography and, therefore, directed opposite to the Eulerian mean flow shown in Figure 4.

variability [S. Jacobs, personal communication, 1984]. This variability could be a consequence of the simulated recirculation along the flanks of the bump just to the south of the ice front. The recirculation time scale indicated by the model is approximately 3 years. Significant alteration of the water mass comprising the warm core is therefore possible below the ice shelf before it emerges back into the open Ross Sea.

The sensible heat flux into the sub-ice shelf region driven by tidal rectification (denoted by HF) is the product of the temperature and the cross ice front flow, integrated over depth and over the length of the ice front

$$HF = \int_\Gamma -\rho c <\underline{u}_1 \cdot \underline{n}\, \theta> D\, dl \qquad (16)$$

where Γ is the contour extending along the ice front, $\underline{n}$ is a unit vector that points out of the sub-ice shelf cavity, $\underline{u}_1$ is the Lagrangian mean flow, c is the heat capacity, ρ is the density, and θ is the depth-averaged temperature. Equation (16) represents only the contribution by the barotropic flow. Heat transport by the baroclinic or density driven flow and by latent heat transport is not determined in this analysis.

To evaluate the sensible heat flux integral given by equation (16), the observed hydrographic structure along the ice front was examined to determine a representative temperature difference $\Delta\theta$ between the inflow and outflow. The two most prominent areas of simulated inflow occur northwest of Roosevelt Island and through McMurdo Sound. The observed depth-averaged seawater temperatures at these two locations imply an average inflow temperature of $-1.6°C$. The outflow temperature is assumed to be the freezing point for the depth of the ice shelf base in the vicinity of the ice front ($-2.1°C$). The approximate temperature difference between inflow and outflow is thus estimated to be $0.5°C$. Uncertainty of $\Delta\theta$ resulting from deviations of the warm core observed off the ice front to the northwest of Roosevelt Island can be estimated to be $0.25°C$ [Jacobs et al., this volume].

The sensible heat flux calculated by using the above figure of $\Delta\theta$, equation (16) and the simulated Lagrangian-mean velocities is 2.8 ±

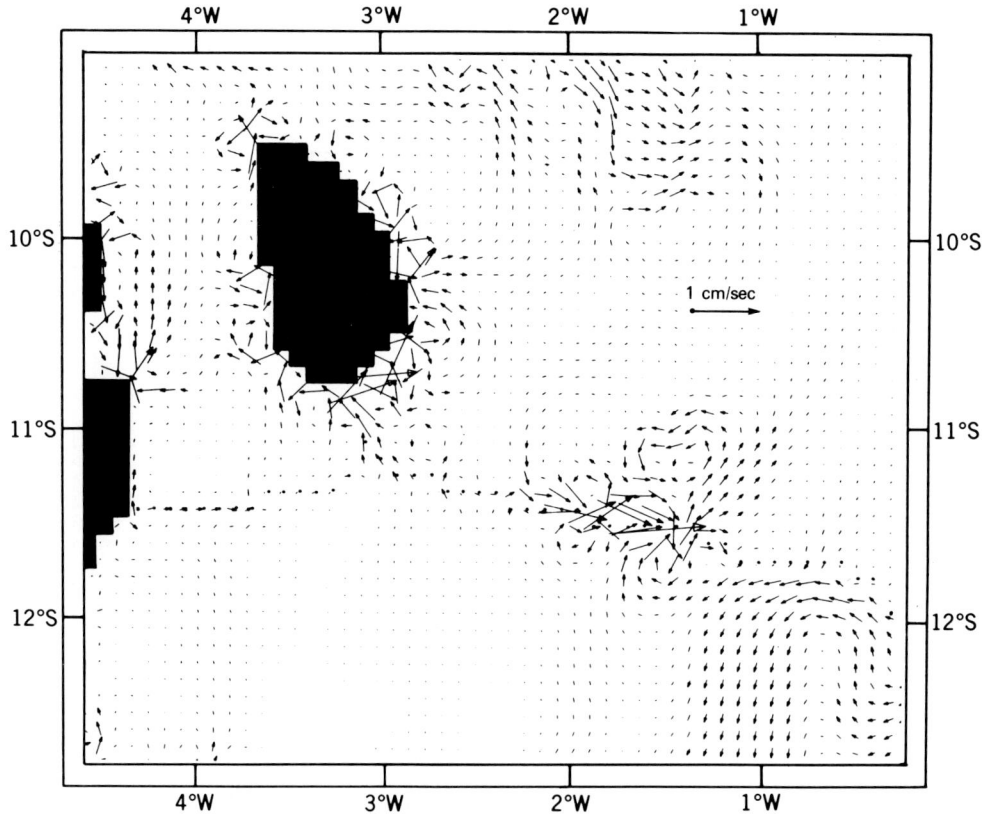

LAGRANGIAN MEAN VELOCITY

Fig. 18. The Lagrangian mean flow, composed of the sum of the Stokes drift and the Eulerian-mean flow. Unlike the Eulerian mean flow, this flow field defines the trajectories of marked water parcels. Heat, salt, geochemical tracers, and similar properties are advected by the Lagrangian mean flow.

1.4×10^{11} W. This result is compared with previously published estimates in Table 2. The average basal melting rate implied by this tidally driven component of the total heat flux is 0.05 ± 0.025 m/yr over the entire 5.8×10^{11} m^2 area of the Ross Ice Shelf. Upward heat conduction through the ice shelf is estimated to be equivalent to approximately 0.02 m/yr basal freezing [Clough and Hansen, 1979], and could significantly reduce the above estimate.

In view of the inefficient penetration of barotropic flow into regions farther than approximately 100 km from the ice front, the estimated cross ice front heat flux is likely to induce melting only in the limited region near the ice front. This region comprises approximately 10% of the total ice shelf area, so the tidally driven basal melting rate will be small unless some other form of circulation that may be caused by tidally driven vertical mixing is active [MacAyeal, 1984a]. Positions along the ice front where Lagrangian mean flow is directed out from below the ice shelf would be expected to exhibit significantly reduced basal melting rates.

The basal melting pattern near the ice front suggested by the fan of inflowing tracer trajectories presented in Figures 20-22 is supported by observations. Temperature-depth profiles of the ice shelf, observed in core holes and through electrical resistivity measurements, indicate basal mass balance conditions [Wexler, 1960; Crary, 1961a, 1961b; Shabtaie and Bently, 1979; Clough and Hansen, 1979; MacAyeal and Thomas, 1979]. Sharp temperature gradients near the base of the ice shelf indicate downward ice advection, with respect to the basal surface, associated with basal melting. The temperature-depth profile at the Little America V station located near the ice front at the Bay of Whales, for example, indicated a basal melting rate exceeding 0.5 m/yr [Wexler, 1960; Crary, 1961a, 1961b]. This rate is consistent with the simulated penetration of oceanic heat below this section of the ice front. Temperature-depth measurements made closer to the ice

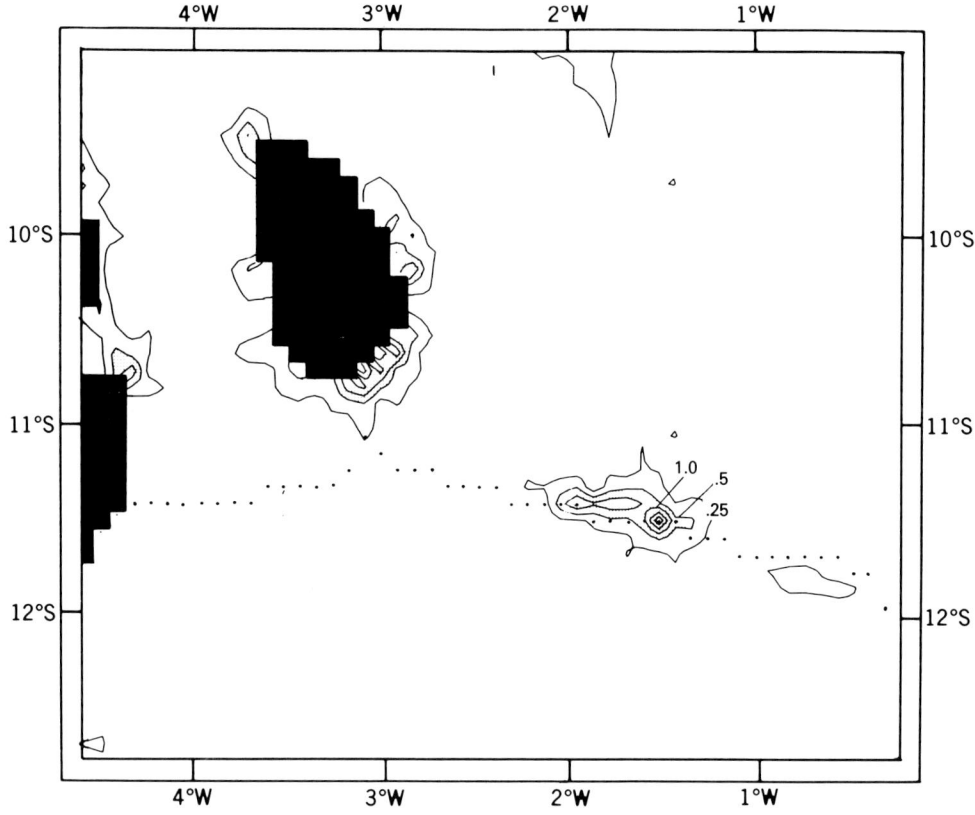

LAGRANGIAN MEAN VELOCITY MAGNITUDE (cm/sec)

Fig. 19. The contours of the Lagrangian mean flow magnitude, indicating that the magnitude of the Lagrangian mean flow is approximately half as great as that of the Eulerian mean flow shown in Figure 7.

front section where trajectories emerge from below the ice shelf indicate zero basal melting [Shabtaie and Bentley, 1979]. The temperature-depth measurement at J9 also indicates near zero basal melting, and the ice core recovered from the J9 site verifies the presence of a thin basal sea ice layer resulting from slow basal freezing [Zotikov et al., 1980]. These indications of near zero melting or freezing are consistent with simulated limited penetration of Lagrangian mean trajectories into this region.

Ice thickness patterns in the vicinity of Ross Island indicate strong basal melting, in agreement with the simulated heat transport through McMurdo Sound [Robin, 1975]. MacDonald and Hatherton [1961] estimate a 0.75 m/yr basal melting rate in this area, based on observed ice advection and snow accumulation. This result is also supported by the mass budget of a numerical ice flow simulation presented in D. R. MacAyeal and R. H. Thomas [unpublished data, 1984].

Radio echo sounding profiles of the Ross Ice Shelf shown in Figure 23 display alternating bands of high- and low-radio wave reflectivity at the ice/water contact that are aligned with the direction of ice shelf flow [Neal, 1979]. Several interpretations of the bands of low reflectivity have been offered, one of which is that the stripes are salt-rich basal sea ice [Neal, 1979; Robin, 1979; Bentley et al., 1979; Shabtaie and Bentley, 1979]. The organization of the apparent basal sea ice into long and narrow bands suggests two important characteristics of the hypothetical basal mass balance that would create them: (1) coastal zones of intense basal freezing must be spatially confined and temporally permanent, and (2) elsewhere, basal melting (or freezing) is insufficient to erase (or overwrite) the delicately structured streaks that extend as "relict" features hundreds of kilometers from their sources (points of actual sea ice deposition) in the southern Ross Sea. If coastal freezing zones were spatially diffuse, broader stripes or splotches of basal sea ice would be produced. Even small amounts of basal melting or freezing would "erase" existing sea ice bands or "cloud" them over

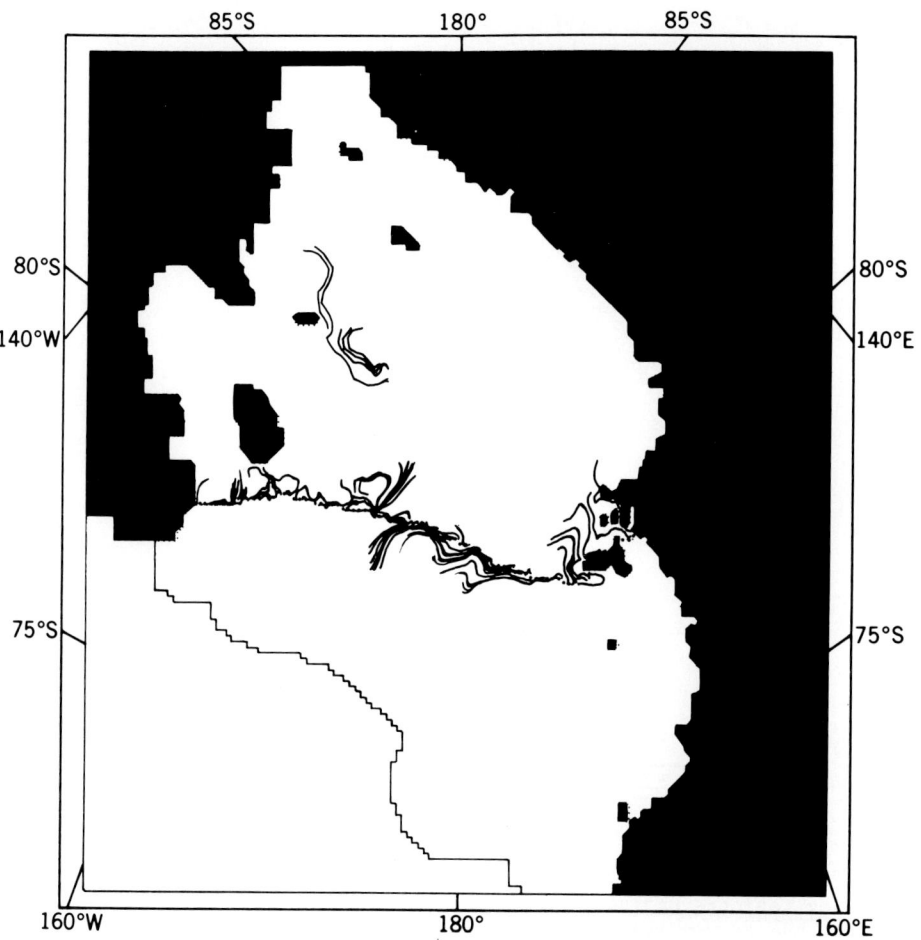

MIXING INTO THE SUB-ICE SHELF REGION: 3 YEARS

Fig. 20. Tracer streaklines emitted from the ice front, displaying how tidal rectification ventilates the sub-ice shelf cavity. After 3 years of advection by the simulated Lagrangian mean flow, tracers have penetrated farthest into the sub-ice shelf cavity near the rectification site northwest of Roosevelt Island and through McMurdo Sound. Heat transport associated with the indicated water-parcel movement will cause approxiamtely 0.5 ± 0.25 m/yr basal melting in the regions penetrated by the tracer trajectories (this region comprises 10% of the total ice shelf area).

with additional sea ice. Additionally, temporal persistence of both the basal freezing zones and the ice shelf flowlines is required to maintain lengthwise continuity of the stripes and their alignment with present-day flowlines. Jezek [1984] and Jezek and Bentley [1984] have investigated apparent deviations of radio-reflectivity features from present-day flowlines and have deduced aspects of transient ice shelf behavior.

Because of its spatial confinement and temporal permanence, tidal rectification may possibly deposit basal sea ice at points where the stripes originate. In the numerical simulation, many vortex pairs were generated by tidal flow past coastal headlands along the Transantarctic Mountains where some of the observed stripes originate. Assuming that the stripes represent relict basal sea ice, these residual eddies are sufficiently confined and permanent to explain the small width and length-wise continuity of the sea ice bands in the western region. The stripes originating from the Siple Coast near the Steershead Ice Rise are broader and more consistent with possible large-scale tidal residuals generated by the seabed topography in this region.

The mechanism proposed here by which basal freezing occurs is based upon the relationship between hydrostatic pressure and the freezing

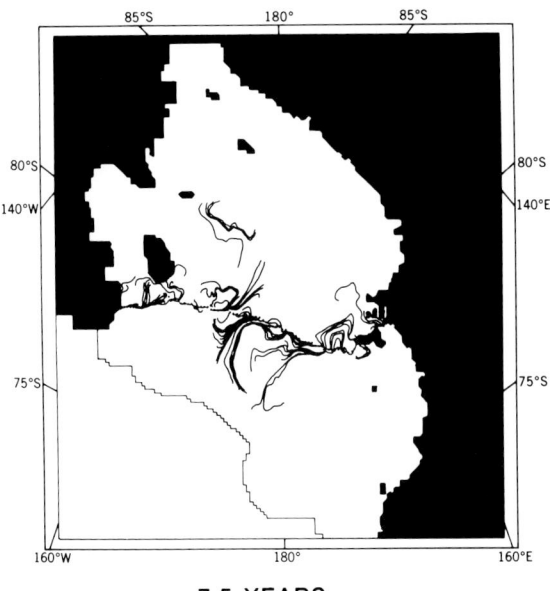

7.5 YEARS

Fig. 21. Same information as in Figure 20, after 7.5 years.

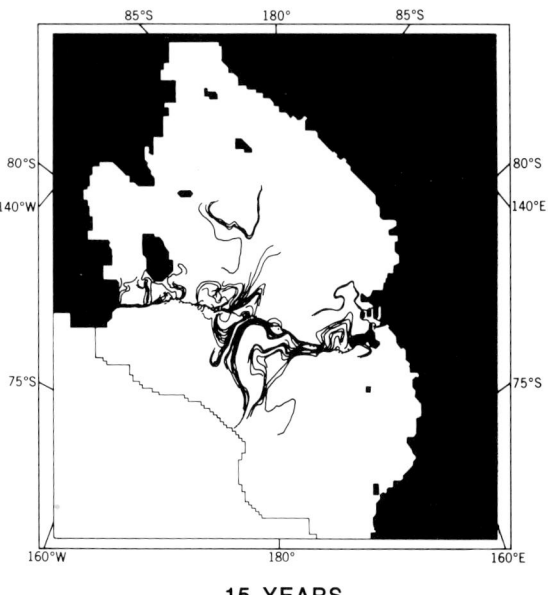

15 YEARS

Fig. 22. Same information as in Figure 20, after 15 years.

temperature. For every 10 m of increasing depth, the freezing temperature, θ_f, is depressed by 7.6×10^{-3}°C [Fujino et al., 1974; see also UNESCO, 1978]. As a result, the temperature at the base of the ice shelf is typically 0.3°C colder than the freezing temperature of water of the same salinity at the sea surfce. Assuming that the seawater in contact with the ice shelf is in chemical and thermodynamic equilibrium with the ice, seawater flow toward decreasing (increasing) ice shelf draft will cause basal freezing (melting). This effect has been proposed as a cause of basal melting and freezing by a number of authors [Gordon, 1974; Doake, 1976; Robin, 1979; Neal, 1979] and has considerable appeal in the present situation because no heat exchange is required between the cavity and the outside ocean.

Conclusion

The leading implication of the analysis presented here is that tidal currents below the ice shelf may trigger large-scale oceanic convections that control the heat and mass exchange between the ice shelf and the ocean. Tidal rectification at the ice front site northwest of Roosevelt Island may, for example, be the process that initially drives WMCO water into the sub-ice cavity. This speculation is based upon the apparent coincidence of the site where WMCO is most often observed with the site corresponding to the strongest simulated tidal rectification. Estimated basal melting associated with tidal rectification is restricted to 10% of the tidal ice shelf area closest to the ice front, and is approximately 0.5 m/year.

TABLE 2. Oceanic Heat Flux Into the Cavity Under the Ice Shelf

Location	Observer	Water Mass	Heat Flux	Ice-Shelf-Averaged Melt
Entire ice front	simulation	HSSW and Warm Core (WMCO)	$2.8 \pm 1.4 \times 10^{11}$ W	0.05 ± 0.025 m/yr
McMurdo Sound	Heath [1977] and others	HSSW	1.3×10^{10} W	0.002 m/yr
Eastern ice front	Jacobs et al. [1979]	WMCO	4.2×10^{11} W	0.08 m/yr

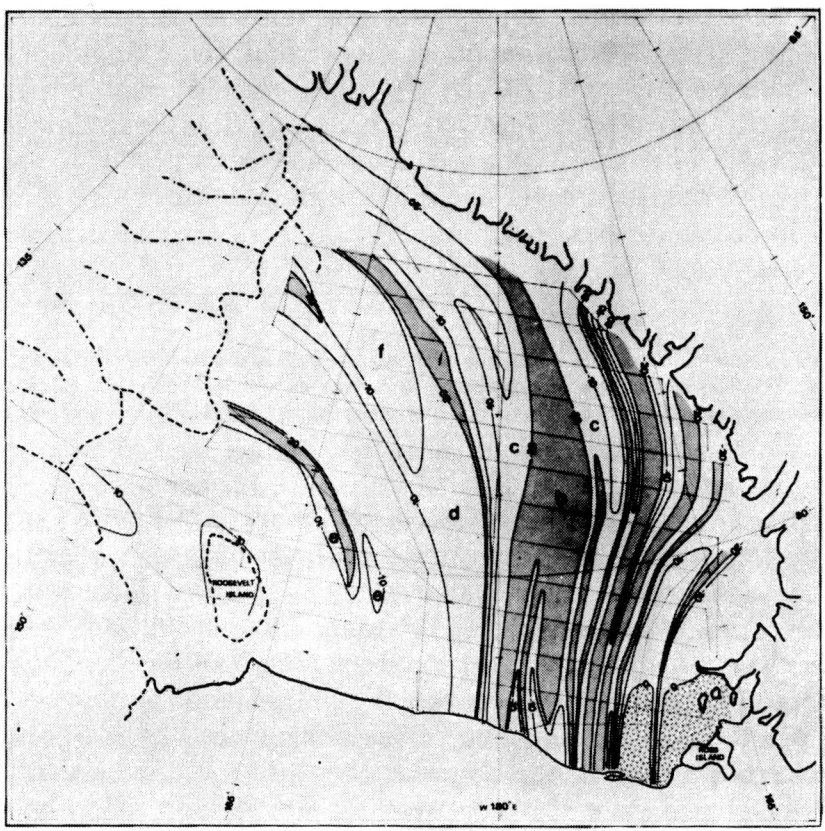

Fig. 23. Reflection coefficient of the ice/water interface, after Figure 4 of Neal (1979). Shaded regions indicate where low-radio wave reflectivity from the ice/water contact was observed. Neal (1979) proposed that these bands of low reflectivity represent deposits of relict basal sea ice that have been swept downstream (with respect to the ice shelf flow) from confined zones of basal freezing. Some of these stripes originate in areas of strong tidal rectification. Perhaps the best example is the one labeled "e" that originates at the tidal rectification site west of Roosevelt Island and northwest of Steershead Ice Rise. Basal freezing caused by tidally driven flow toward shallow ice shelf draft is a possible cause of this and other stripes. The abundance of the observed stripes near the midwestern part of the ice front and the absence elsewhere along the ice front support the basal melting pattern associated with tidally driven cross ice front transport displayed in Figure 22.

As means for ventilating the greater portion of the sub-ice cavity, tidal rectification is very inefficient. The simulation revealed that 100-200 years are required to completely flush the cavity. Geochemical observations suggest that this time scale is an order of magnitude too high, and that thermohaline processes must inevitably ventilate the bulk of the sub-ice cavity.

To further evaluate the possible role of tidal rectification in the Ross Sea, long term field measurements of the currents at the ice front site northwest of Roosevelt Island are recommended. If these measurements indicate the presence of a barotropic westward flow along the ice front that decays to near zero within 10-20 km of the ice front, then further efforts should be made to simulate tidal rectification numerically. Refined simulations should encompass a much finer grid resolution than was possible in this study to avoid the mismatch between grid scale and off-ice-front decay scale of the rectified current.

Acknowledgments. I gratefully acknowledge the encouragement and editorial assistance of S. Jacobs. The computations were performed at the Geophysical Fluid Dynamics Laboratory of Princeton University. Advice and technical assistance were provided by Kirk Bryan and Phil Tunison. Glenda York is appreciated for her editorial assistance.

References

Bentley, C. R., J. W. Clough, K. C. Jezek, and S. Shabtaie, Ice-thickness patterns and the dynamics of the Ross Ice Shelf, Antarctica, J. Glaciol., 24(90), 287-294, 1979.

Bruchhausen, P. M., J. A. Raymond, S. S. Jacobs, A. L. DeVries, E. M. Thorndike, and H. H. DeWitt, Fish, crustaceans, and the sea floor under the Ross Ice Shelf, Science, 203, 449-451, 1979.

Clough, J. W., and B. L. Hansen, The Ross Ice Shelf Project, Science, 203(4379), 433-434, 1979.

Crary, A. P., Glaciological studies at Little America Station, Antarctica, 1957 and 1958, IGY Glaciological Rep. no. 5, 197 pp., Am. Geogr. Soc., New York, 1961a.

Crary, A. P., Glaciological regime at Little America Station, Antarctica, J. Geophys. Res. 66(3), 871-878, 1961b.

Doake, C. S. M., Thermodynamics of the interaction between ice shelves and the sea, Polar Record, 18(112), 37-41, 1976.

Fujino, K., E. L. Lewis, and R. G. Perkin, The freezing point of seawater at pressures up to 100 bars, J. Geophys. Res., 79(12), 1792-1797, 1974.

Gordon, A., RISP oceanographic observations, in RISP Science Plan, pp. 41-58, University of Nebraska, Lincoln, 1974.

Greischar, L. L., and C. R. Bentley, Isostatic equilibrium grounding line between the West Antarctic inland ice sheet and the Ross Ice Shelf, Nature, 283, 651-654, 1980.

Hayes, D. E., and F. J. Davey, A geophysical study of the Ross Sea, Antarctica, Initial Rep. Deep Sea Drill. Proj., 28, 263-278, 1974.

Heath, R. A., Circulation across the ice shelf edge in McMurdo Sound, Antarctica, in Polar Oceans, edited by M. J. Dunbar, pp. 129-149, Arctic Institute of North America, Calgary, Alba., 1977.

Huthnance, J. M., Tidal current asymmetries over the Norfolk Sandbanks, Estuarine Coastal Mar. Sci., 1, 89-99, 1973.

Jacobs, S. S., and W. E. Haines, Oceanographic data in the Ross Sea and along the George V Coast, 1976-1979, Ross Ice Shelf Proj. Tech. Rep. LDGO-82-1, Lamont-Doherty Geol. Obs., Palisades, N.Y., 1982.

Jacobs, S. S., A. F. Amos, and P. M. Bruchhausen, Ross Sea oceanography and Antarctic Bottom Water Formation, Deep Sea Res., 17, 935-962, 1970.

Jacobs, S. S., A. L. Gordon, and J. L. Ardai, Circulation and melting beneath the Ross Ice Shelf, Science, 203, 439-442, 1979.

Jacobs, S. S., R. G. Fairbanks, and Y. Horibe, Origin and Evolution of water masses near the Antarctic continental margin: Evidence from $H_2^{18}O/H_2^{16}O$ ratios in seawater, this volume.

Jezek, K. C., Recent changes in the dynamic condition of the Ross Ice Shelf, Antarctica, J. Geophys. Res., 89(B1), 409-416, 1984.

Jezek, K. C., and C. R. Bentley, A reconsideration of the mass balance of a portion of the Ross Ice Shelf, Antarctica, J. Glaciol., 30(106), 381-384, 1984.

Lipps, J. H., T. E. Ronan, and T. E. DeLaca, Life below the Ross Ice Shelf, Antarctica, Science, 203(4379), 447-449, 1979.

Loder, J. W., Topographic rectification of tidal currents on the sides of Georges Bank, J. Phys. Oceanogr., 10(9), 1399-1416, 1980.

Longuet-Higgins, M. S., On the transport of mass by time-varying ocean currents, Deep Sea Res., 16, 431-447, 1969.

MacAyeal, D. R., Tidal-current rectification and tidal mixing fronts: Controls on the Ross Ice Shelf flow and mass balance, Ph.D. dissertation, 287 pp., Princeton University, June 1983.

MacAyeal, D. R., Thermohaline circulation below the Ross Ice Shelf: A consequence of tidally induced vertical mixing and basal melting, J. Geophys. Res., 89(C1), 607-615, 1984a.

MacAyeal, D. R., Numerical simulations of the Ross Sea Tides, J. Geophys. Res., 89(C1), 597-606, 1984b.

MacAyeal, D. R., and R. H. Thomas, Ross Ice Shelf temperatures support a history of ice-shelf thickening, Nature, 282(5740), 703-705, 1979.

MacDonald, W. J. P., and T. Hatherton, Movement of the Ross Ice Shelf near Scott Base, J. Glaciol., 3(29), 859-866, 1961.

Michel, R. L., T. W. Linick, and P. M. Williams, Tritium and Carbon-14 distributions in sea water from under the Ross Ice Shelf Project ice hole, Science, 203(4379), 445-446, 1979.

Neal, C. S., The dynamics of the Ross Ice Shelf revealed by radio echo-sounding, J. Glaciol., 24(90), 295-319, 1979.

Nihoul, J. C. J., Hydrodynamic models, Modelling of Marine Systems, edited by J. C. J. Nihoul, pp. 41-67, Elsevier Oceanography Ser., No. 10, Elsevier, New York, 1975.

Pedlosky, J., Geophysical Fluid Dynamics, 624 pp., Springer-Verlag, New York, 1979.

Pillsbury, R. D. and S. S. Jacobs, Preliminary results from long-term current meter moorings near the Ross Ice Shelf, this volume.

Ramming, H.G., and Z. Kowalik, Numerical Modelling of Marine Hydrodynamics, Applications to Dynamic Physical Processes, 368 pp., Elsevier Oceanography Ser., No. 26, Elsevier, New York, 1980.

Robin, G. de Q., Ice shelves and ice flow, Nature, 253(5488), 168-171, 1975.

Robin, G. de Q., Formation, flow and disintegration of ice shelves, J. Glaciol., 24(90), 259-272, 1979.

Robinson, I. S., Tidal vorticity and residual

circulation, Deep Sea Res., 28(3), 195-212, 1981.

Schey, H. M., Div. Grad. Curl and All That, 163 pp., W. W. Norton, New York, 1973.

Schwiderski, E. W., On charting global ocean tides, Rev. Geophys. Space Phys., 18(1), 243-268, 1980.

Shabtaie, S., and C. R. Bentley, Investigation of bottom mass-balance rates by electrical resistivity soundings on the Ross Ice Shelf, Antarctica, J. Glaciol., 24(90), 331-344, 1979.

Thomas, R. H., D. R. MacAyeal, D. H. Eilers, and D. R. Gaylord, Glaciological studies on the Ross Ice Shelf, Antarctica, 1973-1978, in The Ross Ice Shelf: Glaciology and Geophysics, edited by C. R. Bentley and D. E. Hayes, Antarct. Res. Ser., vol. 42, pp. 21-53, AGU, Washington, D.C., 1984.

UNESCO, Freezing point of sea water, Eighth report of the Joint Panel of Oceanographic Tables and Standards, Appendix 6, UNESCO Tech. Pap. Mar. Sci., 28, 29-35, 1978.

Wexler, H., Heating and melting of floating ice shelves, J. Glaciol., 3(27), 626-645, 1960.

Williams, R. T., and E. S. Robinson, The ocean tide in the southern Ross Sea, J. Geophys. Res., 85(C11), 6689-6696, 1980.

Williams, R. T., and E. S. Robinson, Flexural waves in the Ross Ice Shelf, J. Geophys. Res., 86(C7), 6643-6648, 1981.

Zimmerman, J. T. F., On the Euler-Lagrange transformation and the Stokes' drift in the presence of oscillatory and residual currents, Deep Sea Res., 26(A), 505-520, 1979.

Zimmerman, J. T. F., Vorticity transfer by tidal currents over an irregular topography, J. Mar. Res., 38(4), 601-630, 1980.

Zimmerman, J. T. F., Dynamics, diffusion and geomorphological significance of tidal residual eddies, Nature, 290, 549-555, 1981.

Zotikov, I. A., V. S. Zagorodnov, and J. V. Raikovsky, Core drilling though the Ross Ice Shelf confirms basal freezing, Science, 207, 1463-1465, 1980.

(Received October 28, 1983; accepted August 22, 1984).

EVOLUTION OF TIDALLY TRIGGERED MELTWATER PLUMES BELOW ICE SHELVES

Douglas R. MacAyeal

Department of the Geophysical Sciences, University of Chicago, Chicago, Illinois 60637

Abstract. Theory suggests that tidally induced vertical mixing and tidal rectification may trigger basal melting in two widely separated regions of the sub-ice cavity in the Ross Sea. Vertical separation of two meltwater masses observed off the Ross Ice Shelf by others reaffirms this suggestion and provides geochemical evidence useful for testing models of sub-ice shelf meltwater plume evolution. A simple model of this sort is used here to examine the idealized evolution of two meltwater plumes originating at 1,000-m depth and at 250-m depth. Results indicate that melting along the plume path driven by turbulent entrainment of ambient seawater strongly controls the net vertical penetration of the plume as it flows along the sloping ice shelf base. Entrainment-driven melting along the plume path is possible under present climatic conditions, but at depths greater than approximately 550 m. Such melting may be possible at all depths, however, if climatic change were to warm the ambient water column by approximately 0.6°C.

Introduction

Jacobs et al. [this volume] observe an isotopic and hydrographic distinction between two meltwater masses in the open Ross Sea and, as a result, have proposed a dual circulation regime below the ice shelf. Salinity contrast between these two masses constrains their vertical separation within the ambient water column. This vertical distinction has prompted Jacobs et al. to refer conveniently to the two masses as "Deep Ice Shelf Water" (DISW) and "Shallow Ice Shelf Water" (SISW).

The DISW mass is observed between 350 m and 500 m along the central and western portions of the Ross ice front (See Fig. 6c of Jacobs et al. [this volume]). Judging from the observed distribution of ice shelf draft [Bentley, et al., 1979] (also shown in MacAyeal [1984a]), DISW is likely to be in contact with the ice-shelf base along the remote southeastern region of the sub-ice cavity. The circulation regime proposed by Jacobs et al. [this volume] leading to DISW production must, therefore, embody deep penetration into the sub-ice cavity without direct influence on the basal melting regime of the thin portion of ice shelf near the ice front. The hydrographic and geochemical composition of DISW measured by Jacobs et al. [this volume] confirms this concept of deep thermohaline circulation in that DISW has the properties expected from a mixture of pure glacial meltwater and High Salinity Shelf Water (HSSW), the predominant water type at the seafloor on the Ross Sea continental shelf.

The SISW mass, in contrast to DISW, is observed at, or just below, the sea surface (see Fig. 6c of Jacobs et al. [this volume]). The ice shelf draft patterns suggest that SISW has been in contact with thin ice near the ice front. Jacobs et al. [this volume] suggest that SISW results predominantly from the influx from the continental slope region of a distinct core of relatively warm water (WMCO) to the central part of the Ross Ice Shelf Barrier.

The numerical tidal simulations discussed in companion papers [MacAyeal, 1984a; MacAyeal, this volume] identify two tidal processes that may generate the two meltwater circulation schemes envisioned by Jacobs et al. [this volume]. The companion study presented in this volume suggests that WMCO influx is driven, in part, by tidal rectification along the ice front. Tidal rectification may thus factor into SISW production. The other companion study [MacAyeal, 1984a] suggests that tidally driven vertical mixing erodes density stratification sufficiently to catalyze basal melting below thick ice shelf in the extreme southeastern part of the sub-ice cavity. Tidally driven mixing may thus factor into DISW production.

As an initial test of the hypothetical link between the simulated tidal processes and the observed meltwaters, the evolution of two idealized meltwater plumes are calculated to see how closely they match the observed meltwater properties. A simple stream tube model [Smith, 1975; Killworth, 1977; Bo Pedersen, 1980; Melling and Lewis, 1982] is adapted here to simulate buoyancy driven flow along the sloping base of an idealized ice shelf. The objective is to determine the depth, salinity and oxygen-isotope depletion at the point where the flow breaks free of the ice shelf base and begins to interleave within the ambient stratified water column. It will be

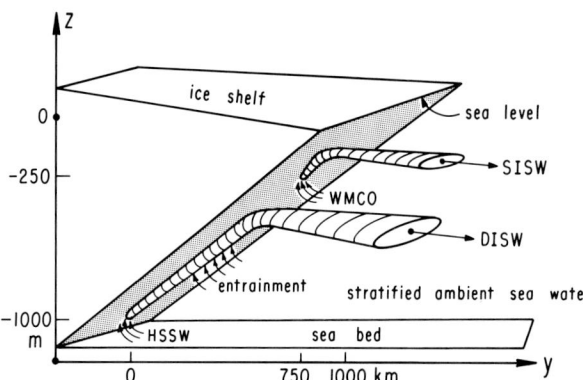

Fig. 1. DISW and SISW production was modeled using this idealized geometry. Meltwaters are assumed to rise along the sloping ice shelf base as buoyant plumes which separate from the ice at the level where their buoyancy is zero. In some simulations presented here, earth rotation introduces a component of plume velocity (not shown) directed along the contours of constant ice shelf draft. In other simulations, rotation effects are suppressed by assuming that a coast, or an inverted basal ice channel, directs the plume along the ice shelf draft gradient. The long plume representing DISW production from HSSW is initiated 1000 km from the ice front where the shelf is 1000 m thick. The short plume representing SISW production from WMCO is initiated 250 km from the ice front where the shelf thickness is 250 m.

assumed that the fluid properties at the point where the plume ceases to interact with the ice are directly comparable to the observed properties of meltwaters in the open Ross Sea.

Model Description

The stream tube model was developed previously to simulate dense water drainage off polar continental shelves into the abyssal ocean basins [Killworth, 1977; Melling and Lewis, 1982]. Here it is adapted to simulate the drainage of positively buoyant meltwater out of an idealized sub-ice shelf cavity. Figure 1 shows schematically how the stream tube model is envisaged in this application. The term "stream tube" refers to fluid flowing within an idealized conduit of stream lines that separate the plume from its surroundings. Properties within the tube are assumed uniform, steady in time, and are characterized by a velocity magnitude u (m s^{-1}), a flow direction angle β (radians) measured (counterclockwise) with respect to the ice shelf draft gradient, a buoyancy Δ (m s^{-2}) a salinity S_p in parts per thousand, a plume thickness h (m), an oxygen isotope concentration c (o/oo), and a cross sectional width w (km). The temperature of the plume is assumed to be in constant equilibrium with the ice at the freezing point θ_f (°C) associated with the local hydrostatic pressure and plume salinity [Fujino et al., 1974; see also Millero, 1978]:

$$\theta_f = -0.036 - 0.0499\, S_p - 0.000112\, S_p^2 + 7.59 \times 10^{-8} \int_0^z \rho g\, dz \qquad (1)$$

where z is the vertical coordinate (zero at sea level and negative downwards), and ρ (kg m^{-3}) is the ambient seawater density. The buoyancy of the plume Δ (m s^{-2}) is assumed uniform transverse to the direction of flow and is given by

$$\Delta = \frac{g}{\rho}(\rho - \rho_p) \qquad (2)$$

where ρ_p (kg m^{-3}) is the plume density. Both the ambient density and plume density are taken as functions of salinity only and are given by

$$\rho = 1000.0 + \beta_0\, S(z) \qquad (3)$$

and

$$\rho_p = 1000.0 + \beta_0\, S_p \qquad (4)$$

where $S(z)$ is the salinity of the ambient water (taken to represent typical Ross Sea conditions) and $\beta_0 = 8.0 \times 10^{-1}$ kg m^{-3}.

Ambient seawater through which the plume ascends is idealized as having uniform density stratification determined by a vertical salinity gradient. The salinity profile taken to represent typical conditions in the Ross Sea is given by

$$S(z) = 34.5 - 6 \times 10^{-4} z \qquad (5)$$

Simplicity of the model requires that the salinity gradient be constant, thus the idealized profile used here departs considerably from that observed [Jacobs et al., 1979]. Nevertheless, this profile represents conditions expected in a deep 1000-m water column composed entirely of HSSW. The Brunt-Väisälä frequency N [Gill, 1982] is constant at 2.17×10^{-3} s^{-1}.

The ambient temperature is assumed to be -1.87°C, the approximate sea surface freezing temperature corrected for the local salinity. This temperature is also based on the assumption that vigorous winter sea ice production and associated vertical convection maintains the sea-surface freezing temperature at all depths. No correction is made here to account for the presence of colder ice shelf meltwater in the ambient water column. The WMCO that penetrates below the Ross Ice Shelf from the continental shelf break [Jacobs et al., this volume] is not explicitly placed within the idealized water column. The WMCO mass is treated implicitly, however, as the initiator of an idealized SISW plume by assuming initial SISW plume conditions equivalent to those produced when all sensible heat flux contained in the WMCO is used to melt ice at 250 m depth. The WMCO mass flux into the sub ice cavity is assumed to be 5×10^3 m^3 s^{-1}. This assumed flux is consistent with the tidal rectification volume fluxes determined by MacAyeal [this volume]. Recent field measurements of currents [Pillsbury and Jacobs, this volume] suggest, however, that the actual WMCO flux may be several times larger.

For simplicity, all basal melting resulting from contact between ambient ocean and the ice where there is no intervening plume is disregarded. This procedure is justified by the likely presence of a neutrally buoyant and insulating basal water film at the freezing point wherever motion does not induce vertical mixing [Gill, 1973].

The ice shelf is idealized as having a constant basal slope $\alpha = 10^{-3}$ in the horizontal y-direction, and is assumed uniform in the horizontal x direction. The grounding line is taken at y = 0, and the ice-shelf base is assumed to intersect the sea surface with a wedgelike ice front at y = 1000 km. Heat conduction upwards through the ice shelf is idealized by assuming a simple linear temperature-depth profile in the ice having the following vertical gradient

$$\frac{\partial \theta_i}{\partial z} = \frac{\theta_s - \theta_f}{H} \quad (6)$$

where $\theta_s = -26°C$ is the assumed average atmospheric temperature [see Thomas et al., 1984], θ_f is the basal freezing temperature (equation (1)) adjusted for pressure and salinity, θ_i is the ice temperature, and H is the local ice shelf thickness calculated from the local ice shelf draft by assuming a constant ice density of 917.0 kg m^{-3}. The treatment of upward heat conduction through the ice shelf is intended to allow the plume buoyancy to change according to the local ablation rates determined by the heat balance at the ice-water contact. In the present analysis, feedback between strong basal melting and increased vertical heat flux in the ice shelf is disregarded. This feedback is likely to reduce basal melting rates whenever the plume flow direction and ice flow direction coincide.

The oxygen-isotope ratio (as defined by Jacobs et al. [this volume]) of the ice shelf c_i is assumed to be constant at a value of -42 o/oo. This value is taken from the observed concentration in the basal ice layer immediately above basal sea ice at the J-9 core hole [Grootes and Stuiver, 1982]. In actuality, oxygen isotope ratios greater (less negative) than this assumed value are expected downstream of basal melting areas where the highly oxygen-18 depleted "Wisconsin-age" ice is ablated. For the present analysis, the simple assumption of a constant ratio is sufficient to determine the evolution of the oxygen-isotope ratio of the plume.

The equations used to simulate plume evolution are taken from Killworth [1977] with several alterations. These equations specify the conservation of mass, momentum, buoyancy, and oxygen-isotope concentration, respectively, and are written

$$\frac{\partial}{\partial \lambda} (Au) - Ewu - \dot{B} = 0 \quad (7)$$

$$A \Delta \sin \alpha \cos \beta - w K u^2 = 0 \quad (8)$$

$$\frac{\Delta}{u^2} \sin \alpha \sin \beta - f/u = 0 \quad (9)$$

$$\frac{\partial}{\partial \lambda} (Au\Delta) + N^2 \sin \alpha \cos \beta \, Au - \Delta_m \dot{B} = 0 \quad (10)$$

and

$$\frac{\partial}{\partial \lambda} (Auc) - \dot{B}c_i He(\dot{B}) - \dot{B}cHe(-\dot{B}) = 0 \quad (11)$$

where E is the nondimensional entrainment parameter, $\dot{B}$ (m^2 s^{-1}) is the basal meltwater production rate per unit length of the stream tube, A = wh is the plume cross section area, K is the nondimensional friction parameter, f = -1.43×10^{-4} s^{-1} is the Coriolis parameter (constant at 80°S), Δ_m is the buoyancy of salt-free meltwater or of the water being deposited as sea ice at a salinity of 5.0, He($\dot{B}$) is the Heaviside function (=0 if $\dot{B}<0$, =1 if $\dot{B}>0$) and λ is the curvilinear coordinate directed along the plume path. For simplicity, the plume width w and thickness h are assumed to have a fixed ratio of 5.0×10^3. This ratio is based on the observed extent of DISW at the ice front [Jacobs et al., 1979], but does not adequately describe SISW [Jacobs et al., this volume] which appears along the entire ice front. The model results are highly sensitive to the parameters E and K. Their values are thus chosen to define a range of sensitivity and to maintain consistency with previous work by Killworth (1977) and Melling and Lewis (1982). A full discussion of these parameters is provided under "Modeling Strategy."

The horizontal coordinates x and y are related to λ according to

$$\frac{\partial x}{\partial \lambda} = -\sin \beta \cos \alpha \quad (12)$$

and

$$\frac{\partial y}{\partial \lambda} = \cos \beta \cos \alpha \quad (13)$$

where β is measured counterclockwise from the y axis and x is positive to the right with respect to the direction of positive y.

The basal meltwater production rate per unit length of the plume $\dot{B}$ (m^3 s^{-1} m^{-1}) is de-

TABLE 1. Plume Simulations

Run	Eo	K	$\dot{B}$	z_c, m	$x(z_c)$, km	$S_p(z_c)$	$\theta_f(z_c)$, °C	$c(z_c)$, o/oo	$w(z_c)$, km	$h(z_c)$, m
					DISW Analog					
1	0.072	0.025	suppressed	-829	0	35.00	-2.568	-0.064	124	25
2	0.072	0.025	calculated	-212	0	34.63	-2.061	-0.203	410	82
3	0.036	0.025	calculated	-262	0	34.66	-2.161	-0.230	233	47
4	0.018	0.025	calculated	-330	0	34.70	-2.210	-0.258	145	29
a	0.072	0.01	calculated	-816	2996	34.94	-2.541	-0.176	112	22
b	0.072	0.1	calculated	-658	954	34.89	-2.417	-0.240	175	35
					SISW Analog					
5	0.072	0.025	suppressed	-54	0	34.53	-1.934	-0.078	120	24
6	0.072	0.025	calculated	-27	0	34.52	-1.912	-0.091	124	25
7	0.036	0.025	calculated	-5	0	34.50	-1.895	-0.141	80	16
8	0.018	0.025	calculated	0	0	34.50	-1.891	-0.219	50	10
c	0.072	0.01	calculated	-119	1405	34.55	-1.985	-0.086	202	40
d	0.072	0.1	calculated	0	270	34.32	-1.880	-0.400	55	11

z_c defined to be level where Δ becomes less than 1.0×10^{-5}.

termined by the entrainment rate of warm ambient water and by the vertical heat flux through the ice. The equation for $\dot{B}$ is

$$\dot{B} = \frac{c_o}{L_o} Ewu (\theta_{(z)} - \theta_f) - \frac{k_i}{L_o \rho} w \frac{(\theta_s - \theta_f)}{H} \quad (14)$$

where $\theta_{(z)}$ is the temperature of the entrained water, $c_o = 4000$ J kg^{-1}°C^{-1} is the heat capacity of sea water at 0°C, $L_o = 3.35 \times 10^5$ J kg^{-1} is the latent heat of melting, and $k_i = 2.4$ W m^{-1} is the conductivity of ice at -12°C. It is important to realize that the basal melting rate given by equation (14) applies only to the area in contact with the plume. Elsewhere, $\dot{B}$ is assumed zero.

The buoyancy of the meltwater Δ_m is given by

$$\Delta_m = \frac{g}{\rho} (\rho - 1000.0) \text{ if } \dot{B} > 0$$

$$\Delta_m = \frac{g}{\rho} (\rho - 1004.0) \text{ if } \dot{B} < 0 \quad (15)$$

where ρ is the ambient density, the density of salt-free meltwater is assumed to be 1000 kg m^{-3}, and the density of water being frozen as basal ice is assumed to be that corresponding to a salinity of 5.0.

The model equations are solved numerically using an Adams-Moulton predictor/corrector scheme given by Young and Gregory [1972]:

$$G_{n+1} = G_n + \frac{\Delta\lambda}{24} [9 F(G_{n+1}^P) + 19 F(G_n) - 5 F(G_{n-1}) + F(G_{n-2})] \quad (16)$$

where

$$G_{n+1}^P = G_n + \frac{\Delta\lambda}{24} [55F(G_n) - 59F(G_{n-1}) + 37F(G_{n-2}) - 9F(G_{n-3})] \quad (17)$$

G is the solution of

$$\frac{\partial G}{\partial \lambda} = F(G) \quad (18)$$

$\Delta\lambda$ is the step-size, and F is a "forcing" function depending on G and other parameters. The first three steps are taken using a Euler forward scheme [Young and Gregory, 1972] to initialize the predictor/corrector sequence.

The model integration is halted when the plume breaks free of the ice-shelf base and begins to interleave within the stratified ambient water column. This criterion is taken to be that point, (y_c, z_c) at which the plume buoyancy Δ is so low that the equations for u and β are no longer valid because inertial terms become comparable to buoyancy terms. This critical plume buoyancy is approximately 10^{-4} m s^{-2}. The results are not sensitive to this choice.

Modeling Strategy

The stream-tube model used here to provide insight into buoyancy-driven flow along the ice shelf base is a highly simplified representation of nature. In addition, the parameters chosen to represent the complex physical processes such as friction and entrainment represent conjectures based on values used in other contexts [e.g., Killworth, 1977; Melling and Lewis, 1982]. Without more detailed observation of the sub-ice regime, modeling studies such as this can at best merely identify logical processes to be examined elsewhere through observation or through more sophisticated modeling techniques.

The objectives of the present modeling exercise must be chosen with the above difficulties in mind. Rather than attempt to tune model parameters to better match observations the objectives here are to explore the range of parameter sensitivity and to identify qualitative rules governing sub-ice plume evolution. Specific tasks to be accomplished here are: (1) to display graphs showing the evolution of S, C, and B as functions of λ, (2) to determine the range of depths at which DISW and SISW analogs break free of the ice shelf and interleave within the ambient water column, and (3) to examine the effects of the earth's rotation. To meet these objectives, numerous simulations were undertaken. A description of these simulations is presented in **Table 1**. Differences among them fall into four categories: (1) initial conditions, (2) parameter values of E and K, (3) artificial suppression of basal melting, and (4) artificial suppression of earth rotation effects.

Two sets of initial conditions are formulated so that both DISW and SISW may be simulated. For DISW production, tidal mixing in the remote portions of the cavity is assumed to cause an initial 5×10^3 m^3 s^{-1} flux of ambient water (idealized HSSW) to equilibrate with the ice shelf at 1000-m depth. This initial flux and that used for SISW production below are arbitrary and based on the assumption that the plume originates as a small flux which subsequently builds through entrainment. The salinity, buoyancy, oxygen-isotope concentration, and temperature of the initial DISW flux are calculated from conservation equations and have values of 34.79, 2.437×10^{-3} m s^{-2}, -0.37 ‰ and -2.76°C, respectively.

For SISW production, tidal rectification near the ice front is assumed to cause an initial 5×10^3 m^3 s^{-1} flux of an idealized WMCO having temperature of -1.0°C and salinity 34.65 to equilibrate with the ice shelf at 250-m depth. The idealized salinity is assigned to WMCO so that it would reside at a depth of 250 m in the idealized water column. In reality, the salinity of WMCO is less than 34.5, but this salinity does not occur at any level in the model's ambient water column. The initial salinity, buoyancy, oxygen-isotope concentration, and temperature of this equilibrated flux are 34.21, 3.45×10^{-3} m s^{-2}, -0.53 ‰, and -2.08°C, respectively. The DISW plume begins at y = 0 and z = -1000 m, and the SISW plume begins at y = 750 km and z = -250 m. These parameters are consistent with the spatial dimensions and velocity scales observed in the Ross Sea.

The nondimensional entrainment parameter E is inversely proportional to the bulk Richardson number and is given by the formula developed by Bo Pedersen [1980] and by Melling and Lewis [1982]:

$$E = E_o K/(h\Delta u^{-2}) \qquad (19)$$

where $E_o = 0.072$. This parameterization allows the entrainment rate to go up when the stability of the interface separating the plume from the environment goes down. This interfacial stability is measured by the bulk Richardson number $h\Delta u^{-2}$. To test sensitivity, the model is run using values of E_o reduced by 1/2 and by 1/4.

The nondimensional friction parameter K is varied in order to evaluate model sensitivity. For simulations in which earth rotation is disregarded, $K = 2.5 \times 10^{-3}$. This choice is consistent with the value used previously to parameterize friction in shallow water tidal models [MacAyeal, 1984b], and falls within the range applicable to coastal circulation regimes summarized by Ramming and Kowalik [1980]. For simulations in which earth rotation deflects the stream tube, K is assigned one of two possible values, 10^{-1} or 10^{-2}. The higher values of K used when rotation effects are embraced are consistent with the values used in studies by Killworth [1977] (who used 1.5×10^{-1}) and Melling and Lewis [1982] (who used 1.0×10^{-2}). These high values are motivated in this study, and in those previous, by the desire to reduce the deflection angle β. Without such a reduction (e.g., leaving K at 2.5×10^{-3}) the stream tube does not exit the sub-ice region without extending too far ($\sim 10^6$ km) along the ice shelf draft contours. The need for high K to reduce β implies that earth rotation effects are counterbalanced by effects associated with coasts, inverted channels in the base of the ice shelf, or with baroclinic instability (meandering). Such effects are not embraced here.

Two regimes are examined in the present study. In one regime, friction and buoyancy balance, and the Coriolis force is assumed balanced by an unspecified force such as a pressure acting on a coast or on the side of an inverted channel in the base of the ice shelf. Optical leveling surveys show ice shelf surface topography indicative of such channels in the southeastern corner of the

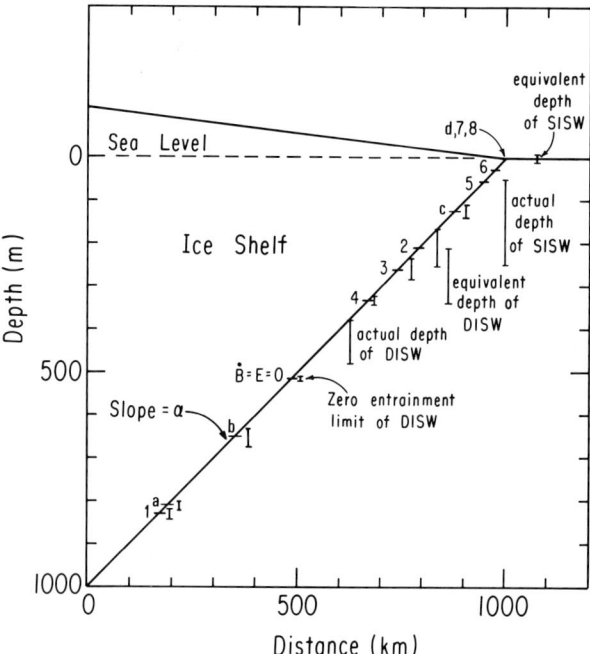

Fig. 2. Levels of plume separation for each model run and the analytic solutions along with the observed depths and equivalent depths of DISW and SISW. The equivalent depth is defined as the depth at which a water mass with its observed temperature and salinity would reside in the idealized ambient water column used in this study. The number and letter identification scheme is summarized in Table 1. In brief, numbered results differ from lettered results by the suppression of earth rotation effects. The results show that rotation effects tend to diminish the vertical penetration of the plume by suppressing entrainment-driven basal melting.

Ross Ice Shelf (near 83°58'S, 160°00'W) [R. Bindschadler, personal communication, 1984]. These channels, however, extend only a limited distance to the north, and are thought to be relict features advected into the ice shelf from grounded ice streams. They may, nevertheless, host meltwater plumes. In this circumstance, β is set to zero, the Richardson number becomes approximately constant at $K/(\sin\alpha \cos\beta)$, and the entrainment parameter E is approximately constant at $E_o \sin\alpha$.

The other regime represents a geostrophic balance between the Coriolis force and buoyancy. This regime is expected to result in predominantly slow flows in which basal ice topography and coasts exert little influence on the deflection of the stream tube. The Richardson number for this regime is increased by a factor of $\cos^{-1}\beta$ where

$$\beta = \tan^{-1}\left(\frac{fh}{ku}\right)$$

and the entrainment parameter is reduced to approximately 10% of the value derived above for the regime in which buoyancy and friction balance.

Analytic Results

Before discussing the numerical solutions, it is instructive to derive an analytic solution valid for zero entrainment and a quasi-static momentum balance between buoyancy and friction (earth rotation disregarded). The buoyancy and velocity derived from equations (10) and (8) are written as

$$\Delta = \Delta(y = y_o) - N^2 \sin\alpha\, y \qquad (20)$$

and

$$u = \left(h\,\frac{\sin\alpha}{k}\,(\Delta(y=y_o) - N^2 \sin\alpha\, y)\right)^{1/2} \qquad (21)$$

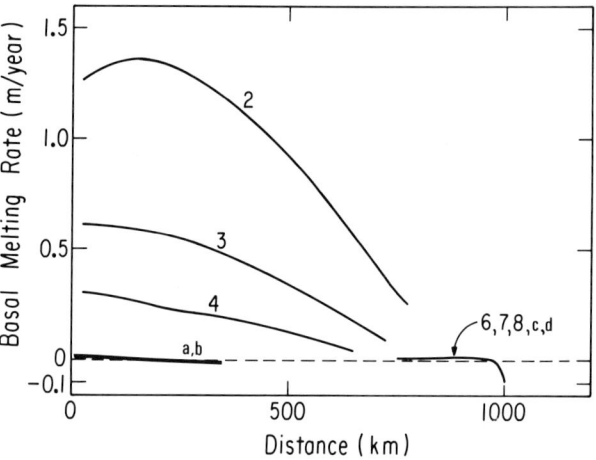

Fig. 3. Melting rates on sections of the ice shelf base with which the plumes are in contact, plotted as functions of y. These ice shelf sections are of the order of 100 km wide. DISW plumes 2, 3, and 4 cause the greatest melting because of their high entrainment rates and the greater temperature difference between the ambient water entrained and the in situ freezing point. These runs demonstrate the effectiveness of entrainment-driven basal melting. The effect of the earth's rotation, however, is to suppress basal melting because the geostrophic balance implies slower flow and less entrainment. The SISW plumes show little basal melting as a result of lower temperature difference between the entrained water and basal ice. High heat flux through the thin idealized ice shelf in contact with the SISW plume also suppresses melting. This suggests that the initial loss of sensible heat from the WMCO dominates near-ice-front melting.

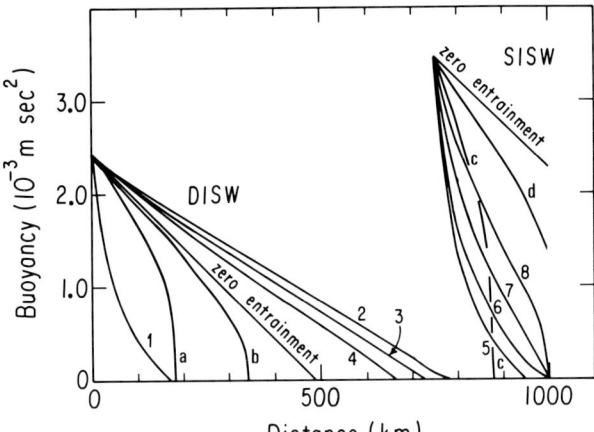

Fig. 4. Plume buoyancy plotted as a function of y. Buoyancy is maintained above the "zero-entrainment" curve for DISW plumes in which earth rotation effects are suppressed. This results from entrainment-driven basal melting. Rotation effects reduce buoyancy below the "zero-entrainment" curve for DISW and SISW plumes. This is because geostrophic balance implies a reduced entrainment rate. Curve a, in which rotation effects buoyancy most strongly, approaches curve 1, which represents the effect of artificial suppression of entrainment-driven basal melting.

where y_o is the initial value of y or the position of the plume source. From this solution the depths at which the DISW and SISW plume analogs break free of the ice shelf base are

$$z_c = \frac{\Delta(y = y_o)}{N^2} = -516 \quad (DISW)$$

$$z_c = 0 \quad (SISW) \qquad (22)$$

The DISW plume rises along the ice shelf base until it reaches the level where its salinity is equal to the ambient salinity. The SISW plume, in contrast, rises until it reaches the sea surface. This is because the SISW plume has an initial salinity lower than that of the ambient water column at any level. These critical depths are displayed in Figure 2. The velocities along the plume paths are plotted in Figure 3.

Break Free Depths

The numerical results shown in Table 1 and Figure 2 indicate that all the DISW plume analogs interleave within the ambient water column at a considerable depth. No reasonable combination of parameters was found to allow the DISW plume to reach the sea surface. The deepest break free depth (z_c) was attained by run 1 for which rotation and basal melting were suppressed, and for which the entrainment had it greatest value. The shallowest break free depth was attained by run 2 for which rotation, but not basal melting, was suppressed. Runs 2 through 4 indicate that increasing entrainment reduces the break free depth. This relationship occurs because greater entrainment incorporates more heat into the plume which subsequently produces more meltwater. Figures 3 and 4 show the basal melting rate (B) and plume buoyancy (Δ), as functions of y, respectively. Entrainment-driven melting is demonstrated by these figures to reduce DISW buoyancy decay, therefore allowing greater vertical penetration.

The effect of earth rotation on the DISW plume is to deepen the break free depth below the zero-entrainment limit determined analytically and shown in Figure 2. As seen in Figure 5, the geostrophic balance of runs a and b (letters denote runs with rotation effects) suppresses the plume velocities. This suppression tends to reduce the entrainment rate which subsequently reduces the basal melting rate (Figure 3). The buoyancy decay with increasing y tends to fall below the zero-entrainment limit, however, because the plume path is not directly along the y axis.

In comparing the break free depth predicted by the model with the observed depth of DISW along the ice front [Jacobs, et al., this volume] it is useful to define an "equivalent observed depth." This definition allows the observed depth to be corrected for the fact that the salinity (density) stratification of the

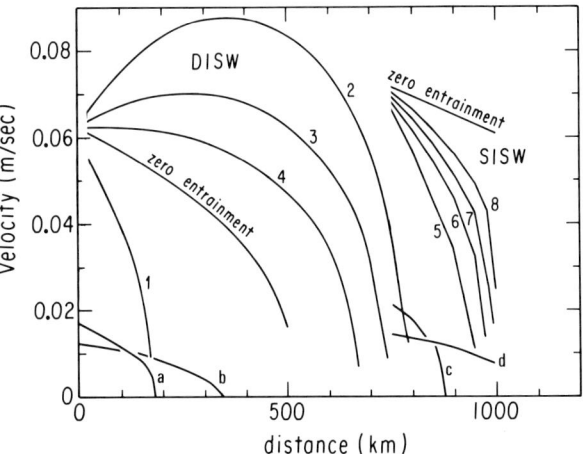

Fig. 5. The plume velocities are plotted as functions of y. Numbered plumes (earth rotation suppressed) accelerate rapidly downstream of their sources to reach a velocity characterized by balance between friction and buoyancy. Lettered plumes move more slowly as is characterized by the geostrophic balance. Velocity maxima associated with plumes 2 and 3 are caused by the effect of plume thickness.

natural water column is not uniform. The observed salinity of DISW is between 34.63 and 34.73; thus if the actual DISW mass were to reside in the idealized sea, it would occupy the depth range between 217 m and 383 m. This falls precisely within the range of break free depths spanned by the DISW plume experiments in which earth rotation effects are ignored. Possible implications of this comparison are: (1) the natural DISW plume is channeled by a coastline, by the ice-shelf topography or by horizontally nonuniform stratification; (2) the natural DISW plume originates much higher in the water column, closer to the depth at which it interleaves within the open ocean.

In contrast with the simulated DISW plume, the simulated SISW plume generally emerges at the sea surface unless entrainment is enhanced by a high value of the entrainment parameter and by suppression of earth rotation effects. Also unlike the DISW plume, sensible heat entrained into the SISW plume is negligible because shallow depth implies that the ambient water column temperature is close to the basal ice temperature. Without entrainment-driven melting along the plume path (Figure 3), entrainment simply forces a more rapid plume buoyancy decay (Figure 4).

The equivalent depth for SISW is sea level in the idealized sea rather than at depth [Jacobs et al., this volume]; thus the tendency for the modeled SISW plume to reach the sea surface before breaking free of the ice is consistent with observation. This agreement may be fortuitous, however, because the substantial idealizations of WMCO influx and ambient temperature may emphasize incorrect natural processes. The SISW experiments demonstrate that basal melting near the ice front is not likely to be sustained by entrained sensible heat from HSSW. The results here confirm that sensible heat flux associated with inflowing WMCO and with seasonal upper-ocean warming is the dominant factor for near-ice-front melting.

Entrainment-Driven Melting

The numerical simulations may be organized into two categories besides those of DISW and SISW production depending on whether the entrainment helps or hinders the maintenance of plume buoyancy. All the SISW simulations, and the DISW simulations in which rotation effects are embraced, show that entrainment increases the buoyancy decay rate above the "zero-entrainment limit" determined analytically (Figure 4). The DISW plume simulations represented by runs 1, 2, and 3 display the opposite effect as a result of entrainment-driven melting. This result suggests that a critical depth exists below which entrainment-driven melting is able to counterbalance the otherwise buoyancy reducing effects of entrainment.

To estimate this critical depth, equation (10) is simplified by assuming that the change of buoyancy flux with λ is mostly due to changing buoyancy rather than changing volume flux ($\Delta \frac{\partial (Au)}{\partial \lambda} \ll Au \frac{\partial \Delta}{\partial \lambda}$):

$$\frac{\partial \Delta}{\partial \lambda} = \frac{1}{Au} \{\dot{B} \Delta_m - (Ewu)\Delta\} - N^2 \sin\alpha \cos\beta \quad (23)$$

The last term on the right hand side of equation (23) expresses the zero-entrainment limit of buoyancy decay that depends entirely on the vertical motion of the plume with respect to the density stratified environment. Depending on whether the first term on the right hand side of equaton (23) is positive or negative, entrainment will either diminish or enhance plume buoyancy decay. If the basal melting rate given by equation (14) is substituted into equation (23) and the vertical heat flux through the ice is disregarded, the difference $\theta(z) - \theta_f$ determines whether $(B\Delta_m - (Ewu)\Delta)$ is positive or negative. The criterion for whether entrainment helps or hinders buoyancy decay is thus

$$\theta_{(z)} - \theta_f \geqslant \frac{\Delta}{\Delta_m} \frac{L_o}{C_o} \quad (24)$$

This criterion can be simplified and expressed in terms of depth by assuming $\theta_{(z)} \cong 1.87°C$, using a representative value $\Delta/\Delta_m \cong 5 \times 10^{-3}$, and by adopting in place of equation (1)

$$\theta_f \cong -1.87 + 7.59 \times 10^{-3} m^{-1} z \quad (25)$$

This alternative criterion is roughly

$$z \leqslant -550 \ m \quad (26)$$

where the value 550 m may be changed depending on the value chosen to represent the ratio Δ/Δ_m.

The implication of this criterion is that only DISW plumes are susceptible to buoyancy enhancement by entrainment-driven melting. The SISW plumes flow through an environment where sensible heat entrainment and melting is too low to counterbalance the additional density attained when plume and environment mix. If climatic change were to warm the ambient temperatures above approximately -1.3°C, the criterion expressed by equation (24) would be met at all depths. SISW plumes would thus induce entrainment-driven melting, and would become more prominent in their effect on the ice shelf mass balance.

Evolution of Plume Geochemistry

Jacobs et al. [this volume] associate meltwaters observed in the open Ross Sea with parent water masses by examining the effects of

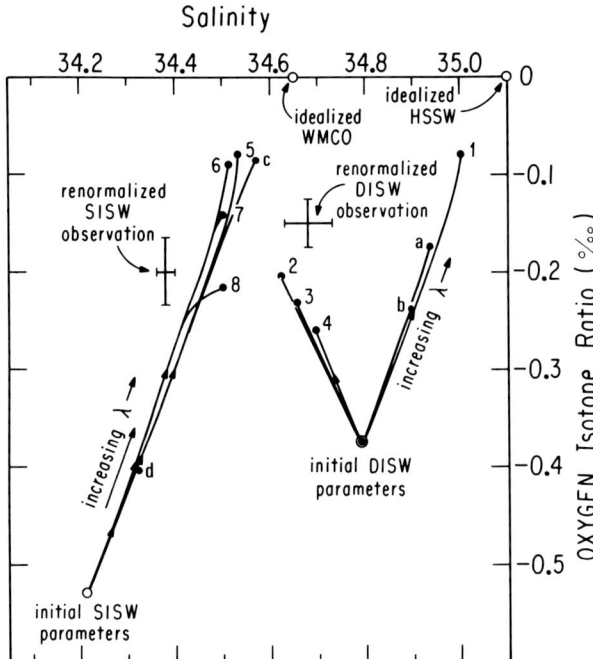

Fig. 6. Plume evolutions expressed as parametric curves in S_p versus c space. Without rotational effects, the DISW plumes reduce their salinity as their oxygen-isotope concentrations increase (become less negative) because entrainment tends to elicit further melting. SISW plumes and DISW plumes effected by earth rotation evolve in the opposite sense because, without strong melting, entrainment always increases their salinity. End points of the parametric curves are compared with the observed S_p and c of DISW and SISW normalized to account for the adjusted S_p and c values assigned to HSSW and WMCO. This renormalization of the observations is required to account for the assumption that the ambient water column has a constant vertical salinity gradient. Without this renormalization the model results and data cannot be directly compared. Best agreement with "observation" is attained for SISW runs with moderate entrainment.

melting on the salinity and oxygen-isotope composition. The diagram of salinity versus oxygen-isotope ratio of the plume shown in Figure 6 indicates the simulated geochemical evolution of the two meltwater masses as they travel from their respective points of origin to where they interleave into the open ocean. For both DISW and SISW plumes, the oxygen-isotope ratios are expressed relative to the initial ratios of the respective WMCO and HSSW input water masses. These input water massses are assumed to have an oxygen-isotope ratio of 0 o/oo rather than the slightly negative values given by Jacobs et al. [this volume]. This renormalization, or recalibration, must be accounted for when comparing Figure 6 with its counterpart in Jacobs et al. [this volume].

For DISW, best agreement between the simulated plume properties at the break free depth and the renormalized observation is for the run having largest entrainment-driven basal melting (run 2). By suppressing this form of melting, earth rotation effects tend to increase the plume salinity along the plume path. Otherwise, meltwater dilution causes the plume salinity to decrease along the plume path.

Entrainment-driven melting is negligible for the simulated SISW plumes; thus plume salinity increases along the plume path for all runs. Best agreement between simulation and renormalized observation is achieved for runs in which entrainment is suppressed by earth rotation or by a small value of the entrainment parameter.

Ice Shelf Mass Balance

The entrainment-sustained basal melting described above has several serious drawbacks which presumably can be corrected in a more realistic study. The most apparent defect is that basal melting is probably too high for the DISW plume and probably too low for the SISW plume. This judgment is based on the tentative perception of the overall mass budget of the Ross Ice Shelf presented by MacAyeal and Thomas [1984]. Modeled and observed ice shelf flow patterns suggest that two areas of melting are required to maintain a steady state ice thickness distribution on the Ross Ice Shelf. These areas are: (1) between the Crary Ice Rise and Siple Coast, and (2) within 100 km of the ice front (especially near the location of the warm core). The modeled DISW plumes, in contrast, elicit a basal melting pattern that persists along the entire plume trajectory. Furthermore, the modeled SISW plumes produce appreciable melting only at their points of origin.

This apparent model defect may, in future studies, be corrected by: (1) allowing feedback between the basal melting rates and the internal ice shelf temperatures, (2) adjusting the temperature and salinity stratifiation of the ambient seawater to conform with the observed hydrography, and (3) allowing the plume to determine, in concert with other factors, the stratification of the ambient water column. The first correction would intensify the vertical heat flux at the ice-ocean contact and halt or diminish basal melting downstream of the plume origin [MacAyeal, 1979]. This effect could account for the apparent lack of melting associated with the possible DISW plume in the central part of the Ross Sea cavity. The second correction would allow greater basal melting along the ice front, where ambient water in summer [Pillsbury and Jacobs, this volume] has a temperature of -1.25°C, or warmer, rather than the colder temperature assumed in the model. The third correction would diminish basal melting associated with the DISW plume because recirculation of meltwater back into the sub-ice cavity would diminish the heat available to be entrained. Additional factors to be addressed in an improved study of meltwater plume evolution are: instabilities, coastal boundary effects, variable ice shelf thickness gradients, variable ice shelf oxygen-isotope concentration, and time-dependent ambient stratification.

Conclusion

Buoyant meltwater plumes flowing along the base of large Antarctic ice shelves interleave within the surrounding ocean at depths determined by the ice shelf draft at the plume source and the degree of entrainment experienced along the plume trajectory. Model studies presented here suggest that DISW plumes, or plumes originating where the ice shelf draft is greater than approximately 500 m, tend to penetrate vertically farther than do SISW plumes. This result is a consequence of basal ice shelf melting forced by the sensible heat flux accompanying plume entrainment. The geostrophic balance tends to mitigate this effect because plume velocity and entrainment rates are reduced when the Coriolis force, rather than friction, balances the buoyancy force.

The capacity for ventilating a typical sub-ice shelf cavity is strongest for the most vigorous DISW plumes in which rotation effects are suppressed and entrainment parameters are set high. Assuming a return flow of HSSW, the outgoing DISW plume represented by run 2 (Table 1), for example, would flush the entire cavity below the Ross Ice Shelf in approximately 2 years. If averaged over the entire ice shelf area, the ablation rate driven by plume entrainment would be approximately 0.2 m/yr.

Such vigorous sub-ice shelf flushing by a DISW plume is not anticipated in the Ross Sea, however, becasue earth rotation effects are likely to suppress necesssary entrainment-driven melting. The rotational effects referred to here would be cancelled by the presence of coast or inverted "channels" in the ice-shelf base. That may explain the relatively vigorous flushing observed by Potter and Paren [this volume] below the George VI Ice Shelf. Comparison between the simulated evolution of plume geochemistry and the hydrographic properties of DISW observed by Jacobs et al. [this volume] suggests that the real DISW plume may originate where the Ross Ice Shelf has a draft of approximately 600 m rather than 1000 m as was assumed in this study. Some areas in the southeastern corner of the ice shelf (near 87°37'S, 166°00'W) have an ice thickness of approximately 600 m and additionally display ice flow convergence requiring greater than 0.5 m/yr melting to maintain steady state [R. Bindschadler, personal communication, 1984].

In contrast to the DISW plumes, SISW plumes caused by WMCO impingement on an ice shelf base at 250 m depth do not have the capacity for entrainment-driven melting. As a result, the geochemical evolution of SISW represents a simple mixture of plume with ambient water. The net basal ablation associated with SISW production is, furthermore, entirely associated with the plume source where sensible heat from the WMCO is lost. Entrainment-driven basal melting similar to that associated with possible DISW plumes could be sustained, however, if the ambient water through which the SISW plume flows had a temperature greater than approximately -1.3°C. Temperatures this warm near the surface of the Ross Sea are not characteristic of the present climatic regime; but could be expected if CO_2 warming were to occur.

Acknowledgments. I sincerely appreciate the scientific advice and editorial assistance of S. Jacobs, several anonymous reviewers, and G. York. This work was supported by a grant from the National Science Foundation (DPP 84-01016).

References

Bentley, C. R., J. W. Clough, K. C. Jezek, and S. Shabtaie, Ice-thickness patterns and the dynamics of the Ross Ice Shelf, Antarctica, J. Glaciol., 24(90), 287-294, 1979.

Bo Pedersen, F., Dense bottom currents in a rotating ocean, J. Hydraul. Div. Am. Soc. Civ. Eng., 106(HY8), 1291-1308, 1980.

Fujino, K., E. L. Lewis, and R. G. Perkin, The freezing point of seawater at pressures up to 100 bars, J. Geophys. Res., 79(12), 1792-1797, 1974.

Gill, A.E., Circulation and bottom water production in the Weddell Sea, Deep Sea Res., 20, 111-140, 1973.

Gill, A.E., Atmosphere-Ocean Dynamics, 662 pp., Academic Press, New York., 1982.

Grootes, P.M., and M. Stuiver, Ross Ice Shelf and Dome C oxygen-isotope analysis, Antarct. J. U.S., 17(5), 76-78, 1982.

Jacobs, S.S., A.L. Gordon, and J.L. Ardai, Jr., Circulation and melting beneath the Ross Ice Shelf, Science, 203(4379), 441-443, 1979.

Jacobs, S.S., R.G. Fairbanks, and Y. Horibe, Origin and evolution of water masses near the Antarctic continental margin: Evidence from $H_2^{18}O/H_2^{16}O$ ratios in seawater, this volume.

Killworth, P.D., Mixing on the Weddell Sea continental slope, Deep Sea Res., 24, 427-448, 1977.

MacAyeal, D.R., Transient Temperature-Depth Profiles of the Ross Ice Shelf, M.Sc. thesis, 116 pp., Univ. of Maine at Orono, May 1979.

MacAyeal, D.R., Thermohaline circulation below the Ross Ice Shelf: A consequence of tidally induced vertical mixing and basal melting, J. Geophys. Res., 89(C1), 597-606, 1984a.

MacAyeal, D.R., Numerical simulations of the Ross Sea tides, J. Geophys. Res., 89(C1), 607-615, 1984b.

MacAyeal, D.R., Tidal rectification below the Ross Ice Shelf, Antarctica, this volume.

MacAyeal, D.R., and R. H. Thomas, The effects of basal melting on the present flow of the Ross Ice Shelf, J. Glaciol., in press, 1984.

Melling, H., and E.L. Lewis, Shelf drainage flows in the Beaufort Sea and their effect on the Arctic Ocean pycnocline, Deep Sea Res., 29(8A), 967-985, 1982.

Millero, F.J., Freezing point of sea water, Eighth Report of the Joint Panel of Oceanographic Tables and Standards, Appendix 6, UNESCO Tech. Pap. Mar. Sci., 28, 29-35,

Pillsbury, R.D. and S.S. Jacobs, Preliminary observations from long-term current meter moorings near the Ross Ice Shelf, Antarctica, this volume.

Potter, J.R., and J.G. Paren, Interaction between ice shelf and ocean in George VI Sound, Antarctica, this volume.

Ramming, H.G., and Z. Kowalik, Numerical Modeling of Marine Hydrodynamics, Applications to Dynamic Physical Processes, Elsevier Oceanography Series No. 26, Elsevier/North Holland, New York, 368 pp., 1980.

Smith, P.C., A stream-tube model for bottom boundary currents in the ocean, Deep Sea Res., 22, 853-873, 1975.

Thomas, H.G., D.R. MacAyeal, D.H. Eilers, and D.R. Gaylord, Glaciological studies on the Ross Ice Shelf, Antarctica, 1973-1978, in The Ross Ice Shelf: Glaciology and Geophysics, Ant. Res. Ser., vol. 40, edited by C.R. Bentley and D.E. Hayes, pp. 21-53, American Geophysical Union, Washington, D.C., 1984.

Young, D.M., and R. T. Gregory, A Survey of Numerical Mathematics, vol. 1, 492 pp., Addison-Wesley, Reading, Pa., 1972.

(Received October 18, 1984; accepted December 27, 1984.)

THE WINTER OCEANOGRAPHY OF MCMURDO SOUND, ANTARCTICA

E. L. Lewis and R. G. Perkin

Frozen Sea Research Group, Institute of Ocean Sciences
Sidney, British Columbia, Canada   V8L 4B2

Abstract. Analysis of current meter and conductivity/temperature/depth (CTD) data have given an overall picture of the winter circulation in McMurdo Sound. As has been shown in other studies, High Salinity Shelf Water enters from the Ross Sea along the east-central side of the sound as far as Cape Royds where it swings west then northwest to exit along the western side. The geostrophic currents relative to 700 dbar are in agreement with the record of a current meter deployed to the west of Cape Royds and indicate a large anticyclonic eddy at that location which produces upwelling and a northward moving current at the Cape. The High Salinity Shelf Water also occupies the deeper levels of McMurdo Sound but its circulation is thought to be sluggish compared to the surface layers. South of Cape Royds, the upper 200 m of the sound are heavily influenced by northward flowing, cold, low-salinity water. This water cannot result from local melting because its salinity is too low to be formed from High Salinity Shelf Water. It is advected from under the Ross Ice Shelf and exits McMurdo Sound on the extreme western side. Water coming from the eastern part of the ice shelf edge is caught up in a relatively complex flow, partially due to the blockage effect of the Erebus Glacier Tongue, and may recirculate under the ice shelf after relieving some of its supercooling in Erebus Bay and along the Hut Point Peninsula. Profiles showing extremely high supercooling near the ice/water interface gave indications of correspondingly high salt fluxes related to the relief of supercooling. Convection from these sites is capable of reaching depths of approximately 200 m, and a new water mass, produced by the relief of supercooling, is shown to exist and take part in mixing processes within the sound.

Introduction

The location of McMurdo Sound and the surrounding landmass is shown in Figure 1. The sound is best considered as a small inlet at the northwest corner of the Ross Ice Shelf bounded on the east by Ross Island. The Ross Ice Shelf, which is the dominant feature in local coastal oceanography, is of the order of 500 m thick and forms the floating "coast" of Antarctica for a distance of nearly 1000 km [Clough and Hansen, 1979]. Its western end is intercepted by Ross Island so that the southern boundary of McMurdo Sound consists of about 60 km of ice shelf. It has been generally assumed that subsurface circulation south of Ross Island is not blocked by the grounded portion of the ice sheet. The surface circulation in McMurdo Sound as studied by Heath [1977] is indicated by the hollow arrows in Figure 1.

The oceanography of the sound has attracted considerable attention over the years, largely due to its proximity to the major logistic centers at McMurdo Station and Scott Base. Previous studies appear to have been of two types: time series of oceanographic profiles made in the immediate vicinity of McMurdo Station by scientists staying for periods of a year or more [Tressler and Ommundsen, 1962; Littlepage, 1965] and spatial surveys using ships involved in summer supply voyages [Heath, 1971; Jacobs et al., 1981]. This last study was particularly directed towards the discussion of thermohaline steps discovered in the vicinity of the Erebus Glacier Tongue (EGT), Figure 1. The present investigation, conducted in October/November 1982, after the end of the winter season, consists of month-long current meter records and a quasi-synoptic survey using helicopter-mounted conductivity/temperature/depth (CTD) equipment.

Figure 2 is a schematic diagram of the distribution of the various water masses in the Ross Sea taking Jacobs et al. [1970] and Jacobs et al. [this volume] as the main sources of information. On the surface lies Antarctic Surface Water (AASW) with salinities less than 34.50 and a large seasonal variation in both salinity and temperature. This layer is usually about 200 m deep and is the water mass most affected by the freezing and melting cycle of the sea ice. Offshore and beneath the surface layer is Circumpolar Deep Water (CDW). This water, with salinity near 34.7 and temperature as high as 1.4°C, mixes with shelf water across a frontal zone at the shelf break to form an intermediate-depth warm layer (WMCO) with a temperature maximum between -1.0° and 0.0°C on the eastern shelf. On the

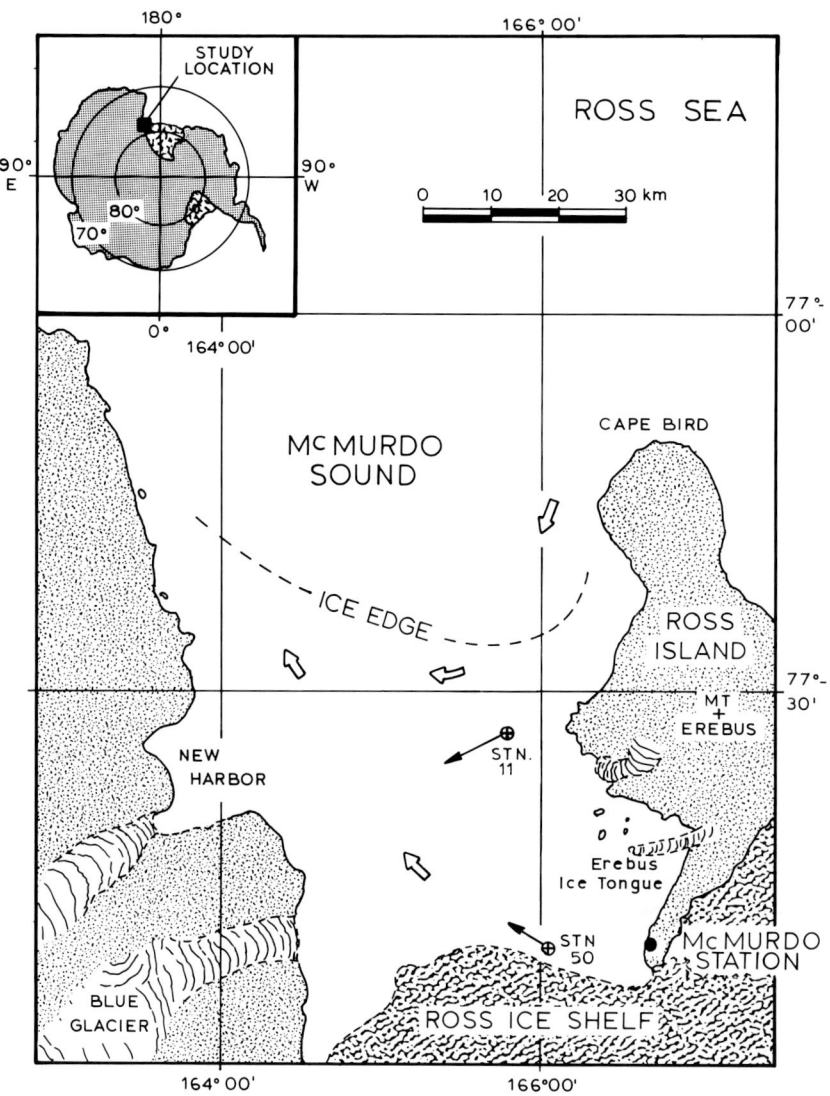

Fig. 1. The location of McMurdo Sound, an inlet at the western end of the Ross Ice Shelf, Antarctica. The hollow arrows indicate water movement as described by Heath [1977]. The vectors at the locations labeled STN 11 and 50 indicate the mean currents measured by us over a period of a month at a depth of 200 m.

continental shelf the circulation of these water masses is cyclonic (clockwise) with a resulting westward flow along the Ross Ice Shelf barrier. On the western side of the Ross Sea lies High Salinity Shelf Water (HSSW) with salinities greater than 34.75 and near the surface freezing temperature, the densest water to be found on the shelf. Leaking out of the northwest Ross Sea Shelf, HSSW is a source for one variety of Antarctic Bottom Water which is formed on the continental slope by mixing with CDW. HSSW is concentrated in the southwest Ross Sea, the location of McMurdo Sound, by a combination of trapping by sills to the north and by the dynamics of the shelf circulation.

Circulation of water beneath the ice shelf results in production of Ice Shelf Water (not shown on Figure 2) with temperatures frequently below -2°C [Jacobs et al., 1979]. Temperatures such as these come about through the pressure dependence of the freezing point which amounts to 0.000753°C/dbar [UNESCO, 1978]. They are substantially below the surface freezing temperature for water of the same salinity, and therefore signify contact with ice at depth - in this case the Ross Ice Shelf. The formation of this very cold water has been discussed in connection with profiles measured near the Filchner Ice Shelf by Foldvik and Kvinge [1974] and Foldvik et al. [this volume]. Any circulation which moves this

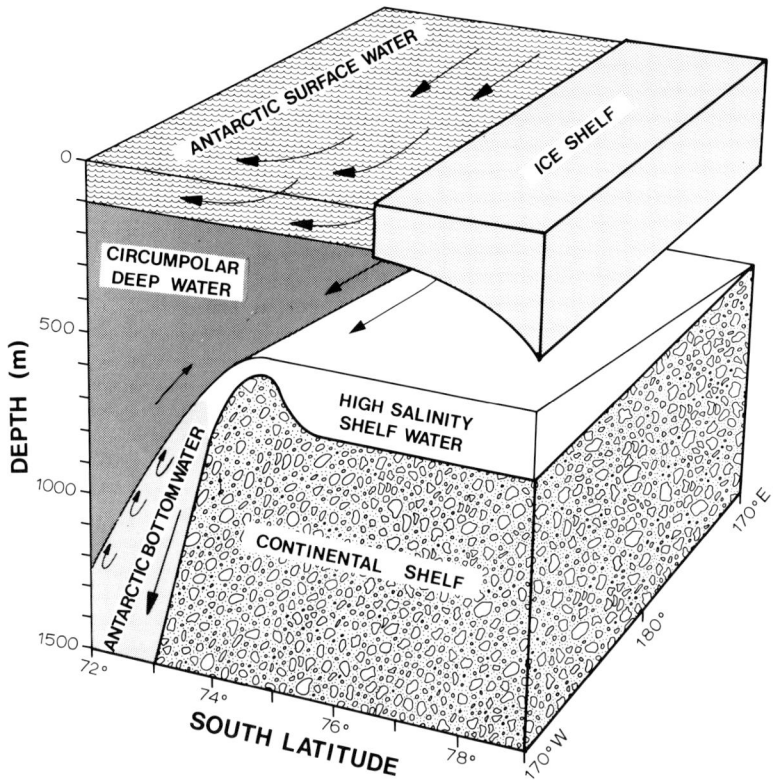

Fig. 2. Schematic diagram of the water masses of the western Ross Sea. High Salinity Shelf Water collects in the southwest regions and occupies most of the water column in McMurdo Sound (see also Jacobs et al. [this volume, Figures 6a and 6b]).

water upward to lower pressures causes it to become supercooled and is thus a potential heat sink for underwater ice formation.

All the above water mass parameters have been taken from summer measurements, and are subject to seasonal variations. The apparent seasonal change of salinity in HSSW is 0.05 to 0.15 and our measurements, taken after the end of winter, represent the highest salinities and lowest temperatures of the year. All the water detected in outer McMurdo Sound at this time falls within the HSSW classification, in contrast with the profiles in Jacobs et al. [1981] showing the presence of AASW and a subsurface warm feature in summer (February).

Observational Program

Data to be considered were obtained using Aanderaa RCM 4 current meters and a Guildline Type 8706 CTD system mounted in a U.S Navy UH-1N twin engine helicopter. The winch, generator, and associated instrument electronics were designed or modified to fit within the helicopter. Upon arrival at a site, a hole was drilled through the sea ice at a specific location with respect to the helicopter skids, a boom was extended and the CTD probe lowered into the sea directly. Our data have accuracies, estimated from daily calibration, of about ±0.01 in salinity, ±0.003°C in temperature, with precisions about half as great. All our stations are numbered on Figure 3 which also shows depth contours. A major concern was to obtain nearly synoptic profiles for the two sections labeled "east-west" and "north-south" in the figure and these were completed in 6 and 3 days, respectively. The first station was taken on October 26, 1982, and the last on November 19, 1982.

The horizontal component of the earth's magnetic field at McMurdo is just adequate to obtain a directional reference for the current meter but special modifications including a new vane design were made to allow the assembly to be put through a 25-cm-diameter hole in the sea ice sheet. Current meters were installed at 200-m depth for a total period of 32 days and the mean vectors representing currents at stations 11 and 50 are shown on Figure 1. Figure 3 gives a rough position of the sea ice edge near the beginning of our field operations (October 26) but it had come south and was much closer to the location of our east-west section stations (about 5 km north of Cape Royds) by the time of departure

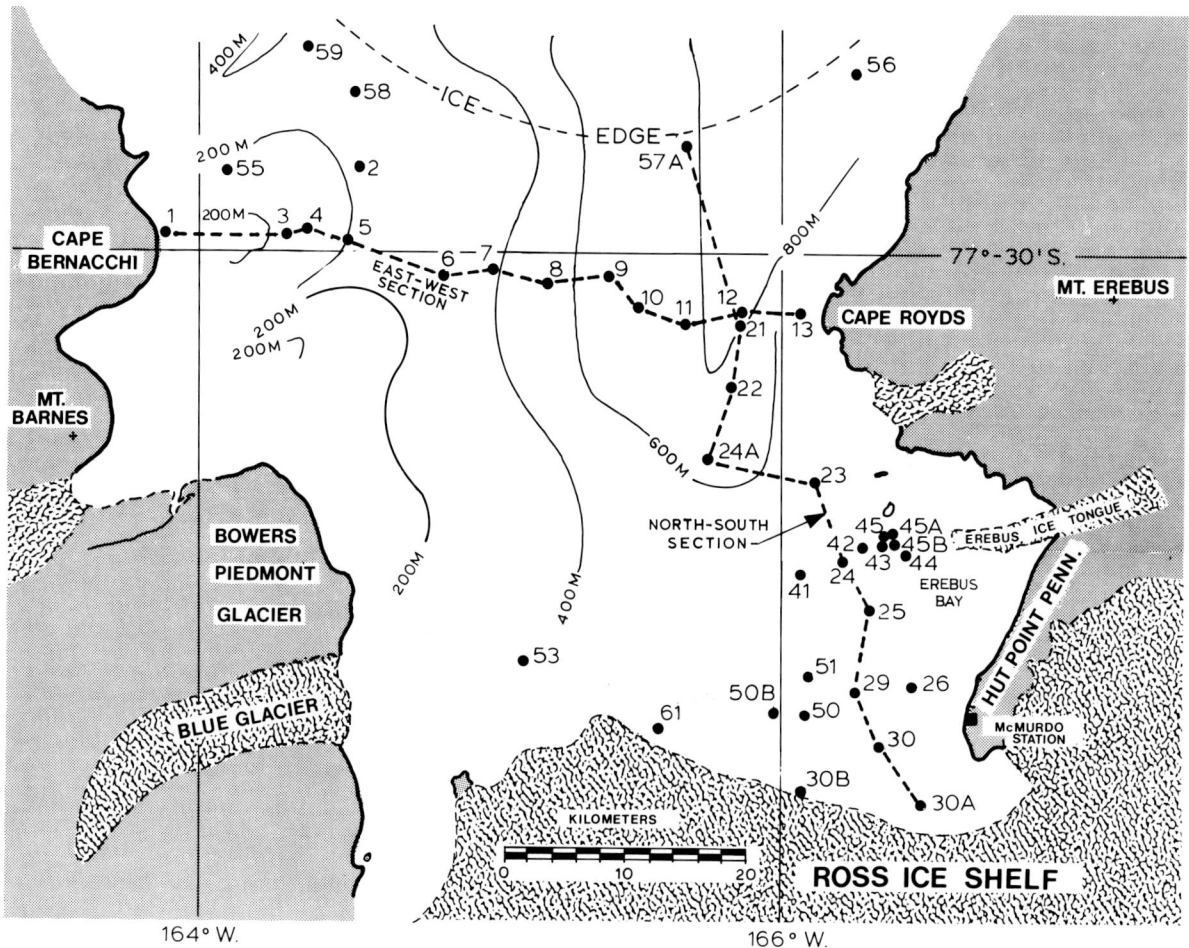

Fig. 3. The location of our oceanographic stations in McMurdo Sound. Also shown are depth contours and the location of the sea ice edge at the beginning of the survey (October 26, 1984). The exact juncture between the Ross Ice Shelf and the sea ice is difficult to discern because thinning of the edge of the shelf ice by melting results in a smooth transition into the contiguous sea ice. It is known to have moved in recent years and the indicated position is approximate.

(November 23). Satellite observations [Zwally et al., 1983, this volume], show that this reduction in the ice cover north of Ross Island is normal.

Thermohaline Properties of the Water

In the following discussion, salinities will be given without units, on the Practical Salinity Scale 1978 as recommended by UNESCO [1981]. Pressures, given in decibars (dbar) can be considered numerically equal to depth in meters within 1% accuracy and variations in density can be considered to follow variations in salinity since the entire range of measured temperatures affects density by an amount equal to the effect of only a 0.0013 change in salinity.

Figure 4 gives profiles of the temperature, potential temperature, and salinity at a station north of Cape Royds where water enters McMurdo Sound from the Ross Sea (station 56, Figure 3). These profiles show very little range in the variations of temperature and salinity compared with the larger range to be found in the summer, [cf. Jacobs et al., 1981]. However, there is a weakly stratified surface layer which extends down to about 50 m. From 50 m to 700 m, the salinity is almost constant, increasing by only 0.016 to a value of 34.794. Below 700 m there is a comparatively rapid salinity increase. Although the entire profile falls within the definition of High Salinity Shelf Water (HSSW), the layering suggests different histories within that classification. Jacobs et al. [this volume] dis-

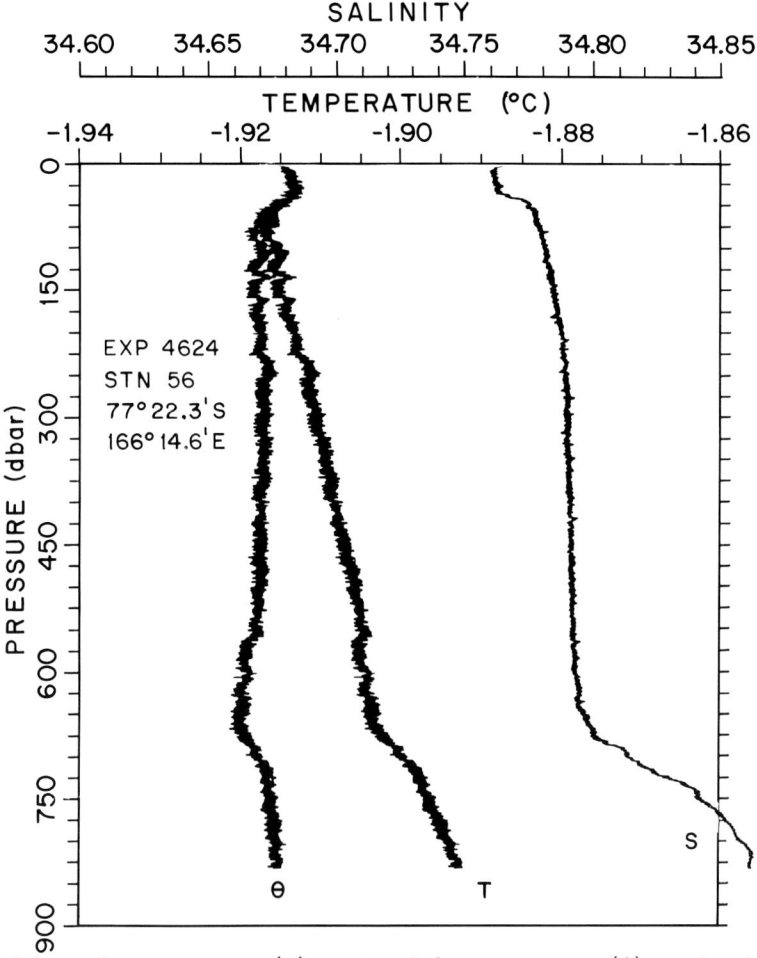

Fig. 4. Profiles of temperature (T), potential temperature (θ), and salinity (s) versus depth for station 56 (see Figure 3). These profiles are thought to be representative of water entering McMurdo Sound from the Ross Sea.

cuss the water masses of the Ross Sea in the context of the additional information provided by oxygen isotope ratios. Using their summer data, they found that HSSW could be derived from other surface waters by a combination of freezing and precipitation. They also identified another variety under the Ross Ice Shelf station J9, which could be derived from the HSSW by a combination of melting at the bottom of the ice shelf and freezing. The details of this process in the Ross Sea are beyond the scope of this paper, and the origin of the water below 700 m cannot be determined from our data. The potential temperature is notable for being almost constant over the whole profile and for being slightly below the surface freezing temperature at the observed salinity. This general supercooling suggests that the entire water column has been affected by contact with ice. The slightly lower potential temperatures near 600 m may result from the circulation of HSSW beneath the Ross Ice

Shelf where its temperature can be further depressed by contact with the ice at high pressure. The observed changes in these properties since the water entered McMurdo Sound will now be interpreted in terms of the circulation.

Contours of potential temperature, salinity, and sigma-t were drawn for the east-west section of stations and these are shown in Figures 5a, 5b and 5c, respectively. The essential constancy of the potential temperature over the greater part of the section is again noticeable. The sigma-t contours, closely following the salinity, give an indication of flow into and out of the sound. Taking the horizontal nature of the isopycnals below 700 m to indicate a level of no motion, we can anticipate an outflow immediately adjacent to Cape Royds and a corresponding inflow in the vicinity of station 10. Between stations 9 and 6, once again, an outflow is to be expected. The lower-salinity water on the

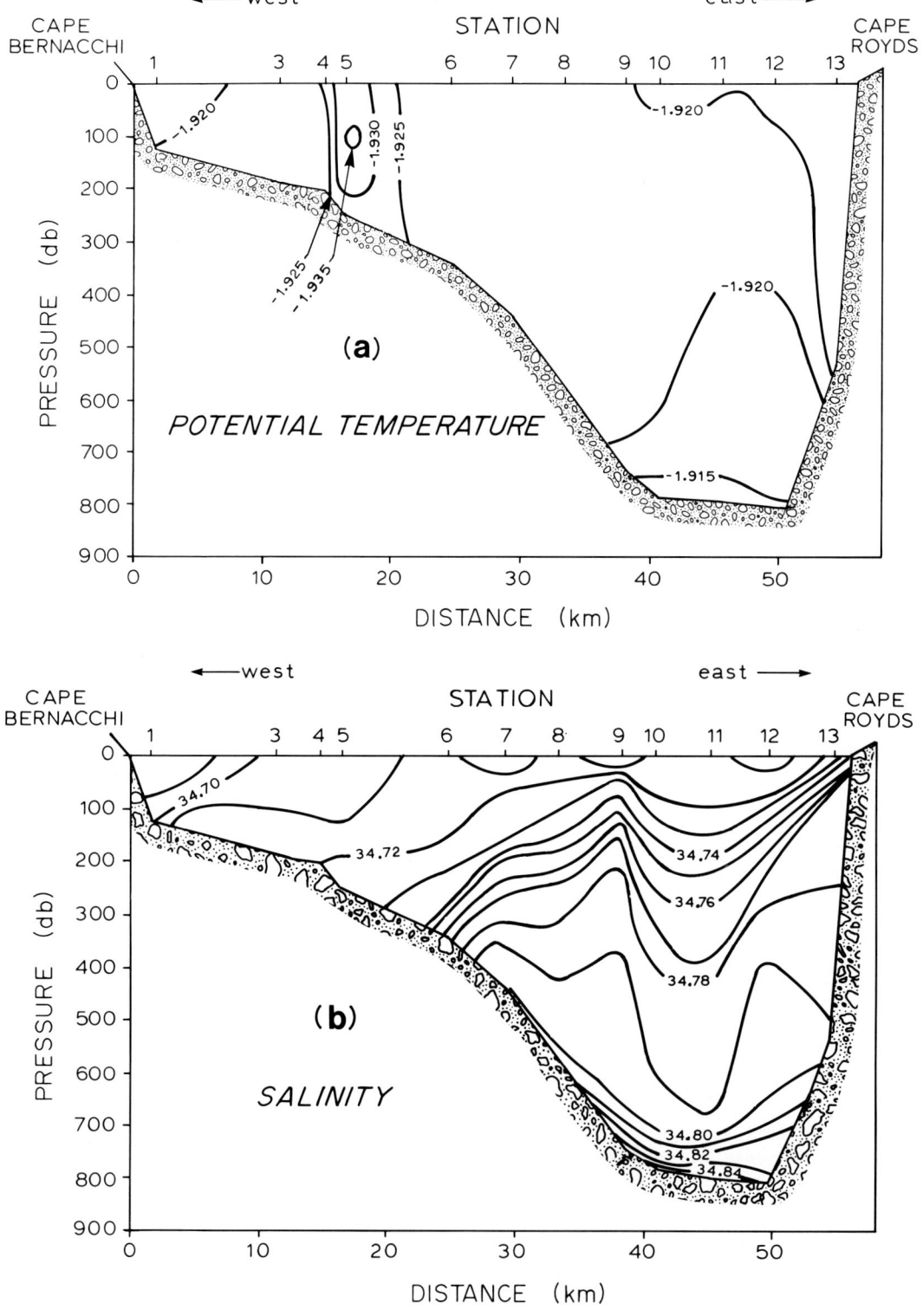

Fig. 5. Contours of (a) potential temperature, (b) salinity, and (c) sigma-t for our east-west section (see Figure 3). There is little variation in potential temperature in the vicinity of Cape Royds. The contours in sigma-t indicate alternating current flow centered around station 11.

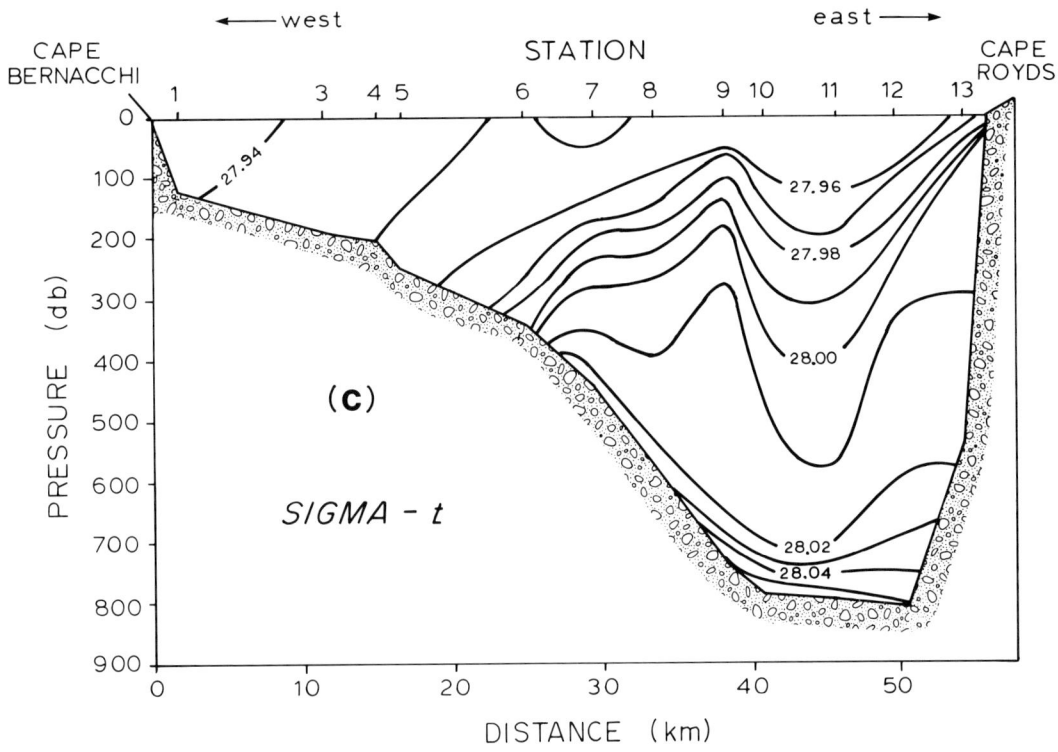

Fig. 5. (continued).

west side indicates a second source of water associated, at station 5, with temperatures considerably below the surface freezing temperature. These low temperatures indicate that water has been in contact with ice considerably below the sea surface and, from the implied circulation pattern, its source is in southern McMurdo Sound.

The internal Rossby radius obtained by consideration of the density profiles at 700 dbars off Cape Royds is estimated to be 4.5 km and Figure 5 shows that this horizontal scale is of the correct order for the observed variations. The station spacing should be adequate to capture the major features of the baroclinic (internal) adjustment.

Figure 6 gives the corresponding contours for the north-south section. The deeper waters of this section have the same potential temperature and salinity characteristics as the water north of Cape Royds and, except in the eastern basin near the EGT, the deep circulation is not expected to be strong. Station 24 appears to be anomalous due to its location significantly off the line joining stations 22 and 23. The data from these stations do fit into a general circulation pattern as will be seen below. In the shallower waters, a major feature of the temperature contours of the north-south section, Figure 6, is the cold intrusion proceeding from the front of the Ross Ice Shelf at a depth of about 100 m. A general decrease in salinity is noted as the shelf ice is approached from the north. Together with the aforementioned cold intrusion this may be attributable to the melting of ice at depth or possibly to the introduction of a new water mass coming from under the shelf ice through a connection south of Ross Island. This is shown more clearly in Figure 7 which shows the surface salinity pattern. Given a general outflow along the western coast, a source of low-salinity (34.70 to 34.72) water entering McMurdo Sound from under the Ross Ice Shelf is indicated.

These inferences regarding circulation are given greater weight when the degree of surface supercooling is plotted (Figure 8). As in the case of the low-salinity water, the source appears to be the ice shelf and the supercooled water appears to exit on the west side of the sound.

To calculate supercooling the UNESCO [1978] formula has been used with the pressure taken as atmospheric. This supercooling has resulted in many observations of underwater ice formation in McMurdo Sound [e.g., Dayton et al., 1969]. In the course of taking these soundings it was observed that large quantities of ice discs would well up in freshly augered holes in the sea ice. Often 5 minutes of continuous bailing was required before the instrument could be lowered without the possibility of being fouled. Divers making local

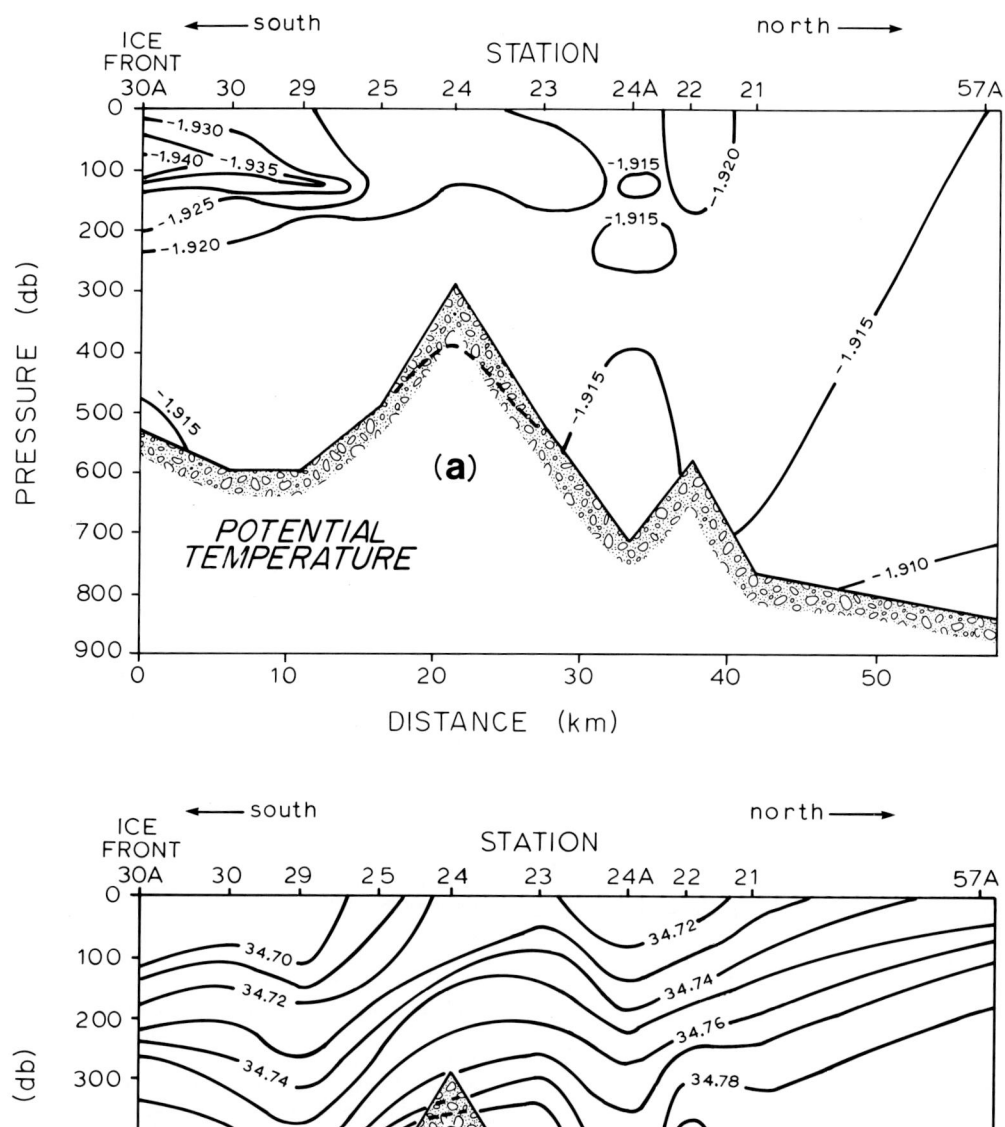

Fig. 6. Contours of (a) potential temperature, (b) salinity, and (c) sigma-t for our north-south section (see Figure 3). Of particular note is the intrusive cold layer coming from the ice front at a depth of about 100 m. The distinct dip in salinity and sigma-t contours at station 24A is an artifact of the position of that station, somewhat to the west of the other stations along this section.

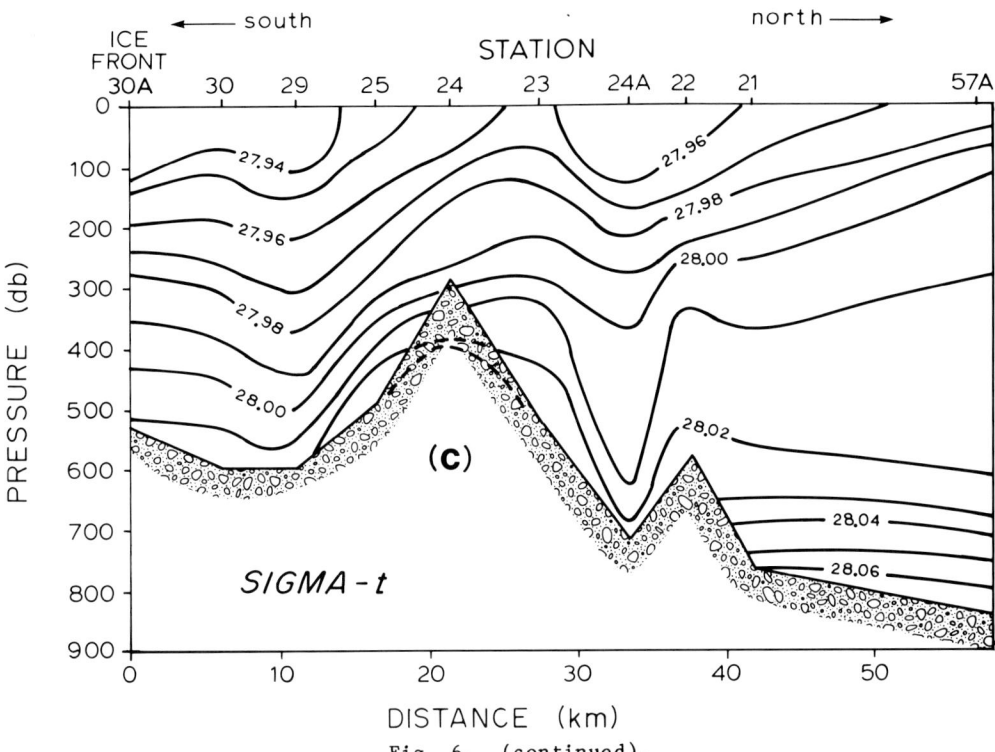

Fig. 6. (continued).

biological studies found the undersurface of the ice to be covered with a thick layer of these loose discs which had formed in the water and floated up to the surface. Figure 9 shows ice formations on the suspension of a current meter which was deployed off Hut Point at 5 m below the ice and ceased to operate 27 hours after installation. Upon recovery, this instrument was found to be completely encased in ice. Finally, it should be noted that the authors' colleague in Antarctica, S. J. Jones (personal communication, 1984) states that approximately 40% of the ice sheet in the vicinity of McMurdo Station consisted of "underwater ice" which has a distinctive crystallographic structure enabling the heat sink for its growth to be identified as supercooling.

In general, Figures 7 and 8 show that surface salinities increase and supercooling decreases as one moves out of the sound, consistent with the picture of colder, fresher water emanating from the ice shelf. It can also be seen from Figures 7 and 8 that maximum supercooling near the east side of the ice shelf appears to be associated with the local salinity maximum and suggests an upwelling of this supercooled water from a depth of approximately 100 m on the nearby north-south section (Figure 6a). A similar salinity maximum seen near the west side of the ice shelf also suggests upwelling, although there are too few data to define the horizontal or temporal variations of either feature. Upwelling is thought to be a result of an anticyclonic circulation pattern centered on station 29 and described later in connection with the dynamic height topography.

Figure 8 and Figure 3 illustrate that station 53 has the greatest amount of surface supercooling. Figure 10 shows the details of this profile near the ice/water interface. Even including pressure effects, the water at this site is supercooled to a depth of 60 m, while the entire profile shows water considerably below the surface freezing temperature to a depth of 290 m. It is difficult to comment on the uppermost part of the profiles because of the local effect of the hole in the 3 m thick ice through which the instrument was lowered. However, below 4 m depth the sensors can be considerd to be well flushed and it is clear that the salinity, and hence density, is decreasing away from the ice sheet while the temperature is rapidly decreasing to a value which makes the seawater supercooled by 0.047°C. It is necessary that the temperature reach the freezing point value, $T_f = -1.906°C$, in the boundary layer immediately in contact with the ice. Although complicated by the effects of the hole, this profile shows the unstable boundary layer next to the ice where supercooling is being relieved with the consequent rejection of salt. Such a situation would require a strong downward convective motion to be associated with the density flux in a region of general upwelling. Assuming

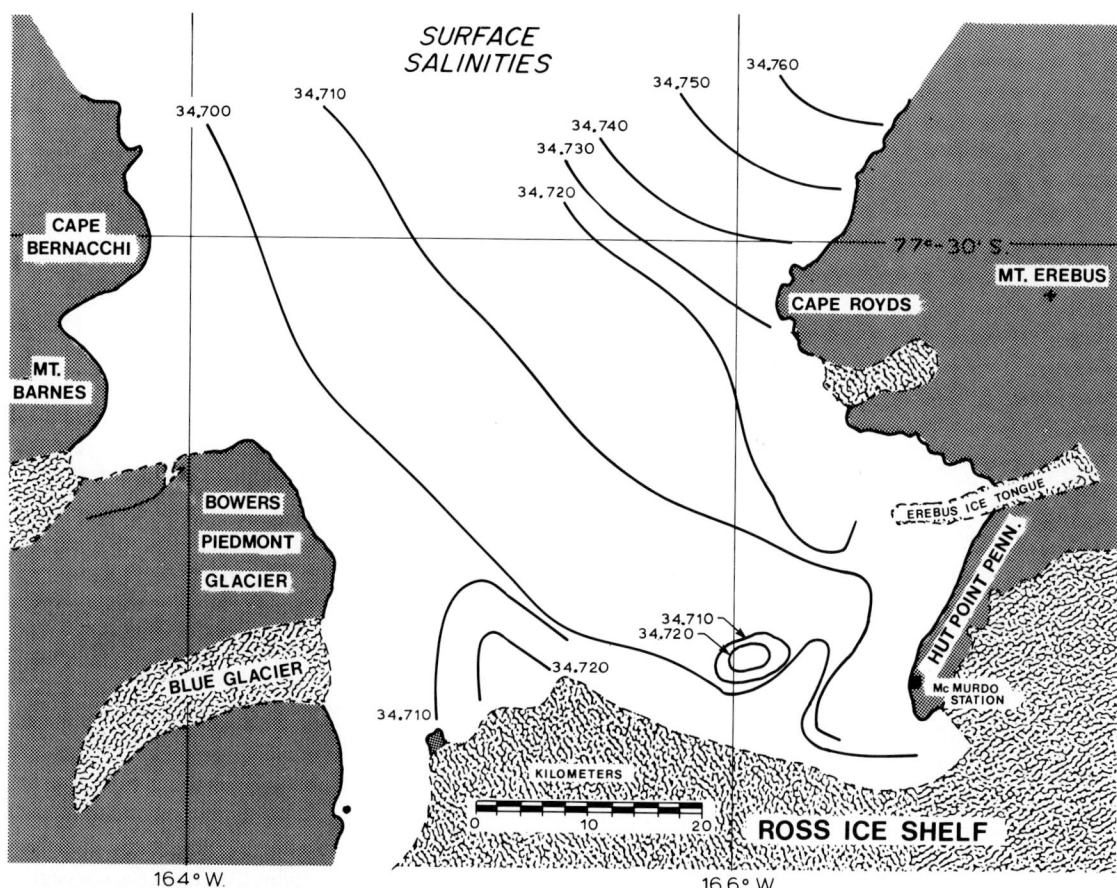

Fig. 7. Near-surface salinities in McMurdo Sound. These would indicate a source of fresher water associated with the Ross Ice Shelf and moving out of the Sound on its western side.

Monin-Obukov similarity [Turner, 1979, p.131], the shape of the profile can be used to calculate a Monin-Obukov length. If currents beneath the ice and the roughness of the ice undersurface were known, a stress could be estimated and combined with the Monin-Obukov length, used to estimate the density and salt fluxes. It is worth noting that the profile at this site is unstable down to 190 m so this convection is capable of reaching great depths. A similar density inversion was found at station 51, the site of the other salinity and supercooling maximum. An outstanding problem requiring further investigation is the horizontal and temporal scales of these convective events and their influence on local water properties.

Tides and Currents

Figure 11a shows the currents recorded at a depth of 200 m at station 11. The main flow is tidal and reaches a maximum speed of about 6 cm/s. On occasions, the velocities were so low that the rotor of the Aanderaa RCM 4 remained stationary. When this occurred the threshold value of 2 cm/s was assigned to the unknown current speed; the vane has a much lower threshold speed and was assumed to be correct. Data from the stalled rotor were excluded from the tidal analysis but were used in the production of the progressive vector diagrams. Figure 11a shows a beat period of 13.6 days between the components of the tide, with basically diurnal variations. The relative frequency section in Figure 11a shows a predominant current direction near 250°.

Figure 11b gives an impression of the behavior of the mean current over the period of observation. The current takes on a direction of 209° from day 300 to day 305, shows very little net motion for 10 days, then takes a set of almost due west from day 315 to day 320 before returning to a direction close to the original 209°. The average current velocity, 1.8 cm/s, has been represented by an arrow on Figure 1 but the situation is one of considerable variability and demonstrates the impossiblity of predicting mean flows from short records.

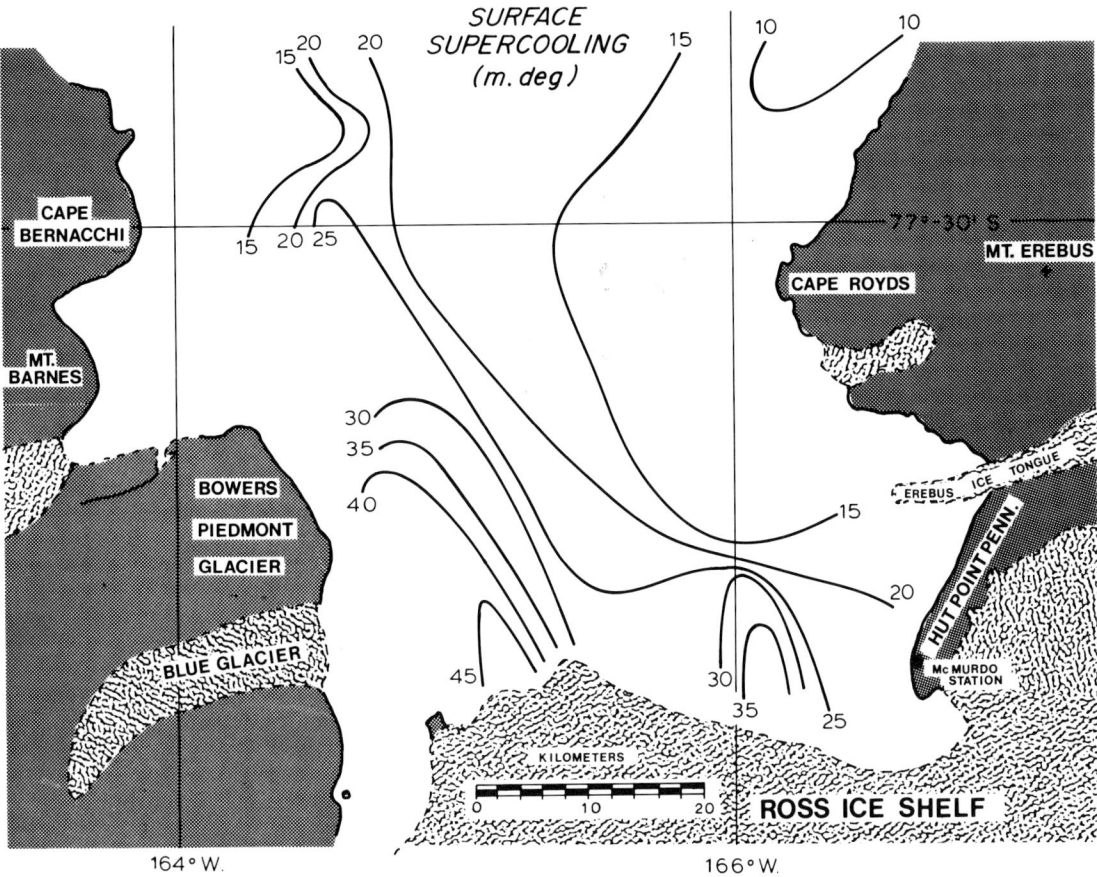

Fig. 8. Contours of surface supercooling beneath the sea ice. These indicate a northward flow from the ice shelf at around 166°W and 165°W. Such supercooling must be produced by water coming into contact with the ice at depth. This figure may therefore be interpreted in conjunction with Figure 7 to indicate water movement from the vicinity of the ice shelf.

Figure 12 presents similar information from station 50 near the ice shelf edge. After day 322 (Figure 12a) noise appears in the record for speed and is due to a loose contact in the instrument. Other parameters measured by the current meter and unaffected by this contact showed changes at this time, notably a small lowering of the temperature. It may well be that enhanced water movement, strumming the line by which the current meter was suspended from the ice, was responsible for the intermittent contact. Figure 12b shows a very considerable change in the mean direction of flow over the period of observation and suggests highly variable water movement. Up until day 300 water movement is essentially perpendicular to the ice shelf edge and afterwards nearly parallel to it.

These current meter records have been analyzed to obtain current ellipses following Foreman [1978]. The main constituents are listed in Table 1 along with the tidal height data from Heath [1971] for McMurdo Sound and Williams and Robinson [1979] for the adjacent Ross Ice Shelf. It can be seen that the diurnal constituents K1 and O1 dominate all the records with the semidiurnal components M2 and S2 taking a secondary role. For McMurdo Station tidal heights (O1 + K1)/(M2 + S2) = 7.3.

Even in these conditions of weak stratification, tidal currents are capable of great complexity and a detailed analysis based on short records from only two current meters is not justified. However, in comparing the phases of the tidal height at McMurdo and on the Ross Ice Shelf to the south, one can see that the diurnal phases and magnitudes are nearly equal whereas the semidiurnal components are considerably out of phase. Similarly, the diurnal tidal currents at stations 50 and 11 are in phase, and the semidiurnals are not. Also, from station 11 to station 50, the ratio of the magnitudes of M2 to S2 changes from 1.6 to 0.75. Therefore, it would seem that the semidiurnal currents at 200 m depth near the ice shelf (station 50) may be modi-

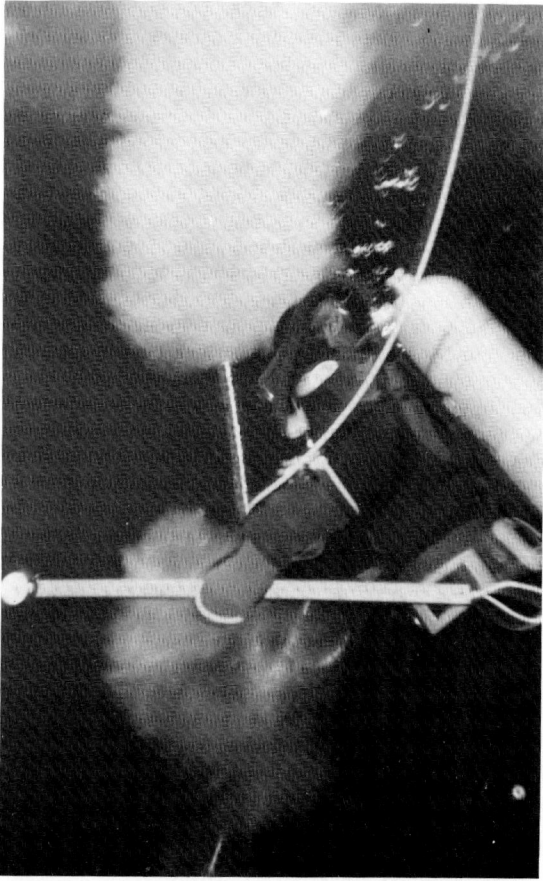

Fig. 9. A photograph of underwater ice on an Aanderaa current meter suspended below the sea ice off Hut Point near McMurdo Station. The current meter has been enveloped by an 80 cm diameter cylinder of ice over a period of 30 days. The heat sink for this growth is the supercooling in the surface layer of seawater in McMurdo Sound.

fied by currents under the ice shelf, possibly driven by a phase difference between the McMurdo Sound and Ross Ice Shelf tidal height constituents.

Figure 13 shows the contours of dynamic height anomaly relative to 300 dbar so as to include as many stations as possible. Some profiles to the west side of the ice shelf edge were recorded about 15 days later than profiles at the other stations, so the contours may include some temporal variation. In areas where stations are sparse, current values obtained from the graph inset on this figure are clearly uncertain. Nevertheless, we consider the overall pattern to be correct. The shoal that exists off the end of the EGT has been emphasized in Figure 13 as it may contribute to the westward current deflection as shown in the figure at that point.

Of considerable interest are the eddy shown off Cape Royds and the circulation inferred in the immediate vicinity of the ice shelf. Stations near the eddy are deep enough to allow geostrophic currents to be referenced to the 700 dbar level. When this is done, the baroclinic component of the surface current is increased about 5 cm/s over that derivable from Figure 13, the direction of the motion being preserved. The eddylike current pattern penetrates to great depth which suggests that it is not a transitory pattern entrained in the mean flow but a quasi-permanent feature related to the local topography. As can be seen from Figure 6, the vertical separation of isopycnal surfaces in the 300- to 600-dbar interval is reduced by about half in moving from outer McMurdo Sound (station 57a) to the area of Cape Royds (station 21). If planetary vorticity dominates the relative vorticity in the outer region, then such a compression of isopycnals means that the water column must acquire positive relative vorticity [Pedlovsky, 1979] consistent with the direction of rotation of the Cape Royds eddy in Figure 13. If the eddy is a permanent feature, a continuous or periodic input of vorticity would be necessary to compensate for its frictional decay. Thus the eddy is probably connected with a transport of water into McMurdo Sound along the eastern side, and is responsible for considerable upwelling at Cape Royds through the tilting of isopycnals (Figure 5c). The location of an Adélie penguin rookery at Cape Royds may reflect the resulting greater local abundance of food. Reports on the productivity of the benthos [Dayton and Oliver, 1977] and the microbial uptake of dissolved organic matter [Hodson et al., 1981] emphasize the higher degree of biological activity in the eastern side of the sound, in contrast to the oligotrophic western side. Such conditions may require the rapid replacement of nutrients in the eastern Sound as would occur by the advection and upwelling of water as described above.

Comparing Figure 13 with Figures 7 and 8, it can be seen that currents inferred from the dynamic height topography are given credence by the presence of low-salinity, cold water in the western half of McMurdo Sound. The outflow of water from under the Ross Ice Shelf in this location, where we have few stations, is further supported by the radioactive tracer data of Michel et al. [1979]. Their measurements of tritium and carbon-14 isotopes tagged water in the western part of the Sound as originating under the Ross Ice Shelf.

Flow patterns along the ice shelf edge can be clarified by potential temperature profiles taken from stations in that vicinity (Figure 14). Stations 30A and 53 represent sites where, according to Figure 13, cold water is flowing out from under the Ross Ice Shelf.

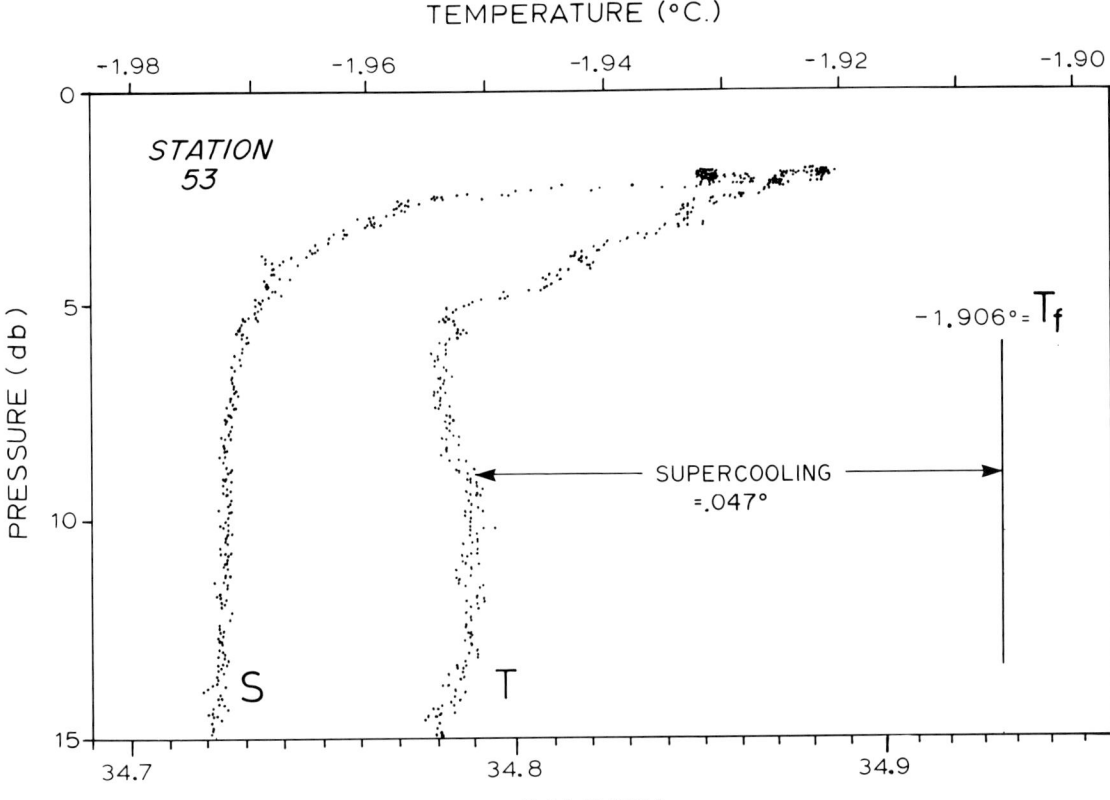

Fig. 10. Supercooling at station 53 on the route of outflow of waters from the Ross Ice Shelf northward into McMurdo Sound. As the sea ice was approximately 3 m thick, the gradients shown in the interval 3-4 m may owe part of their shape to the hole through which the CTD was lowered into the water. Nevertheless, the supercooling of 0.047°C must reduce to zero at the ice/water interface and so reach the -1.906°C freezing point. With the relief of supercooling there must be an increase in salinity which is indeed shown.

Station 61 shows a relatively warm inflow between 75 and 150 m. The salinity of this warmer water is 34.71, too low to be entering from the Ross Sea, and Figure 13 suggests that it may be coming from Erebus Bay, south of the EGT. Other stations, such as 50B, exhibit a combination of both warmer and colder waters but always below the surface freezing temperature.

The temporal variability just north of the ice shelf is illustrated by a time series of profiles taken at station 29 (Figure 15). These profiles are separated by 2 hours. The profiles of potential temperature clearly illustrate short-term changes in the thickness of the cold intrusive layer centered around 125 m. During this time interval, the current meter at station 50, 5 km to the southwest, showed the current to be near the end of a tidal cycle carrying water in that direction. The most probable explanation is that the temperature minimum is being advected back under the ice shelf, thinning from 70 m to 25 m at station 29. Therefore, the water at the eastern ice shelf edge is a combination of relatively low-salinity water masses at a variety of temperatures below the surface freezing point. Since water of this salinity cannot originate in the outer sound, it must originate under the ice shelf, with supercooling relief at the ice/water interface accounting for the temperature variabilty. Details of the flow pattern may be too complex to be resolved by the available data but some of the warmer water is being advected back under the ice shelf. The baroclinic component of the mean flow is such that isopycnals tilt upward toward Hut Point Peninsula, possibly contributing to the high productivity noted there by Hodson et al. [1981].

The Effects of the Erebus Glacier Tongue

A total of seven stations, numbered 41 to 45B were taken in the immediate vicinity of the Erebus Glacier Tongue (EGT); three off the

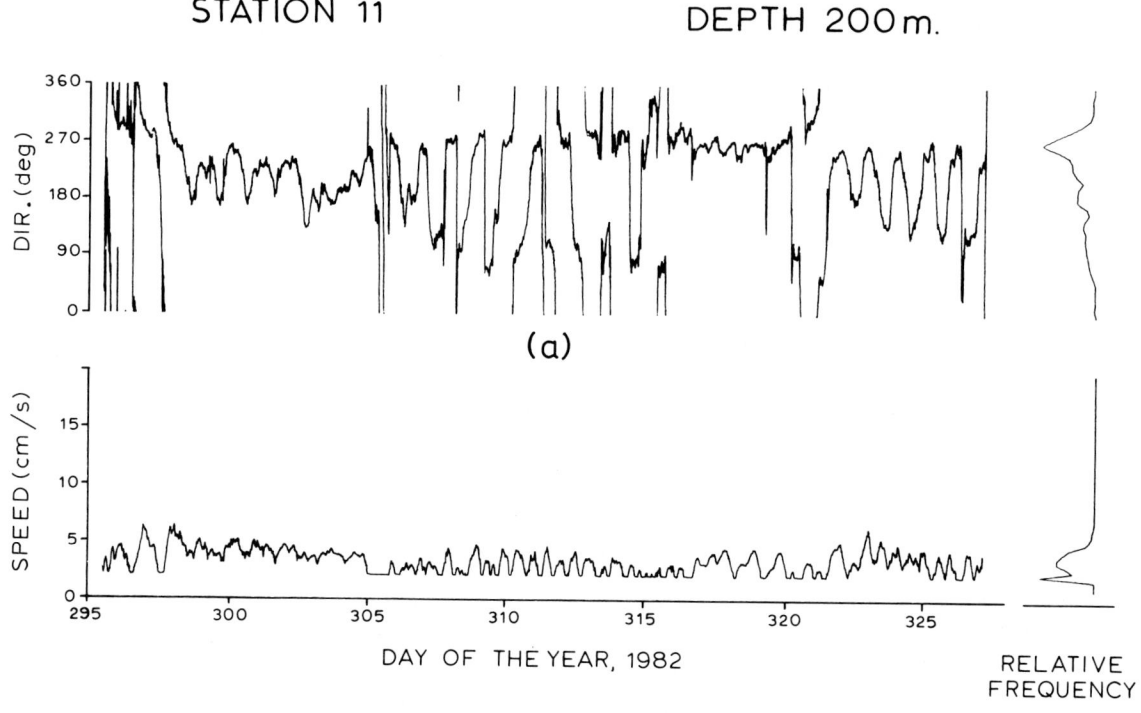

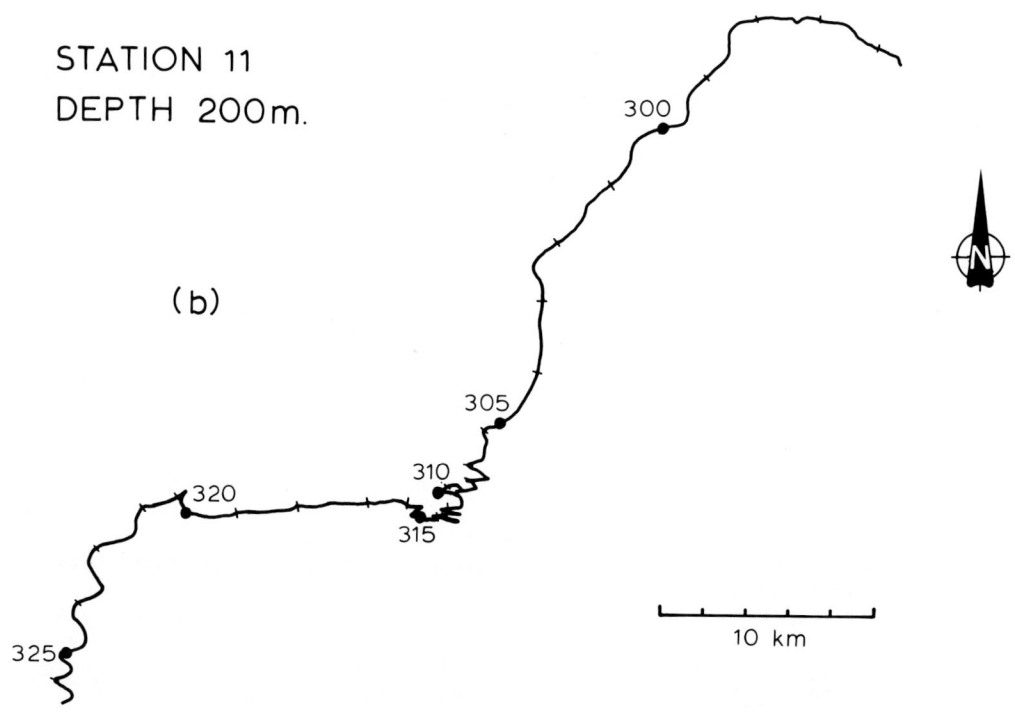

Fig. 11. (a) Current magnitude and direction at 200-m depth at station 11 over a period of a month. (b) The same information expressed as a progressive vector diagram. Note that two distinct directions are taken by the current over the period of measurements.

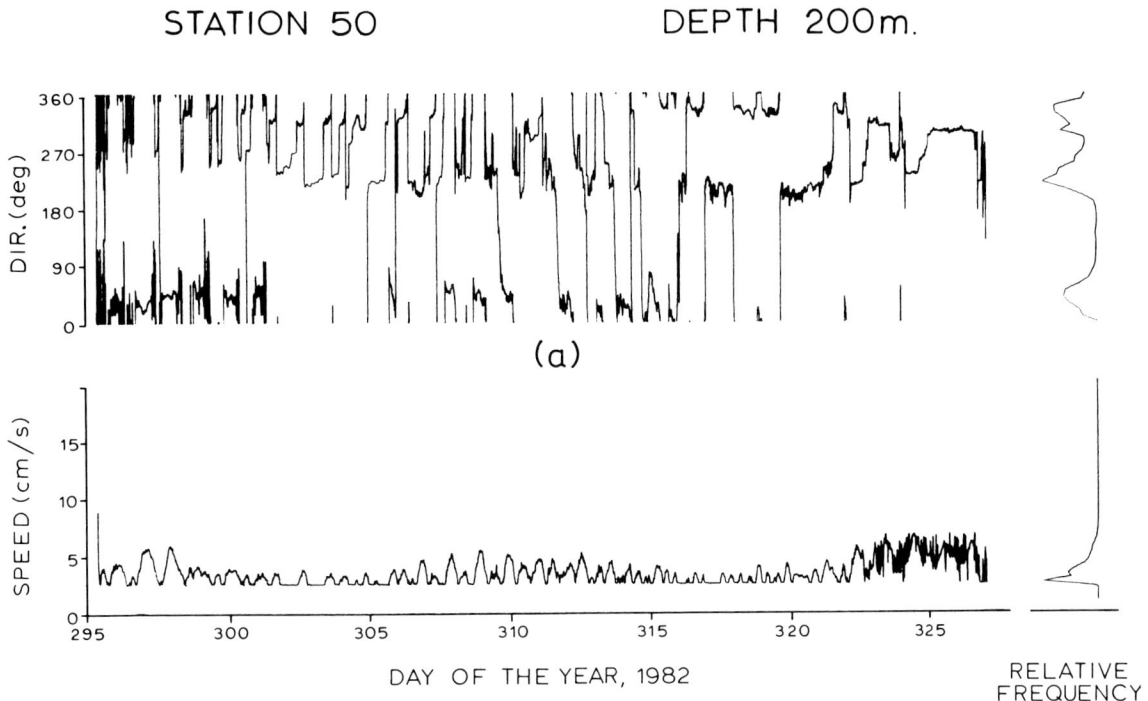

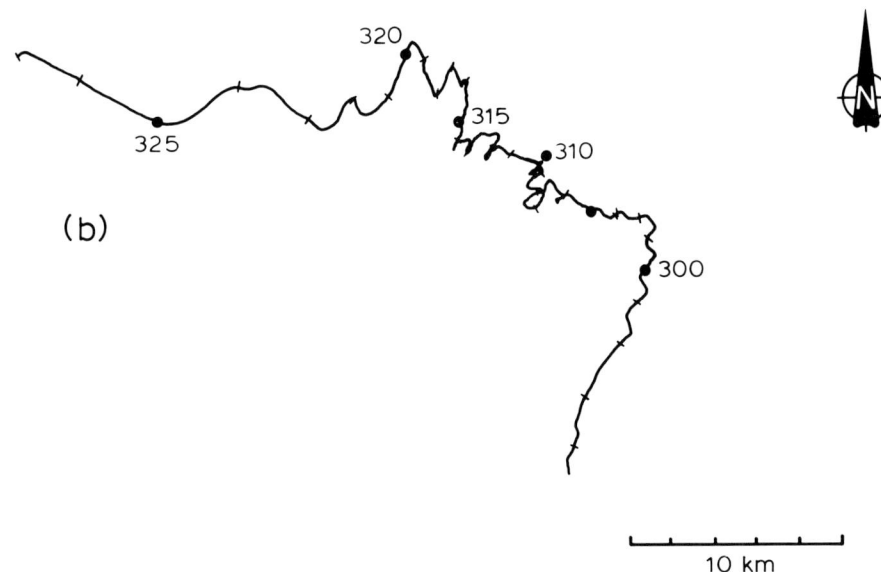

Fig. 12. (a) Current magnitude direction at 200-m depth at station 50 over a period of a month. (b) The same information expressed as a progressive vector diagram. The noise towards the end of the record was caused by a loose contact in the instrument. The current suffered a distinct change in direction over the period of the measurement.

TABLE 1. Parameters for the Tidal Current Ellipses at Stations 11 and 50

Tidal Component	Station 11				Station 50				McMurdo Station		Ross Ice Shelf[a]	
	Major Axis, cm/s	Minor Axis, cm/s	Inclination, deg. T	Phase,[b] deg.	Major Axis, cm/s	Minor Axis, cm/s	Inclination, deg. T	Phase, deg.	Height, cm	Phase, deg.	Height, cm	Phase, deg.
O1	1.9	0.0	163.0	267.2	0.8	0.3	042.6	275.0	21	195	29	196.
P1	0.7	0.0	158.7	297.4	0.5	0.0	033.8	328.0	8	213	10	203.
K1	2.2	0.0	158.7	297.4	1.5	0.0	033.8	328.0	23	212	31	208.
N2									2	263	3	180.
M2	0.8	0.1	001.3	166.6	0.6	0.1	175.3	119.3	4	242	4	340.
S2	0.5	0.0	154.4	217.4	0.8	-0.1	162.6	095.3	2	327	2	190.
K2	0.1	0.0	154.4	217.4	0.2	0.0	162.6	095.3	0.4	082		
M4	0.2	0.0	027.9	230.8	0.2	0.1	021.6	002.5	0.3	270		
MS4	0.2	0.0	174.3	144.5	0.5	0.0	177.1	144.8	0.2	325		

Tidal height data for McMurdo Sound and the Ross Ice Shelf are taken from Heath [1971] and Williams and Robinson [1979] respectively.
[a] 79.6°S, 163.3°E.
[b] All phases with respect to UT.

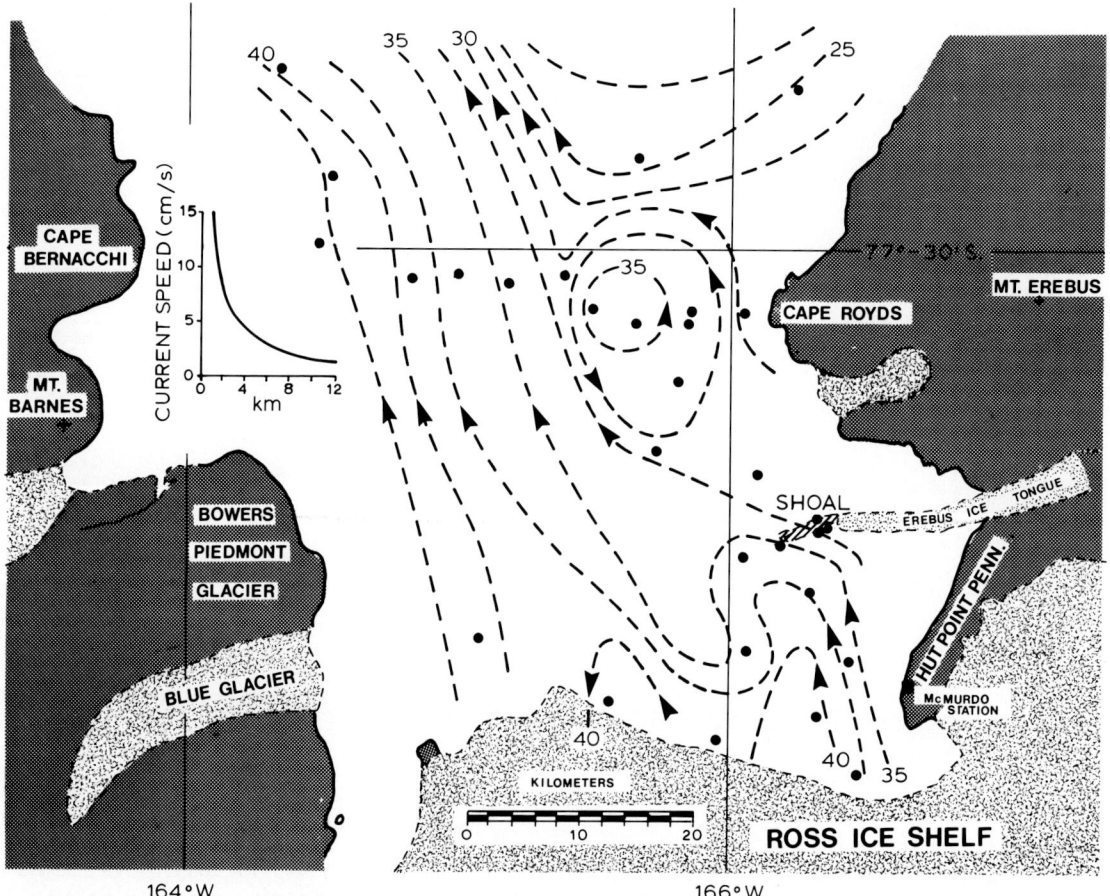

Fig. 13. The circulation pattern in McMurdo Sound as inferred from dynamic height topography in dynamic millimeters relative to 300 dbar. The inset graph gives current speed as a function of separation of the contours.

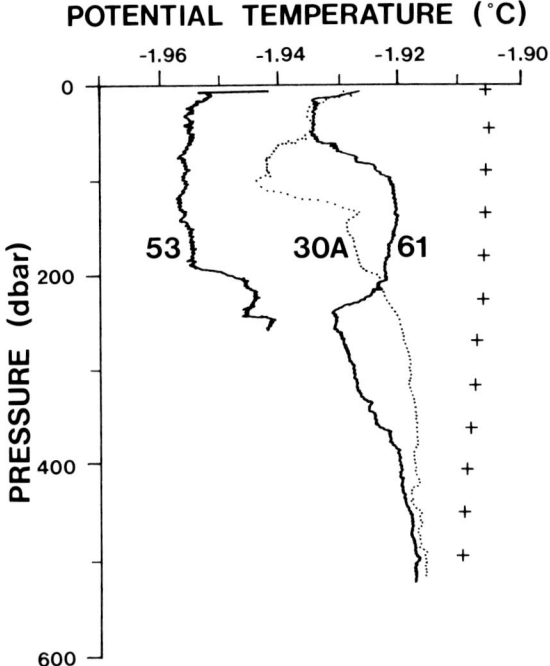

Fig. 14. Potential temperatures at stations in the vicinity of the Ross Ice Shelf edge in McMurdo Sound. Station numbers are given next to each profile: station 53 is on the west side; station 61 is in the middle; and station 30A is on the east side. The surface freezing temperatures are marked with crosses.

tip in a line to the southwest, three off the north side of the tip in a line to the northwest, and one to the south in Erebus Bay. These were positioned to investigate thermohaline steps associated with glacial melting as reported in Jacobs et al. [1981].

At station 43, about 100 m southwest of the tip of EGT, an anomalously large current was detected by a noticeable deflection of the CTD cable. Analysis of the CTD descent rate, usually kept constant by the winch operator, showed that this current first affected the probe at a depth of 150 m. It would have to have been about 50-60 cm/s to influence the descent rate by the amount measured. This condition continued to the bottom, a depth of 324 m. The salinity profiles from the three stations off the northwest side of the tip of the ice tongue are shown in Figure 16. These stations were placed approximately 500 m apart, 45B being closest to the vertical wall of the EGT. The potential temperatures were practically constant at $-1.920 \pm 0.003°C$ so the density profiles are similar to the salinity profiles shown in Figure 16. At stations 45 and 45A, the profiles show inherently unstable water columns both in the gross sense of having thick unstable layers centered at 220 m (45) and 150 m (45A) and in having dozens of fine structure inversions on 5 to 10-m vertical scales. This is obviously an area of strong mixing, no doubt related to the current detected at the tip of the EGT (station 43). Although these profiles are unstable in that any downward disturbance would move a water parcel into a water mass of lesser zero pressure density, the in situ density of the water (with the pressure effect included) increases monotonically with depth. Transient density inversions such as these are not uncommon in areas where high currents cause horizontal interleaving between two water masses.

Some steps are apparent in the salinity profile closest to the ice tongue, station 45B, Figure 16. Jacobs et al. [1981] give an equation for the vertical size of steps created by the convective regime associated with the melting of ice submerged in stratified water. In this case the temperature is practically constant with depth so the temperature difference driving the melting comes almost entirely from the pressure effect on the freezing point. If the far field density gradient is taken from station 23, the calculated thickness of the mixed layers between

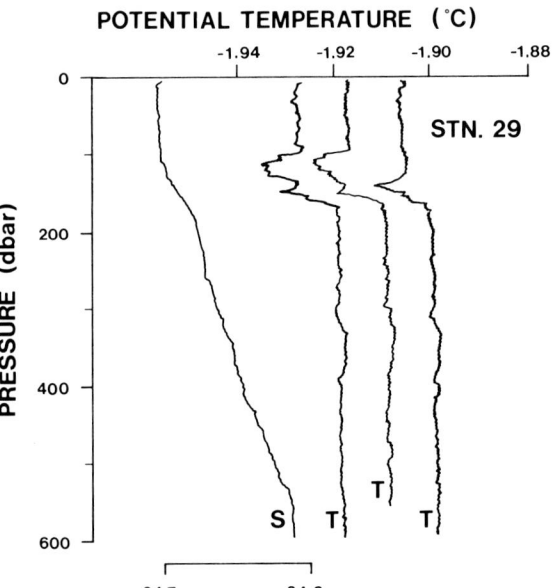

Fig. 15. A time series of profiles of potential temperature taken at station 29 at intervals of approximately 2 hours. The left-hand record gives the correct temperature directly with the subsequent two records each displaced 0.01°C to the right for the sake of clarity. They are included here to indicate the considerable variability of the cold water intrusive layer from the ice shelf occurring at a depth of about 150 m. The salinity profile shown is applicable to all the temperature profiles within ±0.05.

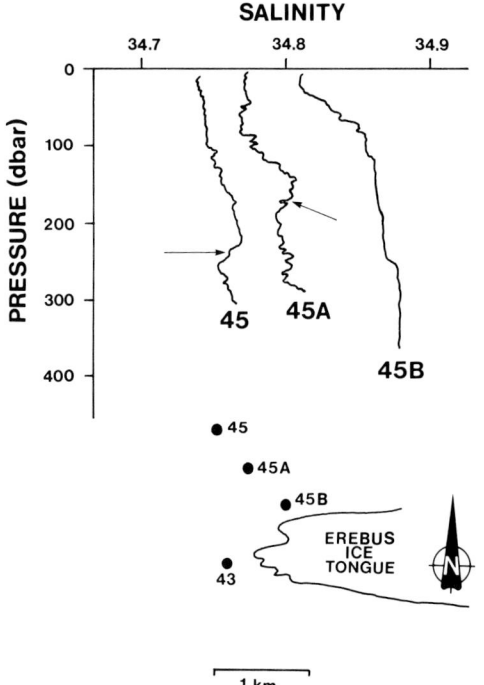

Fig. 16. Salinity profiles at stations 45, 45A, and 45B. Successive profiles are shifted by 0.05 to the right. Arrows point to density inversions. Arguments in the text indicate that this area may be one of very strong currents. Station positions are shown in the sketch.

salinity steps is 130 m. Separation between the strongest steps at station 45B is 135 m, apparently in good agreement with theory. If the gradient Richardson number (the square of the ratio of buoyancy frequency to mean shear) reaches a value less than 0.25, a fluid develops stratified shear flow instabilities which lead to mixing [Turner, 1979]. Under this criterion, the steps described above would be subject to mixing if the shears exceeded 0.013 s^{-1}. The strong currents noticed at station 43 should thus be quite effective in breaking down the steps and the horizontal extent of this convective feature may be limited to relatively quiescent areas in the lee of the ice tongue.

Another interesting aspect of these profiles is the relatively high surface salinity found at station 45. This value of 34.74 is also found at a depth of 250 m in Erebus Bay to the south (station 25), at a depth of 100 m to the north (station 23), and at a depth of 290 m to the west (station 41). Evidently, considerable upwelling accompanies the density inversions previously discussed.

Stations 5, 6, and 7 (Figures 13 and 3) are in an area of outflow from the sound where mixtures between waters coming directly from the west side of the ice shelf and waters coming from the Erebus Bay area would be anticipated. Figure 17, showing the potential temperature and salinity profiles at these stations, lends support to this concept. At station 5 the profile of potential temperature shows a cold intrusive layer centered around 150 m depth. This is well below the surface freezing temperature and can only have originated by contact with ice at depth on the ice shelf. In contrast, station 7 shows no such intrusion, but below about 50 m has salinities higher by 0.02 and temperatures generally higher by about 0.016°C. A possible origin for the water at station 7 is found in Ice Shelf Water that has lost its supercooling, mixed with HSSW by the stong currents and upwelling in the vicinity of the EGT. The profile at station 6 shows a series of interleaving layers alternating between properties of the two different water sources.

This mixing line on the T/S diagram for supercooled water does not necessarily follow a straight line. Should nucleation occur, the temperature-salinity line for this process would modify any mixing line between two adjacent water masses. Taking the nucleation process in isolation, one would expect

$$\partial T/\partial S = L/(c_p S) \sim 2.3\,°C$$

when $S \sim 35$, where L is the latent heat of formation of sea ice, C_p is specific heat of seawater and S is salinity.

Figure 18 is a T/S diagram to illustrate this point. Supercooling greater than 0.005°C

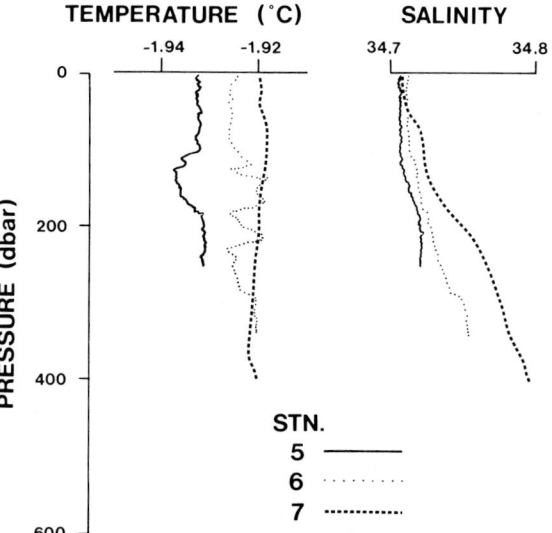

Fig. 17. Potential temperature and salinity profiles from stations in the western part of McMurdo Sound showing interleaving between Ice Shelf Water and modified Ice Shelf Water (Erebus Water).

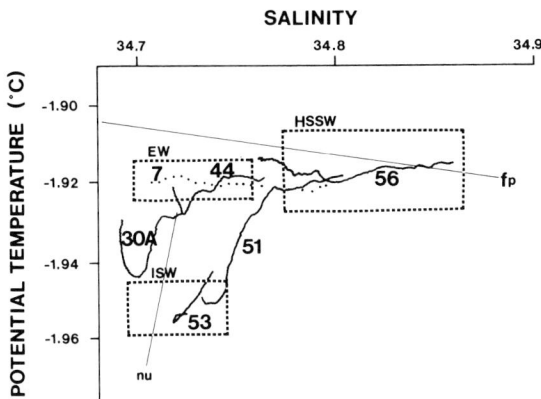

Fig. 18. Temperature-salinity (T/S) diagram of selected stations. Stations 30A and 53 show the properties of Ice Shelf Water (ISW). Station 56 shows High Salinity Shelf Water (HSSW). Where ISW supercooling has been relieved through nucleation along the line labeled "nu", Erebus Water (EW) has been formed. All lines with the same slope are potential nucleation paths. These properties are observed in Erebus Bay (station 44) and at the center of the east-west section (station 7). The line labeled "f_p" gives the zero pressure freezing line for the salinities shown.

cannot normally be attributed to local pressure ridge keels, but indicates that water must have been in contact with glacial ice at depth. Thus the lowest temperatures (stations 51, 53, and 30A) represent Ice Shelf Water (ISW). As previously discussed, station 56, below 400 m, represents HSSW. Station 51 has properties joining the cold ISW to the upper HSSW, the normal mixing condition to be expected at depth where the water is not supercooled in situ. However, a third water mass, here labeled Erebus Water (EW), is present at station 44 and is joined to the ISW by a nucleation line (nu). This water mass is also found at station 7 where it mixes laterally with station 5 water that has come directly from the ice shelf. Since EW is of lower salinity than HSSW, it must be formed from ISW through nucleation. In order to relieve its supercooling, the ISW must rise to a depth where the pressure effect on freezing is reduced and contact with ice can be made. These conditions are met in Erebus Bay where strong upwelling would be associated with the blockage effect of the EGT, the shoal extending from its tip, and the general upward slope of isopycnals toward the east. Currents computed from dynamic heights at stations 29 and 26 near Hut Point indicate a 15 cm/s outflow of Ice Shelf Water in the depth range where the lowest temperatures are found. Anomalously high currents have been reported off Hut Point [Heath, 1977]. Upwelling along Hut Point Peninsula may also result in a relatively intense rate of nucleation there, as evidenced by the ice formation on the current meter (Figure 9).

Conclusions

Water in McMurdo Sound at the end of winter season falls into the High Salinity Shelf Water category. The potential temperature is almost constant from the surface to 800 m depth, except for a local source of cold water which can be traced to the Ross Ice Shelf in the southern reaches of the Sound.

The surface circulation in the sound can be inferred from the dynamic topography relative to 300 dbar supported by current meter measurements at two points. In general, it confirms the earlier results of Heath [1977], but reveals the presence of a large anticyclonic eddy to the west of Cape Royds. This feature extends to about 700 m depth and may be spatially anchored by the topography although variability in the tidally averaged current suggests limited movement. Upwelling produced by this eddy along the coast could be a significant factor in the biological activity asociated with the Cape Royds penguin rookery.

Another anticyclonic surface circulation pattern is inferred between the EGT and the ice shelf, although it could take the form of a meander rather than a closed eddy. This pattern has the effect of inducing water to upwell from under the ice shelf and move along the Hut Point Peninsula. Since the freezing temperature dependence upon pressure causes the ice shelf water to be supercooled at the sea surface, nucleation should be intense along this coastline. A photograph of underwater ice formation has been shown as an example.

The cold Ice Shelf Water was found at the sea surface at two other locations near the ice shelf edge. These were apparently sites of deep-reaching vertical convection resulting from the density instability caused by the rejection of salt during the nucleation process.

The horizontal and temporal scales of these upwelling and downwelling processes are unknown. Instability in the profiles at these two sites could be offset by the buoyancy of entrained ice crystals in the manner proposed by Foldvik and Kvinge [1974]. Attempts to collect ice crystals on a filter and measure the salinity change by melting have given inconclusive results, although it may be possible to establish upper limits on the importance of this effect.

The relief of supercooling has been shown to result in a new water mass. This "Erebus Water", formed mainly in Erebus Bay, leaves McMurdo Sound toward the western side, interleaving with water coming directly from the ice shelf and with HSSW.

Tidal constituents of the current measured at 200-m depth near the Ross Ice Shelf show a pattern of relative magnitudes typical of measurements of tidal height taken on the Ice Shelf much farther to the east. Combined with the pattern of outflow in the surface layers, this indicates that the ice sheet is not completely grounded and that water flows around Ross Island, under the ice, into the southern end of McMurdo Sound. This conclusion is supported by the isotope measurements of Michel et al. [1979] which can be interpreted to indicate that Ice Shelf Water is predominant on the west side of McMurdo Sound.

The presence of a floating ice shelf significantly effects the surrounding sea ice sheet if the density stratification allows water from the melting ice shelf to reach the surface. This is the case in late winter when the stratification in the surface layer is reduced by cooling and by the rejection of salt from the growing sea ice, producing a surface isopycnal layer. If this surface layer reaches the depth of the ice shelf, then large quantities of cold, buoyant water formed at high pressure can reach the surface, where supercooling can be relieved by contact with the ice sheet. In these circumstances preferential melting of deep ice occurs with the overall effect of transferring ("pumping") ice from the base of the ice shelf to the sea ice sheet [Lewis and Perkin, 1983]. An increased sea ice thickness would thus be expected near the ice shelf, with a predominance of underwater ice (platelets) in the lower portion of that sea ice. These characteristics have been noted in McMurdo Sound, and the gentle gradation of ice thickness from the shelf ice into the sea ice also reflects this ice transfer.

Acknowledgments. We thank Ron Cooke and Ernie Sargent for their able assistance in the collection of these data and Bill Buckingham for his help in their reduction. Staff and equipment for the field operation were supplied by our Institute (Canadian Department of Fisheries and Oceans) while the U.S. National Science Foundation supported our work in Antarctica under grant DPP-81-19863 to Columbia University. Our thanks are due to both.

References

Clough, J. W., and B. L. Hansen, The Ross Ice Shelf Project, Science, 203, 433-434, 1979.

Dayton, P. K., and J. S. Oliver, Antarctic soft bottom benthos in oligotrophic and eutrophic environments, Science, 197, 55-58, 1977.

Dayton, P. K., G. A. Robilliard, and A. L. DeVries, Anchor ice formation in McMurdo Sound, Antarctica, and its biological effects, Science, 163, 273-274, 1969.

Foldvik, A., and T. Kvinge, Conditional instability of sea water at the freezing point, Deep Sea Res., 21, 160-174, 1974.

Foldvik, A., T. Gammelsrød, and T. Torresen, Circulation and water masses on the southern Weddell Sea shelf, this volume.

Foreman, M. S. S., Manual for tidal current analysis and prediction, Pac. Mar. Sci. Rep., 78-6 70 pp., Inst. of Ocean Sci., Sidney, B.C., 1978.

Heath, R. A., Circulation and hydrology under the seasonal ice in McMurdo Sound, Antarctica, N.Z. J. Mar. Freshwater Res., 5(3,4) 479-515, 1971.

Heath, R. A., Circulation across the Ice Shelf Edge in McMurdo Sound, Antarctica, in Polar Oceans, Proceedings of the Polar Oceans Conference, edited by M. J. Dunbar, pp. 129-139, McGill University, Montreal, 1977.

Hodson, R. E., F. Azam, A. F. Carlucci, J. A. Fuhrman, D. M. Karl, and O. Holm-Hansen, Microbial uptake of dissolved organic matter in McMurdo Sound, Antarctica, Mar. Biol., 61, 89-94, 1981.

Jacobs, S. S., A. F. Amos, and P. M. Bruchhausen, Ross Sea oceanography and Antarctic Bottom Water formation, Deep Sea Res., 17, 935-962, 1970.

Jacobs, S. S., A. L. Gordon, and J. L. Ardai, Jr., Circulation and melting beneath the Ross Ice Shelf, Science, 203, 439-443, 1979.

Jacobs, S. S., H. E. Huppert, G. Holdsworth, and D. J. Drewry, Thermohaline steps induced by melting of the Erebus Glacier Tongue, J. Geophys. Res., 86(C7), 6547-6555, 1981.

Jacobs, S. S., R. G. Fairbanks, and Y. Horibe, Origin and evolution of water masses near the Antarctic continental margin: Evidence from $H_2^{18}O/H_2^{16}O$ ratios in seawater, this volume.

Lewis, E. L., and R. G. Perkin, Supercooling and energy exchange near the Arctic Ocean surface, J. Geophys. Res., 88(C12), 7681-7685, 1983.

Littlepage, J. L., Oceanographic investigations in McMurdo Sound, Antarctica, in Biology of the Antarctic Seas II, Antarct. Res. Ser., vol. 5, edited by G. A. Llano, pp. 1-37, AGU, Washington, D.C., 1965.

Michel, R. L., T. W. Linick, and P. M. Williams, Tritium and carbon-14 distributions in seawater from under the Ross Ice Shelf Project ice hole, Science, 203, 445-446, 1979.

Pedlosky, J., Geophysical Fluid Dynamics, pp. 1-624, Springer-Verlag, New York, 1979.

Tressler, W. L., and A. M. Ommundsen, Seasonal oceanographic studies in McMurdo Sound, Antarctica, Tech. Rep. TR-125, pp. 1-141, U.S. Navy Hydrographic Office, Washington, D.C., 1962.

Turner, J. S., Buoyancy Effects in Fluids, pp. 1-368, Cambridge University Press, New York, 1979.

Williams, R. T., and E. S. Robinson, Ocean tide and waves beneath the Ross Ice Shelf, Antarctica, Science, 203, 443-445, 1979.

UNESCO, Freezing point of sea water, Eighth report of the Joint Panel of Oceanographic Tables and Standards, Appendix 6, UNESCO Tech. Pap. Mar. Sci., 28, 29-31, 1978.

UNESCO, The Practical Salinity Scale 1978 and the International Equation of State of Seawater 1980, Tenth report of the Joint Panel on Oceanographic Tables and Standards, Annex 1, UNESCO Tech. Pap. Mar. Sci., 36, 1981.

Zwally, H. J., J. C. Comiso, C. L. Parkinson, W. J. Campbell, F. D. Carsey, P. Gloerson, Antarctic Sea Ice, 1973-1976: Satellite Passive Microwave Observations, NASA SP-459, 1983.

Zwally, H. J., J. C. Comiso, and A. L. Gordon, Antarctic offshore leads and polynyas and oceanographic effects, this volume.

(Received March 19, 1984; accepted June 6, 1984).

OBSERVATIONS IN THE BOUNDARY LAYER UNDER THE SEA ICE IN MCMURDO SOUND

W. M. Mitchell[1] and J. A. T. Bye

The Flinders Institute for Atmospheric and Marine Sciences
The Flinders University of South Australia, Bedford Park, South Australia 5042

Abstract. High-resolution observations of current (three components), temperature, and conductivity at two levels just below the seasonal sea ice are presented for two sites in McMurdo Sound, Antarctica, in January 1977. The dynamics of the melting process are found to differ between the two sites. At the eastern site near McMurdo Station, relatively high melting rates occurred due to the southward advection of relatively warm oceanic water, whereas at the western site in the region of oceanic advection from under the Ross Ice Shelf, melting was slight and due to surface intrusions of coastal meltwater probably from the Hobbs Glacier. The frequency spectra indicated a buoyancy subrange for the velocity components and a fine structure range for the density at frequencies greater than the Brunt-Väisälä frequency and the probable existence of internal wave spectra at lower frequencies.

Introduction

Observations of the boundary layer beneath seasonal sea ice were made in McMurdo Sound, Antarctica (Figure 1), with high-resolution instrumentation designed to measure buoyancy and momentum fluxes. The information gathered gives some indication of the rates of melting (and freezing) of the annual sea ice cover. Initially, the instrumentation was intended for use during the Ross Ice Shelf Project [Clough and Hansen, 1979] but problems in maintaining the hole open in the latter stages of the project meant that attention was diverted to the annual sea ice cover of McMurdo Sound.

Two sites were occupied in an effort to obtain some perspective of the spatial variability. The first site, S1 (77°51'11"S, 166°39'11"E), was located approximately 600 m due west of McMurdo Station in Winter Quarters Bay and was occupied from January 7 to 14, 1977. On January 17, 1977, S2 (77°54'27"S, 164°34'48"E), about 300 m offshore and close to the observed meltwater inflow from the Hobbs Glacier, was occupied for 2 days. The water depth at S1 was greater than 100 m, while at S2 it was 98 m. The sea ice cover at both sides was uniform, and there was no evidence of pressure ridges, although cracks and frozen leads occurred adjacent to the shore.

The station positions were fixed by theodolite bearings on landmarks with the aid of U.S. Naval Oceanographic Office [1969] Chart H.O. 6666. At each site, 16 hours of data at two levels were collected for the three components of velocity, temperature and conductivity. An effective sampling interval of 1 s gave approximately 10^5 data points.

Methods

The flux-measuring instrumentation was designed and constructed at the Flinders Institute for Atmospheric and Marine Sciences. It consisted of high-resolution solid state current, temperature, and conductivity sensors. The current vector was determined by three mutually orthogonal pitot tubes which contained transistors that gave an analog output proportional to speed. This instrument is described by Gordon [1981] and is based on a design by Steedman [1972].

Temperature and conductivity were determined by a bead-in-glass thermistor and a single electrode probe, the latter being based on the laboratory instrument described by Mied and Merceret [1974]. There were two arrays of the current, conductivity, and temperature sensors, 2 m apart on a vertical stem, the center of which housed preamplifiers and appropriate bridge electronics. The complete "fluxatron" (Figure 2) was designed to fit through a 20-cm-diameter ice hole. Vertical positioning of the fluxatron could be determined by a pressure sensor, and horizontal orientation by a modified (to compensate for the large angle of dip) Aanderaa compass, both of which were housed in the center stem.

Analog voltages were transmitted via 16 core video cable to the surface, where signal filtering and recording were carried out on chart recorders and analog tapes.

Calibration was performed in the laboratory and in situ. On returning to base, checks were made for instrumental drift and perfor-

---

[1]Now at Department of Applied Mathematics, The University of Adelaide, Adelaide, South Australia, 5001.

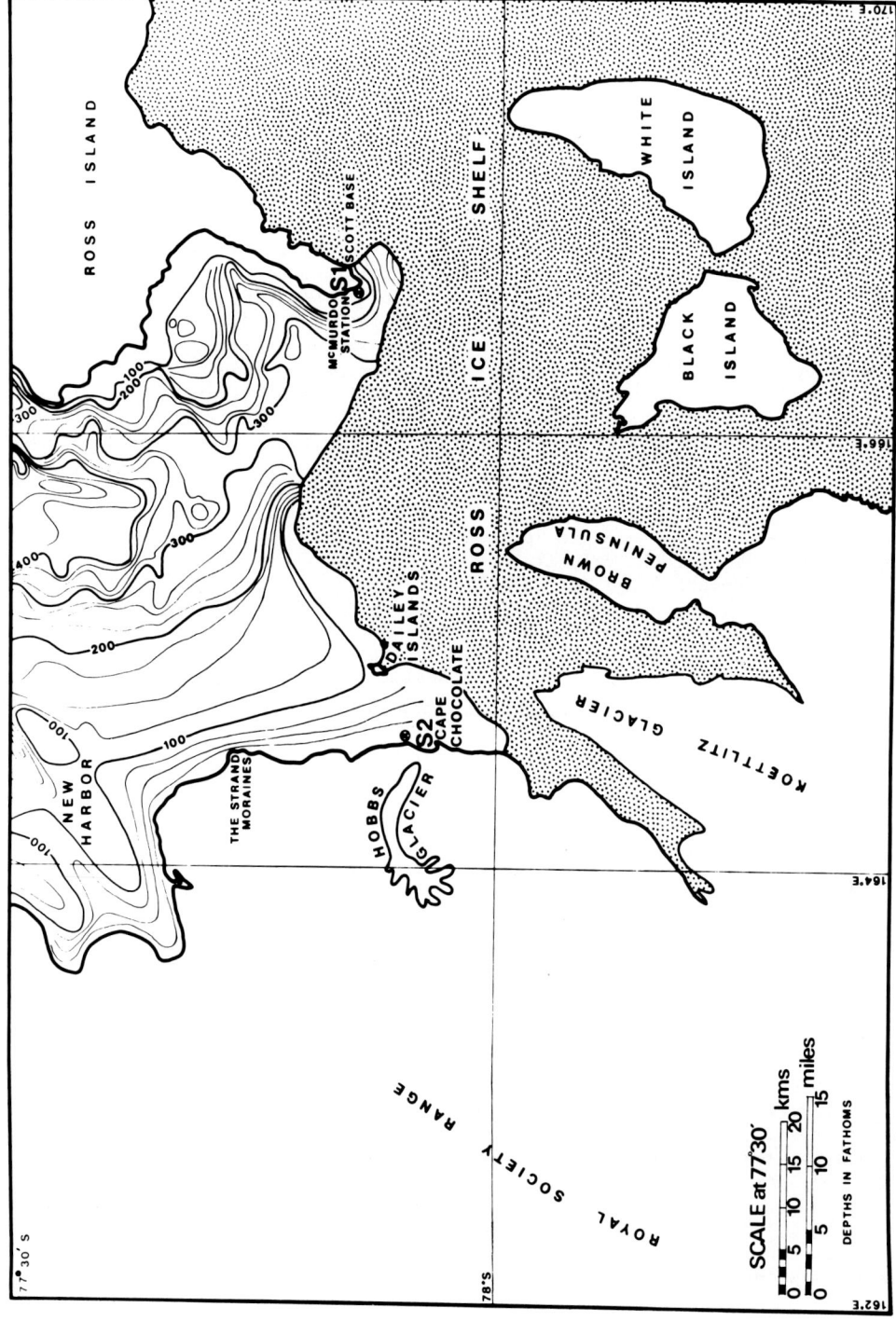

Fig. 1. Bathymetry of McMurdo Sound from U.S. Naval Oceanographic Office [1969] Chart H.O. 6666.

Fig. 2. Flux-measuring instrumentation: the fluxatron.

mance. Particular care had to be taken with the conductivity sensor, since the measurements involved detecting minute resistivity changes and some chemical deterioration of the platinum-tipped electrodes was unavoidable. Table 1 shows the frequency response, absolute accuracy, and resolution of each sensor when coupled to the recording instruments. After recording, the analog data were digitized, filtered again to remove discretizing errors, then spectrally analyzed.

At each site an access hole 30 cm in diameter was drilled with an auger. Over this hole a tripod was erected, and the instrumentation was lowered into the hole by winch. Once clear of the hole at the lower boundary, the fluxatron was raised until a rubber flange fitted to its top was flush with the underside, effectively sealing the hole and minimizing disruption to the boundary layer flow. When positioned in this manner, the orientation of the fluxatron to the vertical was observed to be $0 \pm 5°$, and the centers of the top and bottom arrays were 31 cm and 231 cm below the ice-seawater boundary.

Measurements of the salinity and temperature profiles were collected simultaneously with the flux observations. These were obtained by a Hamon Bridge. The sensor was lowered through another hole drilled in the vicinity of the access hole. By this means, not only was information on the entire water column to 100 m obtained, but in situ checks on the absolute values of conductivity and temperature as ascertained by the flux instruments could be made.

Ice thicknesses were also recorded at intervals during the data runs. A steel rod attached at its center to a measuring tape was lowered through each access hole until clear

TABLE 1. Characteristics of Fluxatron Sensors

Instrument	Frequency Response	Absolute Accuracy	Resolution
Current meter[a]	0.3 Hz	0.04-0.1 cm/s	0.01 cm/s
Salinity meter	0.6 Hz	0.04×10^{-3}	0.02×10^{-4}
Temperature meter	0.6 Hz	0.02°C	0.0005°C

[a]Threshold current is 0.01 cm s^{-1}.

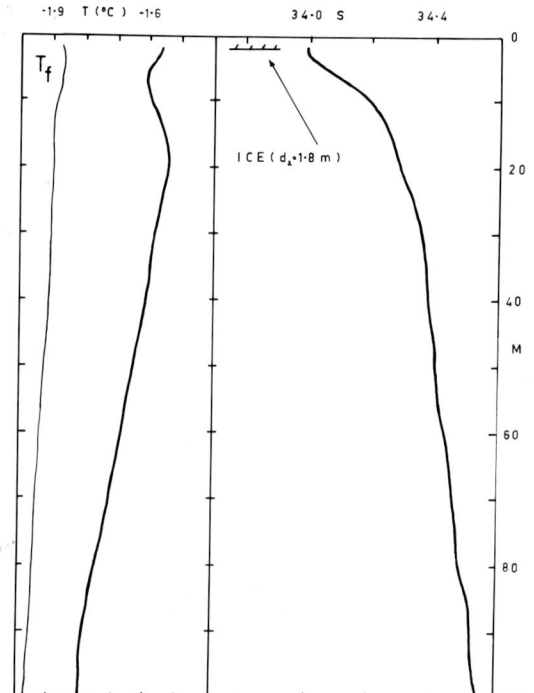

Fig. 3. Vertical profiles of mean temperature and salinity at S1. Freezing temperture (T_f) after Millero [1978].

of the underside, then raised until flush with the interface. This procedure was repeated several times until the measurements of ice thickness (d_1) agreed to within 1 cm. A useful check on these estimates was to measure the distance (d_2) between the water level in the access hole and the bottom surface of the ice. If it is assumed that the densities of the ice and the water remain the same, and that the ice is floating in isostatic equilibrium, the ratio d_1/d_2 should be constant. The salinities of the ice cores obtained from the access holes were also measured.

Results

Site Environment Data

At S1, ambient thermohaline conditions in the upper 5 m were temperatures of about -1.56°C with variations of ± 0.05°C and salinities ranging ± 0.04 about a mean of 34.01. The mean temperature profile at greater depths shows a maximum between 15 and 20 m, below which there is a steady decrease to 100 m. The mean salinity profile has a strong positive gradient in the upper 10 to 15 m, and then a uniform increase to 100 m (Figure 3). There is a significant difference between the observed temperature and the in situ freezing point (T_f) at all depths.

By contrast, at S2 (Figure 4) the upper 5 m exhibited a strong temperature and salinity gradient, while the water column between 10 and 98 was uniform, with a mean temperature and salinity of -1.9°C and 34.48, respectively. The difference between the in situ freezing point (T_f) and the observed temperature is small in comparison with that observed at S1. Note that pressure effects cause the slight decrease with depth of the freezing temperature [Millero, 1978].

At both of these near-shore sites, the water column salinities are lower than shown by Lewis and Perkin [this volume] for an east-west transect of winter (October/November) stations across McMurdo Sound at 77°30'S. However, they also show nearly homogeneous water in the upper 100 m on the west side of the sound, and a well-developed halocline in the top 100 m on the east side.

Ice thickness measurements showed that there was a significantly higher mean melting rate at S1 than at S2 (29 mm d^{-1} compared to 1 mm d^{-1}). Cores of the ice cover at S1 exhibited salinities between 6.5 and 4.3 and a thickness of about 2 m. At S2, the thickness was about 2.5 m with salinities in the range of 0 to 3.4. The lower salinity of the sea ice at S2 could be attributed to older ice there. The presence of freshwater runoff from the nearby Hobbs Glacier may cause the very low near-surface salinity at S2.

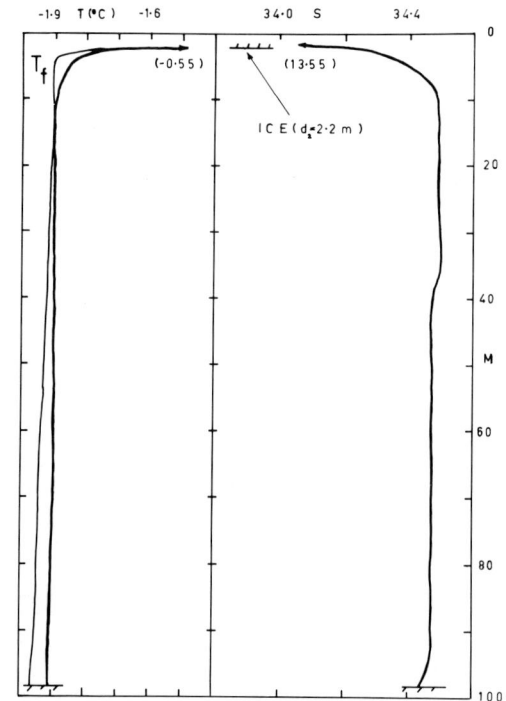

Fig. 4. Vertical profiles of mean temperature and salinity at S2.

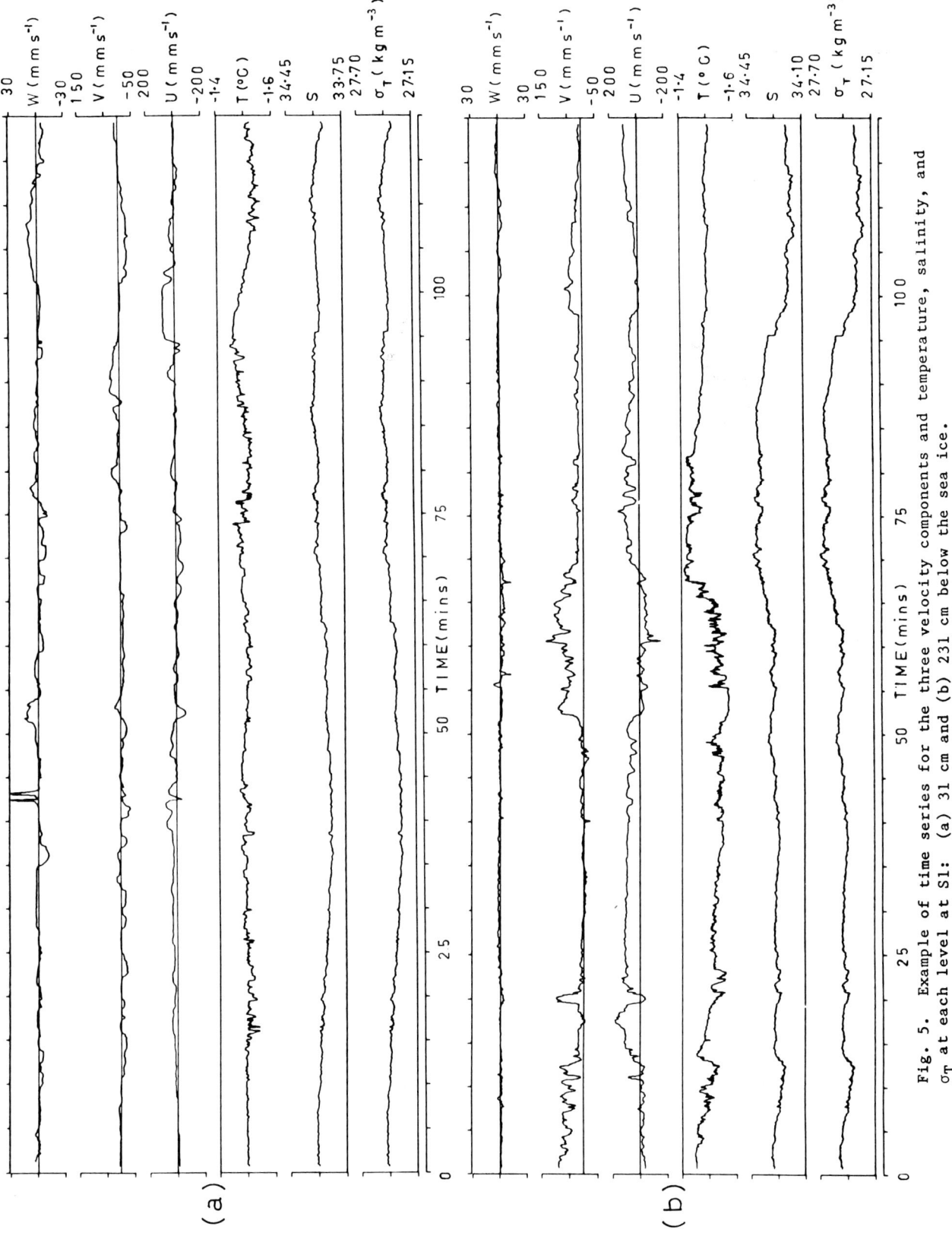

Fig. 5. Example of time series for the three velocity components and temperature, salinity, and σ_T at each level at S1: (a) 31 cm and (b) 231 cm below the sea ice.

High-Resolution Observations

Figures 5 and 6 are two examples of the records obtained with the high-resolution instruments at S1 and S2, respectively. The salinity and density records were calculated from the standard relationship between conductivity and temperature. The records demonstrate many interesting features. At S2, the vertical velocity component shows relatively long periods of "updraughts" and "downdraughts" (at times 5 to 25, 45 to 55, 75 to 90, and 113 to 132 min) at the upper level. These events are largely confined to the upper array. At the lower level at S1, there is a change of about 90° in direction of the horizontal current after 17 min. Noticeable at both sites is the anisotropic nature of the flow, with fluctuations in current speed being usually much larger in the horizontal than in the vertical and with the vertical currents being more variable at the upper than at the lower level. The horizontal velocity fluctuations are also very large in comparison to the mean speed.

Density fluctuations were mainly a result of salinity variations. Good examples of this are at 70 and 77 min in the S2 upper level and at 87 min in the lower level records. At S2 there are several very large excursions in density (at 30 and 70 min at the upper level and 84, 110, and 133 min at the lower level).

The relatively large fluctuations of horizontal velocity superimposed on a small mean value presents a problem in the interpretation of spectra. Nearly all spectral theories utilize the horizontal wave number found by assuming a turbulent structure to be imbedded in a relatively large mean flow. This hypothesis cannot be justified for many of the velocity records obtained at S1 and S2.

Normalized frequency spectra therefore are presented (Figures 7-10) in which spectral quantities have been normalized with respect to their value at the frequency 0.08 Hz, then multiplied by their respective frequency, so that a -n frequency dependence is represented by a line of slope 1-n. As Figures 7, 8, and 9 show, there is a frequency dependence of -3 for all velocity spectra at the higher frequencies contained in these records. The range of this frequency dependence is possibly somewhat less for the vertical component ($\gtrsim 3 \times 10^{-2}$ Hz) than for the two horizontal components ($\gtrsim 2 \times 10^{-2}$ Hz). At the lower frequencies the spectral densities for all the velocity components are only weakly dependent on frequency. For the density spectra the mean slope is about $-2\frac{1}{2}$ for the higher frequencies ($\gtrsim 2 \times 10^{-2}$ Hz), decreasing to about $-1\frac{1}{2}$ for the lower frequencies ($\lesssim 2 \times 10^{-2}$ Hz). The temperature and salinity spectra show a similar behavior to the density spectra. There is no evidence of the Kolmogorov universal equilibrium ranges at high frequencies. Table 2 shows the vertical heat and salt fluxes at each level together with the boundary fluxes for the observed rates of melting (29 mm d^{-1} at S1; 1 mm d^{-1} at S2). Records 1 and 4 correspond to Figures 5 and 6, respectively.

The components of apparent slope (X, Y) of the bottom of the ice have been calculated by solving the simultaneous pairs of equations

$$\bar{u}_1 X + \bar{v}_1 Y = \bar{w}_1$$

$$\bar{u}_2 X + \bar{v}_2 Y = \bar{w}_2$$

in which $\bar{u}_1$ and $\bar{u}_2$ are the observed mean current vectors at the two levels and define a plane. Vertical turbulent flux estimate (1) in Table 2 is the component normal to this plane. Vertical turbulent flux estimate (2) has been obtained by high-pass filtering the records with a cutoff frequency of 1.67×10^{-3} Hz; ie., contributions due to fluctuations of period $\geqslant$ 10 min have been suppressed. This cutoff frequency is similar to the lowest frequency of the frequency spectra (Figures 7-10). The cross-correlation coefficients for the high pass records are defined by

$$R_T = \frac{\overline{w'T'}}{(\overline{w'^2})^{1/2} (\overline{T'^2})^{1/2}}$$

$$R_S = \frac{\overline{w'S'}}{(\overline{w'^2})^{1/2} (\overline{S'^2})^{1/2}}$$

where a prime denotes a fluctuation, and the boundary fluxes have been calculated using the formulae

$$F_T = \rho M L$$

$$F_S = M(\rho \bar{S} - \rho_I \bar{S}_I)$$

where L is the latent heat of melting of ice, ρ and S are the water density and salinity, respectively, and ρ_I and $\bar{S}_I$ are the ice density and salinity, respectively, at the ice-water interface. At S1, the bottom slope estimates are all less than the error in positioning of the fluxatron which corresponds to a slope of $\pm$ 0.1, and hence conditions are essentially indistinguishable from a flat ice bottom. It is clear, however, that the high-pass fluxes are significantly smaller than the normal fluxes and somewhat smaller than the boundary fluxes. We conclude that an important part of the melting fluxes is contained in fluctuations of periods $\geqslant$ 10 min. The length of our records (1-2 hours), however, precludes an accurate evaluation owing to problems in stationarity. Figure 5 indeed

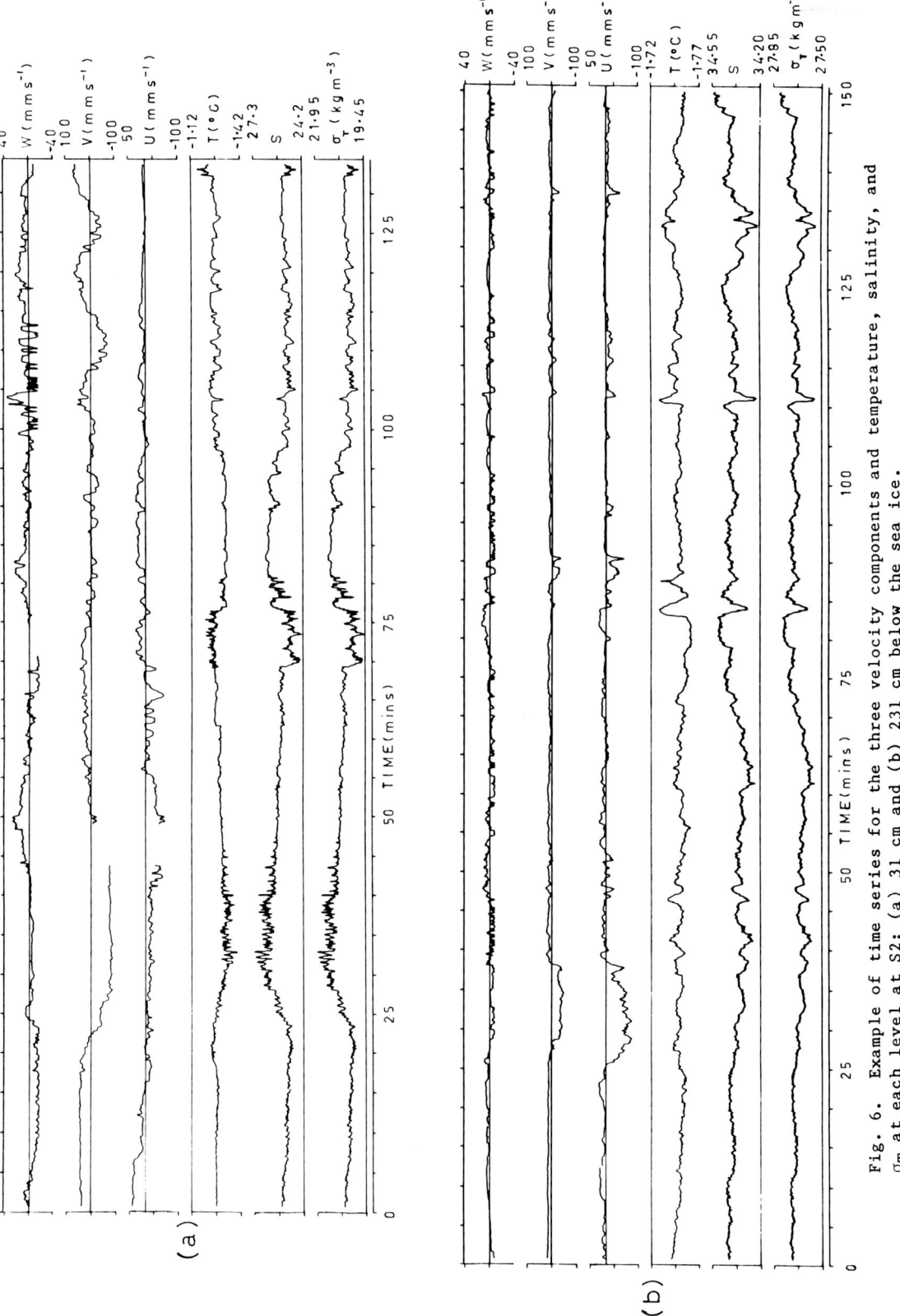

Fig. 6. Example of time series for the three velocity components and temperature, salinity, and σ_T at each level at S2: (a) 31 cm and (b) 231 cm below the sea ice.

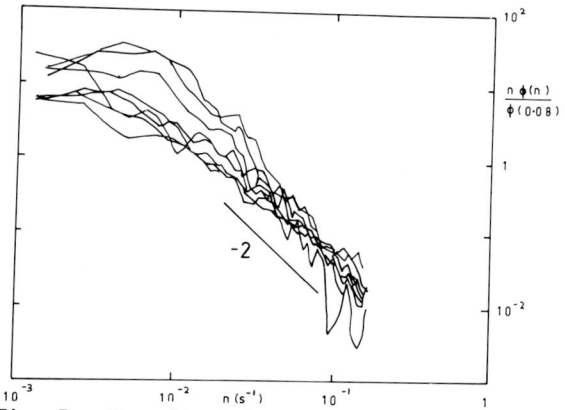

Fig. 7. Normalized longitudinal velocity frequency spectra.

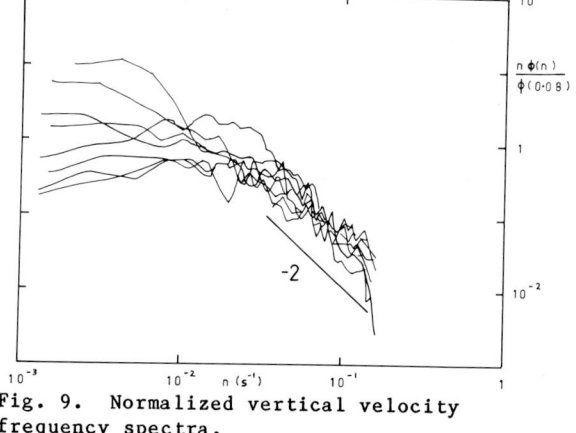

Fig. 9. Normalized vertical velocity frequency spectra.

shows a clear pattern of long-term variability in which w is nearly, but not exactly, in quadrature with T and S. At S2 the apparent bottom slope is very large and corresponds to an angle to the horizontal of $\sim 30°$. This result is almost certainly an artifact of the analysis, since Figure 6 shows that w_1 has an important long-term variability, and its mean value ($\bar{w}_1 = 1.3$ mm s^{-1}) is large. The normal turbulent fluxes accordingly are also very large and must be regarded as misleading. The high-pass fluxes are also greater than the boundary fluxes, in contrast to S1. We attribute this to the presence of a highly turbulent, horizontally inhomogeneous flowfield.

Discussion

The spectral results will be interpreted in terms of the basic concepts of the generation of turbulence by internal waves and its extraction by buoyancy forces. The classical theory [Lumley, 1964; Phillips, 1966] of the buoyancy subrange in which the extraction of energy occurs predicts that the scalar energy density should be proportional to the scalar wavenumber to the power of -3, and to the square of the Brunt-Väisälä frequency (N). The fundamental result from internal wave theory is that the maximum wave frequency is $\sim N$. Hence one may expect a system in which N marks a division in frequency between dynamics primarily controlled by internal waves and by turbulence (for the energy spectra) or fine structure (for the scalar spectra). Phillips [1966] notes that for this division to occur the energy density of the large-scale internal waves must be large enough to generate forced modes of similar amplitudes. Our observations are in qualitative agreement with this picture, provided that the frequency spectrum of the buoyancy subrange has a -3 frequency dependence in analogy with the wave number spectrum. This conclusion is based on the following observed spectral properties.

1. A change in slope of the energy spectra

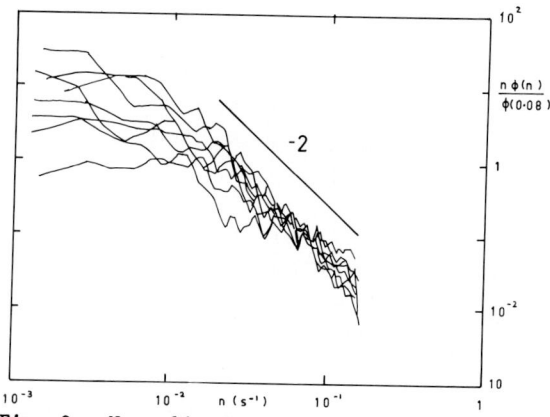

Fig. 8. Normalized transverse velocity frequency spectra.

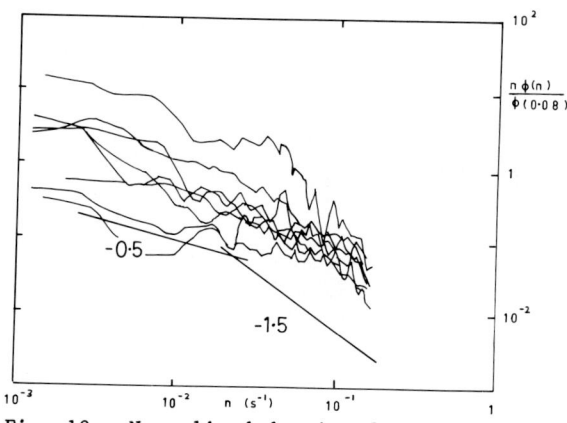

Fig. 10. Normalized density frequency spectra.

occurs at $\sim 2 \times 10^{-2}$ Hz, which is just greater than the maximum Brunt-Väisälä frequency (N $\sim 1 \times 10^{-2}$ Hz) in the upper 10 m at S1 (Figure 3). At S2, N has a similar value at this depth (Figure 4).

2. The energy spectra for each of the velocity components exhibit a -3 frequency dependence, and a significant anisotropy [Gordon, 1981] between the vertical and horizontal components.

3. There appears to be no systematic difference between sites in energy densities in the -3 range. This result (which may be fortuitous) is consistent with the existence of similar values of N at each site. The selection of N at S2 which predicts this coincidence of energy densities is possibly somewhat biased. A Brunt-Väisälä frequency based on the vertical density gradient of the thin, very stable surface layer (N $\sim 2 \times 10^{-1}$ Hz), which is probably only a very localized phenomenon, would, however, lead to a prediction of energy densities at S2 much greater than at S1, and this was not observed.

4. Energy spectral plateaus occur at frequencies ≲ N which suggest the approach to a saturated internal wave field.

5. The scalar spectra also show a change in slope at N $\sim 10^{-2}$ Hz, and our slopes agree with the slopes obtained by Levine and Irish [1981] for moored temperature spectra in which the slopes attributed to internal waves and fine structure were -1.5 and -2.5, respectively.

In summary, our spectral observations appear to be consistent with the existence of energetic buoyancy and fine structure subranges at both sites. It is interesting that velocity and temperature frequency spectra with similar slopes have been obtained in the contrasting environment of a stable layer at 700 m in the tropical atmosphere [Berman, 1976]. A discussion of recent observations of the buoyancy subrange and their interpretation in the upper ocean is given by Gargett et al. [1981].

On a larger scale at S1 the properties of the upper 100 m of the water column are typical of the summer melting regime near McMurdo Station [Tressler and Ommundsen, 1962], which has been interpreted in terms of warm advection from the north [Heath, 1971] that leads to the annual breakout of seasonal sea ice. Our data suggest that a minor part of the heat and salt fluxes associated with the melting process occurs in the frequency band of the fine structure/internal waves represented in Figures 7-10 and a major part through correlations at periods ⩾ 10 min.

Thus we may speculate that the existence of a vigorous internal wave field on the shallow thermocline which occurs under the ice during the melting process enchances the rate of melting. In other words, rapid melting results through a positive feedback process in which initial melting gives rise to a shallow

TABLE 2. Vertical Turbulent and Boundary Fluxes at S1 and S2

Site	Record No.	Record Length s	Vertical Turbulent Fluxes				Boundary Fluxes		Horizontal Speed mm s^{-1}		Boundary Slope	Flux Correlation Coefficients (2)	
			$C_p\overline{T'w'}$ Wm^{-2}		$\overline{S'w'}$ mg m^{-2} s^{-1}		F_T Wm^{-2}	F_S mg m^{-2} s^{-1}	Q_1	Q_2	$B \times 10^{-3}$	R_T	R_S
			(1)	(2)	(1)	(2)							
S1	1	6919	-94	-19	-11	12	6	1	11	53	11	0.02 0.14	0.11 0.03
	2	3259	225	38	-49	38	--	--	5	12	55	--	--
	3	6702	1	-14	-3	-32	1	-0	40	61	14	-0.05 -0.11	0.18 -0.01
S2	4	7037 8764	580	-7	-3180	57	-11	2	4	6	600	-- 0.15	-- -0.11

$B = (X^2 + Y^2)^{1/2}$, $Q_1 = (\overline{u_1^2} + \overline{v_1^2})^{1/2}$, $Q_2 = (\overline{u_2^2} + \overline{v_2^2})^{1/2}$, and C_p is the specific heat at constant pressure of seawater. The entry pairs refer to the upper (31 cm) and lower (231 cm) levels, respectively. (1) refers to the normal fluxes and (2) to the high-pass fluxes (cutoff frequency 1.67×10^{-3} Hz). The differences in record lengths between levels are due to gaps in the data. Dashes indicate records of insufficient 'gap free' length for high-pass analysis. See Figures 5 and 6 for Record Nos. 1 and 4, respectively.

pycnocline which favors the generation of internal waves. The internal waves bring warmer, sub-boundary layer water into contact with the ice, increase the melting, and strengthen the pycnocline. We may call this process the "oceanic melting regime," since an essential feature is warm advection from the adjacent ice-free ocean.

At S2, on the other hand, the water column below 10 m is almost uniform with a temperature very close to the 10-m freezing temperature. This is consistent with an origin under the Ross Ice Shelf [Gilmour, 1975]. The relatively fresh and warm surface water is probably due to irregular penetrations of meltwater from the nearby Hobbs Glacier as surface intrusions which advect water at in situ freezing temperatures. This surface inflow raises the temperature of the sea ice, and melting occurs if the temperature attains the melting point. Thus we speculate that the instantaneous melting front lies essentially along the line where the temperatures of the sea ice and the surface intrusion are identical. Seaward of this line (in the absence of warm oceanic advection), melting does not occur. This process gives rise to the inshore coastal leads and may be called the "continental melting regime," since it is initiated by run-off from the neighboring continent.

Our sets of measurements, by good fortune, appear to monitor both these regimes.

Acknowledgments. The encouragement of Peter Schwerdtfeger during all phases of the project is gratefully acknowledged and also the work of Scott Gordon, Max Whittington, and Allan Suskin without which these observations could not have been made. The project was supported by an Australian Research Grants Committee grant 15066, and facilities at McMurdo Sound were graciously made available through the National Science Foundation.

References

Berman, E. A., Measurements of temperature and downwind spectra in the "buoyant subrange," J. Atmos. Sci., 23, 495-498, 1976.

Clough, J. W., and B. L. Hansen, The Ross Ice Shelf Project, Science, 203, 433-434, 1979.

Gargett, A. E., P. J. Hendricks, T. B. Sanford, T. R. Osborn, and A. J. Williams, A composite spectrum of vertical shear in the upper ocean, J. Phys. Oceanogr., 11, 1258-1271, 1981.

Gilmour, A. E., McMurdo Sound hydrological observations, 1972-1973, N.Z. J. Mar. Freshwater Res., 9, 75-95, 1975.

Gordon, S. H., Boundary layer flow under an Antarctic ice sheet, Res. Rep. 37, Flinders Inst. for Atmos. and Mar. Sci., 144 pp. Flinders Univ. of South Australia, Bedford Park, South Australia, 1981.

Heath, R. A., Circulation and hydrology under the seasonal ice in McMurdo Sound, Antarctica, N.Z. J. Mar. Freshwater Res., 5, 497-515, 1971.

Levine, M. D., and J. D. Irish, A statistical description of temperature finestructure in the presence of internal waves, J. Phys. Oceanogr., 11, 676-691, 1981.

Lewis, E. L., and R. G. Perkin, The winter oceanography of McMurdo Sound, Antarctica, this volume.

Lumley, J. L., The spectrum of nearly inertial turbulence in a stably stratified fluid, J. Atmos. Sci., 21, 99-102, 1964.

Mied, R. P., and F. J. Merceret, Jr., The construction of a simple conductivity probe, Internal report, 27 pp., Dept. of Mech., Johns Hopkins Univ., Baltimore, Md., 1974.

Millero, F. J., Freezing point of sea water, Eighth Report of the Joint Panel of Oceanographic Tables and Standards, Appendix 6, UNESCO Tech. Pap. Mar. Sci., 28, 29-35, 1978.

Phillips, O. M., The Dynamics of the Upper Ocean, 336 pp., Cambridge University Press, New York, 1966.

Steedman, R. K., A solid state oceanographic current meter, J. Phys. E., 5, 1157-1162, 1972.

Tressler, W. L., and A. M. Ommundsen, Seasonal oceanographic studies in McMurdo Sound, Antarctica, Tech Rep. TR-125, 141 pp., U. S. Navy Hydrographic Office, Suitland, Md., 1962.

U.S. Naval Oceanographic Office, Chart H.O. 6666, 4th ed., Suitland, Md., 1969.

(Received August 27, 1984; accepted December 27, 1984).

A RECURRING, ATMOSPHERICALLY FORCED POLYNYA IN TERRA NOVA BAY

Dennis D. Kurtz and David H. Bromwich

Institute of Polar Studies, The Ohio State University, Columbus, Ohio 43210

Abstract. The Terra Nova Bay polynya is a large, stable, annually recurring feature in the western Ross Sea which markedly influences sea ice dynamics and physical oceanography in that region. Strong, persistent katabatic winds which blow far offshore, and blockage of northward drifting sea ice by the Drygalski Ice Tongue are both necessary for polynya existence. Secondary factors, such as katabatic and synoptic wind interactions, and seawater depth, characteristics and circulation also affect the polynya. However, it is the absence of this katabatic-blocking combination along other windy coasts that explains why similar polynyas do not form there. Direct though limited measurements provide strong confirmation that the katabatic conditions previously inferred from qualitative historical accounts occur each year. The winter wind regime at Inexpressible Island is similar to that at coastal Port Martin, the second windiest location in the Antarctic. Thermal infrared satellite images contain ample evidence that the katabatic winds can extend well beyond Terra Nova Bay. In addition, the images suggest that "anomalous" longevity is a common feature of katabatic drainage through the Transantarctic Mountains onto the Ross Ice Shelf. Seasonal and winter time series satellite data document the Drygalski Ice Tongue blocking effect, and suggest that decreases in polynya area reflect rapid sea ice freezing in response to local weakening of katabatic wind action. The latter finding means that synoptic forcing is important only during periods of major polynya expansion, and explains why this polynya's areal fluctuations are weakly correlated with the zonal component of the surface geostrophic wind in the western Ross Sea. Sensible heat in northward flowing High Salinity Shelf Water (HSSW) and solar energy locally stored during summer can supply only a small fraction of the annual energy loss from the polynya to the atmosphere; the remainder comes from latent heat released by the freezing of seawater. Allowing for these energy sources, we calculate cumulative annual ice production of ~ 60 m from the surface energy balance. This estimate is an upper limit on possible ice production because of the uncertain impact upon spatially averaged energy fluxes of new ice formed in the polynya margins. Ice production in Terra Nova Bay amounts to 10% of the total formed over the Ross Sea continental shelf. Brine rejected during surface freezing of seawater may play a key role in maintaining the HSSW.

Introduction

The complexities of wintertime sea ice dynamics and its influence on air-sea energy exchange, water column characteristics, sea ice mass balance, and biological productivity are being increasingly appreciated. Coastal polynyas, located within winter pack ice limits, are of particular importance because of their intricate interactions with the atmosphere, ocean, and surrounding pack ice. They produce a significant fraction of the annual sea ice surrounding Antarctica and, via the formation of saline shelf water, play an important role in the production of Antarctic Bottom Water [Zwally et al., this volume; Cavalieri and Martin, this volume]. Coastal areas of open water have been noted for many years in the Ross Sea [Priestley, 1914; U.S. Naval Hydrographic Office, 1960; Stonehouse, 1967; Streten, 1973], but detailed observations during the polar night have only been available since the advent of infrared and microwave satellite imagery. O'Connor and Bromwich [unpublished manuscript] have shown that polynyas can form on the east and west sides of Ross Island near Cape Crozier and Cape Royds as a result of the deflection of southerly barrier winds by the steep volcanic island. Zwally et al. [this volume] studied a polynya that forms along the front of the Ross Ice Shelf to the east of Cape Crozier and demonstrated that it fluctuates in response to synoptic scale winds. Such investigations provide an approximate idea of the governing physical mechanisms, but suffer from the absence of an accurate depiction of the wind and temperature fields affecting the open water. For example, katabatic winds draining through the Transantarctic Mountains also appear to influence the polynya studied by Zwally et al. [this volume] which will be discussed later. Detailed analysis of the formation and fluc-

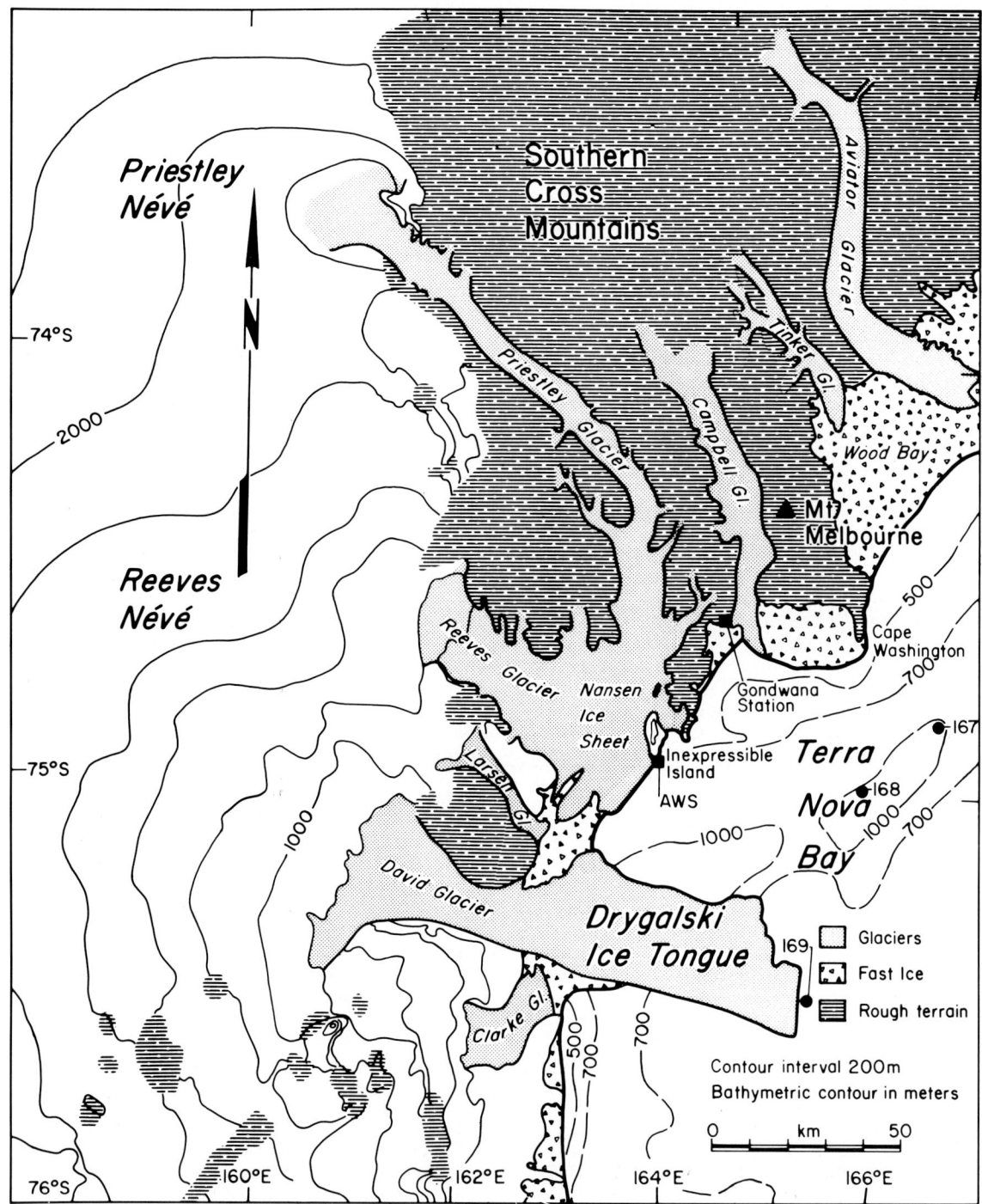

Fig. 1. Neighborhood of Terra Nova Bay. Solid circles denote locations of hydrographic profiles in Jacobs and Haines [1982].

tuations of a prominent, recurring coastal polynya permits the important physical mechanisms common to all these features to be identified. The regional impact of these atmospheric heat sources and oceanographic heat sinks can then be assessed.

The polynya in Terra Nova Bay (Figures 1 and 2) is a dynamic and thermally anomalous feature in the western Ross Sea. Though cartographically an embayment bordered on the south by the Drygalski Ice Tongue, and on the west and north by the Victoria Land coast, the

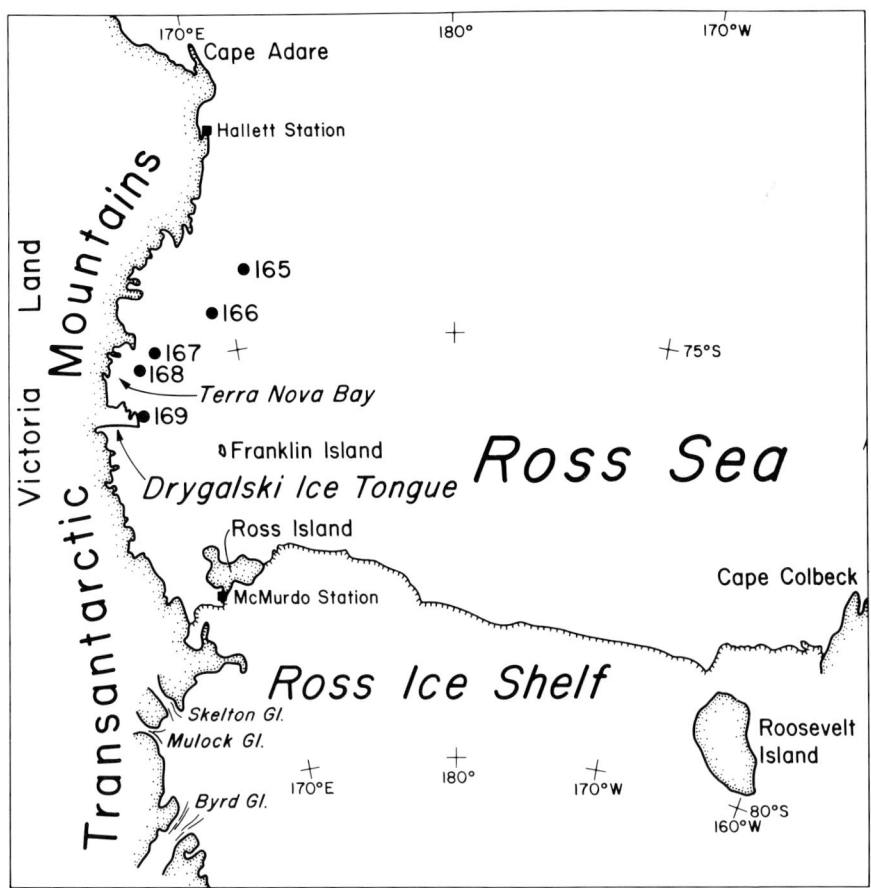

Fig. 2. Regional map. Filled circles give positions of hydrographic stations.

bay is not oceanographically separated below the approximate 100-m draft of the Drygalski Ice Tongue from the open Ross Sea (see bathymetry, Figure 1). Wintertime open water in Terra Nova Bay was first reported by the men of Scott's Northern Party who noted that part of the bay was ice free throughout the winter of 1912 [Priestley, 1962]. Knapp [1972] first reported this using satellite imagery, and hypothesized that it was one of a class of polynyas that form when strong winds associated with passing cyclones blow ice away from the lee of fixed barriers. Our comparison of polynya area fluctuations and synoptic wind conditions [Kurtz and Bromwich, 1983] indicates that this is unlikely. While the area of open water in Terra Nova Bay probably does fluctuate in response to synoptic interactions with polynya forcing mechanisms, the persistence of open water throughout the winter, regardless of synoptic conditions, indicates that it is not formed by passing cyclones. Szekielda [1974] later noted the polynya and attributed its presence to submarine volcanic activity and/or upwelling of warm deep waters which prevented sea ice from freezing. Bottom sediment samples and sonar data collected in Terra Nova Bay during Deep Freeze 80 [Anderson and Kurtz, 1980] indicate no submarine volcanic activity. No exceptionally warm deep waters have been sampled in this area [Jacobs and Haines, 1982] and we have shown [Bromwich and Kurtz, 1984] that none need be present. The polynya can be formed and maintained through the action of the "plateau wind" endured by Scott's Northern Party, and believed by them to have kept the bay partly ice free.

In 1979, the polynya, and a surrounding region which we interpret to have consisted of thin or loosely consolidated pack ice, occupied roughly 25,000 km^2 (Figures 3a and 3c). Mean polynya area was 1300 km^2, varying from near zero to 5000 km^2. A nucleus of open water was virtually always present offshore from Inexpressible Island and the outlet of the Reeves Glacier (Figures 1 and 3a). At its maximum extent the entire bay was ice free, with the seaward polynya boundary being as far east as the eastern tip of the Drygalski Ice Tongue. The area of thin or loose pack extended east and northeast of the polynya, and was increasingly colder and more consolidated farther from the polynya (Figure 3a). The designations "thin or loose pack ice" are opera-

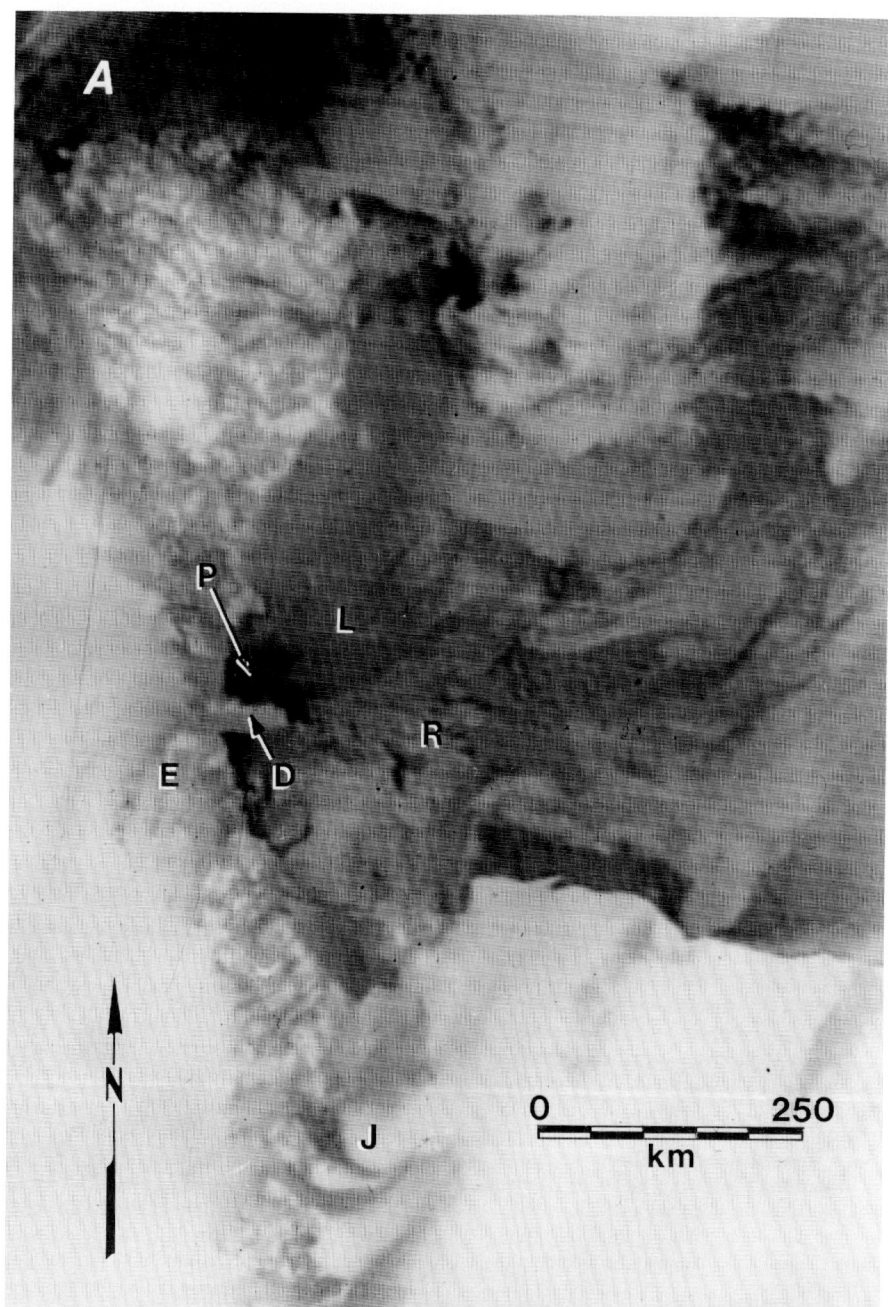

Fig. 3. Thermal infrared satellite (Defense Meteorological Satellite Program) images of the Western Ross Sea. Light tones indicate cold temperatures, dark tones are warm. Symbols are identical throughout; D, Drygalski Ice Tongue; E, thermal signature of katabatic winds; L, zone of unconlidated pack ice; P, polynya; R, Ross Sea pack ice; J, katabatic jet. (a) April 19, 1979; autumn polynya. (b) July 22, 1978; winter polynya (partially obscured by a thin synoptic scale cloud field). (c) August 10, 1979; late winter polynya.

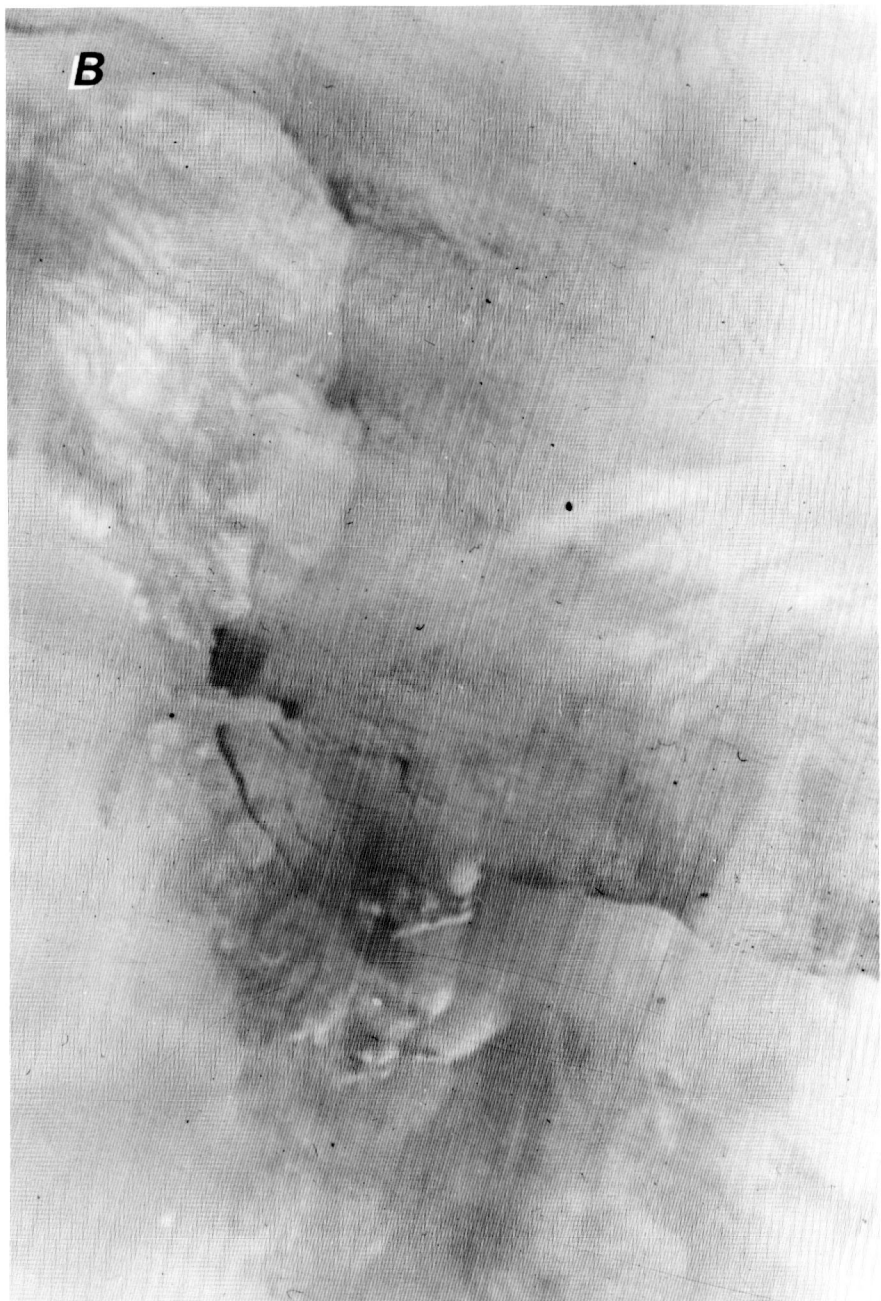

Fig. 3. (continued)

tional ones. The thermal signature of this zone is consistent with the presence of a thin ice-covered sea surface but the visible network of fractures and leads characterizing the Ross Sea pack ice field is absent. Based upon the accumulating evidence that katabatic winds can blow far offshore (discussed later), we interpret this region as being due to katabatic winds which, while not able to maintain essentially open water conditions, prevent sea ice consolidation. Observed variations in the shape, size, and boundary locations of the zone of loose pack during winter 1979, and its thermal continuity with the eastern edge of the polynya, suggest sea ice present there formed in the polynya and was advected eastward by katabatic winds. The region appears physically continuous with the Ross Sea pack ice, though a band of cold, perhaps thick, sea ice is often present between them [Szekielda,

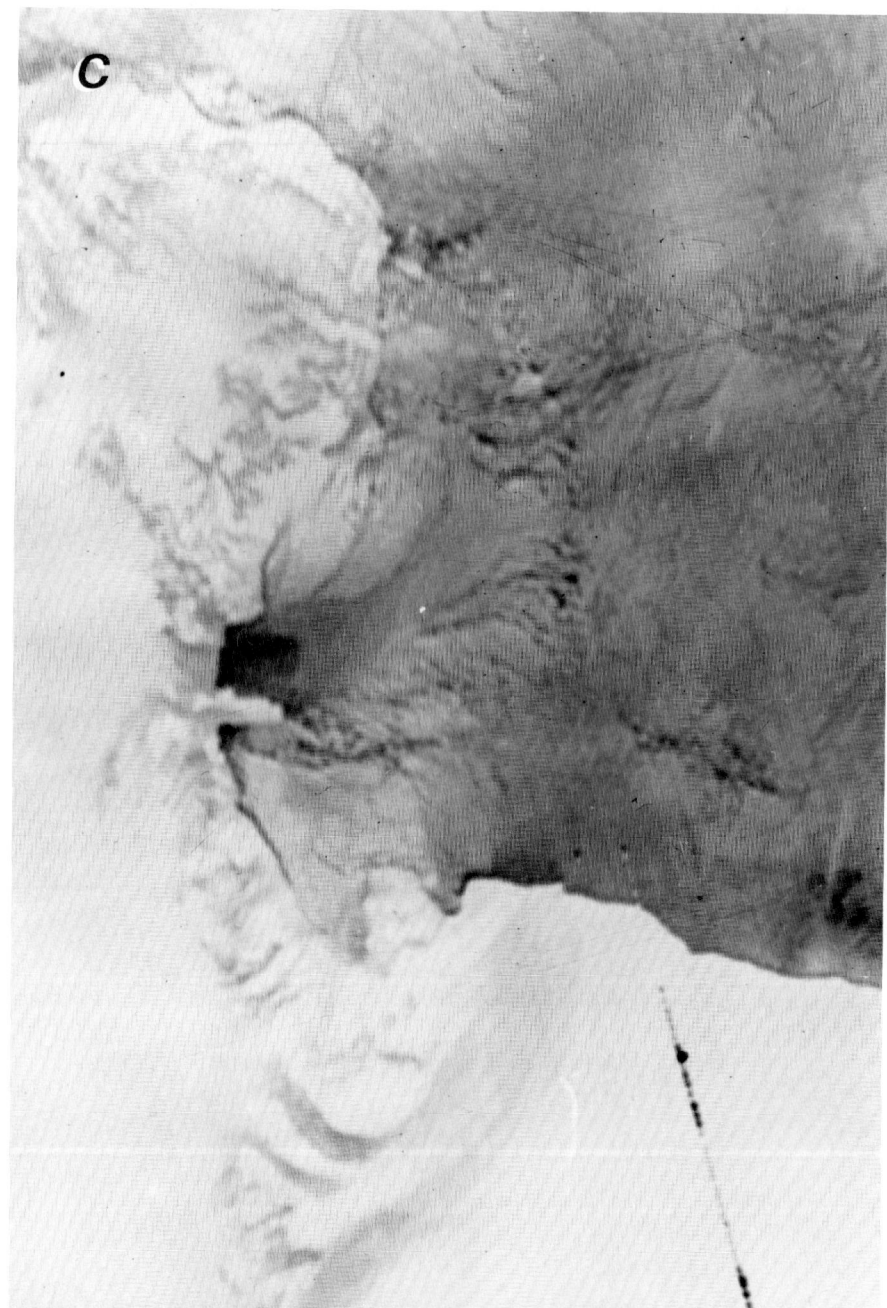

Fig. 3. (continued)

1974; Kurtz and Bromwich, 1983]. This band separates the polynya and surrounding regions from the Ross Sea, and may consist of ice from the zone of loose pack that accumulates at the edge of the Ross Sea pack ice field.

We have formulated a multifaceted conceptual model concerning the processes that create and maintain this polynya. On site measurements needed to test various aspects of this thesis are currently being obtained. At present, our understanding relies principally on evaluation of historical records and infrared satellite images, constrained where possible by independent meteorological and glaciological observations. Salient factors in developing this model are that the relevant Antarctic meteorological processes and glacial features recur annually, or are stable on decadal time scales. For example, winter surface winds on the polar plateau are assumed to blow in the same patterns from year to year, and throughout the winter season. (Seasonal terms are defined as follows: winter refers to the period without sunshine (early May to mid-August); summer, with 24 hours of daylight, extends from early November to mid-February; spring and autumn encompass the transitional periods.) This stability reflects the dominant influences of terrain slope and winter surface temperature inversion upon surface airflow over the ice sheet [Parish, 1982]. Such considerations enable us to utilize numerical simulations and data from different years to support our contentions. This conceptual framework reflects an interdisciplinary synthesis of several lines of evidence.

The previously elucidated model [Bromwich and Kurtz, 1982, 1984; Kurtz and Bromwich, 1983] is summarized, discussed, and updated here. In particular, direct though limited measurements of the katabatic outflow in Terra Nova Bay show close agreement with earlier inferences from qualitative sources. Also, several lines of evidence suggest that persistence of katabatic winds far beyond the slope break is a common feature of airflow through the Transantarctic Mountains. The seasonal cycles and shorter-term fluctuations of the polynya size are examined with the primary aims of demonstrating that satellite images record the blocking effect of the Drygalski Ice Tongue and that synoptic forcing plays a key role only during periods of substantial polynya expansion. At all other times the behavior of the katabatic wind is the dominating mechanism. Finally, annual ice production in the polynya and the resulting salinization of shelf waters in the western Ross Sea are estimated.

Polynya Forcing

The Terra Nova Bay polynya is thought to be formed and maintained through the combined action of (1) strong, persistent offshore katabatic winds which prevent sea ice from consolidating in the bay, and (2) the blocking action of Drygalski Ice Tongue which prevents existing sea ice from entering Terra Nova Bay from the south. Each of these influences is necessary, though insufficient without the other, and absence of this combination along other windy coasts (e.g. Cape Denison) may in part explain the absence of similar polynyas there.

Katabatic Winds

Existence and seasonal occurrence of katabatic winds in Terra Nova Bay are inferred from several lines of evidence. Numerical modeling of winter surface winds in East Antarctica predicts air confluence and drainage into Terra Nova Bay [Parish, 1982]. Calculations are based upon integrated boundary layer equations which express a steady state balance between the pressure gradient force generated by the sloped surface temperature inversion, the Coriolis effect, and friction. Explicit mass conservation was not required, so only a qualitative picture of the time-averaged drainage mass flux is provided. Parish [1981, 1982] has argued that the prerequisite for strong, persistent coastal katabatic winds is a large cold air reservoir which discharges through a relatively narrow outlet. Surface airflow draining roughly 3% of East Antarctica appears to blow principally down the ~ 15-km-wide Reeves Glacier valley. The simulation is supported strongly by the accounts of Scott's Northern Party who, during the winter of 1912, observed strong, persistent katabatic winds 34 km to the east of that valley exit [Bromwich and Kurtz, 1982], and by sastrugi orientations on the ice sheet west of Terra Nova Bay [David and Priestley, 1914; Stuart and Heine, 1961]. Sastrugi are wind-formed snow surface ridges that are aligned with the prevailing wind direction. Their orientations agree closely with Parish's streamline pattern, and in particular demonstrate that the simulated position of the drainage divide between airflow into Terra Nova Bay and flow to Cape Denison is well located [Bromwich and Kurtz, 1984].

An automatic weather station (AWS) was installed at an elevation of 78 m on the southern tip of Inexpressible Island on February 4, 1984 (Figure 1). It operated flawlessly until April 19, 1984, when transmission stopped abruptly. Very strong katabatic winds were measured during most of the period. Vector average surface winds for February, March, and April were from 295° at 14.1 m s^{-1}, 295° at 18.4 m s^{-1}, and 304° at 16.4 m s^{-1}. Directional constancies were 0.97, 0.99, and 0.98, respectively. The short record suggests that the katabatic conditions encountered by Scott's Northern Party during the 1912 winter [Bromwich and Kurtz, 1982] recur each year; Bromwich and Kurtz [1984] used a historically-derived coastal resultant wind of 293° at 15 m s^{-1} in their scaling analysis of polynya processes. Wind conditions at Inexpressible Island are similar to those prevailing at Port Martin in Adélie Land [Parish, 1981]. This is remarkable because Port Martin sits at the base of the ice slope whereas Inexpressible Island is some 34 km to the east of the Reeves Glacier exit. Because the airflow slows under friction as it crosses the flat Nansen Ice Sheet [Bromwich and Kurtz, 1984], winter wind speeds at the foot of the Reeves Glacier valley must rival those at Cape Denison, a region renowned for the presence of intense katabatic winds [Mawson, 1969].

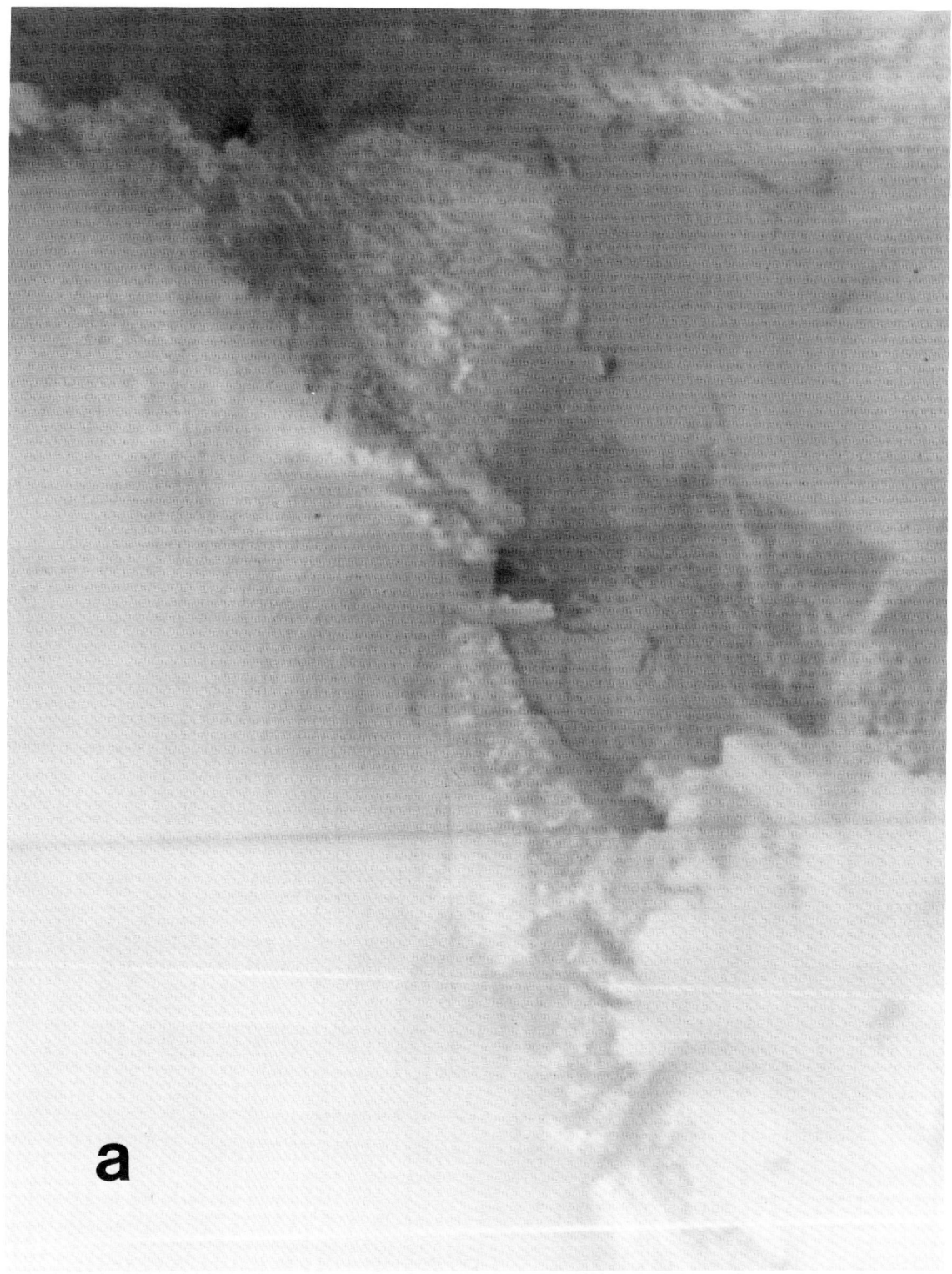

Fig. 4. Satellite infrared sequence illustrating polynya expansion during May 1979; see text for discussion. Symbols are same as Figure 3. (a) May 16. (b) May 21. (c) May 23. (d) May 24.

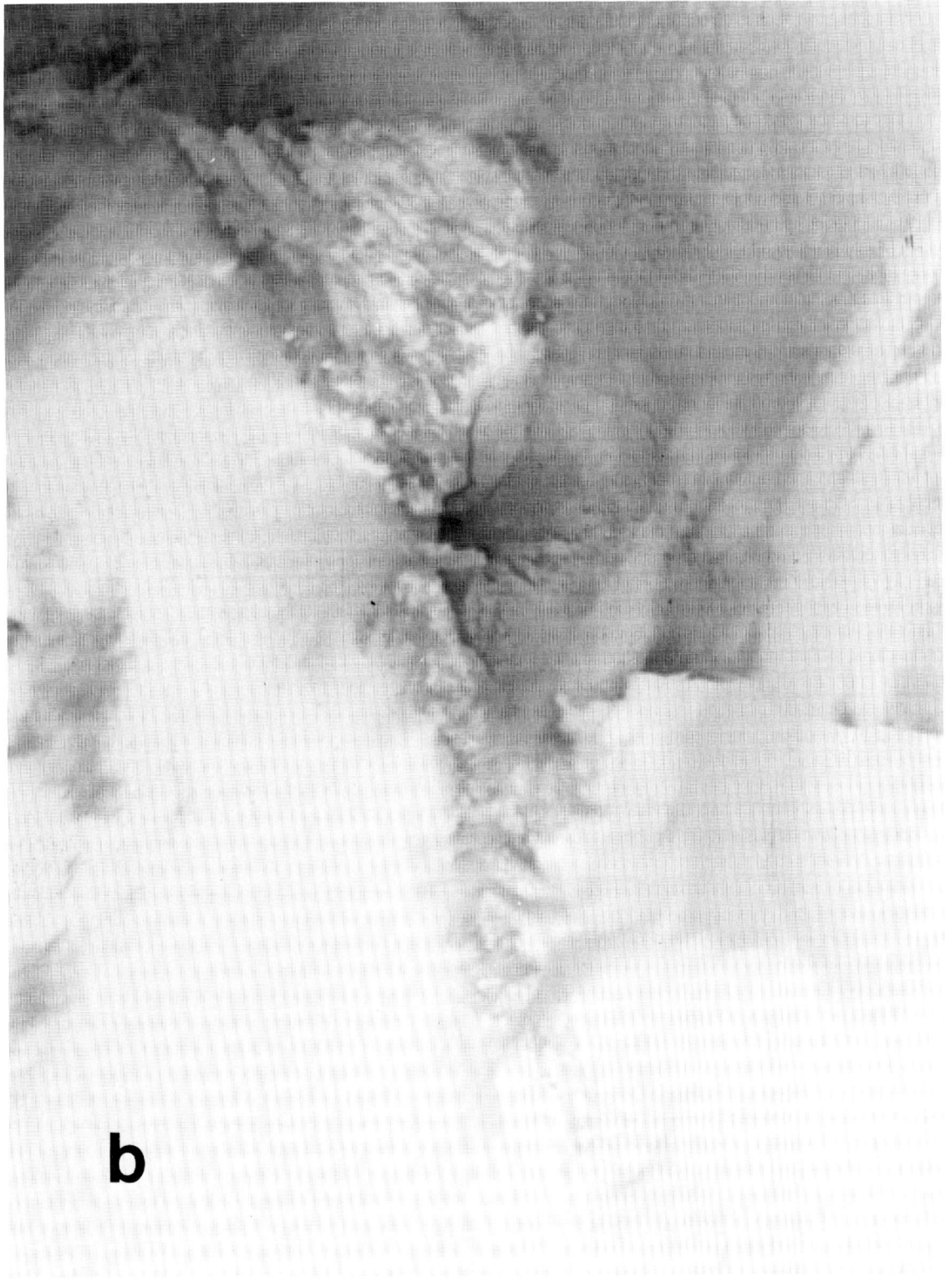

Fig. 4. (continued)

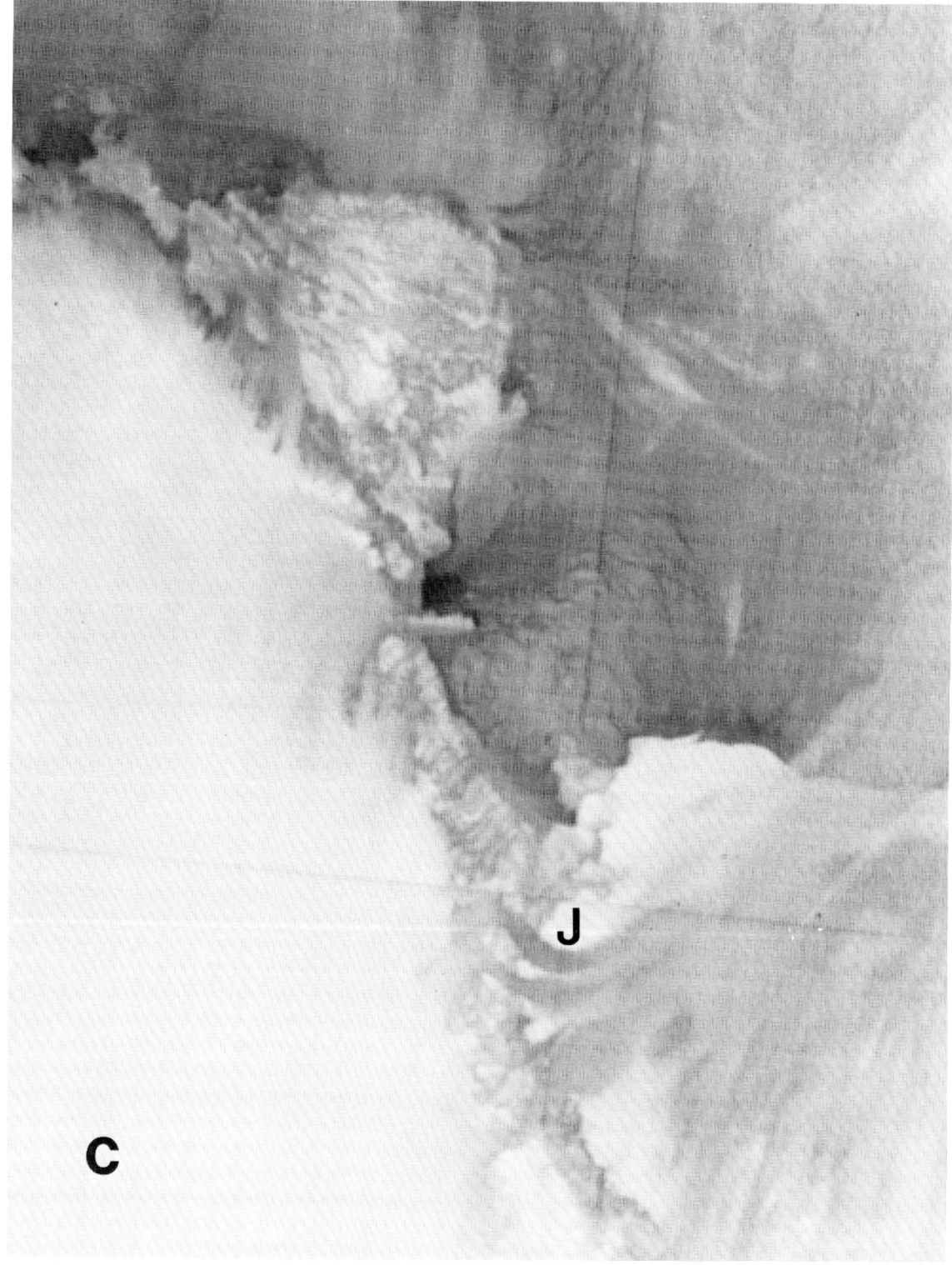

Fig. 4. (continued)

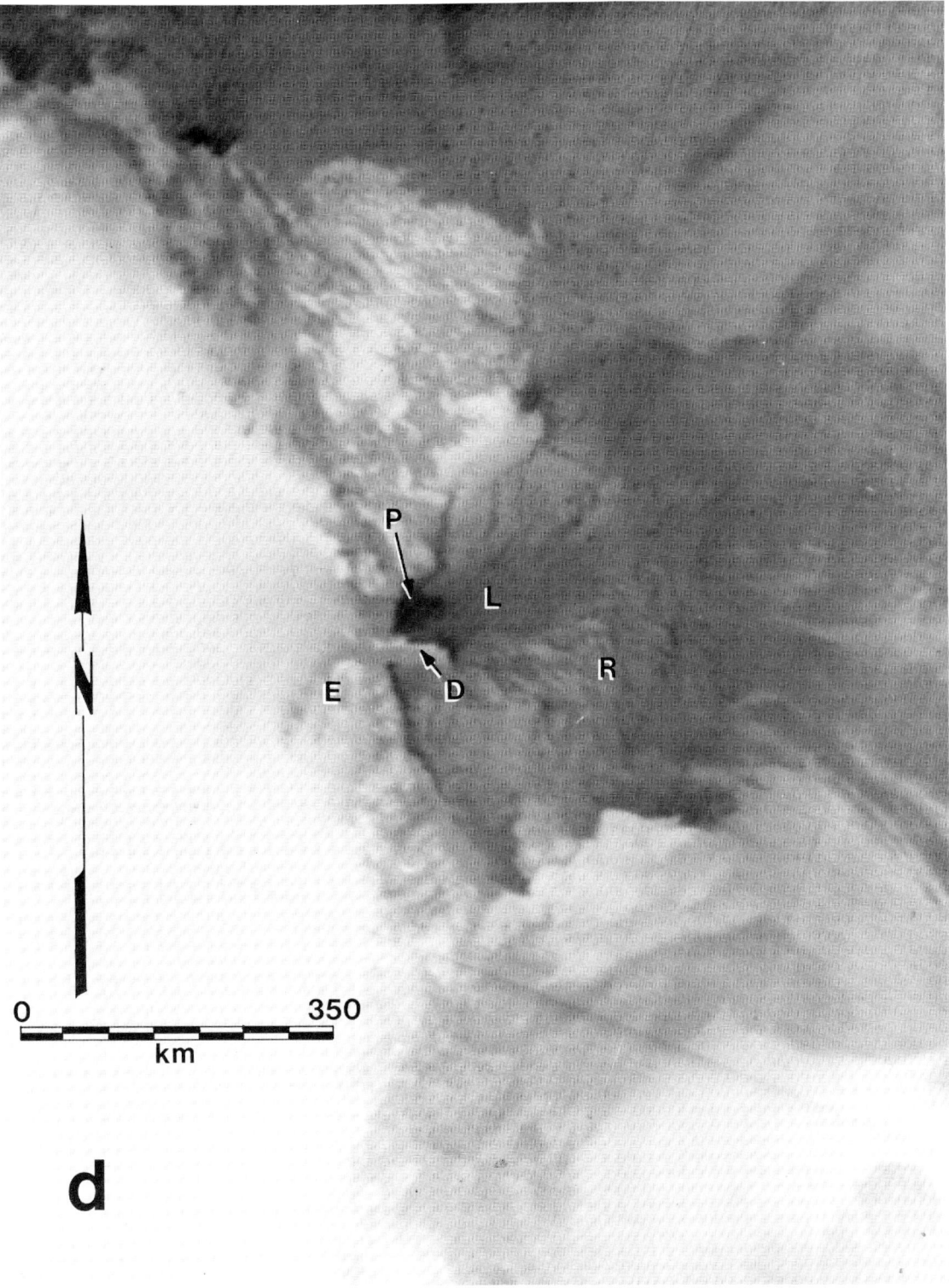

Fig. 4. (continued)

An infrared signature (labeled by E in Figures 3 and 4) thought to reflect the descent of adiabatically warming (drift-bearing?) air was described by Kurtz and Bromwich [1983]. This signature is the warmest feature in and around the western Ross Sea apart from the polynya itself and is present in satellite images throughout the winter. It can be divided into two general regions. The cooler (lighter), fainter area on the polar plateau probably represents the converging air currents modeled by Parish [1982]. The warmer (darker) region adjacent to the coast is most pronounced in the Reeves Glacier valley and adjacent Nansen Ice Sheet. Landsat images and air photographs reveal that the ice surface is very ablated there [Bromwich and Kurtz, 1984]. The coastal signature is very similar to that generated by strong valley winds in Alaska [Marvill and Jayaweera, 1975] and probably occurs for the same reason; i.e., strong (katabatic) winds thoroughly mix the near surface air and cause the ice surface temperature to be significantly higher than that in more quiescent adjacent areas. The extent and borders of this warm area are consistent with observed and inferred paths of katabatic winds. Low relative humidities, generated by dry adiabatic descent of air from the plateau, coupled with high persistent winter wind speeds favor substantial evaporation from the ice surface, consistent with the ablated surface observed during summer.

The apparent persistence of katabatic outflow across Terra Nova Bay contrasts sharply with observations elsewhere along the East Antarctic coast. There katabatic winds typically die out within 10-20 km of the foot of the terrain slope [Ball, 1957; Tauber, 1960; Weller, 1969; Schwerdtfeger, 1970]. This is an important consideration because we infer that the katabatic jet forms and maintains the open water. Rapid dissipation has been attributed to turbulent instability close to the foot of the terrain slope, where the wind experiences an abrupt increase in depth [Ball, 1956]. Visible, drift-bearing eddies ("whirlies") and walls of drift snow associated with this turbulence were commonly observed at Cape Denison [Mawson, 1969]. Such features were not remarked upon by Scott's Northern Party though their absence in the Terra Nova Bay vicinity has not been demonstrated. Simulations indicate that the strong katabatic winds which reach Inexpressible Island can be maintained seaward far enough to account for ice accumulation in the satellite-observed band of ice along the eastern polynya boundary [Bromwich and Kurtz, 1984].

Thermal infrared images provide substantial evidence that persistence of katabatic flows far beyond the slope break is a common feature of drainage through the Transantarctic Mountains. Thermal "plumes" emanating from the Skelton, Mulock, and Byrd glaciers (Figure 2) can be seen on many of the images in Figures 3 and 4 (labeled by J), and occurred very frequently throughout the 1979 winter. These signatures have previously been observed during January 1973 [Swithinbank, 1973] and October 1973 [Godin, 1977], and thus appear to be climatological features on the flat Ross Ice Shelf. The plumes may appear because the emission temperature near the top of a turbulent drift-bearing [Godin, 1977] katabatic layer is much higher than that of the quiescent snow surface outside of, but adjacent to, the katabatic airstream. This explanation is made more plausible by the strong winter mean surface inversion strength ($\sim 10°C$) inferred by Schwerdtfeger [1984, p. 83] for the northwestern part of the Ross Ice Shelf. It is to be noted that typically the plumes from the Skelton and Mulock glaciers exhibit the leftward turning of pseudo-inertial flow while the Byrd Glacier plume does not. Further analysis of the nature and dynamics of these airflows is clearly needed. Finally, Figure 3c shows that these plumes can merge into a broad air stream and reach the Ross Sea; this has been observed on other images. This observation suggests that these airflows may play a role in the formation in this area of the polynya that was studied by Zwally et al. [this volume].

Automatic weather station observations from Franklin Island, about 200 km southeast of Terra Nova Bay (Figure 2), also suggest that katabatic winds can blow for considerable distances over the ocean. Throughout 1982 and 1983, northwesterly winds were frequently measured (10-40% of 3-hourly observations, depending upon the month). This recurring feature is not easily explained in terms of known synoptic events; prevailing winds in the western Ross Sea are southerlies [Taljaard et al., 1969; Kurtz and Bromwich, 1983]. Similar persistent winds were experienced in June 1915, as the Aurora drifted past Franklin Island, leading Wordie [1921] to suggest that they came from the David Glacier, which feeds the Drygalski Ice Tongue. The frequent presence near Franklin Island of a katabatic jet from the David Glacier would be consistent with the sustained southeast movement of large ice floes near the Drygalski Ice Tongue, as observed by Ahlnaes and Jayaweera [1983] during the 1981-1982 austral summer. Other explanations are possible; Savage and Stearns [1984] believe that the wind regime at Franklin Island is governed solely by synoptic events. Research is underway to verify this apparently remarkable example of the offshore longevity of katabatic flow.

Bromwich and Kurtz [1984] have argued that katabatic conditions in Terra Nova Bay are best explained as boras, with the air blowing downslope being denser than the air at sea level. Bora-type conditions do not require a favorable synoptic pressure gradient. We have estimated the mean winter surface wind speed at Inexpressible Island to be 15 m s^{-1}, and the katabatic layer depth to be at least 390 m (the maximum elevation of the island). Both values are based upon observations during 1912 by Priestley [1913] (see Bromwich and Kurtz [1982]). As indicated previously, 1984 data from the AWS on Inexpressible Island are in excellent agreement with our wind speed estimate. We use this value and assume an initial 450 m katabatic layer depth to model katabatic trajectories across the bay, and to scale rates of dissipation of the horizontal density difference between the jet and air into which it flows. Details of these calculations are provided in Bromwich and Kurtz [1984]; principal assumptions are summarized below. We assume the layer-average wind speed and surface speed to be equal, and neglect fluxes of sensible and latent heat when computing trajectories of cold air as it crosses the bay. Winds are assumed to issue from the Reeves Glacier valley where wind depth begins to increase from 300 m (a typical value at Cape Denison) to the assumed depth of 450 m at Inexpressible Island; wind speed decreases over the same distance from a simulated 23 m s^{-1} (representative for Cape Denison) to 15 m s^{-1}. Regardless of synoptic conditions, there is relatively little difference in jet trajectories within 20-30 km of the western shore of Terra Nova Bay, a typical polynya width. Geostrophic winds from the east and south are most favorable for maintenance of the jet; such winds occurred 90% of the time during winter 1979 [Kurtz and Bromwich, 1983]. In the presence of southerly or easterly geostrophic winds, trajectories swing round to the northeast, consistent with the usual orientation of the thermal signature of this airflow. Offshore winds in the bay do not prevent sea ice from forming; they prevent it from consolidating as a thick pack ice cover and advect it out of the bay. During calm intervals in 1912, sea ice was observed to freeze rapidly, but was rapidly dispersed when winds resumed.

Surface Energy Balance

An approximation of wintertime energy exchanges across the air-sea interface can be obtained by considering an average surface wind speed of 13 m s^{-1} over the bay (Tables 1 and 2). Energy exchanges during summer were computed using lower wind speeds based upon in situ observations taken by D. Skinner and 1984 AWS data. These calculations yield large winter heat fluxes which are a substantial fraction of the solar constant (1376 W m^{-2}) but even so permit a horizontal density contrast between the katabatic jet and sea level air to persist for distances comparable to typical polynya widths. For example, Bromwich and Kurtz [1984] note that for a layer 500 m deep, an initial horizontal layer-average buoyancy deficit of 2K, and a sensible heat flux of 630 W m^{-2} into the layer, there is no density contrast once the air has traversed 29 km of open water. Assuming a combined sensible and latent heat flux of 820 W m^{-2}, implying complete water vapor condensation, the density contrast has disappeared after the jet has traversed 22 km of open water. Bromwich and Kurtz [1984] note that frazil ice, which forms in the polynya and supplies the energy lost through the water surface, is probably herded by wind-generated Langmuir circulations into long rows of grease ice parallel to the katabatic wind direction. Martin and Kauffman [1981] find that as the wind forcing diminishes the grease ice tends to consolidate into pancake ice. Surface heat transfer from the water to the air is little affected by grease ice at the interface but is significantly reduced from its open water value by pancake ice. Thus the surface energy balance in the more distal parts of the polynya and zone of loose pack is probably a complicated function of the types and quantities of sea ice present. On the other hand reduced surface heat fluxes will allow the katabatic jet to maintain its negative buoyancy for greater distances offshore.

The winter water column in Terra Nova Bay probably consists entirely of High Salinity Shelf Water (HSSW, previously called Ross Sea Shelf Water) [Jacobs et al., 1970; Jacobs et al., this volume]. With this water mass being nearly isothermal at the sea surface freezing point ($\sim -1.9°C$) the only significant energy source available to supply the large winter heat losses presented in Table 1 is latent heat release associated with ice formation. Detailed discussion of the available energy supplies and the required ice production rates is presented in the physical oceanography section. Brine formation during evaporation and sea ice formation induces haline convection that, along with wind-driven mixing, could extend to great depths [Lewis and Walker, 1970; Killworth, 1983], affecting much of the ~ 1000-m water column.

Surface energy balance calculations can also be used to assess polynya area estimates obtained from thermal infrared satellite images. Free atmospheric subsidence associated with entrainment into the katabatic air mass tends to dissipate clouds immediately above the boundary layer [compare Tauber, 1960]. This considerably enhances the chances for viewing the winter polynya on satellite images. (This phenomenon was responsible for the comparatively good weather in Terra Nova Bay during January 1983 [Mortimer, 1983] at a time when strong katabatic winds at Inexpressible Island were observed frequently from

TABLE 1. Calculated Surface Energy Balance (Watts per Square Meter) of Open Water in Western Terra Nova Bay

Month	SW	LW	R	H	LE	G	Evaporation Corresponding to LE cm d^{-1}
March	63	-37	26	379	176	-529	0.61
April	15	-40	-25	403	179	-607	0.62
May	0	-48	-47	575	192	-814	0.66
June	0	-46	-46	601	188	-835	0.65
July	0	-50	-50	574	192	-816	0.66
August	6	-48	-42	625	194	-861	0.67
September	36	-42	-6	615	193	-814	0.67
October	117	-41	76	391	173	-488	0.60
November	247	-44	203	218	128	-143	0.44
December	278	-39	239	39	55	+145	0.19
January	257	-40	217	37	45	+135	0.16
February	164	-36	128	124	111	-107	0.38

Variable	Description

Sellers [1965]

SW	absorbed shortwave radiation = $(Q+q)_o (1-\alpha)(1 - 0.16n - 0.38n^2)$
$(Q+q)_o$	total incident shortwave radiation at the ground on a horizontal surface under clear skies
α	water surface albedo
n	total fractional cloud cover

Gordon [1981], Reed [1976]

LW	net longwave absorption = $-\varepsilon\sigma T_s^4 (0.254 - 0.00495 e)(1-0.7n)$
ε	emissivity of sea water = 0.97
σ	Stefan-Boltzmann constant = 5.67×10^{-8} Wm^{-2} K^{-4}
e	vapor pressure of the surface air in millibars
T_s	sea surface temperature in Kelvins
R	net radiation = SW + LW

Liu et al. [1979]

H	upward flux of sensible heat = $\rho C_p C_H V_t (T_s - T)$
ρ	density of air
C_p	specific heat of air at constant pressure = 1005 J kg^{-1} K^{-1}
C_H	transfer coefficient for sensible heat = 1.25×10^{-3} March-October; 1.39×10^{-3} November, February; 1.73×10^{-3} December, January
V	surface wind speed
T	surface air temperature in Kelvins

Liu et al. [1979]

LE	$\rho L C_E V (q_s - q)$ = upward flux of latent heat
L	latent heat of vaporization = 2.502×10^6 J kg^{-1}
C_E	transfer coefficient for water vapor = 1.33×10^{-3} March-October; 1.41×10^{-3} November, February; 1.78×10^{-3} December, January
q_s	saturated specific humidity at T_s
q	specific humidity of the surface air
G	R - H - LE = energy flux gained by the water column from surface energy exchanges

Sign convention [Munn, 1966]: A surface energy gain by radiation is positive. Turbulent fluxes of sensible and latent heat are positive when directed upward. Heat transfer to the surface as a result of advection or storage change in the water column is negative.

TABLE 2. Input Data for Surface Energy Balance Calculations in Table 1

Month	$(Q+q)_o$, W m^{-2}	α	n	T_s, °C	V, m s^{-1}	T, °C	T_d, °C	P_s, mbar
March	106	0.13	0.72	-1.9	13	-19.0	-29.0	989.6
April	24	0.15	0.66	-1.9	13	-20.0	-30.0	991.0
May	0	0.23	0.52	-1.9	13	-27.0	-35.0	991.4
June	0	0.23	0.56	-1.9	13	-28.0	-32.0	992.0
July	0	0.23	0.48	-1.9	13	-27.0	-35.0	989.0
August	9	0.23	0.52	-1.9	13	-29.0	-36.0	989.8
September	58	0.16	0.64	-1.9	13	-28.7	-35.0	986.1
October	178	0.11	0.64	-1.9	13	-19.6	-27.0	983.1
November	349	0.09	0.58	-1.2	10	-13.0	-20.0	988.8
December	423	0.09	0.67	-0.5	5	-4.0	-10.0	993.4
January	381	0.09	0.64	-0.5	5	-3.8	-7.8	999.1
February	270	0.10	0.74	-1.2	10	-8.0	-16.0	991.9

Variable	Description	Source
V	Assumed values	
T_s	Assumed values	
$(Q + q)_o$	Interpolated to 75°S	Rusin [1964]
α	Applies at 70° latitude	Budyko [1964]
n	Observed values at McMurdo Station	Schwerdtfeger [1970]
T	Monthly climatological values at 75°S, 165°E	Taljaard et al. [1969]
T_d (dew point temperature)		
P_s (sea level pressure)		

Gondwana Station (F. Tessensohn, personal communication, 1983); farther north, low clouds and fog persisted along the coast.) As the katabatic air mass crosses the open water, evaporation can saturate the cold winter air and lead to fog and/or cloud formation within the boundary layer. This point can be demonstrated by considering the atmospheric water budget. Following Bromwich and Kurtz [1984] a layer-average July wind speed of 13 m s^{-1}, a katabatic depth of 500 m and a 25-km-wide polynya are assumed. With an initial relative humidity of 65%, which is a representative average for Mirny Station during katabatic episodes [Rusin, 1964], the July evaporation rate given in Table 1 will saturate (relative to ice) the katabatic layer well before the eastern boundary of the open water is reached. Because this calculation is sensitive to the assumed temperatures and initial relative humidities, it only demonstrates that condensation can occur frequently. Fog has a similar thermal signature to thin ice; attempts were made to exclude regions with these characteristics from the polynya domains defined by Kurtz and Bromwich [1983]. It is thus conceivable that the polynya area has been underestimated.

Radiation conditions on the polar plateau during late summer play a role in polynya formation. Schwerdtfeger [1977] notes that the southern polar summer is very short, lasting for only about 30 days between mid-December and mid-January. This is followed by rapid cooling and concomitant strengthening of the surface temperature inversion [Schwerdtfeger, 1970]. This situation is responsible for the early resumption of strong, persistent katabatic winds in Terra Nova Bay before sea ice begins to form. D. Skinner observed strong katabatic winds (speeds to 40 m s^{-1}) at Inexpressible Island on 7 of the 8 days from January 10-17, 1983. The onset of these winds in mid-February was noted by Scott's Northern Party in 1912 and measured by the AWS in 1984. Thus sea ice cannot consolidate prior to the onset of strong winds; otherwise, a solid ice cover might form that could not be dispersed. Similarly, in spring the continued presence of marked offshore winds maintains open bay waters.

During late winter and spring (August-

October) the bay water column probably contains the least sensible heat of any season. Despite daily increasing solar insolation, the net surface energy balance remains negative (Table 1). Overall katabatic outflow starts to decrease while sea surface temperature remains low. Rapid sea ice freezing takes place when winds diminish or cease, and it is then that the bay is likely to freeze over [Wright and Priestley, 1922]. Katabatic outflow persists to some degree year-round, however, and increasing winds destroy the thin ice cover.

Sea Ice Drift

Pack ice in the western Ross Sea probably drifts northward along the coast during winter, given the northward surface currents there [U.S. Navy Hydrographic Office, 1960; Priestley, 1974], and the winter drift track of the Aurora [Wordie, 1921]. As noted by Kurtz and Bromwich [1983], the Ross Sea within about 200 km of the Victoria Land coast is apparently under the influence of a marked barrier wind regime. Such winds arise because the steep Transantarctic Mountains deflect westward moving cold air masses [Schwerdtfeger, 1979a], and will induce rapid ice movement to the north [Schwerdtfeger, 1979b]. Observations from the Sierra Nevada Mountains (T.R. Parish, personal communication, 1984) show that relatively weak surface temperature inversions, such as those given by Phillpot and Zillman [1970] for the western Ross Sea, can result in pronounced barrier winds. A numerical simulation of the winter drift of Ross Sea pack ice [Baranov et al., 1977] also finds large drift rates in the western Ross Sea. Their calculation partly accounts for the influence of barrier winds because the broadscale structure of the surface pressure field has been established from many years of simultaneous pressure observations at McMurdo and Hallett Stations (Figure 2). Bromwich and Kurtz [1984] pointed out that katabatic winds from the Reeves Glacier and barrier winds in the western Ross Sea can co-exist if the shallow, dense katabatic airstream undercuts the much deeper and less dense barrier wind layer.

Orientation of the Drygalski Ice Tongue perpendicular to this drift direction results in ice blockage that is essential for maintaining wintertime open water in Terra Nova Bay. Blockage is also indicated by east-west trending pressure ridges south of the ice tongue [David and Priestley, 1914], and observations by early explorers of ice floes being deflected eastward around this feature [Priestley, 1974]. Infrared satellite images indicate that, during winter 1979, the eastern border of the polynya did not extend seaward of the extremity of the Drygalski Ice Tongue. This observation reflects the ice tongue's control on winter polynya width, because katabatic winds probably extend much farther seaward than its eastern end.

We postulate annual recurrence of this polynya. The major influences, persistent katabatic outflow through the Reeves Glacier valley and ice tongue blockage of northward drifting sea ice, are recurring, stable phenomena. Satellite observations of the polynya spanning several years, open water in Terra Nova Bay in 1912, and careful evaluation of nearly a decade of synoptic ice charts [Fleet Weather Facility, 1975, 1977, 1979; Naval Polar Oceanography Center, 1981, 1983] support this interpretation.

Polynya Character: Seasonal and Short-Term Behavior

General characteristics of the Terra Nova Bay polynya that persist through the winter are visible in Figures 3a-3c and 4a-4d. From April through October the bay is the warmest (darkest tones) region in the western Ross Sea. West of the polynya (P), a warm signature (E) of descending (drift-bearing?) katabatic winds is visible. In most images a definite thermal (wind) convergence into the Reeves and David Glacier valleys can be seen. The zone of loose pack (L), clearly defined in Figures 4a-4d, extends east and northeast of the polynya. Fractures similar to those visible elsewhere in the Ross Sea are not present in this zone because of katabatic wind action, and pack ice surrounding it is markedly colder (lighter tones). The often diffuse boundary between open water and loose pack ice may indicate the presence of either thin ice or cloud/fog within the boundary layer. Fractures and genetically related leads form as the Ross Sea pack ice shifts in response to regional wind stresses.

The autumn to late winter sequence from 1978 and 1979 (Figure 3) demonstrates seasonal polynya stability. Sea ice begins freezing in the western Ross Sea in late February and early March; ice melting begins in December, though some ice persists there throughout the year. Winter sunset in Terra Nova Bay occurs on May 5, with no sunrise until August 10. Figure 3a (April 19, 1979) illustrates the autumn polynya, which is large in this image, ~ 3000 km^2. The Ross Sea pack ice field has consolidated and definite polynya boundaries have formed; Streten's [1983] data suggest that the 1979 polynya may have appeared atypically late in autumn. The eastern polynya boundary sometimes extends seaward of the tip of the Drygalski Ice Tongue during autumn, as it may also do in late spring. Coastal leads such as that south of the ice tongue in Figure 3a formed throughout the winter but were typically not this large (Figure 3b). The triangular shape of this particular lead (compared with linear fractures) suggests influ-

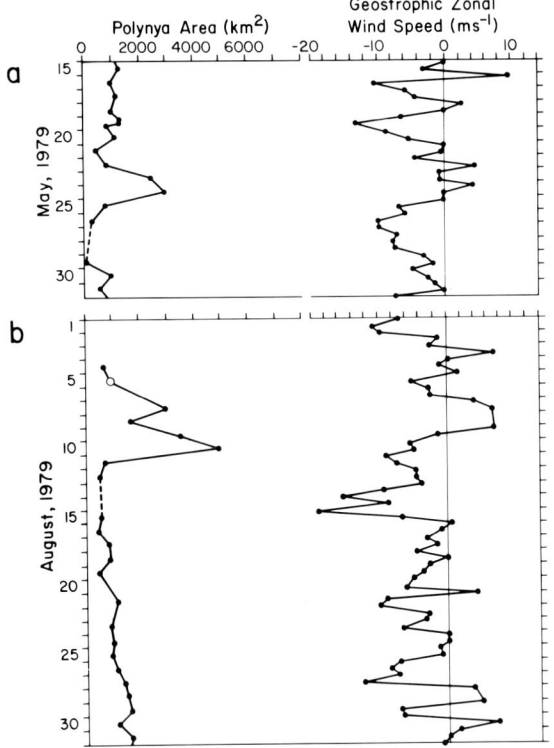

Fig. 5. Time series of polynya area (from satellite images) and zonal geostrophic wind (from synoptic maps). (a) May 15-31, 1979. (b) August, 1979. Positive zonal winds are westerlies and negative values are easterlies. Polynya time series is dashed when data are missing for 2 or 3 consecutive days and omitted when the gap exceeds 3 days. Open circles indicate cloudy images.

ence of offshore winds and probably indicates that new ice there has yet to consolidate.

The winter polynya is shown in Figure 3b during a period of expansion. Synoptic conditions led to eastward movement of the entire Ross Sea pack ice field as evidenced by leads south and east of the Drygalski Ice Tongue. Mean polynya area during winter is ~ 1000 km^2. The polynya probably did not disappear during 1979; some open water was visible near the Reeves Glacier outlet on all clear images. The presence of this nucleus of open water reflects the persistence of katabatic winds in that region, and has been observed on satellite images from other years.

Figure 3c exemplifies the polynya in late winter and early spring. The region was relatively warm, but polynya fluctuations were greatest during that time (Figure 5b). As spring progressed, the polynya tended to be larger than average; but there were brief (24-36 hours) periods when the bay was almost totally ice-covered. It persisted until mid-January 1980 [Naval Polar Oceanography Center, 1981] by which time the Ross Sea pack ice had dispersed and the polynya lost its identity. Pronounced katabatic outflow probably continues well into the summer.

An important change occurring through the season involves the zone of unconsolidated pack ice (L), and further demonstrates the Drygalski Ice Tongue blocking effect. The southern border of the region of loose pack shifted in orientation from approximately east-west during autumn (Figure 3a) (i.e. almost parallel to the ice tongue), to roughly northeast-southwest in late winter and spring (Figure 3c), pivoting around the northeastern corner of the ice tongue. The boundary between loose pack and northward moving consolidated pack probably defines a transition between the region where katabatic action prevents sea ice accumulation and those areas where it does not. In early autumn the boundary accurately reflects the extent and magnitude of these winds. Persistence of the zone of loose pack throughout the winter indicates the persistence of katabatic winds. Wind trajectories are not expected to undergo systematic seasonal variation, however, so progressive changes in trajectories cannot account for the boundary shift. As winter progresses, location of the boundary is determined less by the position of the katabatic jet, and more by the presence of northward drifting pack ice in the Ross Sea. Ross Sea pack ice is deflected around the Drygalski Ice Tongue and is probably not significantly affected by katabatic wind action. The western edge of the consolidated pack, which is regulated by the ice tongue, may serve as a barrier against which ice advected eastward from the bay accumulates, forming the loose pack-consolidated pack ice boundary. Interaction between polynya forcing agents which maintain the loose pack field and the northward drifting Ross Sea pack ice lead to a seasonally shifting border, whose satellite-observable position and movements record ice tongue blocking.

The May 1979 sequence (Figures 4a-4d) records an expanding polynya during winter. Figure 4a (May 16) depicts average winter conditions; open water area is ~ 1000 km^2. Figures 4b-4d record polynya expansion to ~ 3000 km^2 on May 24. The time series comparing polynya area with zonal geostrophic wind (Figure 5a) displays a correlation that is characteristic of most but not all of the period of detailed study (Figure 5b), i.e., interaction between katabatic and synoptic winds leads to a tripling of polynya area in 2-3 days. The thermal signature of descending katabatic winds (E) persisted unchanged throughout this period, and it was in conjunction with favorable synoptic conditions (i.e., strong westerly and/or weak easterly zonal geostrophic winds) that the area of open water increased. This is further indicated on May 24 by simultaneous opening of coastal leads due to regional wind stresses on the Ross Sea pack and a cyclone centered over the Ross Ice Shelf (Figure 4d).

Polynya area fluctuated quasi-periodically in 1979, with a period of 15-20 days [Kurtz and Bromwich, 1983]. Analysis of the 1979 time series suggests that decreases in polynya area are more rapid than increases. Of seven cycles, expansion of polynya area to double or triple the average size required ~ 2 days. For the four observable cases of these seven, the mean time for the polynya area to decrease to average (~ 1000 km^2) was ≤ 1 day, typically occurring in the interval from one available image to the next. Relative speeds of expansion and closing may indicate the nature of physical processes occurring during those events. Opening most likely reflects eastward advection of sea ice by enhanced katabatic winds due either to a synoptic pressure gradient which largely offsets the frictional dissipation or (possibly) to an atypically large supply of cold surface air on the plateau. Two mechanisms (either individually or in combination) could account for decreases in polynya area. First, westward advection of sea ice from the Ross Sea could fill Terra Nova Bay. In general, the time required for this would be somewhat longer than for eastward advection because katabatic winds in the western bay would oppose this transport. Geostrophic winds with strong easterly components were present during some closing events, but were not strong enough to account for the relatively rapid decreases in polynya area in the presence of offshore katabatic winds. A second mechanism involves freezing of bay waters following cessation or weakening of katabatic winds. This is expected to occur rapidly [Bromwich and Kurtz, 1984] and was observed to happen in only a few hours in 1912 when the winds ceased [Priestley, 1914]. Decrease in polynya area via this mechanism need not indicate cessation of katabatic winds. Rather, it may reflect a modification of the katabatic trajectory, leading to a local decrease in wind intensity and sea surface freezing in marginal regions of the polynya. Though tentative and requiring further documentation, observations from 1979 suggest that the polynya returns to its mean size by the latter mechanism.

Cavalieri and Martin [this volume] have used satellite passive microwave radiance data to estimate the area of open water within six East Antarctic coastal polynyas during the 1979 winter. By comparing 4-day composites of these data with concurrent surface wind speed observations from the nearest manned coastal station, they obtained cross-correlation coefficients which mostly ranged from 0.5 to 0.75. By contrast, Kurtz and Bromwich [1983] obtained a much lower correlation (0.30) between daily estimates of polynya area in Terra Nova Bay (from thermal infrared satellite imagery) and zonal geostrophic winds in the western Ross Sea. When these data were averaged over the same 4-day intervals used by Cavalieri and Martin, the correlation coefficient increased only slightly to 0.36.

All the polynyas shown by Cavalieri and Martin [this volume] appear to be synoptically forced. Those to the west of Dibble Iceberg Tongue, Dalton Iceberg Tongue, Casey Station, Bowman Island, and Shackleton Ice Shelf are in places where open water would be expected according to the model of Knapp [1972] and these locations are not subject to enhanced katabatic outflow [Parish, 1982]. Although the polynya west of the Mertz Glacier Tongue is adjacent to a region of intense katabatic winds [Parish, 1981] that open water also appears to be generated by passing cyclones. Katabatic winds in that area seem to dissipate within a few kilometers of the coast [Ball, 1957]. No substantial area of open water at the coastline has been observed on thermal infrared images [Bromwich and Kurtz, 1982] or is indicated by the Cavalieri and Martin [this volume] data (see their Figure 9a).

Cavalieri and Martin [this volume] used 6-hourly wind observations from Dumont D'Urville, Casey, and Mirny stations. Only at Casey is the surface wind regime governed almost entirely by the synoptic pressure gradient [Bromwich, 1976]. The pronounced influence of synoptic processes on the katabatic regimes at Dumont D'Urville and Mirny stations is discussed by Cavalieri and Martin [this volume]. Thus the synoptic winds generating the adjacent coastal polynyas are well represented by the Casey observations, but less satisfactorily by the Mirny and Dumont D'Urville measurements.

The basis for the apparent discrepancy can be readily identified. As argued above, behavior of the katabatic jet is the key factor governing fluctuations of the Terra Nova Bay polynya, synoptic processes only being of major importance during periods of expansion. A low correlation between polynya area and zonal geostrophic wind is thus to be expected. Conversely, Cavalieri and Martin [this volume] correlated variables which are directly related to the primary forcing (coastal surface winds at the three stations) and to the sea ice response (polynya area). The highest correlations were obtained for the Casey-polynya comparison where wind observations provide the best estimate of synoptic forcing.

Physical Oceanographic Effects of the Polynya

Circulation along the west coast of the Ross Sea is probably characterized by net northward movement of water at all depths [Jacobs et al., 1970], flow rates are uncertain.

Except during summer the water column consists wholly of High Salinity Shelf Water (HSSW), most of which appears to recirculate on the continental shelf [Jacobs et al., this volume]. Density increases with depth are influenced principally by increasing salinity (HSSW salinity ranges from about 34.75 to 34.90); temperature is nearly constant at the surface freezing point of seawater. Sea ice meltwater forms a summer layer of warmer, less saline water which is transformed into HSSW during autumn.

Calculations presented in Table 1 reveal a large energy loss from the ocean to the atmosphere. We now examine how much of this energy could be supplied from the sensible heat content of water masses flowing northward along the Victoria Land coast. The amount of sensible heat available at the polynya surface is taken to be the sum of the sensible heat contained in two volumes of surface water, the temperatures of which are presumed to be lowered to the surface freezing point while in the polynya. Surface water is defined as that water overlying the HSSW in summer. It occupies approximately the top 250 m of the water column and contains almost all of the sensible heat. The first volume (B) consists of the surface water in the polynya (B = 3.25×10^{11} m^3), and the second (F_t) is the volume of surface water advected northward into the polynya. F_t is calculated from

$$F_t = D L_p c t \qquad (1)$$

where D is the water depth (250 m); L_p is the typical east-west dimension of the polynya ($\sim 2.5 \times 10^4$ m); c is the depth-averaged current speed of surface water passing through D x L_p. The time (t) that surface waters contain sensible heat after the start of energy loss from the Terra Nova Bay water column is estimated to be $\sim$ 100 days. This interval extends from about mid-February (Table 1) to the end of May when complete transformation of the water column to HSSW was observed at McMurdo Sound during 1961 [Littlepage, 1965].

Water movements in the western Ross Sea are somewhat uncertain, so we have allowed the depth-averaged current speed (c) to range from 2 to 8 cm s^{-1}. The Aurora broke away from her winter moorings at Ross Island on May 6, 1915 and drifted generally northward with the pack along the Victoria Land coast, passing the latitude of Cape Adare around August 8, 1915 [Wordie, 1921]. Prior to early August when very high drift rates were experienced, the average speed was 7.6 cm s^{-1}, which we will take as an upper bound for the mean surface current in the vicinity of Terra Nova Bay. Gordon et al. [1981] infer a small vertical current shear in the dynamically-similar western Weddell Sea [Killworth, 1974]. Current vectors between 200 and 300 m along the front of the Ross Ice Shelf average around 5-10 cm s^{-1} and show little vertical variation [Pillsbury and Jacobs, this volume]. Considerably stronger surface currents have been inferred there [U.S. Navy Hydrographic Office, 1957] but are highly variable and may be significantly influenced by the ice shelf [compare Amos, 1982].

The observed drift of large ice floes near Drygalski Ice Tongue during December 1981 and January 1982 [Ahlnaes and Jayaweera, 1983] suggests complex water motions. As discussed earlier, their floes A and D to the east and south of the ice tongue we believe may have been influenced by a katabatic outflow from the David Glacier. This airflow could induce a net drift to the southeast and may not be in equilibrium with the large scale atmospheric pressure field. Their floes B and C, to the east of Terra Nova Bay and north of floes A and D, moved clockwise around circular paths of $\sim$ 40 km diameter with an average speed of 2 cm s^{-1}. Assuming that floes B and C were away from the influence of the weaker and more intermittent summer katabatic winds from the Reeves Glacier, those drifts could have been caused by transient or semi-permanent eddies [Ahlnaes and Jayaweera, 1983]. To allow for varying degrees of local water recirculation, computations have also been done with c set equal to 6, 4, and 2 cm s^{-1}. Data from a year-long current meter mooring in Terra Nova Bay during 1984 [D. Pillsbury, personal communication, 1984] may provide a better estimate for mean flow.

The amount of sensible heat delivered to the polynya surface (H_S) can be expressed as

$$H_S = (Q_B B + 0.5 Q_A F_t)/A \qquad (2)$$

where Q_B and Q_A are the summer heat contents per m^3 of bay waters and northward flowing surface waters respectively, and A is the mean polynya area (1.3×10^9 m^2). The sensible heat of the latter surface water is presumed to decrease linearly to zero over the interval t due to sea ice formation to the south of Terra Nova Bay. Thus the average heat content is 0.5 Q_A. In the absence of data to the contrary, $Q_A = Q_B$ is adopted as a useful first approximation. Thus,

$$H_S = Q_A (B + 0.5 F_t)/A \qquad (3)$$

Heat contents were calculated from hydrographic station data collected in the western Ross Sea during February 1978 [Jacobs and Haines, 1982]. Station locations are given in Figures 1 and 2. Differences between observed potential temperatures and salinity-dependent sea surface freezing points [Fujino et al., 1974; see also UNESCO, 1978] were computed every 10-m to 100-m depth and at 25 m intervals to 250 m. The resulting depth-averaged sensible-heat

TABLE 3. Sensible-Heat Content and Salinity Deficit of Surface Water (Upper 250 m) at Hydrographic Stations in the Western Ross Sea During February 1978

Station	Heat Content, kJ m^{-3}	Salinity Deficit
165	1590	0.090
166	619	0.127
167	703	0.123
168	737	0.057
169	100	0.074
Average of 166-168	686 (=Q_A)	0.102

contents are listed in Table 3. Similar calculations for the entire water column indicate that, as expected, most (85-90%) of the sensible heat is contained in the upper 250 m. Station 165, north of the Terra Nova Bay region, may reflect different processes than at the stations near the bay (166-168). Station 169, located very close to the Drygalski Ice Tongue and perhaps influenced by it, is much colder than the others. The average heat content of the three stations near the bay is used to estimate Q_A because their similar heat contents and surface water profiles may reflect the regional character of surface water in this part of the Ross Sea. The resulting H_S values are listed in Table 4.

Assuming that the heat losses calculated in Table 1 reflect average conditions, the amount of ice formation that alone would be needed to balance G can readily be found (Table 5). Antarctic frazil ice, formed under turbulent sea surface conditions, has salinities ranging from 5.0 to ≥ 10.0 [Gow et al., 1982], with the most saline ice being produced at the highest freezing rates. Therefore, we use a latent heat of fusion of 251.4 kJ kg^{-1} (L_f), corresponding to an ice salinity of 10.0, and an ice density of 925 kg m^{-3} (ρ_i) [Schumacher et al., 1983] to calculate ice production. The heat obtainable from cooling surface waters to the local freezing point can also be expressed in terms of equivalent ice thickness. Comparison of Tables 4 and 5 shows that even with the shortest polynya residence times the sensible heat of surface waters could supply only 8% of the needed energy. If the net energy gained by the water column in December and January (Table 1) was completely recycled in the polynya, this would offset only 3.2 m of ice production. Thus, almost all of the energy must be supplied by latent heat associated with the phase change of water from liquid to solid. The Terra Nova Bay polynya is an example of the latent heat type described by Zwally et al. [this volume] for the Antarctic and by Dunbar [1981] for the Canadian Arctic.

A cumulative annual ice production of ~ 60 m is between one and two orders of magnitude greater than the mean thickness of ice formed in the Ross Sea, and is possible because the katabatic wind continually drives the ice away from the shoreline and prevents the formation of an insulating ice cover. Schumacher et al. [1983] used salinity data to infer that 5 m of sea ice formed in a synoptically forced Arctic polynya during the roughly 30 days it was present in the 1980-1981 winter. This suggests that the ice production rates calculated in Table 5 are reasonable. Schumacher et al. [1983] also found that maximum ice production took place at moderate temperatures (~ -10°C). At colder temperatures (~ -20°C) approaching those characteristic of Terra Nova Bay, the high instantaneous ice production rapidly decreased the polynya area and shut off the ice formation. However, the efficient transport (induced by the katabatic

TABLE 4. Heat Available to the Polynya Surface From the Sensible-Heat Content of Surface Waters

c, cm s^{-1}	c, m d^{-1}	Polynya Residence Time, years	F_t, 10^{11} m^3	H_s, 10^5 kJ m^{-2}	Equivalent Sea Ice Thickness, m
2	1728	0.082	10.8	4.6	2.0
4	3456	0.041	21.6	7.4	3.2
6	5184	0.027	32.4	10.3	4.4
8	6912	0.021	43.2	13.1	5.6

F_t from Equation (1); H_s from Equation (3).
Equivalent sea ice thickness is the amount which would release latent heat equivalent to H_s. It is calculated from $H_s/(\rho_i L_f)$.

TABLE 5. Ice Production in the Polynya

Month	G,[a] $W\ m^{-2}$	Ice,[b] $cm\ d^{-1}$	Ice, m	Ratio[c]
March	-529	19.7	6.1	0.46
April	-607	22.6	6.8	0.44
May	-814	30.2	9.4	0.33
June	-835	31.0	9.3	0.31
July	-816	30.3	9.4	0.33
August	-861	32.0	9.9	0.31
September	-814	30.2	9.1	0.31
October	-488	18.1	5.6	0.44
November	-143	5.3	1.6	0.59
December	+145	---	---	---
January	+135	---	---	---
February	-107	4.0	1.1	0.90
Total			68.3	

[a] From Table 1.
[b] Ice production $(cm\ d^{-1}) = (|G| \times 24 \times 3600 \times 100)/(\rho_i\ L_f \times 1000)$ with G in $W\ m^{-2}$, L_f in $kJ\ kg^{-1}$, and ρ_i in $kg\ m^{-3}$.
[c] Ratio of ice production needed to supply latent heat losses (to the atmosphere) to that needed for sensible heat (LE/H) (from Table 1).

wind) of frazil and grease ice into the Ross Sea and northward advection of this ice by the prevailing surface current prevents substantial ice buildup in Terra Nova Bay. Table 5 also shows that evaporation plays a significant but not dominant role in the winter ice production. By contrast, Ledenev [1963] concluded that sensible- and latent-heat fluxes are equally important for ice production and formation of cold, saline water masses along the East Antarctic coast. The difference is due to the much lower temperatures in Terra Nova Bay. For a given wind speed the sensible-heat flux increases steadily as the temperature falls but the latent-heat flux becomes nearly constant, being dependent on the (fixed) saturated specific humidity at the freezing point of seawater.

Sixty meters of ice formation over the 1300 km^2 polynya amounts to a total of 78 km^3, or enough to cover 39,000 km^2 with ice 2 m thick. This area represents ~ 10% of the Ross Sea lying to the southwest of a line running from Cape Adare to Cape Colbeck. Given the present uncertainties as to the quantitative impact of thin ice upon the surface energy balance, 78 km^3 must be regarded as an upper limit to annual ice formation in Terra Nova Bay. As noted earlier, smaller energy losses through the polynya surface allow the katabatic jet to retain its identity for greater distances offshore (as suggested by satellite images) and thus to distribute significant ice formation over a wide area of the western Ross Sea. The calculated ice production in Terra Nova Bay can be compared with values obtained by Cavalieri and Martin [this volume] for six East Antarctic coastal polynyas between late June and September 30, 1979. During this interval Terra Nova Bay is estimated to have produced ~ 40 km^3, while Cavalieri and Martin [this volume] calculated amounts between 5 and 36 km^3 (their values have been transformed to an ice salinity of 10.0).

During ice production at the polynya surface, about one third of the salt is incorporated into the ice and advected out of the bay by the katabatic wind while the majority forms a high-density brine that may convect to the bottom of the ~ 1000-m water column. The resulting depth-averaged salinity increase of a water parcel passing through Terra Nova Bay (ΔS) can be estimated from

$$\Delta S = \frac{I\ S_i\ R_p}{v} \quad (4)$$

The annual rate of sea ice formation (I, $m^3\ yr^{-1}$), and the mass of solids released for each cubic meter of ice formed (S_i, $kg\ m^{-3}$) determine the total quantity of salt released. The speed at which the water parcel moves through the polynya (expressed here as parcel residence time in the polynya (R_p, years)) and the volume of water in which the salinity increase is distributed at any given time (v, m^3) (i.e., 1 m^2 times depth of haline mixing) determine the volume of water affected. Based upon the surface energy balance I equals ~ 60 $m^3\ yr^{-1}$; S_i is taken to equal 25 kg solids released per cubic meter sea ice formed. This corresponds to frazil ice salinities of ~ 10.0 (as before), formed from HSSW of salinity ~ 35.0. Two depths of haline mixing are examined, 250 m and 1000 m ($v = 250\ m^3$, 1000 m^3, respectively). The 250 m depth is approximately the thickness of the fresher, warmer, mixed surface water. This represents a known minimum depth of haline mixing as the HSSW column is homogenized during winter. The maximum depth of haline mixing is taken to be the mean water depth beneath the polynya, ~ 1000 m. Polynya residence times are taken to be the same as those in Table 4.

Table 6 gives modeled increases in salinity for the various circumstances. It is assumed that all salinity enhancement takes place in the polynya, and no attempt is made to estimate the (probably considerable) lateral spreading and mixing of salt along isopycnal surfaces [e.g. Carmack and Killworth, 1978]. Salinization values in Table 6 are yearly totals; they are not intended to reflect summer sea ice melting or likely seasonal variations in ice production, polynya forcing mechanisms,

TABLE 6. Depth-Averaged Salinity Increase
During HSSW Transit of the Polynya

Residence Time R_p, years	Salinity Increase ΔS (kg m^{-3}) from Equation (4)	
	$v = 250$ m^3	$v = 1000$ m^3
0.082	0.49	0.12
0.041	0.25	0.06
0.027	0.16	0.04
0.021	0.13	0.03

current speeds, and depths of mixing. The salinization values for mixing to the bottom of Terra Nova Bay are significantly smaller than the 0.26 - 0.44 salinity enhancement calculated by Zwally et al. [this volume] for the polynya along the northwestern margin of the Ross Ice Shelf.

The magnitude of two neglected salinity effects can readily be determined. Following arguments used to scale the sensible heat available in autumn surface waters, the approximate surface freezing (FR, m of ice) needed in Terra Nova Bay to convert the fresher surface water to HSSW is given by

$$FR = \frac{\rho_w (B + 0.5 F_t) S_d}{A S_i} \quad (5)$$

The product of the density of water (ρ_w, kg m^{-3}), the volume of surface water that is present within or passes through the polynya in 100 days (B or F_t), and its salinity deficit (S_d or $0.5 S_d$, kg of salt per kg of water) determines the mass of salt needed. S_d is the difference between the layer-average summer surface water salinity (calculated from February station data given by Jacobs and Haines [1982]) and the presumed HSSW value. The latter is approximated by extrapolating deep water salinities to 125 m. As before, results for stations 166-168 are composited (Table 3), and the deficit for surface waters entering the bay is supposed to linearly decrease to zero by the end of May. Even with the shortest polynya residence time (R_p = 0.021 year, F_t = 43.2 x 10^{11} m^3, Table 4), the required salt is provided by ~ 8 m of ice formation. The salt left behind by the annual freshwater evaporation (Table 1) is equal to that rejected by ~ 3 m of surface freezing. The net impact of these partly offsetting effects does not substantially alter the results in Table 6.

Cavalieri and Martin [this volume], Zwally et al. [this volume], and Ledenev [1963] have noted the qualitative association between areas with highly saline, cold summer shelf water and recurring East Antarctic coastal polynyas. With the present uncertainties concerning water movements in the western Ross Sea, no attempt is made to use the few available oceanographic measurements to verify the salinity enhancement inferred for the Terra Nova Bay polynya. However, the values presented in Table 6 indicate that salinization in and around the bay is an active process during winter and that a regional signal of this should be detectable.

An estimate of the potential importance of the Terra Nova Bay polynya for the salt budget of the Ross Sea can be provided. From salt flux considerations, Jacobs et al. [this volume] estimate that at least 1.9 m of sea ice annually forms over the Antarctic continental shelf, and that half of this melts locally while the remainder is advected off the shelf. The calculated ice production in Terra Nova Bay would amount to 0.20 m if spread over the entire Ross Sea continental shelf, about 10% of the annual total. Calculations by Zwally et al. [this volume] imply that the polynya adjacent to the Ross Ice Shelf dominates the Ross Sea salt budget; ice production there is equivalent to covering the continental shelf with 1.6 - 2.8 m of sea ice. However, as noted by Jacobs et al. [this volume], the Terra Nova Bay polynya probably plays a very important role in the salinization of HSSW which may amount to 20% of all shelf water in the Ross Sea. Up to one half of the annual salt addition to this water mass could be due to surface energy exchanges in Terra Nova Bay.

Conclusions

The polynya in Terra Nova Bay is a stable feature, recurring annually, persisting throughout the winter, and affecting several thousand square kilometers of the western Ross Sea. Decadal constancy of polynya driving forces enables us to utilize spatially widespread data collected over many years to formulate and substantiate ideas regarding polynya processes. Ongoing AWS monitoring of the katabatic outflow should allow extensive testing and refinement of these concepts. A key future task is to determine the relative mixture of thin ice and open water at the polynya surface as a function of distance offshore, and thus to better evaluate the energy fluxes from the polynya to the atmosphere and to define their role in the offshore longevity of the katabatic wind.

Polynya presence affects physical oceanographic processes. The sustained, cold wind blowing over the bay removes substantial energy from the water column. Because little sensible heat is available, most of the energy transfer must be supplied by latent heat released during seawater freezing. The katabatic wind continually advects newly formed ice away from the shoreline and keeps the water and atmosphere in direct contact for continued ice formation. Surface heat budget calculations predict that a cumulative ice thickness of up to 60 m could be produced each year. This is equivalent to ~ 10% of the sea ice annually formed over the continental shelf in the Ross Sea. Some 60% of the available salt forms a high density brine that descends into and mixes with the High Salinity Shelf Water; the remainder is incorporated into the newly formed ice and is carried away in near-surface layers. This salinity enhancement at depth may partly account for the salinity field that Killworth [1974] attempted to model [Jacobs et al., this volume]. Consideration that polynya effects may be superimposed on the regional density distribution may improve his simulation.

The combination of strong, persistent katabatic winds and ice tongue blocking is necessary to form the Terra Nova Bay polynya. Polynyas can also be formed, in the lee of suitably oriented obstacles, by strong winds associated with passing cyclones [Knapp, 1972]. Loewe [1956] reports that sea ice conditions off Adélie Land are variable even under the influence of similar wind and temperature conditions. The presence of open water and fast ice along this coast in separate years when the average katabatic wind speed was 19 m s^{-1} demonstrates once again that even very strong offshore winds are not always a sufficient condition for formation and maintenance of large open water areas during winter. Winds off Adélie Land appear to dissipate within a few kilometers of the shore [Ball 1956, 1957], and do not influence sea ice motions far from the coast. These strong offshore winds may lead to frequent coastal leads or early ice breakup, such as the sea ice retreat along the Adélie Coast inferred from passive microwave imagery by Zwally et al. [1983].

Acknowledgments. Thermal infrared satellite images presented here were obtained from the Space Science and Engineering Center, University of Wisconsin-Madison and the World Data Center-A for Glaciology in Boulder, Colorado. The Defense Meteorological Satellite Program collection now resides at the latter site. S. Jacobs was the source of much scientific inspiration; J. Ardai and G. Crocker deployed the AWS; D.N.B. Skinner and F. Tessensohn provided essential in situ data. All this assistance is gratefully acknowledged. This study was supported by NSF grant DPP-83-14613 to D.H.B. Contribution 526 of the Institute of Polar Studies, The Ohio State University.

References

Ahlnaes, K., and K. Jayaweera, Sea-ice studies in Ross Sea, Antarctica, using NOAA - satellite imagery, Proceedings Seventh International Conference on Port and Ocean Engineering Under Arctic Conditions, Vol. 1, pp. 42-51, Technical Research Center, Espoo, Finland, 1983.

Amos, A.F., Physical oceanography of the southwestern Ross Sea, January 1982, Antarct. J. U.S., 17, 146-148, 1982.

Anderson, J.B., and D.D. Kurtz, USCGC Glacier Deep Freeze 80, Antarct. J. U.S., 15, 114-117, 1980.

Ball, F.K., The theory of strong katabatic winds, Aust. J. Phys., 9, 373-386, 1956.

Ball, F.K., The katabatic winds of Adélie Land and King George V Land, Tellus, 9, 201-208, 1957.

Baranov, G.I., V.O. Ivchenko, M.I. Maslouskii, A.F. Treshnikov, and D. E. Kheisin, Wind drift of Antarctic sea ice, (Transl. from Russian), Probl. Arctic Antarctic, 47, 136-159, Amerind, New Delhi, 1977.

Bromwich, D. H., Boundary layer characteristics of the Wilkes Ice Cap, M.Sc. thesis, 115 pp., Univ. of Melbourne, Australia, 1976.

Bromwich, D.H., and D.D. Kurtz, Experiences of Scott's Northern Party: Evidence for a relationship between winter katabatic winds and the Terra Nova Bay polynya, Polar Rec., 21, 137-146, 1982.

Bromwich, D.H., and D.D. Kurtz, Katabatic wind forcing of the Terra Nova Bay polynya, J. Geophys. Res., 89, 3561-3572, 1984.

Budyko, M.I., Guide to the atlas of the heat balance of the earth, WB/T-106, translated from Russian by I.A. Donehoo, U.S. Weather Bureau, Washington, D.C., 1964.

Carmack, E. C., and P. D. Killworth, Formation and interleaving of abyssal water masses off Wilkes Land, Antarctica, Deep Sea Res., 25, 357-369, 1978.

Cavalieri, D. J., and S. Martin, A passive microwave study of polynyas along the Antarctic Wilkes Land coast, this volume.

David, T.W.E., and R.E. Priestley, Geology, Vol. I, Glaciology, Physiography, Stratigraphy and Tectonic Geology of South Victoria Land, British Antarctic Expedition 1907-1909, Reports on the Scientific Investigations, 319 pp., Heinemann, London, 1914.

Dunbar, M. J., Physical causes and biological significance of polynyas and other open water in sea ice, in Polynyas in the Canadian Arctic, Occ. Pap. 45, pp. 29-43, Canadian Wildlife Service, Ottawa, 1981.

Fleet Weather Facility, Antarctic Ice Charts, 1973-1974, Suitland, Md., 1975.

Fleet Weather Facility, Antarctic Ice Charts, 1975-1976, Suitland, Md., 1977.

Fleet Weather Facility, Antarctic Ice Charts, 1977-1978, Suitland, Md., 1979.

Fujino, K., E. L. Lewis, and R. G. Perkin, The freezing point of seawater at pressures up to 100 bars, J. Geophys. Res., 79, 1792-1797, 1974.

Godin, R. H., An investigation of synoptic and associated mesoscale patterns leading to significant weather days at McMurdo Station, M.S. thesis, Naval Postgraduate School, Monterey, Calif., 1977.

Gordon, A.L., Seasonality of the southern ocean sea ice, J. Geophys. Res., 86, 4193-4197, 1981.

Gordon, A.L., D.G. Martinson, and H.W. Taylor, The wind-driven circulation in the Weddell-Enderby Basin, Deep Sea Res., 28A, 151-163, 1981.

Gow, A.J., S.F. Ackley, W.F. Weeks, and J.W. Govoni, Physical and structural characteristics of Antarctic sea ice, Ann. Glaciol., 3, 113-117, 1982.

Jacobs, S.S., and W.E. Haines, Oceanographic data in the Ross Sea and along George V coast 1976-1979, Ross Ice Shelf Proj., Tech. Rep. LDGO-82-1, 505 pp., Lamont-Doherty Geological Observatory, Palisades, N.Y., 1982.

Jacobs, S.S., A.F. Amos, and P.M. Bruchhausen, Ross Sea oceanography and Antarctic Bottom Water formation, Deep Sea Res., 17, 935-962, 1970.

Jacobs, S.S., R.G. Fairbanks, and Y. Horibe, Origin and evolution of water masses near the Antarctic continental margin: Evidence from H_2O^{18}/H_2O^{16} ratios in seawater, this volume.

Killworth, P.D., A baroclinic model of motions on Antarctic continental shelves, Deep Sea Res., 21, 815-837, 1974.

Killworth, P.D., Deep convection in the world ocean, Rev. Geophys. Space Phys., 21, 1-26, 1983.

Knapp, W.W., Satellite observations of large polynyas in polar waters, in Sea Ice, edited by T. Karlsson, National Research Council, Reykjavik, Iceland, pp. 201-212, 1972.

Kurtz, D.D., and D.H. Bromwich, Satellite observed behavior of the Terra Nova Bay polynya, J. Geophys. Res., 88, 9717-9722, 1983.

Ledenev, V. G., Influence of evaporation on the formation of cold Antarctic water, Sov. Antarct. Exped. Inf. Bull., Engl. Transl., 5(1), 50-52, 1963.

Lewis, E.L., and E.R. Walker, The water structure under a growing ice sheet, J. Geophys. Res., 75, 6836-6845, 1970.

Littlepage, J.L., Oceanographic investigations in McMurdo Sound, Antarctica, in Biology of the Antarctic Seas II, Antarct. Res. Ser., vol. 5, edited by G. A. Llano, pp. 1-37, AGU, Washington, D.C., 1965.

Liu, W.T., K.B. Katsaros, and J.A. Businger, Bulk parameterization of air-sea exchanges of heat and water vapor including the molecular constraints at the interface, J. Atmos. Sci., 36, 1722-1735, 1979.

Loewe, F., Contributions to the glaciology of the Antarctic, J. Glaciol., 2, 657-665, 1956.

Marvill, S., and K.O.L.F. Jayaweera, Investigations of strong valley winds in Alaska using satellite infrared imagery, Mon. Weather Rev., 103, 1129-1136, 1975.

Martin, S., and P. Kauffman, A field and laboratory study of wave damping by grease ice, J. Glaciol., 27, 283-313, 1981.

Mawson, D., The Home of the Blizzard, vol. 1, 349 pp., vol. 2, 338 pp., Greenwood Press, New York, 1969.

Mortimer, G., GANOVEX III 1982/83, N.Z. Antarct. Rec., 5, 54-60, 1983.

Munn, R.E., Descriptive Micro-Meteorology, 245 pp., Academic Press, New York, 1966.

Naval Polar Oceanography Center, Antarctic Ice Charts, 1979-1980, Suitland, Md., 1981.

Naval Polar Oceanography Center, Antarctic Ice Charts, 1981-1982, Suitland, Md., 1983.

Parish, T.R., The katabatic winds of Cape Denison and Port Martin, Polar Rec., 20, 525-532, 1981.

Parish, T.R., Surface airflow over East Antarctica, Mon. Weather Rev., 110, 84-90, 1982.

Phillpot, H.R., and J. W. Zillman, The surface temperature inversion over the Antarctic continent, J. Geophys. Res., 75, 4161-4169, 1970.

Pillsbury, R.D., and S.S. Jacobs, Preliminary observations from long-term current meter moorings near the Ross Ice Shelf, Antarctica, this volume.

Priestley, R.E., General diary, 1 January 1912-Feburary 1913, Doc. MS298/6/2, Scott Polar Research Institute, Cambridge, England, 1913.

Priestley, R.E., Work and adventures of the Northern Party of Captain Scott's Antarctic Expedition, 1910-1913, Geogr. J., 43, 1-14, 1914.

Priestley, R.E., Scott's Northern Party, Geogr. J., 128, 129-140, 1962.

Priestley, R.E., Antarctic Adventure, 382 pp., McClelland and Stewart, Toronto, Ont., 1974.

Reed, R.K., On estimation of net longwave radiation from the oceans, J. Geophys. Res., 81, 5793-5794, 1976.

Rusin, N.P., Meteorological and Radiational Regime of Antarctica, Transl. from Russian, Israel Program for Scientific Translations, Jerusalem, 1964.

Savage, M. L., and C. R. Stearns, The climate in the vicinity of Ross Island, Antarctica, Antarct. J. U.S., 19(5), 1984.

Schumacher, J.D., K. Aagaard, C.H. Pease, and R.B. Tripp, Effects of a shelf polynya on flow and water properties in the northern Bering Sea, J. Geophys. Res., 88, 2723-2732, 1983.

Schwerdtfeger, W., The climate of the Antarctic, in Climate of the Polar Regions, World Survey of Climatology, vol. 14, edited by H.E. Landsberg, Elsevier, New York, pp. 253-355, 1970.

Schwerdtfeger, W., Temperature regime of the South Pole: results of 20 years' observations at Amundsen-Scott Station, Antarct. J. U.S., 12, 156-159, 1977.

Schwerdtfeger, W., Problems and suggestions concerning synoptic and climatic data for high southern latitudes, in Collection of Contributions Presented at CPM Sessions, IAGA/IAMAP Assembly, Seattle, Washington, September 1977, edited by M. Kuhn, pp. 84-100, NCAR, Boulder, Colo., 1979a.

Schwerdtfeger, W., Meteorological aspects of the drift of ice from the Weddell Sea toward the mid-latitude westerlies, J. Geophys. Res., 84, 6321-6328, 1979b.

Schwerdtfeger, W., Weather and Climate of the Antarctic, Developments in Atmospheric Science, vol. 15, 261 pp. Elsevier, New York, 1984.

Sellers, W.D., Physical Climatology, University of Chicago Press, Chicago, Ill., 272 pp., 1965.

Stonehouse, B., Occurrence and effects of open water in McMurdo Sound, Antarctica during winter and early spring, Polar Rec., 84, 13, 775-778, 1967.

Streten, N. A., Satellite observations of the summer decay of Antarctic sea ice, Arch. Meteorol. Geophys. Bioklimatol., Ser. A, 22, 119-134, 1973.

Streten, N. A., Antarctic sea ice and related atmospheric circulation during FGGE, Arch. Meteorol. Geophys. Bioklimatol., Ser. A, 32, 231-246, 1983.

Stuart, A.W., and A.J. Heine, Glaciological work of the 1959-60 U.S. Victoria Land traverse, J. Glaciol., 13, 997-1002, 1961.

Swithinbank, C., Higher resolution satellite pictures, Polar Rec., 16, 739-741, 1973.

Szekielda, K.H., The hot spot in the Ross Sea; Upwelling during wintertime, Tethys, 6, 105-110, 1974.

Taljaard, J.J., H. van Loon, H.L. Crutcher, and R.L. Jenne, Climate of the upper air, part 1 - Southern Hemisphere, vol. 1, Temperature, dew points, and heights at selected pressure levels, Rep. NAVAIR-50-1C-55, 140 pp., Naval Weather Service Command, Washington, D.C., 1969.

Tauber, G.M., Characteristics of Antarctic katabatic winds, in Antarctic Meteorology, pp. 52-64, Pergamon, New York, 1960.

UNESCO, Freezing point of seawater, Appendix 6, Eighth report of the joint panel on oceanographic tables and standards, Tech. Papers in Mar. Sci., 28, 29-35, 1978.

U.S. Navy Hydrographic Office, Oceanographic Atlas of the Polar Seas, Part I, Antarctic, H.O. Publ. 705, 70 pp., U.S. Govt. Printing Office, Washington, D.C., 1957.

U.S. Navy Hydrographic Office, Sailing Directions for Antarctica, H.O. Publ. 22, 2nd ed., 432 pp., U.S. Govt. Printing Office, Washington, D.C., 1960.

Weller, G.E., A meridional surface wind speed profile in MacRobertson Land, Antarctica, Pure Appl. Geophys., 77, 193-200, 1969.

Wordie, J.M., The Ross Sea drift of the "Aurora" in 1915-1916, Geogr. J., 58, 219-224, 1921.

Wright, C.S., and R.E. Priestley, Glaciology, British (Terra Nova) Antarctic Expedition, 1910-1913, 581 pp., Harrison and Sons, London, 1922.

Zwally, H.J., J.C. Comiso, C.L. Parkinson, W.J. Campbell, F.D. Carsey, and P. Gloersen, Antarctic sea ice, 1973-1976: Satellite passive-microwave observations, NASA, SP-459, 1983.

Zwally, H.J., J.C. Comiso, and A.L. Gordon, Antarctic offshore leads and polynyas and oceanographic effects, this volume.

(Received March 12, 1984;
accepted December 28, 1984.)

ANTARCTIC OFFSHORE LEADS AND POLYNYAS AND OCEANOGRAPHIC EFFECTS

H. Jay Zwally and J. C. Comiso

Laboratory for Oceans, Goddard Space Flight Center, Greenbelt, Maryland 20771

A. L. Gordon

Lamont-Doherty Geological Observatory of Columbia University, Palisades, New York 10964

Abstract. Extensive areas of open water are located within the Antarctic sea ice pack in a near-shore zone of several hundred kilometers, as well as the marginal ice zone near the sea ice edge. The time-series of satellite passive microwave observations during 1974 provides quantitative values of the area of open water in the coastal zone and details of the opening and closing of polynyas on daily time scales. Along the coast, numerous polynya areas are observed. Sixteen study areas located over the continental shelf are analyzed to provide time series of the maximum brightness temperature (T_B), minimum T_B, average T_B, standard deviation of T_B within each study area, and the derived area of open water. Examination of the synoptic pressure maps in the Ross Sea indicates that the intermittent formation of a polynya near the ice shelf front is strongly influenced by the synoptic winds. Other polynya areas appear to be located offshore of major outlet glaciers that are locations of enhanced katabatic winds. In all cases, the intermittent increases in open water during the polynya events are superimposed on a significant background of near-shore open water, which averages about 19% during the winter period from March 17 through November 11 (days 76-315). In some locations, more open water is observed during the winter period than in the summer days November 12 through March 16 (316-75). The principal heat source in the near-shore zone during the winter is deduced to be latent heat from ice production as ice divergence continually produces new areas of open water. The oceanic heat from cooling of the shelf water column by itself would maintain open water for a period less than 30 days. The resulting ice production within the latent-heat polynyas and leads significantly contributes to the sea ice cover, which is generally driven from the continental shelf toward the deep ocean. Estimates of the rate of ice production along with the observed values of open water area are used to calculate the resultant salinization of the Antarctic shelf waters, giving increases in shelf water salinity of many tenths of parts per thousand during the winter months. Although uncertainties remain regarding the amount of ice exported from the shelves and the shelf-slope exchange of water required to maintain steady state shelf water salinity, it is concluded that ice formation in leads and polynyas over the shelf is likely to be a primary factor in the production of saline shelf water and ultimately in bottom water formation.

Introduction

The Antarctic sea ice pack is known to contain substantial areas of open water during all seasons (e.g., Priestley [1913] (see Bromwich and Kurtz [1982]); Zwally and Gloersen [1977]; Bromwich and Kurtz [1982]; Carsey [1980]; and Zwally et al. [1983]). The size, distribution, and duration of the open water areas within the ice pack are highly variable as the sea ice grows, deforms, and melts in response to atmospheric, oceanic, and radiative forcings. Principal factors influencing the Antarctic sea ice dynamics are the absence of geographical constraints to the northern edge of the ice pack and the strong near-shore drainage winds from the ice sheet [Kurtz and Bromwich, 1983; Parish, 1982]. The wind-induced divergence of the sea ice substantially reduces the time-averaged sea ice concentrations, particularly in the near-shore (inner) and marginal (outer) zones of the ice pack. During much of the year, ice production is rapid in the open water and thin ice areas that are continually formed near the coast. Net divergence of the near-shore ice is generally accompanied by a northerly transport of ice, which contributes to the expansion of the ice edge during the growth season and provides ice for melting at the outer edge [Hibler and Ackley, 1983]. Following the period of maximum winter ice extent, for example, the near-shore zone is expected to be a region of net ice production, and the marginal zone a region of net ice dissipation.

The recent realization of the extensive distribution of open water within the ice pack

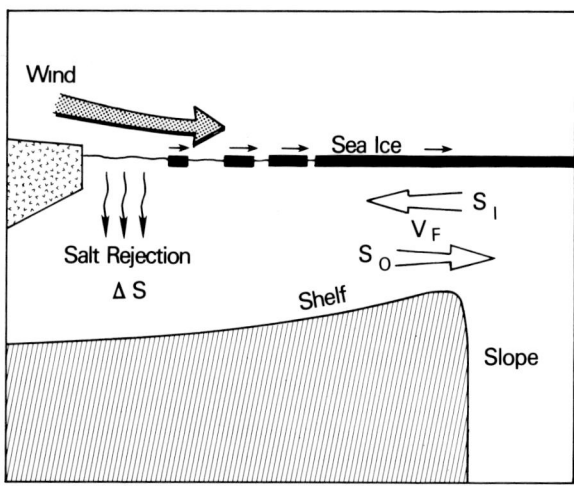

Fig. 1. Schematic diagram of ice production and salt rejection processes in near-shore regions of wind-induced divergence. V_F is the volume exchange rate between the inflow water with salinity S_I and outflow with salinity S_O. Other factors influencing the salt balance are discussed in the text.

is largely a result of year-round observations of the polar regions by passive microwave imagers flown on NASA research satellites over the last decade. The largest area of open water observed within the Antarctic sea ice pack is the Weddell polynya, which had approximately 2×10^5 km^2 of essentially ice-free ocean in some years (e.g. Zwally and Gloersen [1977]; Carsey [1980]; Martinson et al. [1981]; Parkinson [1983]). Except for the Weddell polynya, the sizes of the open water areas within the pack during winter are normally smaller than the 30-km resolution of the passive microwave imagers, and therefore the open water appears in the data as areas of reduced ice concentration.

In general, ice insulates the ocean, thereby sharply reducing the exchange of heat with the atmosphere. In the polynyas and leads created by ice deformation, the oceanic heat loss may be 10 to 100 times above that of the surrounding ice-covered surface (e.g., Maykut [1978]). The oceanic heat flux can be derived from two sources: (1) sensible heat of the sea water as it cools to the freezing point and (2) latent heat of fusion released during ice formation. In some situations during the winter period (when the heat flux is from the ocean to the atmosphere), open water may be maintained over a large polynya with little or no ice production, in which case the principal source of heat is therefore sensible heat from the ocean. In other situations where open water occurs in leads and small polynyas, the principal source of heat is latent heat from ice production. Therefore, open water features can be classified as primarily one of two types, either a sensible-heat or a latent-heat feature. These two types of features have significantly different effects on the ocean water. In the case of a sensible-heat feature, the open ocean continually cools to supply the heat flux to the atmosphere, as in the 1974-1976 Weddell polynyas during which the ocean cooled significantly to 2500 m depth [Gordon, 1982]. In the case of a latent-heat feature, where the sea ice is continually removed mechanically by the wind and the sensible heat from the ocean is inadequate to maintain an average open-water condition, new ice is continually created. The latter case leads to the production of large quantities of sea ice, which are subsequently carried to the surrounding waters.

Most of the polynyas described in the literature are probably of the latent-heat type, (e.g., the Saint Lawrence Island polynya in the Bering Sea, [Schumacher et al. 1983]; or the North Water polynya of Baffin Bay [Dey, 1980]). Within the southern ocean, sensible-heat polynyas are more likely in the deep-ocean region, where a large reservoir of relatively warm water (above 0°C) is found immediately below the shallow and weak permanent pycnocline. Upwelling and entrainment of this heat into the mixed layer is an important factor in the mixed-layer heat budget [Gordon, 1981; Gordon and Huber, 1984].

Over the continental shelf, the water column has an average temperature of -1.43°C (Table 1 from Carmack [1977]) and hence stores little heat above the freezing point. More sensible heat is available over the outer shelf where intrusions of slope water are more prevalent. Over the inner shelf, the water column is colder as the cold shelf water mass occupies a greater percentage of the depth and intrusions of slope water are significantly attenuated. Considering only the sensible-heat flux from open water to the atmosphere, a heat flux of 500 W/m^2 [Maykut, 1978; Schumacher et al., 1983], for example, would cool a 518 meter thick (mean depth [Carmack, 1977]) water column by 0.02°C/d. At this rate, the average shelf water column could sustain open water for only 23 days, not a full winter. Although an ocean circulation pattern that transfers all heat from the shelf waters into a restricted coastal region is conceivable, the sensible-heat flux required to sustain open water for the entire winter would cool the entire shelf volume to the freezing point, which is an unlikely condition since early summer data do not differ significantly from late summer data below the seasonal pycnocline. Therefore, the near-shore areas of open water are considered to be mostly latent-heat features, and thus, principal areas of ice production and salt rejection as shown schematically in Figure 1.

In this paper, the areas of open water in selected coastal regions and their temporal variability are described using passive microwave observations from the Nimbus 5 Electrically Scanning Microwave Radiometer (ESMR). The information on sea ice concentration from the microwave data set is much greater in spatial and temporal detail than other data in the sea ice zone, such as wind and temperature data. Although quantitative surface wind data are not available, examination of the time series of passive microwave data along with the synoptic sea level pressure maps in the Ross Sea, for example, indicates patterns of increases and decreases in sea ice concentration that are consistent with the surface wind fields deduced from the pressure maps. Quantitative values of the total area of open water derived from the passive microwave data for 16 coastal regions are used to estimate the resultant ice production and wintertime salinization of the shelf waters along the Antarctic coast.

Satellite Observations of Polynyas and Leads

Examples of the mean monthly sea ice concentrations derived from satellite microwave data are shown in Plate 1 using methods described below. Reduced ice concentrations in these time-averaged data are apparent in both summer and winter, particularly in the marginal ice zone and the near-shore zone. Open water (C < 14%) is observed where the Weddell polynya is forming in July 1974. The reduced ice concentrations near-shore during winter are most evident in locations of strong drainage winds such as the vicinity of the Ninnis and Mertz outlet glaciers around 148°E [Parish, 1982]. Sixteen study areas, which were chosen to enclose mostly shelf regions along the Antarctic coast, are outlined in Plate 1. The size and location of each area are given in Table 2. The locations of some of the study areas are based on the occurrence of significant offshore polynyas during the winter, and others are based on the absence of such polynyas. An interesting contrast occurs in some areas, such as the Shackleton area at 99°E and the Larsen at 59°W, where more open water is usually observed during winter than in summer (Plate 1). Variations of the open water areas on shorter time scales are described in the following analysis of daily and 3-day average brightness temperatures in the study areas.

Derived Sea Ice Concentration and Error Analysis

The observation and quantification of the amount of open water within the ice pack with passive microwave remote sensing is possible because of the large contrast between the emissivity of water and that of sea ice. The emissivities are approximately 0.44 for ice-free ocean and 0.92 for first-year sea ice at the ESMR wavelength of 1.55 cm. Thus, while the field-of-view of the satellite measurements is coarse (∼30x30 km), the relative areal coverage of open water and sea ice within a field of view can be derived.

In the simplest formulation, the observed microwave brightness temperature is a linear combination of the brightness temperatures of open water and sea ice, in proportion to the areal coverage of each within a field of view. Therefore, using the radiative transfer formulation and approximations described in Comiso and Zwally [1982] and Gloersen et al. [1974], the ice concentration (C) is

$$C(\%) = \frac{(T_B - T_0) \times 100}{\varepsilon_I T_{eff} - T_0} \qquad (1)$$

where T_B is the measured brightness temperature, T_0 is the observed brightness temperature of ice-free ocean, and $\varepsilon_I T_{eff}$ is the observed brightness temperature of 100% ice cover. The ε_I is the microwave emissivity of sea ice, and T_{eff} is the temperature of the ice adjusted to incorporate the atmospheric effects as described in Comiso and Zwally [1982] and Zwally et al. [1983]. Similarly, T_0, which is approximately 135 K under average conditions, is equal to the microwave emissivity of seawater times its physical temperature, plus secondary terms due to atmospheric emissions. Inclusion of the atmospheric effects in T_0 and T_{eff} as approximations retains the simple linear relationship between concentration and observed brightness temperature. A typical value of $\varepsilon_I T_{eff}$ is 235 K, which varies with either the emissivity or physical temperature of the ice.

The emissivity of sea ice in the southern ocean has been observed to have little variation, except during spring and summer when surface melt causes variations in emissivity from 0.85 to about 0.95 [Comiso et al., 1984]. Thus, since most of the data analyzed in this paper are from winter observations, a constant ice emissivity of 0.92 is assumed in the analysis. However, a significant localized effect occurs in the Bellingshausen-Amundsen Seas where a lower emissivity of the old ice surviving the summer melt lowers the derived ice concentration values for those areas by about 10%. Whereas the calculations of the ice concentrations in Plate 1 used climatological air surface temperatures to estimate T_{eff}, in this study the variations in the physical temperature of the ice from one study area to another and over the seasonal cycle are neglected by using an average value of 235 K for the maximum brightness temperature ($\varepsilon_I T_{eff}$) corresponding to 100% ice cover in equation (1).

TABLE 1. Error Analysis of Open Water Observations

	T_B^{MAX}	σ_M	T_B	δW_1 (%)	Area (Pixels)	Overlap (%)	δW_2 (%)	δW (%)
Riiser-Larsen	233.6	6.4	215.2	1.1	10 x 5	8.6	2.6	2.8
Syowa	231.7	5.4	221.8	3.0	3 x 9	11.2	3.4	4.5
Amery	233.8	3.7	215.8	1.0	19 x 6	6.7	2.0	2.2
Prydz	228.2	3.9	208.1	5.3	7 x 6	5.6	1.7	5.6
Shackleton	234.0	4.9	207.1	1.0	9 x 7	4.3	1.3	1.6
Casey	234.6	4.4	217.7	0.3	9 x 7	4.9	1.5	1.5
Sabrina	236.6	4.6	223.6	-1.4	9 x 7	5.8	1.7	2.2
Adélie	228.9	3.7	209.6	4.9	7 x 8	3.9	1.2	5.0
George V	236.2	2.9	225.6	-1.1	9 x 8	3.9	1.2	1.6
Ross	230.9	2.2	212.8	2.2	9 x 18	3.7	1.0	2.4
Ruppert	226.7	3.2	203.1	6.2	3 x 18	11.0	3.3	7.0
Amundsen	234.1	4.0	211.8	0.7	13 x 10	2.7	0.8	1.1
Bellingshausen	236.8	5.1	220.9	-1.5	23 x 11	2.9	0.9	1.7
Larsen	235.5	5.1	214.8	-0.4	13 x 15	2.9	0.9	1.0
Weddell	237.9	2.7	229.0	-2.6	16 x 24	2.0	0.6	2.7
Halley	233.3	3.2	224.8	1.6	18 x 8	4.4	1.3	2.1

Using $\varepsilon_I = 0.92$, the value of 235 K for $\varepsilon_I T_{eff}$ implies an ice temperature of about 255 K and an air temperature of about 250 K.

The error in the derived ice concentration, or open water fraction (W = 100 - C), that is caused by variations in $\varepsilon_I T_{eff}$ is estimated with the following equation derived from equation (1)

$$\delta W(\%) = \frac{C \varepsilon_I T_{eff}}{\varepsilon_I T_{eff} - T_0} \simeq 0.007 \, C \, \delta T_A \qquad (2)$$

Using $\varepsilon_I = 0.92$, $(\varepsilon_I T_{eff} - T_0) = 100$ K, and $T_{eff} \simeq T_I = 0.75 \, T_A + 67.8$, where T_I is the temperature at the snow/ice interface and T_A is the surface air temperature [Zwally et al., 1983]. For example, if T_A were 255 K in an area where 10% open water is calculated using $\varepsilon_I T_{eff} = 235$ K, then δT_A is 5 K and the correct calculated value would be 13% open water using $\varepsilon_I T_{eff} = 238.2$ K. Therefore, the calculated amount of open water is too small in places that are physically warmer than 250 K and too large in places that are colder. Such temperature variations may be on the order of 5 to 10 K during the winter period (days 76 to 315) resulting in open water errors of 3 to 6%, but the average error over the winter is smaller. The time-series of the maximum observed brightness temperature (T_B^{MAX}), which corresponds to 100% ice cover in most study areas during the winter months, also indicate the possible errors due to using $\varepsilon_I T_{eff}$ fixed at 235 K. If the observed T_B^{MAX} is both representative of 100% ice cover and less than 235 K (note the 100% line in Figures 4, 5, and 9 through 23), then the calculated open water would be too large (or too small if T_B^{MAX} is greater than 235 K). To estimate the actual error due to using a fixed value of 235 K, it is assumed that T_B^{MAX} averaged over the midwinter (days 100 to 290) is representative of 100% ice cover and deviations (δT_B) of the average, $<T_B^{MAX}>$, from 235 K cause errors. Actual deviations are expected to be due to variations in ice emissivity and temperature, which cause real errors, and to the intermittent occurrence of less than 100% ice cover throughout the study area, which does not cause real errors in the calculated open water. Therefore, the error estimates obtained here are probably upper limits. The standard deviation σ_M of T_B^{MAX} is an indication of both the short-term variability of the error during the season and departures from 100% ice cover in all pixels. The values of $<T_B^{MAX}>$, the standard deviation σ_M, the midwinter average of the T_B averaged over each study area $<\bar{T}_B>$, and the estimate of the error $<\delta W_1>$ in the winter average open water are given in Table 1 using $<\delta W_1> \simeq C \delta T_B$, where

$$C \simeq \bar{T}_B - 135 \text{ and } \delta T_B = <T_B^{MAX}> - 235 \, .$$

The values of $<\delta W_1>$ are mostly less than 5%.

Another source of error in the estimate of the open water in each study area is the overlap of some coastline ocean pixels on the ice shelf or ice sheet. The effect of the contaminated coastline pixels on the A_W value is limited by the ratio of coastline pixels to the total ocean pixels in each study area. The binary map used to differentiate land areas from ocean areas in the analysis of the data was constructed such that, if the center of the data element is over ocean, the data element is classified as ocean. Therefore, the percentage contamination by ice shelf or ice sheet could range from 0 to about 50% in each coastline pixel. Assuming 25% contamination of the coastline pixels, the areal contamination ranges from about 2% to 11% of the total ocean pixel area as shown in Table 1. The actual effect on the calculated open water area also depends on the winter brightness temperatures inside the coastline compared to those outside the coastline. The T_B inside the coastline ranges from about 155 K on the Larsen Ice Shelf to about 185 K on the Ross Ice Shelf to about 215 K on parts of the East Antarctic ice sheet margin. For a T_B inside the coastline of 185 K, which falsely implies 50% open water, and a value of T_B = 215 K outside the coastline, the possible contamination error ranges from δW_2 = 0.6% to δW_2 = 3.3%, corresponding to the areal contaminations of 2% and 11% respectively. Consequently, the estimated overall error $<\delta W>$ in the calculated winter average open water area, including deviations of $\varepsilon_I T_{eff}$ from 235 K and the pixel contamination, is less than 5% (e.g., $<W>$ = 19% ± 5%) in all areas except Prydz and Ruppert where it is 5.6% and 7.0%, respectively.

Another source of some uncertainty is the microwave emissivity of very thin ice under about 5 cm (new ice and dark nilas). While unconsolidated frazil ice, grease ice, and other new ice forms probably have microwave emissivities more like open water than first-year ice, thin ice of more than about 5-cm thickness (light nilas, gray ice, and gray-white ice) has an emissivity close to that of first-year ice. Therefore, some fraction of the areas of very thin ice under about 5 cm, including, in particular, areas of unconsolidated frazil ice and grease ice, may be included in the ESMR derived values of open water. Consequently, the area of totally ice-free water may be slightly less than the total area of open water indicated by the EMSR. However, this uncertainity is not significant for many purposes such as the estimation of heat fluxes and ice production because of the similarity of these quantities for both open water and very thin ice.

Both daily averages and 3-day averages of brightness temperatures are analyzed. The daily averages give better temporal resolution, while the 3-day averages give better spatial resolution because the three-day maps used only the center of the observation swath (1360 km versus 2500 km). Each map element is approximately 30 x 30 km, which along with other mapping characteristics and associated uncertainities are described in more detail in Zwally and Gloersen [1977], Zwally et al. [1983], and Comiso and Zwally [1984]. The daily average data thus enable the observation and identification of shorter-term polynyas and give better measure of their variability.

However, most of the analyses presented in this paper use the 3-day average maps. When only one date is used to identify a 3-day average map, it is the first day of the 3-day period. Although the 3-day averages may include as many as about 35 observations in each map element distributed over the 3 days, in a few cases, some map elements may have only one or no observations due to the irregular data gaps.

Ross Sea Region

A good example of a recurrent polynya area is in the Ross Sea near the ice shelf front (Plate 2). Three sequences of color-coded images illustrate how the brightness temperature changes from one 3-day period to another. The first set of images (top two) shows the most pronounced change in brightness temperature, which is attributed to a substantial increase in the amount of open water during the 6-day interval from days 100-102 to 106-108. The overall ice condition recovered to a previous state after a few days, but a major change in ice cover occurred again 30 days later as shown in the second set, though not as pronounced as the first. Another significant change in ice cover occurred after another 30 days, but not as large as in either of the first two events. Sea ice concentration changes, similar in magnitude to the third event, occurred a few more times during the year.

The short-term nature of the changes in the Ross Sea suggests that the polynya events are principally driven by winds. Presumably, the reduced concentration is caused by divergence and ice transport from the area. Figure 2 shows transects of brightness temperatures along 179°E across the Ross Sea study area before, during, and after the April polynya event. Notable changes of the ice cover along the transect, including the ice edge, accompanied the change in ice cover in the vicinity of the shelf (e.g., see plot for Julian days 106 and 109). The increase in the sea ice concentration in the outer part of the pack and the slightly higher than normal rate of expansion of the ice edge during the polynya event are further indications of a synoptic-scale wind-driven effect, which is examined in more detail below. The T_B of the ice shelf and the continental ice sheet, in contrast, are relatively constant, and respond slowly to changes in the surface physical temperature (except under melting conditions). The size and location of some of the satellite data elements (pixels) representing the transects in Figure 2, are shown schematically in Figure 3. One pixel (43) is contaminated by contributions from the ice shelf, and the brightness temperature of this pixel did not decrease as much as those farther away from the boundary

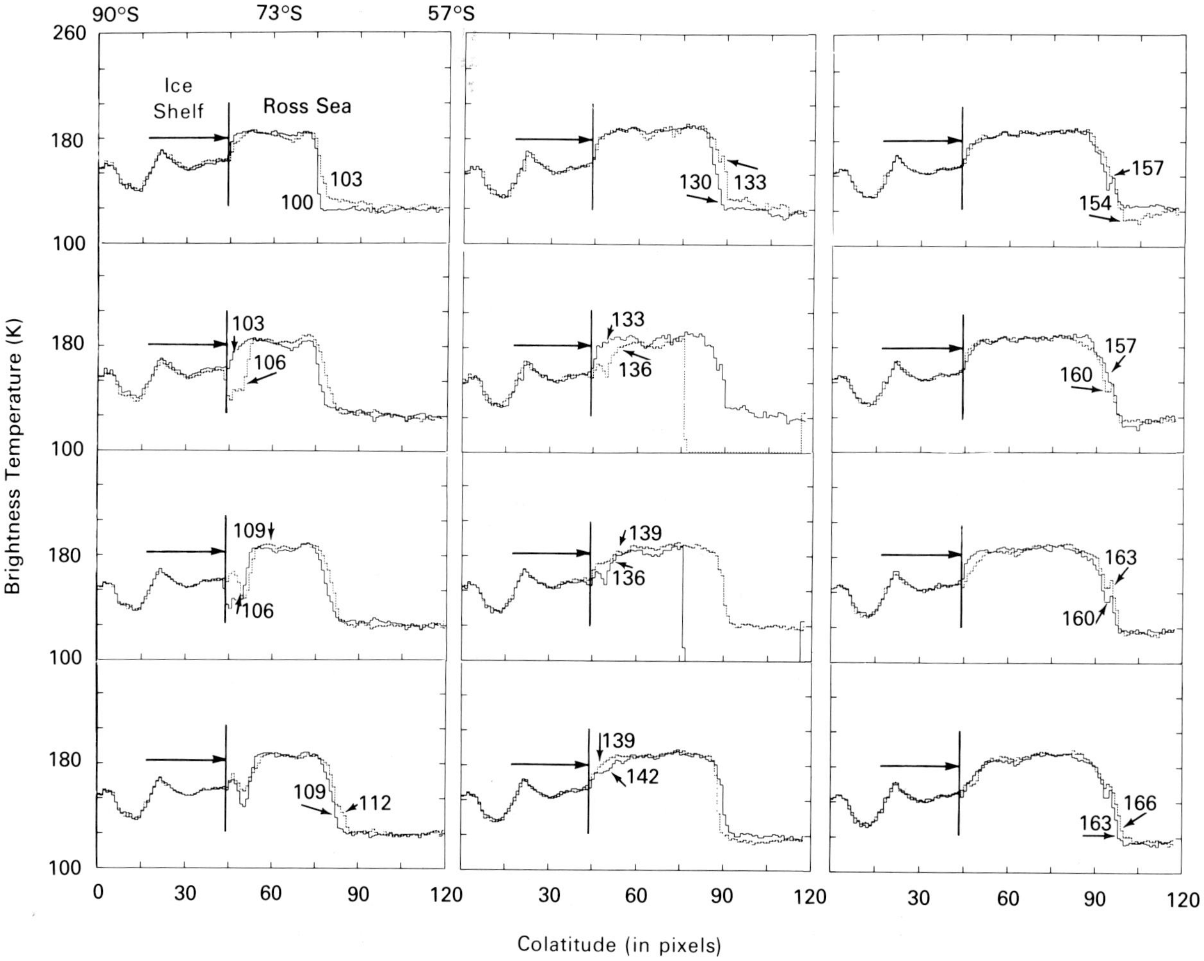

Fig. 2. Latitudinal transects of brightness temperatures along the Ross Ice Shelf and the Ross Sea during occurrences of the Ross Sea polynya.

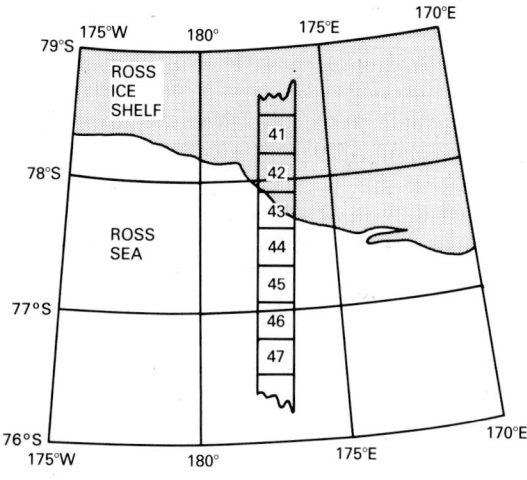

Fig. 3. Pixel (30 km x 30 km) locations along the transect crossing the ice shelf in Fig. 2.

during the occurrence of the polynya. The brightness temperature of the ice shelf in this pixel is substantially less sensitive to the atmospheric forcing.

Figures 4 and 5 show time-series of observed maximum and minimum brightness temperatures (T_B^{MAX} and T_B^{MIN}) in the Ross Sea study area using daily and 3-day averages of all good data in 1974. These time series are obtained by examining the T_B of each of the 123 map elements over the ocean part of the study area for each of the maps. Between January 1 and February 24 (Julian days 1 to 55), the maximum brightness temperature is about 185 K, indicating a maximum ice concentration within the study area of about 50% during summer. As previously noted, the series of T_B^{MAX} are useful for establishing the T_B of 100% ice cover, which is assumed to be found in at least one pixel in most of the maps during winter. From March 11 until November 21 (70-

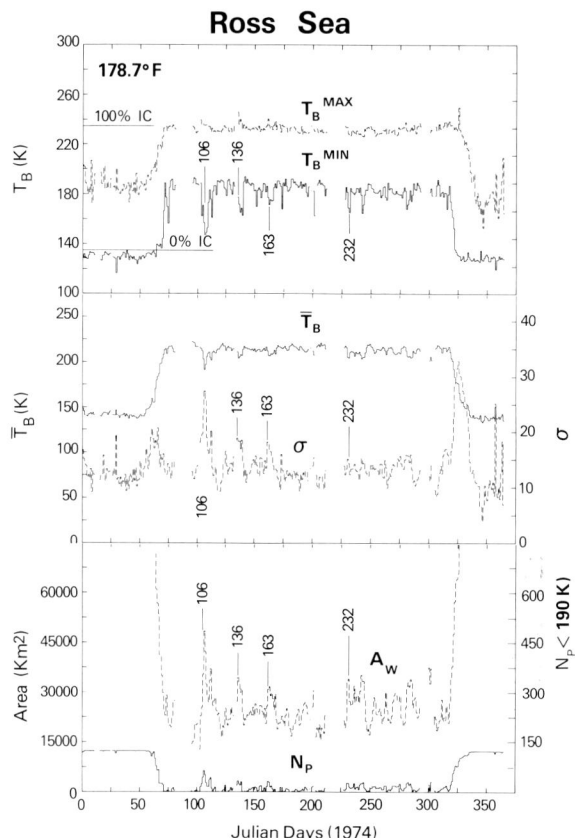

Fig. 4. Time distributions of maximum and minimum brightness temperatures (T_B^{MAX}), (T_B^{MIN}), mean ($\bar{T}_B$), standard deviations σ, number of pixels (N_P) with $T_B < 190$ K, and areal extents of open water (A_W) in the Ross Sea study area using 1-day average data.

325), the T_B^{MAX} is within a few degrees of 235 K. The small variation in the T_B^{MAX} during winter indicates that consolidated ice persisted in some parts of the area throughout this time.

The distribution of minimum T_B, on the other hand, shows a near constant value of 135 K in the summer, which is consistent with the presence of at least one ice-free data element in the study area. The T_B^{MIN} increases abruptly to about 180 K in the fall and remains approximately constant at this value throughout winter, except for the dips around April 16 (106), May 16 (136), and June 12 (163) (and a few smaller ones) consistent with the observed changes shown in Plate 2. In the 1-day average time series on April 17 (107), the T_B^{MIN} is 146 K, indicating the presence of about 89% open water in some part of the study area. In the 3-day average time series, the T_B^{MIN} is 156 K on April 16-18 (106-108), corresponding to 79% open water in at least one of the time-averaged map elements. The lower value of T_B^{MIN} on the one-day average for day 107, compared to the three-day average for 106-108, indicates that the nearly ice-free ocean (89% open water) occurred for only a brief time at the height of this event. Such intermittent formation of open water reduces the average ice-concentration in the coastal regions as shown, for example, in the monthly average in Plate 1.

The variability of the ice cover is quantified by the standard deviation (σ) of the T_B within the study area as plotted in Figures 4 and 5. The average brightness temperature ($\bar{T}_B$) distribution is very similar to that of T_B^{MIN}, but the dips observed in the average T_B series are not as large as in the minima. During the summer, when the whole study area is occupied by a uniform mixture of water and a low concentration of ice floes, and in late winter when the area is predominantly consolidated ice, the standard deviation is smallest and least variable in its magnitude. The standard deviation increases in early fall around March 1 (60), when the study area is no

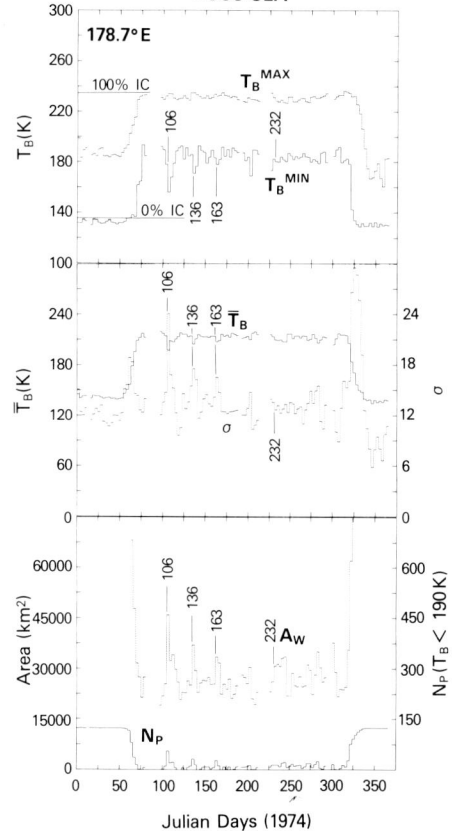

Fig. 5. Time distributions of maximum, minimum, and mean brightness temperatures, standard deviations, number of pixels < 190 K, and areal extents of open water in the Ross Sea study area using 3-day average data.

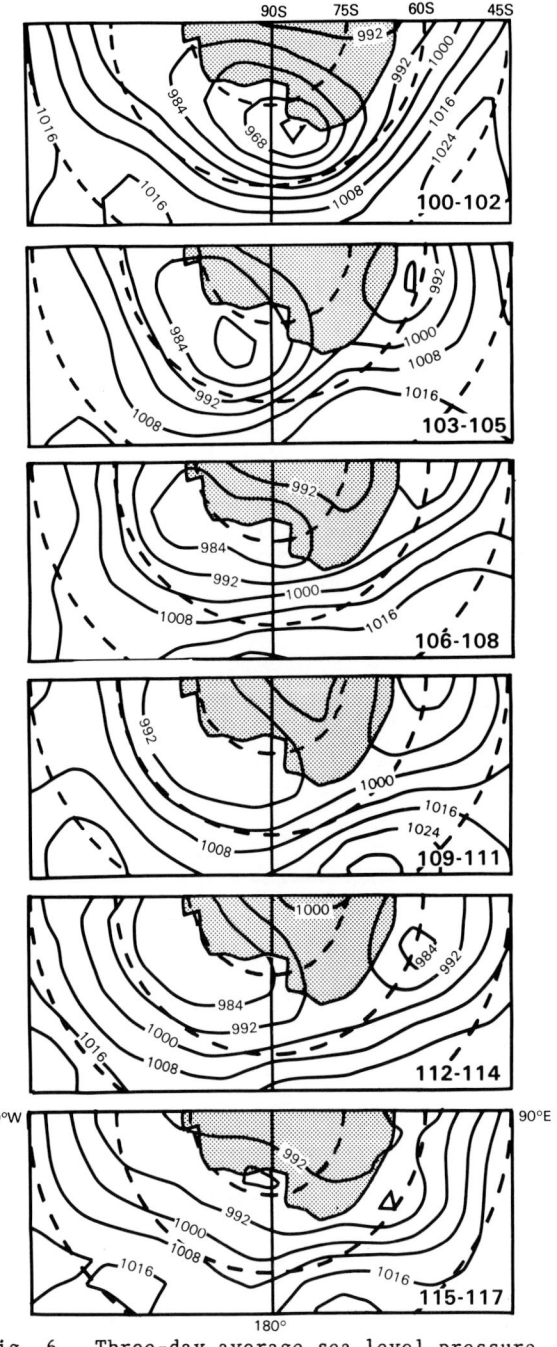

Fig. 6. Three-day average sea level pressure contours from Julian date 100 to 117, 1974.

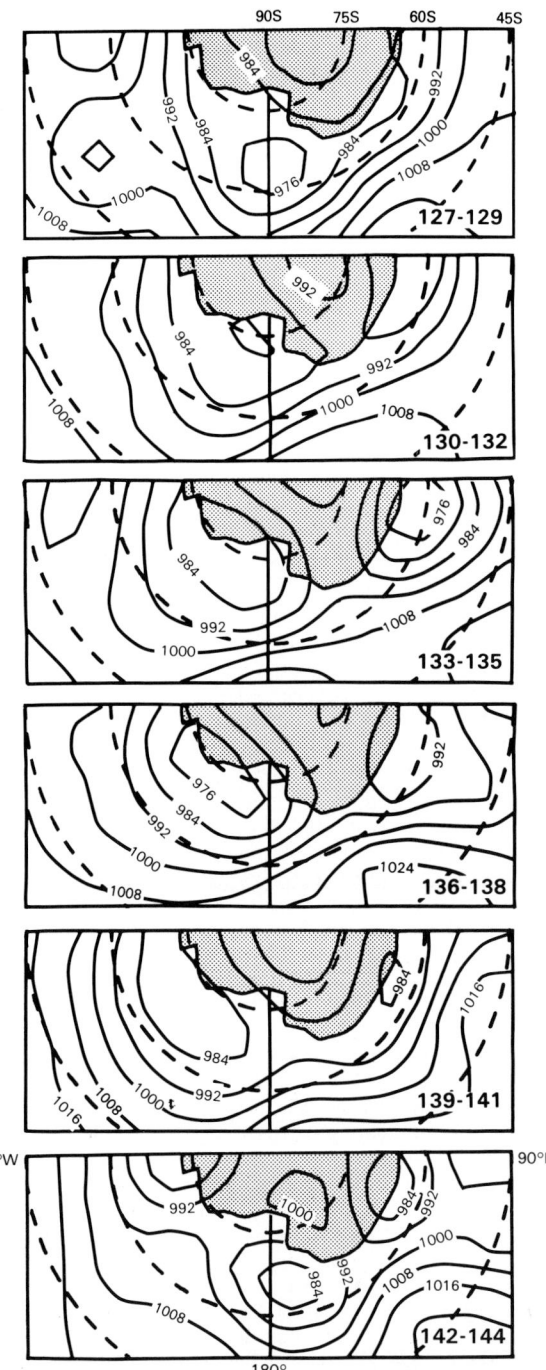

Fig. 7. Three-day average sea level pressure contours from Julian date 127 to 144, 1974.

longer uniform because of the introduction of newly frozen ice, and in the spring around November 21 (325) when substantial melting and breakup occur within the pack. The standard deviation also increases during the polynya events, indicating enhanced ice dynamics.

The areal extent and persistence of the polynya events, as well as the background of nonresolved open leads and polynyas, is quantified by applying equation (1) to each pixel to calculate the area of open water. The results for the Ross Ice Shelf study area are shown in Figures 4 and 5. The peak value of the total area of open water (A_W) in the study

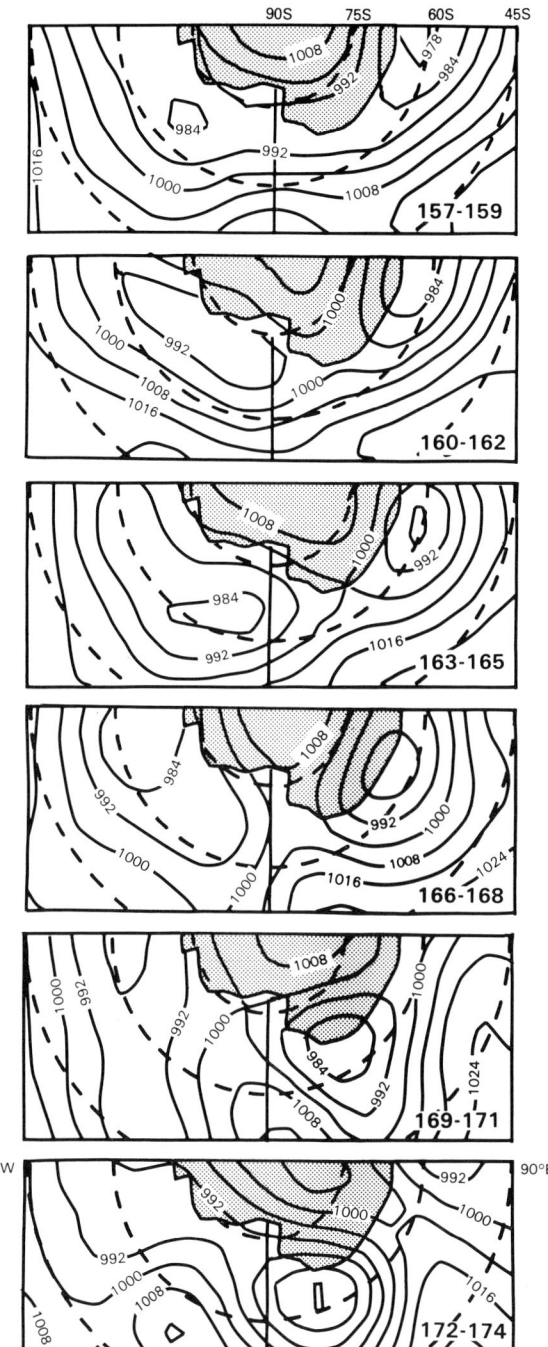

Fig. 8. Three-day average sea level pressure contours from Julian date 157 to 174, 1974.

events also tend to be smaller. However, the averages over 3 or more days are the same in each data set, and the average amount of open water during the winter in the Ross Sea study area is approximately constant at 27,000 km^2. The envelope of the maximum open water in the polynya events decreases to a minimum around mid-July (day 200) when the data indicate that the ice pack is most stable.

The role of synoptic wind in the formation of the Ross Ice Shelf polynya is illustrated by a qualitative analysis of 3-day average pressure contours in Figures 6, 7, and 8 before, during, and after the three polynya events shown in Plate 2. The data were interpolated from the Australian data set that had been prepared for studies of the influence of the wind fields on the position and variability of the ice edge during several seasons of growth and decay [Cavalieri and Parkinson, 1981; Parkinson and Cavalieri, 1982].

The cyclone system, centered near 170°E on days 100-102, migrated to the east to near 150°W on days 103-105 and further to the east to near 135°E on days 109-111. The cyclone position is also shown by the harmonic analysis [Cavalieri and Parkinson, 1981] of the pressure between 60° and 70°S, which is the nominal position of the circumpolar trough in the atmosphere. The increase in open water during the polynya event peaking on days 106-108 coincides with the shift of winds from easterly to southeasterly in front of the ice shelf as indicated by the pressure maps. Since the direction of ice drift is approximately in the direction of the surface winds, the open water during the polynya event is presumably caused by an enhanced northward ice drift. The more rapid increase of the open water during the event compared to the decay is also consistent with the rapid eastward movement of the strong cyclonic system, which was followed by a slower decay of the cyclone in its eastward position over the Amundsen Sea. Similar migrations of weaker cyclonic systems coincide with the weaker polynya events on days 130-132 and 160-162 as shown in Figures 7 and 8.

Changes in the T_B distribution over the ice pack indicate convergence or divergence in response to the wind field. For example, increased T_B indicates a more compact ice pack due to ice convergence or freezing. A small increase in the maximum T_B during the polynya events is evident in the T_B series (Figures 4 and 5) and on the T_B maps (Plate 2) in the eastern portion of the study area. During event 106-108, in particular, the ice in the southeastern part of the Ross Sea appears to be converging in response to the northerly winds at the same time the southeasterly winds in the southwestern part of the Ross Sea are producing open water. Similar changes in the ice edge are also related to the surface

area is 48,500 km^2 in the 1-day average for April 17 (107) of the first polynya event and 45,870 km^2 in the 3-day average for April 16-18 (106-108). Thus, the peak values of open water are larger in the 1 day than the 3-day averages, and the minima between the polynya

winds. For example, the marked expansion of the ice edge by about 250 km around 170°E from days 100-102 to days 106-108 coincides with strong northerly winds that enhance ice growth and northward ice advection.

Other Study Regions

The set of parameters

$$T_B^{MAX}, T_B^{MIN}, \bar{T}_B, \sigma, A_W, \text{ and } N_P$$

in Figures 9 to 23 describe various features of the open water areas and the polynya events for the other 15 study areas. As previously noted, T_B^{MAX} approximately indicates the maximum ice concentration in any of the 30 x 30 km pixels in the study area, and T_B^{MIN} usually indicates the minimum ice concentration. However, T_B^{MIN} is not always an accurate measure of the minimum ice concentration, because the pixel in which T_B^{MIN} is observed may be located partially (or completely due to coastline inaccuracies) over continental ice, which has brightness temperatures in the range of 150 K to 250 K. Because the T_B inside the coastline varies slowly during winter in response to the changing physical temperature, short-term decreases in T_B^{MIN} during winter do indicate polynya events nevertheless. Such events may be located in only one pixel, however, whereas increases in A_W represent the total change in open-water coverage over the entire study area. For example, a large drop in T_B^{MIN} with a small rise in A_W indicates a localized event within the study area. Changes in N_P (the number of pixels with $T_B < 190$ K) are also an indication of the spatial extent of an event. Changes in $\bar{T}_B$ and A_W are inversely proportional to each other, and σ is an indication of the variation of the open water around a mean value over the study area. As discussed below, substantial differences in the seasonal variations in these parameters from one study area to another are observed. In some areas, the approximate length of the ice growth season can be deduced from T_B^{MIN}, but in several areas T_B^{MIN} and the amount of sea ice cover are greater in summer than in winter.

The summer ice conditions are significantly different in the various study areas as can be seen in Plate 1, Figure 5, and Figures 9-23. In the George V and Larsen areas, the minimum T_B do not indicate ice-free water in any pixel at any time during the year. In the Syowa, Shackleton, Sabrina, and Ruppert areas, the minimum T_B indicates low concentrations in at least one pixel, but not totally ice-free water even in the summer. The other 10 areas all show some ice-free pixels during summer.

The maximum T_B is lowest in late summer indicating the absence of 100% ice cover at the end of the melt season in all study areas except the Weddell. Some of the maximum T_B distributions tend to increase slightly in value from fall to midwinter, indicating a higher probability of finding highly consolidated ice later in the growth season. Slight increases in maximum T_B above 235 K during spring are due to the known increase in the emissivity of the surface during the onset of melt [e.g., Zwally and Gloersen, 1977; Comiso, 1983].

The variability in the ice cover over each of the study areas, as indicated by σ, is in most cases greatest during the spring and summer breakup, with a tendency to decrease to a low value in late summer followed by an increase as the ice pack expands in early fall. There is also a tendency in many areas for σ to decrease to a minimum in midwinter indicating a more uniform ice cover, as for example in the Casey and Halley areas (Figures 14 and 23). The most notable exception to this pattern is exhibited in the Larsen area where the is largest in midwinter. In some areas such as Amery, Prydz, Casey, Sabrina, George V, and Ross, the has frequent abrupt changes due to dynamic activity of the ice pack. Also, in some of these cases, the abrupt changes in σ have associated abrupt changes in A_W as in the Prydz, George V, and Ross, but in others the concurrent changes in A_W are less marked due to differences in the way the open water is distributed across the study areas.

The most significant parameter is A_W, which is also most useful for identifying polynya events during which the area of open water increases markedly. In most areas, A_W is largest in late summer and decreases to a minimum in midwinter to late winter when the ice pack is expected to be mostly consolidated in the coastal regions. The most notable exception to this pattern of seasonal variation in A_W occurs in the Larsen area where the maximum A_W occurs in fall and the minimum A_W in summer. The A_W in the Larsen area is actually large throughout the winter, decreasing steadily from its maximum to minimum over a 9-month interval. The time variation of A_W in the Shackleton area is also similar to that in the Larsen.

In general, the production of open water in coastal areas is dependent on the atmospheric and oceanic forcings and the strength of the ice pack. The increase in thickness and consolidation of the ice pack in the coastal regions during the growth season may be a principal factor in the general decline in the open water exhibited in most areas during the winter months. As previously discussed, the production of open water in coastal regions during winter months is mainly wind driven. Therefore, the greater amounts of open water in the Larsen and Shackleton areas during winter than summer indicate a significantly stronger wind forcing of the ice pack in winter, whereas the wind and ocean current forc-

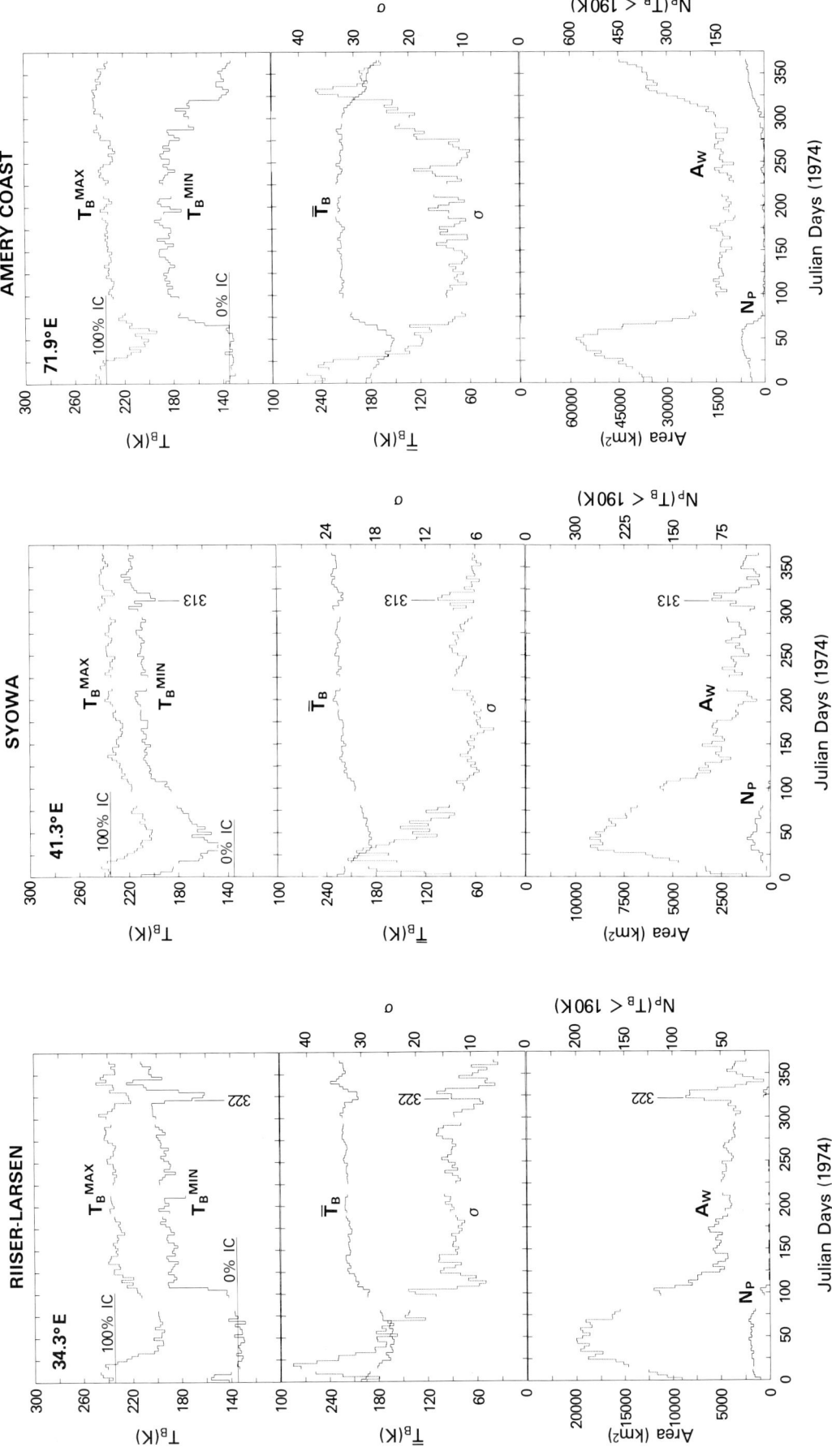

Fig. 9. Time distributions of maximum, minimum, and mean brightness temperatures, standard deviations, number of pixels ≤ 190 K, and areal extents of open water in the Riiser-Larsen Ice Shelf study area using 3-day average data.

Fig. 10. Time distributions of maximum, minimum, and mean brightness temperatures, standard deviations, number of pixels ≤ 190 K, and areal extents of open water near the Syowa study area using 3-day average data.

Fig. 11. Time distributions of maximum, minimum, and mean brightness temperatures, standard deviations, number of pixels ≤ 190 K, and areal extents of open water in the Amery Ice Shelf study area using 3-day average data.

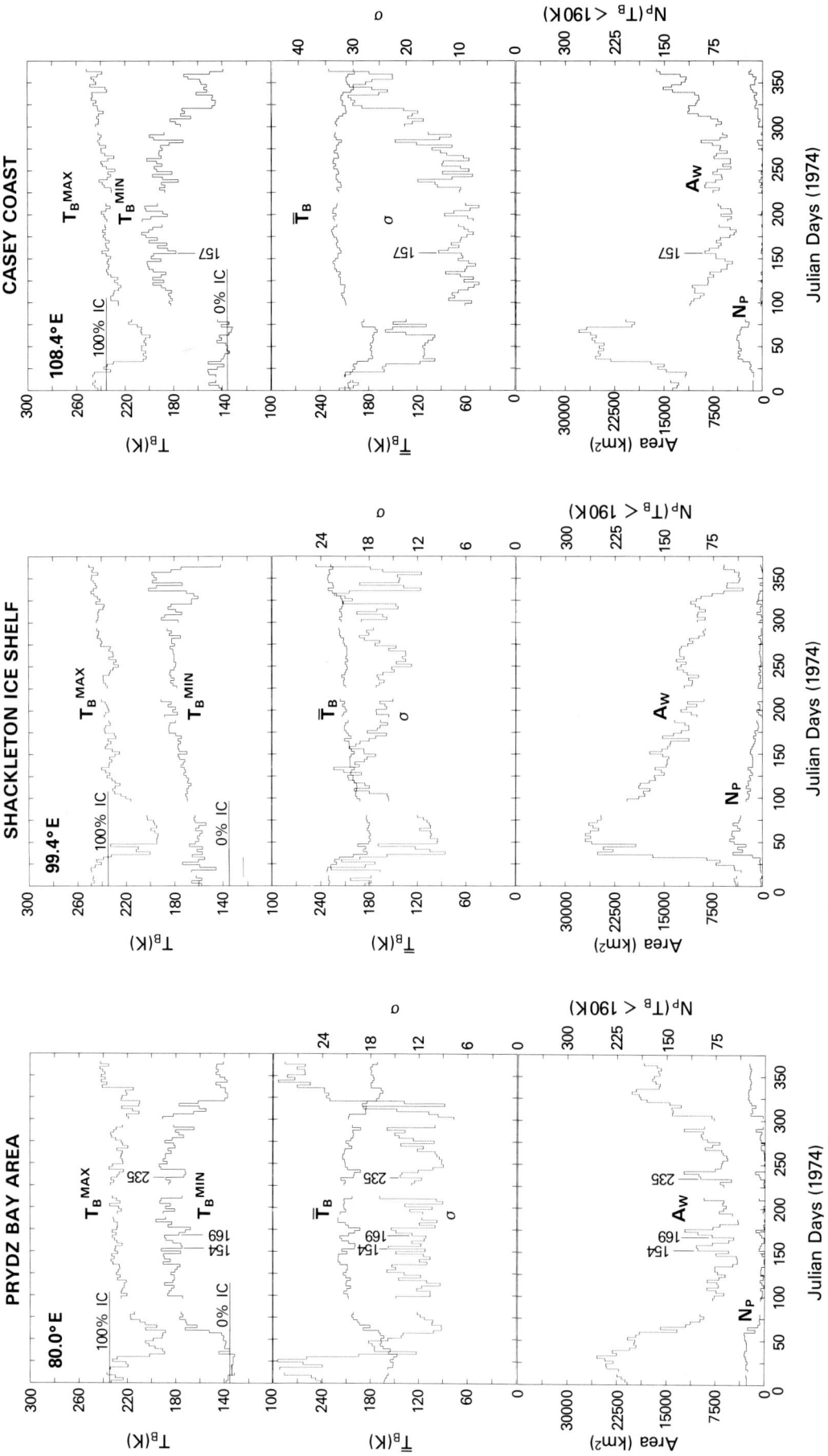

Fig. 12. Time distributions of maximum, minimum, and mean brightness temperatures, standard deviations, number of pixels ≤ 190 K, and areal extents of open water in the Prydz Bay study area using 3-day average data.

Fig. 13. Time distributions of maximum, minimum, and mean brightness temperatures, standard deviations, number of pixels ≤ 190 K, and areal extents of open water in the Shackleton Ice Shelf study area using 3-day average data.

Fig. 14. Time distributions of maximum, minimum, and mean brightness temperatures, standard deviations, number of pixels ≤ 190, and areal extents of open water in the Casey study area using 3-day average data.

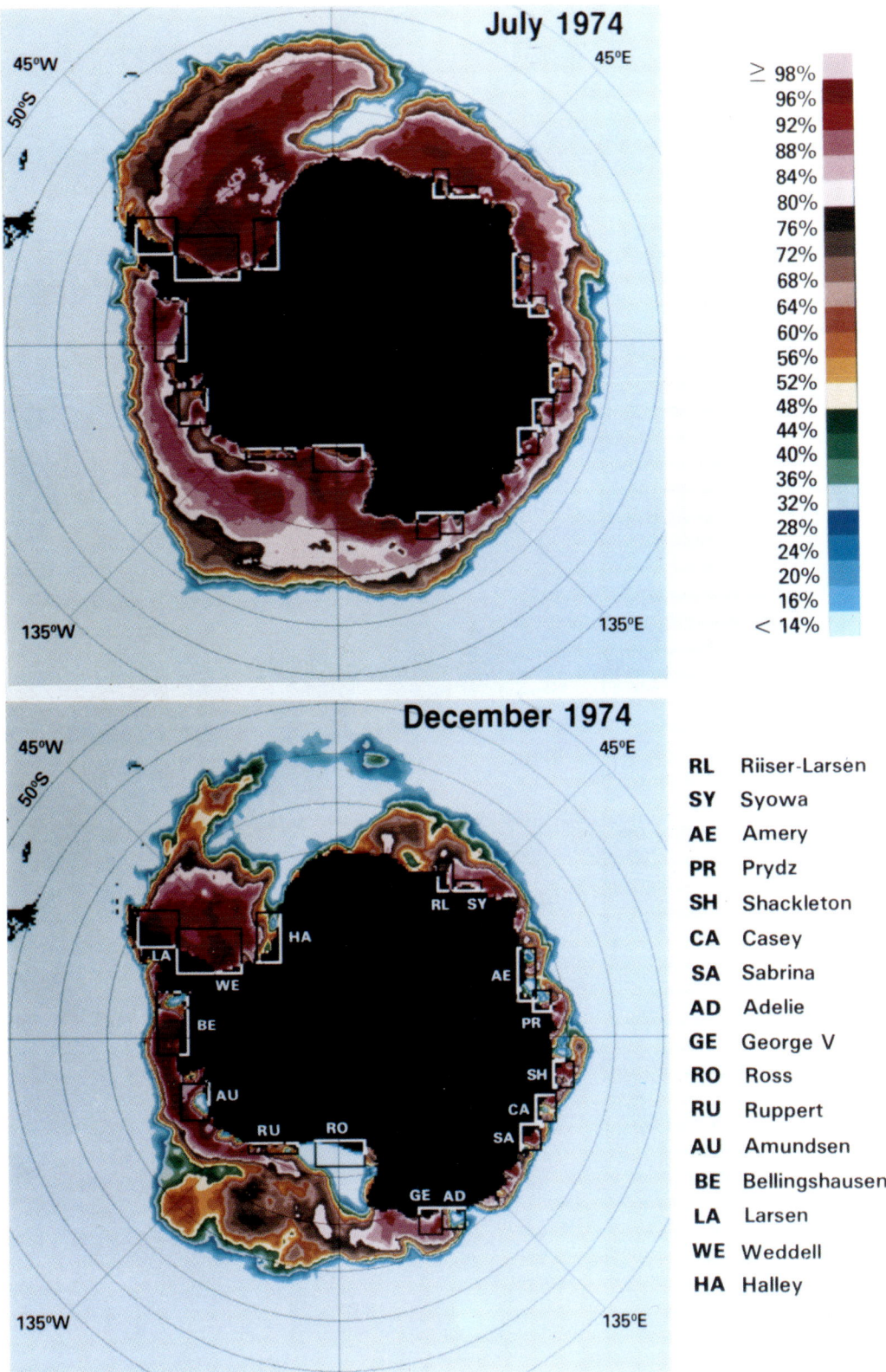

Plate 1. Sea ice concentration map inferred from Nimbus 5 ESMR for July 1974 and December 1974 (from Zwally et al. [1983] with study areas outlined).

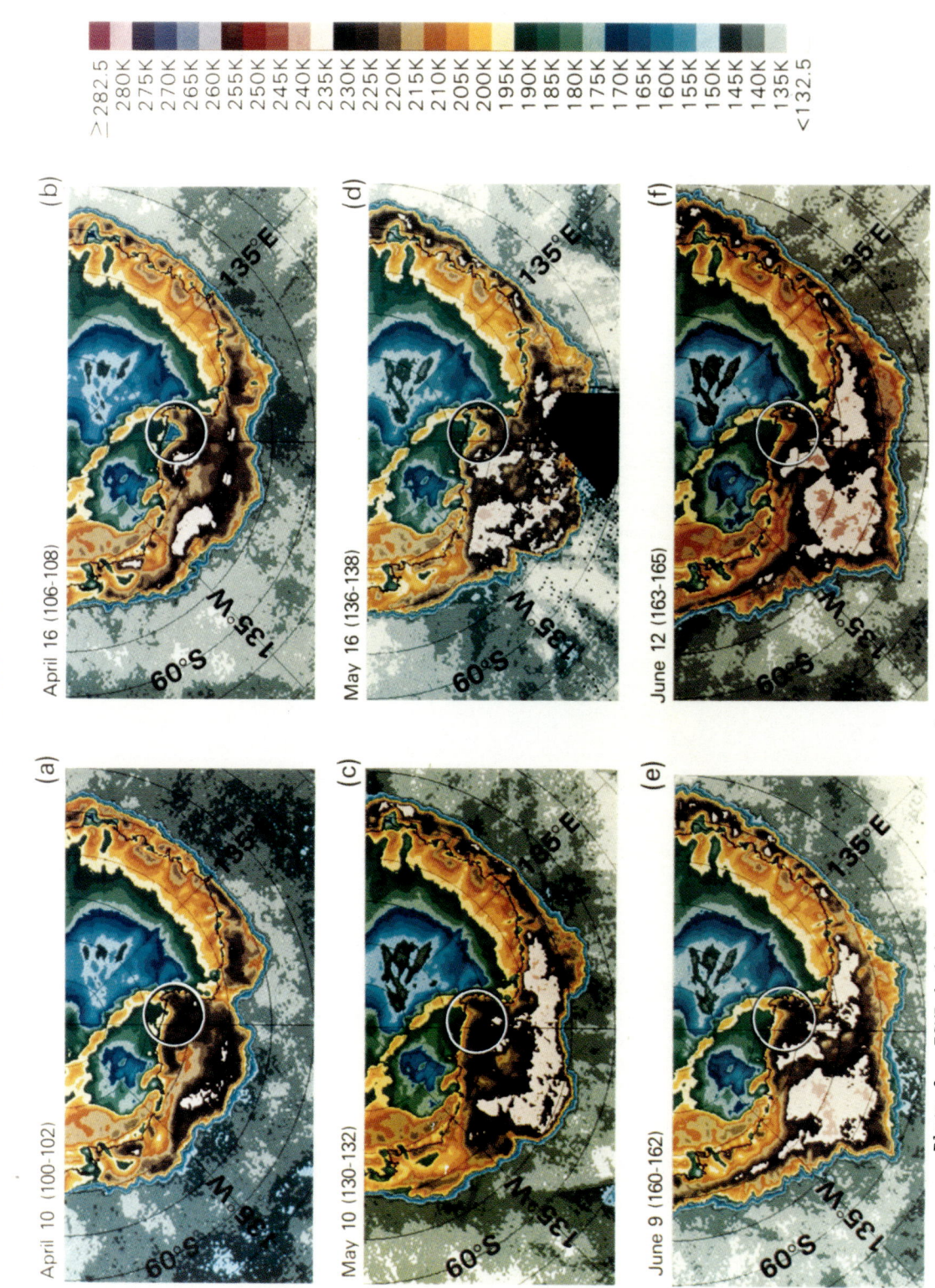

Plate 2. ESMR brightness temperature color-coded maps before and during the major occurences of the Ross Sea polynya.

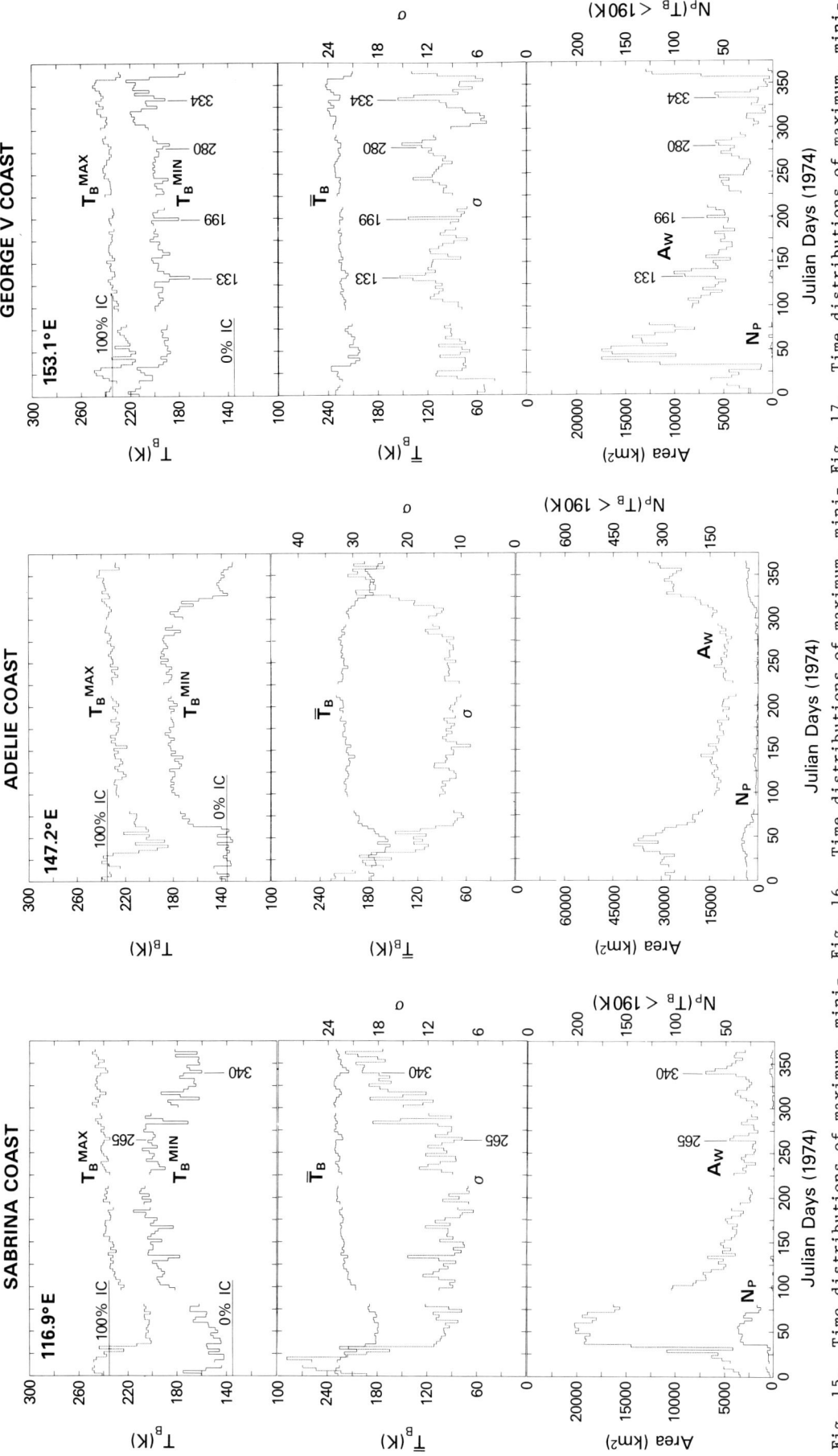

Fig. 15. Time distributions of maximum, minimum, and mean brightness temperatures, standard deviations, number of pixels ≤ 190K, and areal extents of open water in the Sabrina Coast study area using 3-day average data.

Fig. 16. Time distributions of maximum, minimum, and mean brightness temperatures, standard deviations, number of pixels ≤ 190 K, and areal extents of open water in the Adélie Coast study area using 3-day average data.

Fig. 17. Time distributions of maximum, minimum, and mean brightness temperatures, standard deviations, number of pixels ≤ 190 K, and areal extents of open water in the George V Coast study area using 3-day average data.

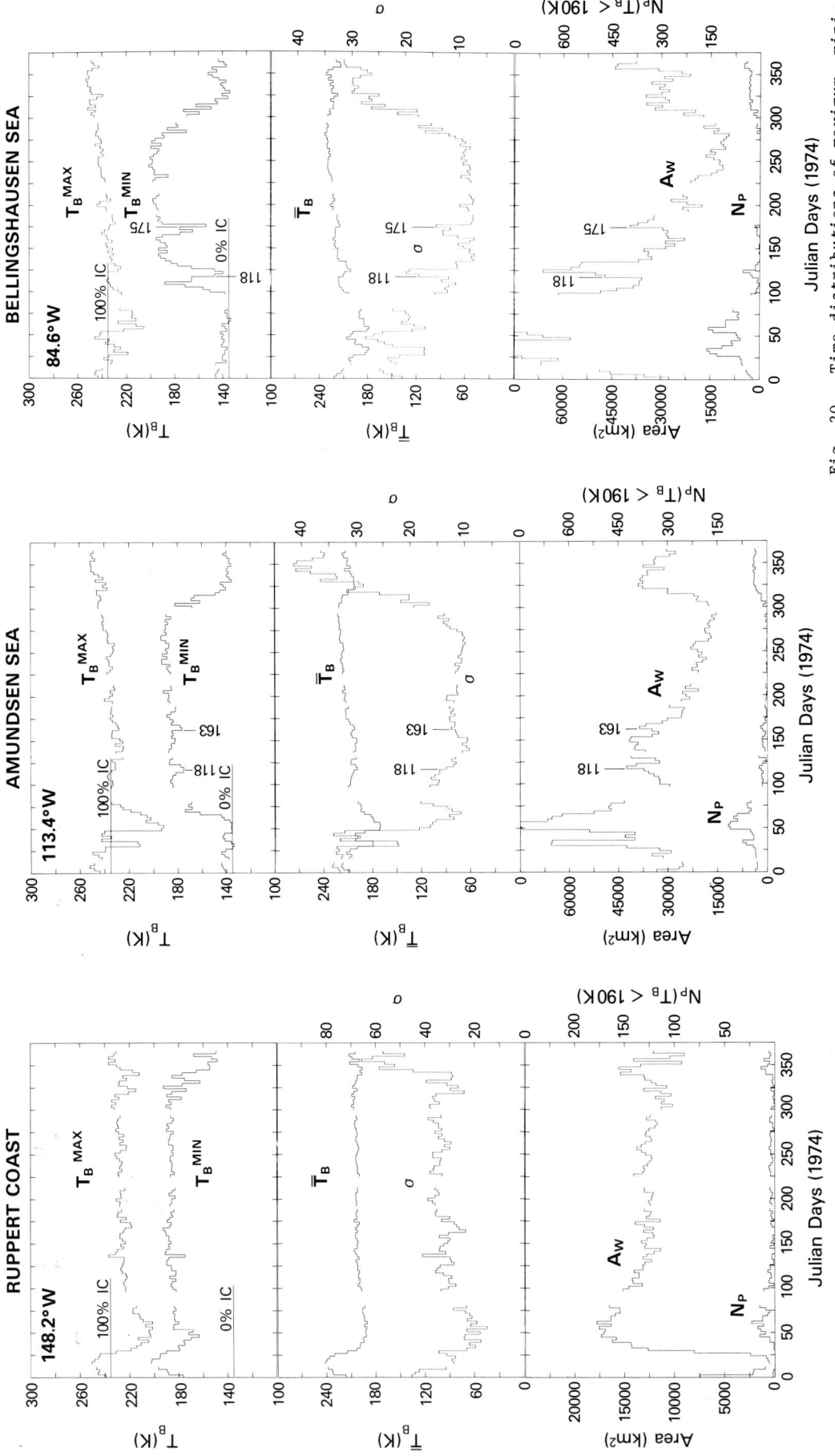

Fig. 18. Time distributions of maximum, minimum, and mean brightness temperatures, standard deviations, number of pixels ≤ 190 K, and areal extents of open water in the Ruppert Coast study area using 3-day average data.

Fig. 19. Time distributions of maximum, minimum, and mean brightness temperatures, standard deviations, number of pixels ≤ 190 K, and areal extents of open water in the Amundsen Sea study area using 3-day average data.

Fig. 20. Time distributions of maximum, minimum, and mean brightness temperatures, standard deviations, number of pixels ≤ 190 K, and areal extents of open water in the Bellingshausen Sea study area using 3-day average data.

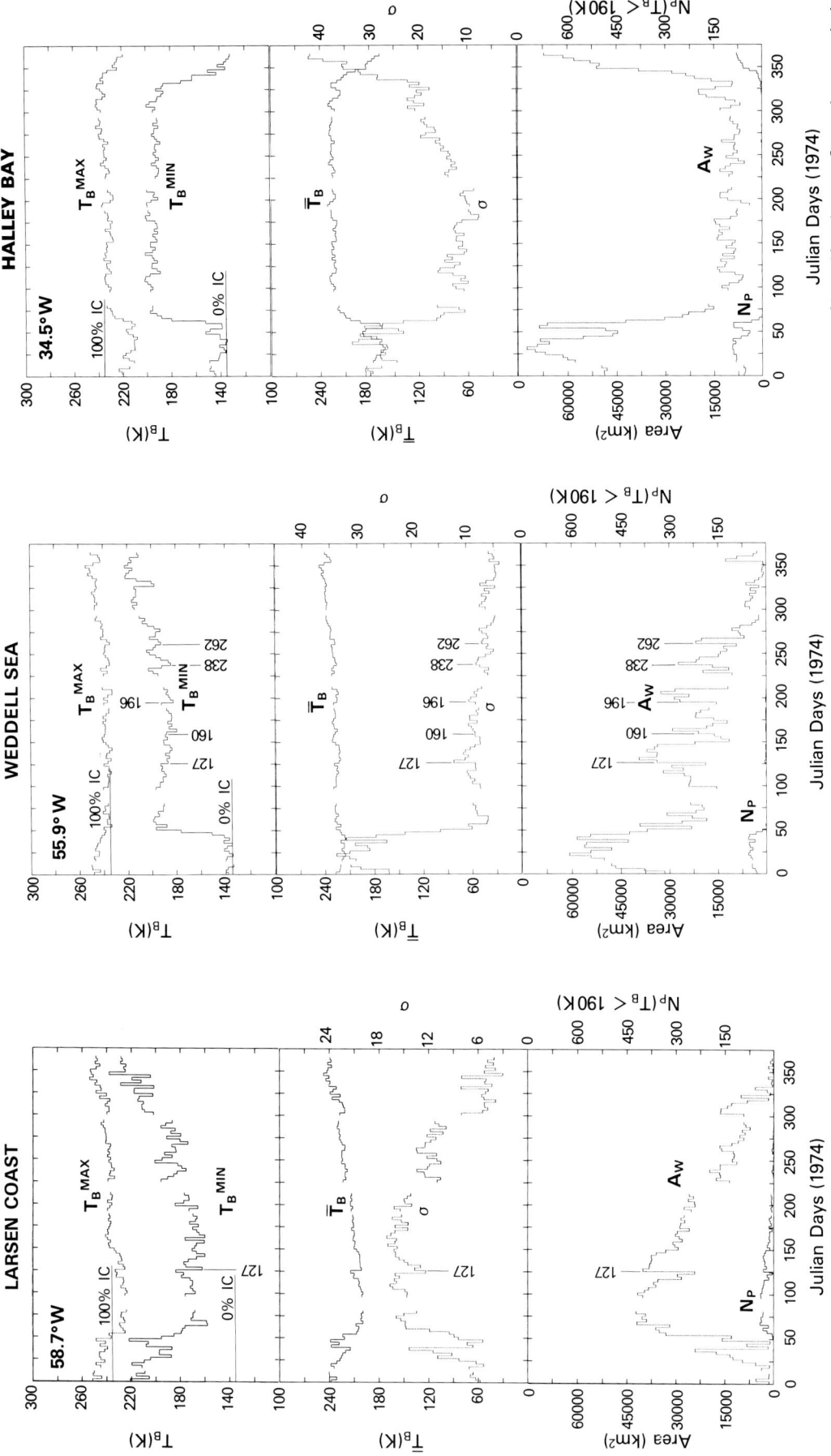

Fig. 21. Time distributions of maximum, minimum, and mean brightness temperatures, standard deviations, number of pixels ≤ 190 K, and areal extents of open water in the Larsen Ice Shelf study area using 3-day average data.

Fig. 22. Time distributions of maximum, minimum, and mean brightness temperatures, standard deviations, number of pixels ≤ 190 K, and areal extents of open water in the Weddell Sea study area using 3-day average data.

Fig. 23. Time distributions of maximum, minimum, and mean brightness temperatures, standard deviations, number of pixels ≤ 190 K, and areal extents of open water in the Halley Bay study area using 3-day average data.

ing in these two areas in summer are apparently favorable for maintaining the observed consolidated ice cover.

Some of the more notable polynya events, in addition to those previously discussed in the Ross area are: in the Riiser-Larsen beginning around day 322; Syowa around day 313; Prydz around days 154, 169, and 235; Casey around day 157; Sabrina around days 265 and 340; George V around days 133, 199, 280, and 334; Amundsen around days 118 and 163; Larsen near day 127; and the Weddell around days 127, 160, 196, 238, and 262. The duration of these events is typically about 9-18 days, which may be associated with changes in the synoptic wind forcing as discussed for the Ross area. However, examination of the pressure maps in the Weddell Sea region did not show similar relations with the ice cover changes in the Larsen and Weddell, possibly because of the barrier influence of the Antarctic Peninsula [Parish, 1983] or inadequacies of the pressure maps. During the event in the Weddell area commencing on day 127, the large (50 x 90 km) tabular iceberg, which was grounded in front of the Ronne Ice Shelf near 77°S and 50°W, became ungrounded and moved 100 km northward to its position in Plate 1 [Zwally and Gloersen, 1977]. During these events in the Weddell Sea, the lowest sea ice concentrations are observed within about 100 km along the ice shelf front. The movement of the iceberg and the reduced ice concentrations are indicative of strong southerly winds. In the Bellingshausen, the increased open water after days 118 and 175 is caused by a retreat of the ice edge into the study area.

The Adélie and the George V areas show some interesting contrasts. The Adélie is located offshore of the Ninnis and Mertz glaciers and has tongues of reduced ice concentration extending over 100 km offshore in winter (Plate 1). The open water is greater in the Adélie area than the George V, but is less variable as shown by changes in both σ and A_W in Figures 16 and 17. The open water offshore from the glaciers is persistent and extends farther offshore than the typical distance of several tens of kilometers over which katabatic winds are usually substantially diminished [e.g., Bromwich and Kurtz, 1984]. Therefore, the variations observed in the coastal open water in these areas suggest that the Adélie is a location of persistent winds at substantial distances offshore from the glaciers, and the George V is a location of more localized and intermittent wind forcing.

In the following sections, the observed A_W is used to estimate the ice production and salinization of the shelf waters. As noted, in some areas the ice-growth season in the coastal regions can be inferred from the temporal changes in T_B^{MIN}, T_B^{MAX}, and the other parameters, but not in all areas as for ex-

amples, George V, Larsen, and Ruppert. Consequently, the growth (or winter) season is estimated for all areas to be from days 76 through 315 inclusive (240 days) and the melt (or summer) to be days 316 through 75 (115 days). The frequency distributions of the ratio of A_W to the total ocean area (A_T) for the winter and summer periods in each of the study areas are shown in Figures 24 and 25. All areas have more open water in summer than in winter except the Larsen and Shackleton. The average open water fraction in winter $(<A_W>/A_T)$ ranges from about 7% in the Weddell to about 25% in the Prydz, Shackleton, Adélie, Amundsen, Ross, and Ruppert (Table 2). These values of open water for the coastal zone are significantly larger than the values in the central part of the ice pack, but are close to the average over the entire pack including the marginal zone where the ice is most divergent. Figures 24 and 25 also show how variable the open water is during the summer and winter periods. In some places, like Amery and Halley, the open water area in winter does not vary much, whereas, in some other areas like Amundsen and Larsen, A_W changes substantially from early winter to late winter. The width of the distributions is also affected by the occurrences of polynyas. To quantify this variability of A_W/A_T, a Gaussian function with a linear background was fitted to each winter distribution by methods of least squares, and the standard deviation (σ_A) of the Gaussian for each study area is given in Table 2.

Oceanic Effect of Offshore Leads and Polynyas

In regions of sea ice and open water such as the Antarctic shelf that do not have significant heat storage in the water column, heat flux into the atmosphere will be derived primarily from the latent heat of fusion. For this situation, estimates of heat flux and daily ice production have been made by Schumacher et al. [1983] for the Saint Lawrence Island polynya in the Bering Sea, based upon changes of water column salinity. For the Antarctic shelf, estimates have been made by Cavalieri and Martin [this volume] based on meteorological data from nearby coastal stations. Schumacher et al. (1983) determine a heat flux of 535 W m^{-2} with an ice production of 0.17 m/d. Cavalieri and Martin [this volume] find an average 90-day ice production of 9 m (their Table 4), which is equivalent to 0.10 m/d, and their average heat flux is 300 W m^{-2}.

The brine rejection associated with ice production within the open water areas increases the water column salinity over what is expected from ice growth through bottom accretion alone. The salinity increase due solely

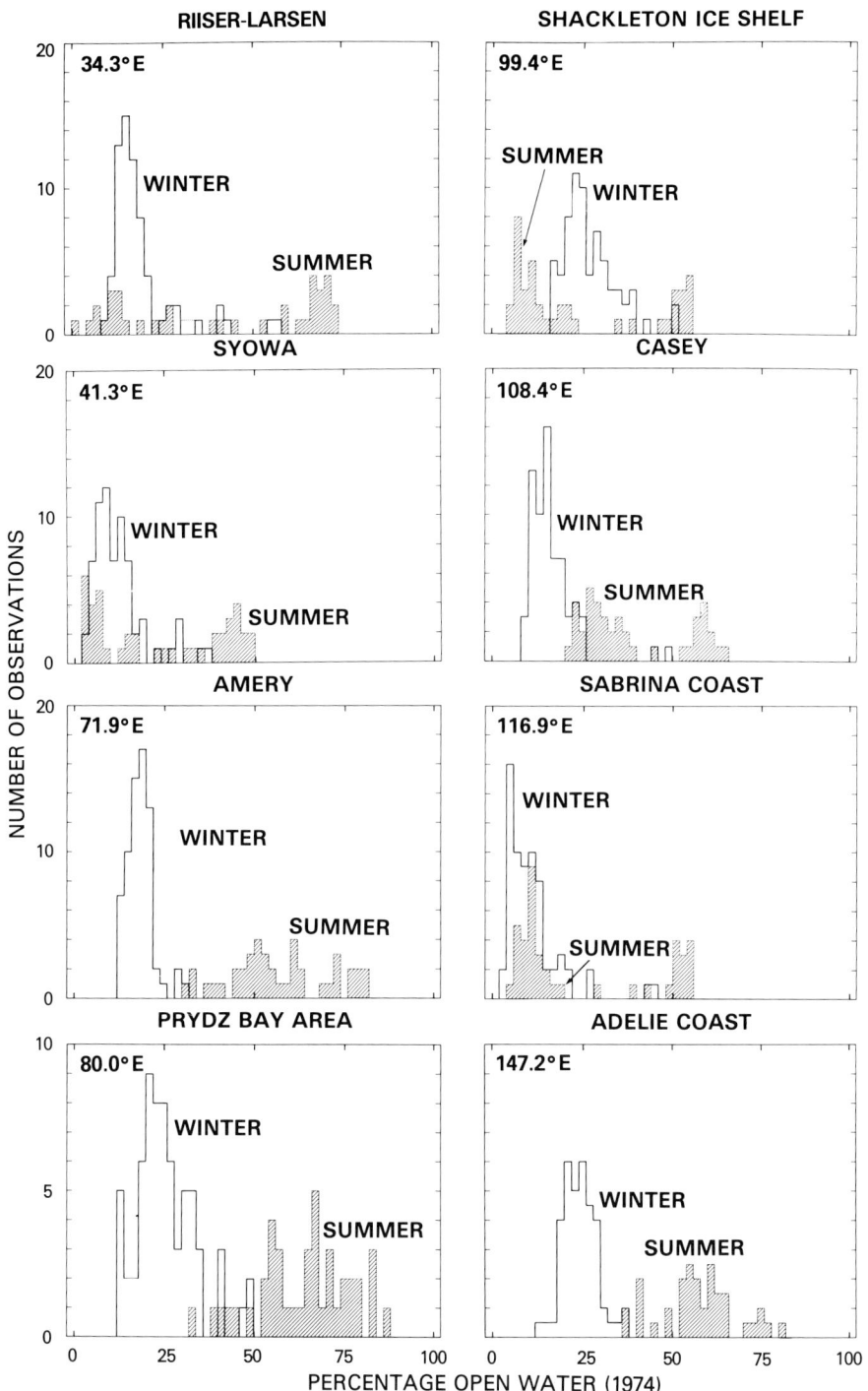

Fig. 24. Frequency distributions of the ratio of open water area to total area in each of the study regions from 34°E to 147.5°E.

to new ice growth in open water areas is estimated for each study area by

$$\Delta S = \frac{s \, T \, A \, R_i}{0.1 \, h} \quad (3)$$

where ΔS is the salinization of the total volume of shelf water within the study region, scaled by the 0.1 to units of $°/oo$, s is the salt rejection in g cm^{-2} for each meter of sea ice formation (a value of 2.5 g cm^{-2} is used,

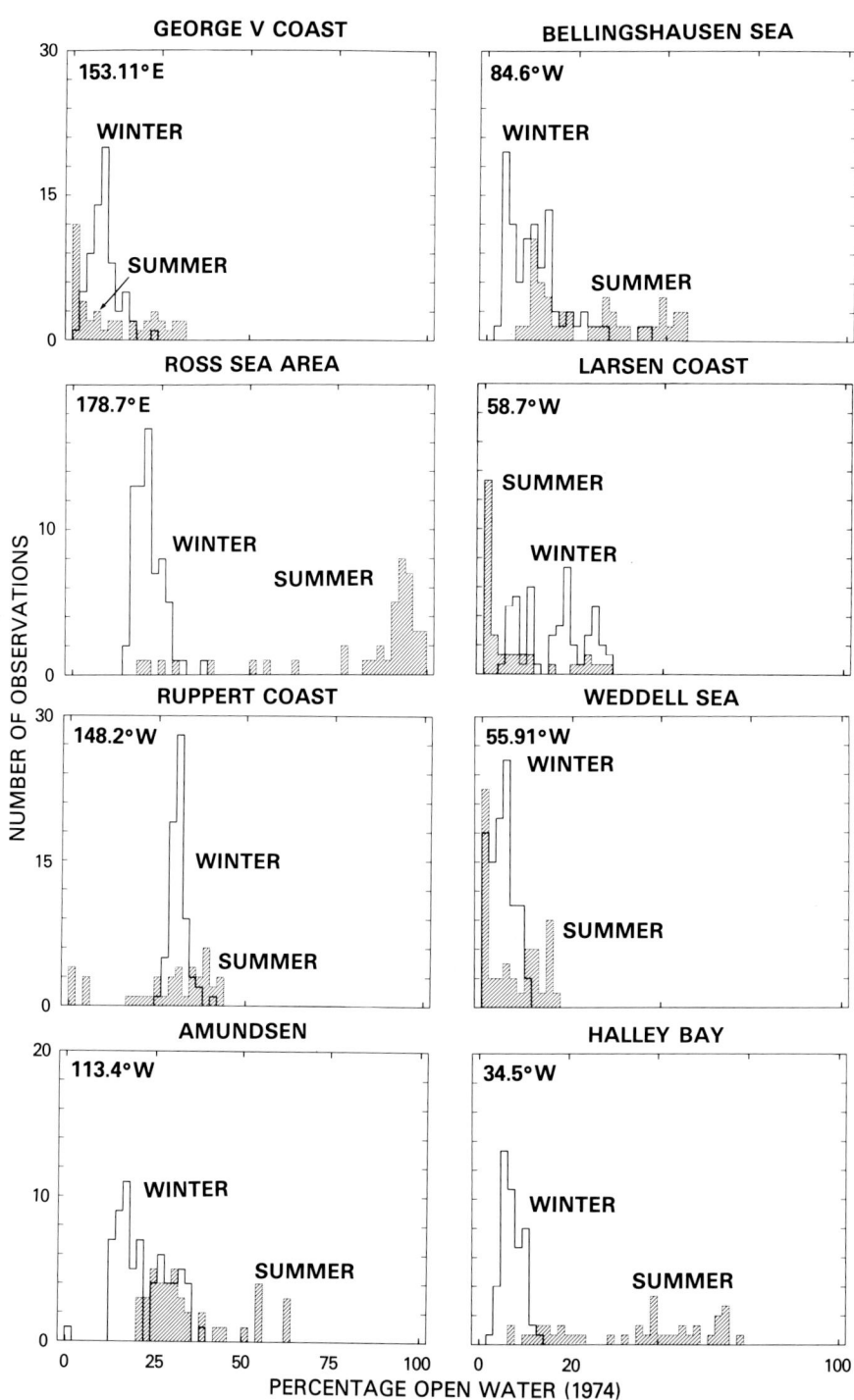

Fig. 25. Frequency distributions of the ratio of open water area to total area in each of the study regions from 153.1°E to 34.5°W.

assuming the newly formed ice has a salinity of 10.0, after Schumacher et al. [1983] and Cavalieri and Martin [this volume]. Also, Wakatsuchi and Ono [1983] report salt rejection values of 1.1 to 3.8 g cm^{-2} per meter of ice for ice growth rates ranging from 0.06 to 0.13 m/d), T is the duration in days of the winter period, (taken from day 76 through day 315, a total of 240 days), A is the observed ratio of open water to total surface area of

TABLE 2. Salinization of Water Volume Within Each Region
Due to Sea Ice Formation with Open Water Fraction

Study Area	Midpoint Lat.	Midpoint Long.	Total Area A_T (km^2)	Open Water Ratio[a] A_W/A_T (%)	σ_A	Salinization[b] (°/oo) low	Salinization[b] (°/oo) high
Riiser-Larsen	69.0°S	34.3°E	27,260	0.201	0.04	0.233	0.396
Syowa	68.2°	41.3°E	18,740	0.134	0.06	0.155	0.264
Amery	68.2°	71.9°E	70,210	0.194	0.04	0.225	0.382
Prydz	67.3°	80.0°E	28,860	0.269	0.08	0.312	0.530
Shackleton	65.0°	99.4°E	47,580	0.281	0.06	0.325	0.553
Casey	66.0°	108.4°E	42,390	0.175	0.04	0.203	0.345
Sabrina	66.4°	116.9°E	36,010	0.121	0.05	0.140	0.238
Adélie	67.0°	147.5°E	47,390	0.254	0.06	0.294	0.500
George V	67.9°	153.1°E	53,250	0.098	0.03	0.114	0.193
Ross	77.5°	178.7°E	121,000	0.223	0.04	0.258	0.439
Ruppert	76.0°	148.2°W	40,120	0.319	0.02	0.369	0.628
Amundsen	72.9°	113.4°W	117,600	0.229	0.12	0.265	0.451
Bellingshausen	71.8°	84.6°W	197,100	0.147	0.06	0.170	0.289
Larsen	67.3°	58.7°W	119,300	0.203	0.12	0.235	0.400
Weddell	73.1°	55.9°W	283,800	0.066	0.04	0.076	0.130
Halley	76.7°	34.5°W	98,900	0.103	0.04	0.119	0.203
Average				0.189	0.06	0.218	0.371

[a] Ratio of winter-average open water area to total ocean area.
[b] From equation 3, with the values for R_i = 0.10 m/d, and h, s, and T given in the text, then $\Delta S = 1.158\ A_W/A_T$. The range of salinization values corresponds to R_i ranging from low, 0.10 m/d, to high, 0.17 m/d.

each study region (the value given in Table 2 represents the average ratio $<A_W>/A_T$ for the 240-day winter period), R_i is the rate of ice formation in meters per day within the open water area (two values, 0.10 and 0.17 m/d, are used for the Table 2 determinations to represent the range of estimates), and h is the water column thickness for the study regions in meters (an average shelf depth of 518 m is used after Carmack [1977]).

Table 2 gives the ΔS values calculated for each of the 16 study regions. The average percentage of observed open water is 19% which gives a significant salinity enhancement of 0.220 to 0.374. Carmack [1977] shows that the largest volume of modified deep water within the shelf water column has a salinity of 34.4. If modified deep water is the initial ingredient of the cold (near freezing point) shelf water, an increase of 0.2 would in 1 year boost its salinity over 34.6, which is considered to be the minimum salinity required for formation of Antarctic Bottom Water.

Comparison of the calculated ΔS values with the measured salinity of near-freezing point shelf water versus longitude (Figure 5 of Gordon [1974]) shows that both values are high along the coast of East Antarctica. In the Weddell region, the calculated ΔS values are relatively low, yet the measured shelf-water salinity is high; thus other factors must also be important in determining shelf-water salinity. Equation (2) is, of course, a gross simplification and does not take into account redistribution of salt by circulation, variations in the value of the water column thickness, or the salinity of the initial water type. Nevertheless, ice generation within the 19% open water area on the shelf appears to be a major factor in raising the density of shelf water.

Discussion

Analysis of the microwave observations provides new information on the extent and variability of the open water in the Antarctic coastal regions. The temporal variability of the open water is observed to be large on daily to seasonal time scales. Spatial variability of the open water is also observed to be large on scales ranging upward from the 30-km resolution of the microwave sensor. Although 3-day averages are found to be adequate to resolve most of the major changes in open water described here as polynya events, the daily averages show more temporal detail and larger fluctuations between minima and maxima. On

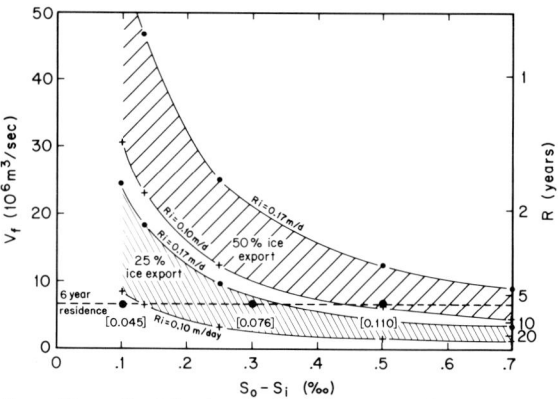

Fig. 26. Shelf-slope water mass volume exchange rate, V_F (10^6 m^3/sec), as a function of the salinity difference between the inflow (S_i) and outflow (S_o) water (see Figure 1), and the percentage of sea ice which forms on the shelf but is exported and melts seaward of the shelf (25 and 50% exported). The shelf water residence time in years is shown at the right. The calculations are carried out for R_i of 0.10 and 0.17 m/d (see text). The values in brackets along the 6-year residence line refer to the required R_i value, with sea ice export of 50%, which would yield that residence time.

monthly time scales, the polynya events are not resolved and are only evident as reduced average sea ice concentrations in the monthly average maps. The openings and closings of individual leads and polynyas are not resolved either in time or space, but the time-average maps represent the average open water during each period of observation.

While one or more 30 km x 30 km pixels often indicate 100% ice cover, the average concentration in most pixels is usually less than 100%. On spatial scales of the 16 study areas, the temporal variations in open water area occur above a significant background of open water. The persistent open water accounts for most of the winter-season average open water, which ranges from 7% to 32% in the various study areas. The results are considered to be representative of the shelf regions around Antarctica, because the study areas represent both regions with consolidated ice packs during winter and regions with significantly reduced ice concentrations or polynya events, as can be noted from the calculated open water values and a visual comparison of the ice concentrations along the coast in Plate 1.

The existence of substantial areas of open water in the Antarctic nearshore sea ice zone during winter has several important consequences. The flux of sensible and latent heat to the atmosphere is much greater from the open water areas than through a consolidated ice pack. Consequently, the growth of new ice in the open water areas is much greater than the growth at the underside of the ice in ice-covered areas. Also, the growth of new ice and nilas, followed by growth of thin young ice, enhances the total seasonal sea ice cover and the amount of salt rejected during the growth season. Overall, more ice is formed in the areas of reduced ice concentration in the coastal zone, the atmosphere is warmed more than it would be over a consolidated ice pack, and the salt rejected during freezing significantly increases the salinity and density of the water in the coastal regions.

The maintenance of open water over a given area of the ocean under freezing conditions requires an export of ice to balance the area covered by new ice. Also, the average shelf-water salinity is a result of the balance between salt rejection during ice growth and the introduction of fresher water from sea ice melt in summer, glacial run-off, excess precipitation over evaporation, and the exchange of water over the continental slope. The fraction of ice exported and the exchange of water between the shelf and deep ocean are major factors in determining the average salinity of the shelf waters, in addition to the amount of ice formed.

What is the contribution to the total sea ice cover of the southern ocean from ice formation within offshore open water? Ice formation rates of 0.10 to 0.17 m/d over a 240-day winter period would produce 24 to 41 meters of ice. Using 19% as a typical open water value for the shelf region (Table 2), ice production would then amount to 4.6 to 7.8 m if spread over the entire shelf. Those effective production rates would be about 15% lower if open water on the shelf were computed from an area-weighted average ratio. Sea ice production by bottom accretion in the other 81% of the area represents additional total ice volume. Thin ice (<0.30 m) grows much faster than thick ice, but not as fast as the initial ice growth in open water. If the spring sea ice cover of the shelf region ranges between 1.5 and 2.5 m [Jacobs et al., this volume], some ice must thicken by bottom accretion while other ice is exported to the north. Spread uniformly over the entire circumpolar belt of seasonal ice cover (about 20 x 10^6 km^2), ice formation in the offshore leads and polynyas would amount to 0.4 to 0.6 m, or approximately 50% of the mean ice thickness [Clarke and Ackley, 1984]. However, the ice formed at high-growth rates in the open water is redistributed in a complex manner as the ice pack diverges to maintain reduced ice concentrations. Although the open water adjacent to Antarctica is clearly a region of significant ice production compared to the total mass of the Antarctica seasonal ice cover, the actual distribution of the ice is determined by the ice drift after formation in the open water areas.

Salinization of the shelf water column by

ice production within the offshore leads and polynyas increases the likelihood of shelf water attaining density values sufficient to result in deep-reaching convective plumes over the continental slope [Killworth, 1983]. The total salt rejection associated with 4.6 to 7.8 m of sea ice over the 2.473×10^6 km^2 of shelf area [Carmack, 1977] is 2.8 to 4.8×10^{17} g/yr. A steady state shelf water salinity is maintained by the inflow of lower salinity water, but not all rejected salt needs to be balanced by shelf-slope water mass exchange because some sea ice melts on the shelf. Sea ice that forms and melts on the shelf has no net effect on shelf water salinity, although it can cause partitioning of salt. In addition, some of the salt is used to mix with the incoming fresh water due to excess precipitation over evaporation and glacial melt. Jacobs et al. [this volume] suggest a net fresh water input of 51 cm/yr to each cm^2 of shelf area. Conversion of this fresh water into average shelf water salinity (taken as 34.5) requires 0.43×10^{17} g/yr of salt.

The shelf-slope water volume exchange rate (V_f, Figure 1), required to balance net sea ice salt rejection by ice exported from the shelf depends on the difference between the inflow salinity (S_i) and outflow salinity (S_o) across the shelf-slope boundary. Adjusting for freshwater input and using the value for S_o-S_i of 0.134 with a sea ice export fraction of 50%, as suggested by Jacobs et al. (this volume), leads to a shelf-slope exchange rate of 23 to 47×10^6 m^3/sec, corresponding to residence times from 1.8 to 0.9 years (Figure 26). A residence time of 6 years (Jacobs et al., this volume), would require an ice export fraction of only 25% and a S_o-S_i difference ranging from 0.13 to 0.35. While these larger salinity differences are possible (if we simply invoke the ΔS calculated by equation 3 and given in Table 1), the export fraction of 25% seems too small.

A large source of uncertainty among these parameters is the rate of sea ice formation (R_i). For low R_i values ($\cong 0.03$ m/d) and 50% sea ice export fraction, sea ice production would only compensate for the fresh water introduction and V_f would be zero. If the 0.10 m/d R_i were valid for only the 90 days studied by Cavalieri and Martin [this volume] and R_i were zero for the rest of the 240-day winter period, then the average growth rate for the full period would be 0.038 m/d. With a 50% export fraction and S_o-S_i of 0.134, V_f would be 2.5×10^6 m/s and the residence time would be 16 years. Alternatively, using the same export fraction (50%) and S_o-S_i value (0.134), R_i would have to be 0.051 m/d in order to yield a 6-year residence time.

Despite uncertainties in R_i, sea ice export fraction, and other parameters affecting the shelf water salinity, it is obvious that offshore leads and polynyas play an important role in southern ocean sea ice generation and salinization of the shelf water. Regional differences in the observed open water areas are large, and interannual variability may also be high. Further studies of the distribution of open water along the Antarctic coastline and of other related variables are needed to improve our knowledge of the quantitative oceanographic effects of these features.

Acknowledgments. The authors thank Steve Schweinfurth and Nancy Aschenbach of Science Applications Research, and Richard Johnson of Computer Sciences Corporation for programing help. The efforts of H.J.Z. and J.C.C. in this project were supported by NASA's Oceanic Processes Program. Lamont-Doherty Geological Observatory contribution 3792. A.L.G. support was derived from NSF grant DPP 81-19863.

References

Bromwich, D. H., and D. D. Kurtz, Experiences of Scott's Northern Party, Evidence for a relationship between winter katabatic winds and the Terra Nova Bay polynya, Polar Rec., 21, 137-146, 1982.

Bromwich, D. H., and D. D. Kurtz, Katabatic wind forcing of the Terra Nova Bay polynya, J. Geophys. Res., 89, 3561-3572, 1984.

Carsey, F. D., Microwave observation of the Weddell Polynya, Mon. Weather Rev., 108, 2032-2044, 1980.

Carmack E. C., Water characteristics of the southern ocean south of the polar front, A Voyage of Discovery, George Deacon 70th Anniversary Volume, edited by M. Azd, pp. 15-37, Pergamon Press, New York, 1977.

Cavalieri, D. J., and S. Martin, A passive microwave study of polynyas along the Antarctic Wilkes Land Coast, this volume.

Cavalieri, D. J., and C. L. Parkinson, Large-scale variations in observed Antarctic sea ice extent and associated atmospheric circulation, Mon. Weather Rev., 109, 2323-2336, 1981.

Clarke, D. B., and S. F. Ackley, Sea ice structure and biological activity in the Antarctic Marginal Ice Zone, J. Geophys. Res., 89, 2087-2095, 1984.

Comiso, J. C., Sea ice effective microwave emissivities inferred from microwave and infrared observations, J. Geophys. Res., 88, 7686-7704, 1983.

Comiso, J. C., and H. J. Zwally, Antarctic sea ice concentrations inferred from Nimbus 5 ESMR and Landsat imagery, J. Geophys Res., 87, 836-844, 1982.

Comiso, J. C., and H. J. Zwally, Concentration gradients and growth/decay characteristics of the seasonal sea ice cover, J. Geophys. Res., 89, 8081-8103, 1984.

Comiso, J. C., S. F. Ackley, and A. L. Gordon,

Antarctic sea ice microwave signatures and their correlation with in situ ice observations, J. Geophys. Res., 89, 662-672, 1984.

Dey, B., Applications of satellite thermal infrared images for monitoring North Water during the periods of polar darkness, Mon. Weather Rev., 93, 425-438, 1980.

Foster, T. D., and Carmack E. C., Frontal Zone mixing and Antarctic Bottom Water formation in the Southern Weddell Sea, Deep Sea Res., 23, 301-317, 1976.

Gloersen, P., T. T. Wilheit, T. C. Chang, W. Nordberg, and W. J. Campbell, Microwave maps of the polar ice of the earth, Bull. Amer. Meteorol. Soc., 55, 1442-1448, 1974.

Gordon, A. L., Varieties and variability of Antarctic Bottom Water, in Processees de Formation des Eaux Oceaniques Prefondes, Colloq. Int. CNRS, no. 215, pp. 33-47, Centre National de la Recherche Scientifique, Paris, 1974.

Gordon, A. L., Seasonality of southern ocean sea ice, J. Geophys. Res., 86, 4193-4197, 1981.

Gordon, A. L., Weddell deep water variability, J. Mar. Res., 40, 199-217, 1982.

Gordon, A. L., and B. A. Huber, Thermohaline stratification below the southern ocean sea ice, J. Geophys. Res., 89, 641-648, 1984.

Hibler, W., and S. F. Ackley, Numerical simulations of the Weddell Sea pack ice, J. Geophys. Res., 88, 2873-2887, 1983.

Jacobs, S., R. Fairbanks, and Y. Horibe, Origin and evolution of water masses near the Antarctic continental margin: evidence from $H_2^{18}O/H_2^{16}O$ ratios in sea water, this volume.

Killworth, P. D., On "chimney" formations in the ocean, J. Phys. Oceanogr., 9, 531-554, 1979.

Killworth, P. D., Deep convection in the world oceans, Rev. Geophys. Space Phys., 21, 1-26, 1983.

Kurtz, D. D., and D. H. Bromwich, Satellite observed behavior of the Terra Nova Bay polynya, J. Geophys. Res., 88, 9717-9722, 1983.

Martinson, D. G., P. D. Kilworth, and A. L. Gordon, A convective model for the Weddell Polynya, J. Phys. Oceanogr., 11, 406-488, 1981.

Maykut, G. A., Energy exchange over young sea ice in the central Arctic, J. Geophys. Res., 88, 3646-3657, 1978.

Parish, T. R., The katabatic winds of Cape Denison and Port Martin, Polar Rec., 20, 525-532, 1981.

Parish, T. R., Surface airflow over east Antarctica, Mon. Weather Rev., 110, 84-90, 1982.

Parish, T. R., The influence of the Antarctic Peninsula on the wind field over the western Weddell Sea, J. Geophys. Res., 88, 2684-2692, 1983.

Parkinson, C. L., On the formation of the Weddell polynya, J. Phys. Oceangr., 13, 501-511, 1983.

Parkinson, C. L., and D. J. Cavalieri, Sea ice variations in the southern ocean, 1973-1975, Ann. Glaciol., 3, 249-254, 1982.

Priestley, R. E., General diary, 1 January 1912-February 1913, Doc. MS298/6/2, Scott Polar Research Institute, Cambridge, England, 1913.

Schumacher, J. D., K. Aagaard, C. H. Pease, and R. B. Tripp, Effects of a shelf polynya on flow and water properties in the northern Bering Sea, J. Geophys. Res., 88, 2723-2732, 1983.

Wakatshuchi, M., and N. Ono, Measurements of salinity and volume of brine excluded from growing sea ice, J. Geophys. Res., 88, 2943-2951, 1983.

Zwally, H. J., and P. Gloersen, Passive microwave images of the polar regions and research applications, Polar Rec., 18, 431-450, 1977.

Zwally, H. J., J. C. Comiso, C. L. Parkinson, W. Campbell, F. Carsey, and P. Gloersen, Antarctic sea ice satellite observations: 1973-1976, NASA Spec. Publ. SP459, 1983.

(Received July 9, 1984;
accepted December 26, 1984.)

A PASSIVE MICROWAVE STUDY OF POLYNYAS ALONG THE ANTARCTIC WILKES LAND COAST

Donald J. Cavalieri

Laboratory for Oceans, NASA Goddard Space Flight Center, Greenbelt, Maryland 20771

Seelye Martin

School of Oceanography, University of Washington, Seattle, Washington 98194

Abstract. Satellite passive microwave radiance data are used to derive the open water area of six polynyas located along the Antarctic Wilkes Land coast for a 3-month period during the austral 1979 winter. The polynyas lie to the west of the following features: the Mertz Glacier Tongue, Dibble and Dalton Iceberg Tongues, Cape Poinsett, Bowman Island, and the western edge of the Shackleton Ice Shelf. A comparison of the temporal variability of the six polynyas over the 3 winter months with a sequence of surface pressure maps shows that the polynyas are largely influenced by the synoptic events. Correlation between the polynya open water areas and wind speeds at the nearest coastal weather station (Dumont d'Urville, Casey, Mirny) to each feature is for most cases significant at the 95% confidence level and as high as 0.75 for the 3-month period. The station data also allowed calculation of the total heat transfer, ice production, and amount of salt rejected to the ocean for each of the polynyas. From these calculations, we find an excellent correlation with earlier oceanographic observations of the salinity on the continental shelf; polynyas with large areas and high ice production, such as the Shackleton and Mertz polynyas, correspond to saline water, while polynyas with small areas and low ice production, such as the Dalton polynya, correspond to less saline water. These results strongly suggest that the coastal polynyas are the sources of the brine which generates the dense shelf water, and thus contribute to the formation of Antarctic bottom water.

1. Introduction

The Scanning Multichannel Microwave Radiometer (SMMR) on the Nimbus 7 spacecraft is used to explore the behavior of six Antarctic coastal polynyas between 90°E and 150°E longitude during the austral winter of 1979. The six polynyas are found adjacent to the following geographic features: the Mertz Glacier Tongue, Dibble and Dalton Iceberg Tongues, Cape Poinsett, Bowman Island, and the Shackleton Ice Shelf (Figure 1).

The satellite observations are used to determine quantitatively the open water area of each polynya as a function of time over the 3-month period. The variability of the polynyas is first related to the general meteorological conditions through a sequence of synoptic weather charts, then correlated with weather station data. The station data are further used to calculate total heat flux, ice production, and salt production for each polynya. Finally, a comparison of the satellite data with earlier in situ oceanographic observations relates the areal extent and salt production rates of the polynyas to the continental shelf seawater salinity.

Much of our recent knowledge of the spatial extent and temporal behavior of Antarctic polynyas derives from observational studies using satellite data. Previous investigations used infrared and visible imagery to study the nature of polynyas [Knapp, 1967 and 1972; Streten, 1973; Kurtz and Bromwich, 1983; and this volume]. For two of our study regions, the Dalton Iceberg Tongue and the Shackleton Ice Shelf, Knapp [1972] observed large areas of open water on visible and infrared Nimbus imagery and showed that the behavior of these polynyas is closely associated with the prevailing synoptic conditions. In the present study, we use passive microwave data which have the advantage of providing information even during periods of darkness and cloud cover. Although limited to spatial resolutions of approximately 15 x 15 km^2, the microwave data provide an integrated measure of the areal fraction of open water and ice cover for each instrument field of view through simple linear algorithms. Infrared and visible satellite imagery provide a higher spatial resolution (approximately 3 km), but are constrained to cloud free periods and require visual interpretation.

The importance of coastal polynyas in polar regions during winter results from the large ocean-to-atmosphere heat flux, which creates both a large ice growth rate in the polynya

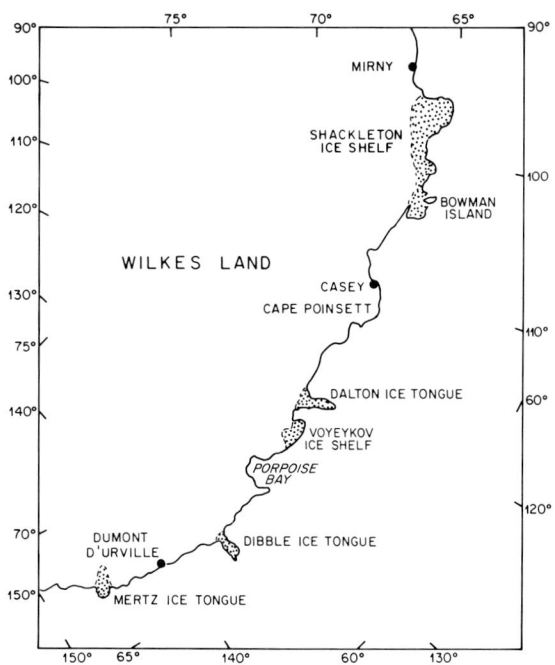

Fig. 1. Map of Antarctic Wilkes Land.

and a large salt flux to the underlying ocean. These semipermanent regions of open water are maintained by winds which sweep the pack ice away from the coast. Several authors speculate [Gordon, 1974; Gordon and Tchernia, 1974; Aagaard et al., 1981; Zwally et al., this volume] that polynyas located on continental shelves in the polar regions play an important role in the formation of Atlantic and Antarctic bottom water. More specifically, for the Antarctic, Gordon [1974] proposes that the formation of Antarctic bottom water is modulated by variations in the westward flowing coastal current. Gordon's [1974] model invokes sea ice production as the most likely mechanism for increasing the salinity of Antarctic shelf water and predicts that bottom water forms in specific regions along the shelf. These include the western and northern periphery of the Weddell Basin, the Adélie Coast, and the Shackleton and Amery Ice Shelf regions. A goal of the present study is to compare the amount of salt rejected as computed from a combination of satellite imagery and station data for each polynya with Gordon's [1974] shelf salinity distribution along the Wilkes Land coast.

In the following section, we describe the ice formation characteristics of previously studied polynyas, while in section 3, we present the satellite observations, the algorithm used to calculate polynya open water areas, and the weather data. Then, in section 4, we discuss the open water data derived from the SMMR and describe the correlation of these data with the synoptic maps and station winds. Finally, in section 5, we present our heat flux, and ice and salt production calculations, and show that the polynyas produce on average a 10-m thickness of ice, compared with a 1-m thickness ice growth away from the polynyas.

2. Polynya Morphology

Polynyas are large regions of open water in the pack ice which frequently forms adjacent to coasts, where the prevailing winter winds force the pack ice away from the coast. This forcing of the coastal polynya by cold, winter winds means that the polynya is also a region of high ice production. The shape of a polynya, which may be approximated by a rectangle or an ellipse, is in contrast to the long, broken linear shape of a lead. Many of these polynyas have been mapped by Knapp [1972] and Streten [1973] along the Antarctic coast, a region particularly susceptible to the formation of polynyas due to the numerous north-south oriented coastal projections which are imbedded in a generally circumpolar atmospheric flow.

From both aircraft overflights of Bering Sea polynyas and laboratory experiments, Martin and Kauffman [1981] studied the characteristics of new ice formation in polynyas. They show that initial ice growth begins with the formation of small discoid frazil ice crystals generated by wind waves in the polynya. These crystals form in the water interior and float to the surface, where a wind-generated Langmuir circulation herds the frazil crystals together into long plumes of grease ice running parallel to the surface winds. Their laboratory experiments also show that, given a wave-agitated grease ice layer with a cold wind blowing over it, the grease ice surface temperature is only reduced by 0.01°C below the seawater freezing point. Thus, the heat transfer from the seawater to the air is reduced only slightly from its open water value by the grease ice presence. As the wave agitation weakens, the grease ice solidifies into pancake ice, and the heat flux is reduced. Unpublished field observations by S. Martin [1983] indicate that the grease ice within the Langmuir plume also undergoes a transition with time from pure grease ice to a mixture of grease and pancake ice.

Examples of grease ice plumes are illustrated in Figures 2a and 2b, which show two aircraft images from NASA Convair-990 overflights of Saint Lawrence Island and Saint Matthew Island in the Bering Sea at an altitude of 9.2 km on February 18, 1983, 2330 UT. At the time of the overflight, an ice weather station near Saint Matthew Island reported a wind speed of 12.5 m s^{-1} from 032° at a 2 m-height and an air temperature of -17°C. The wind had

Fig. 2a. Photomosaic of the Saint Lawrence Island polynya from a NASA Convair-990 overflight on February 18, 1983, at a 9.2-km altitude. The field of view of each frame is 13.8 x 13.8 km^2.

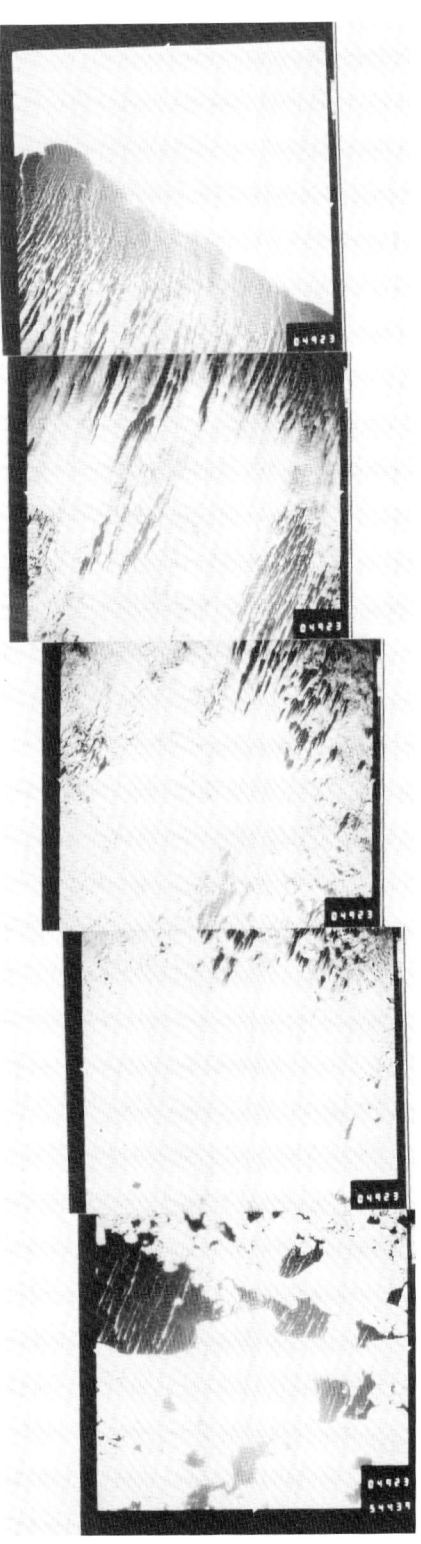

Fig. 2b. Photomosaic of the Saint Matthew Island polynya from a NASA Convair-990 overflight on February 18, 1983, at a 9.2-km altitude. The field of view of each frame measures 13.8 km on a side.

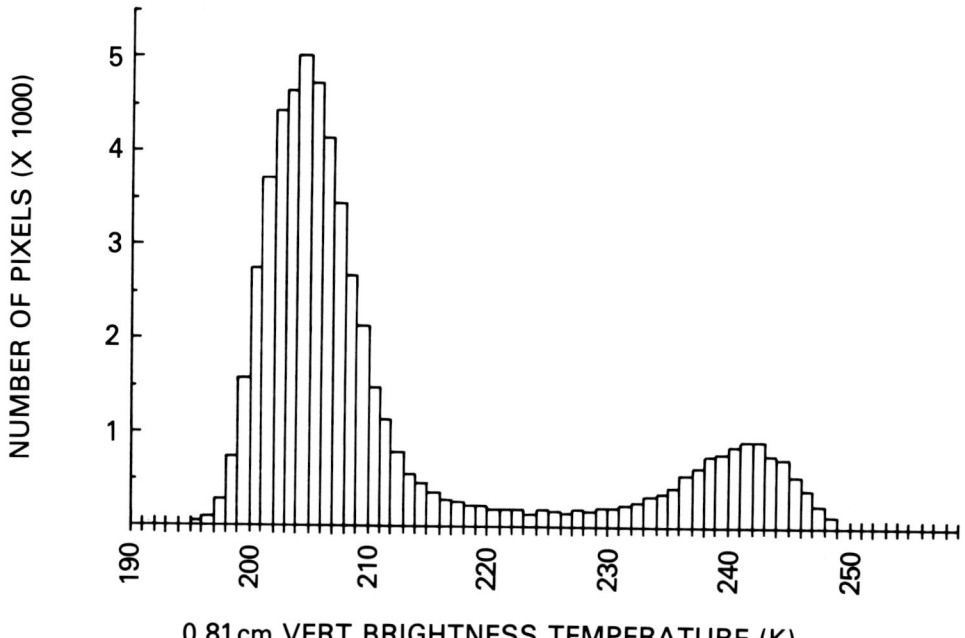

Fig. 3. Histogram of the vertically polarized 0.81-cm SMMR brightness temperatures for the 4-day period beginning on day 242 from the Wilkes Land coastal region.

been steady from the northeast at speeds above 10 m s^{-1} since February 15. Using the heat flux algorithm described in section 5, these temperatures and wind speeds yield a heat flux of 400 W m^{-2} and an ice growth rate of 0.12 m d^{-1}. Since the photographs were taken from an altitude of 9.2 km, with a camera field of view of 73°, the individual frames measure 13.8 km on a side, so that coincidently, each photographic frame is only slightly smaller than one SMMR image pixel. The Saint Lawrence image shows the striated pattern of grease ice plumes in the open water extending about 15 km downwind of the island. The Saint Matthew Island polynya image shows similar plumes, with open water again extending about 15-23 km downwind of the island and with additional open water farther south. The variability of ice concentration within the polynyas shown in these images is probably similar in form to what occurs in the Antarctic polynyas.

3. Data

3.1. Satellite Data

The SMMR on the Nimbus 7 satellite launched in October 1978 measures microwave radiances at five wavelengths (0.81, 1.43, 1.67, 2.81, and 4.55 cm) at both horizontal and vertical polarizations. A summary of the instrument operating characteristics is given by Gloersen and Barath [1977]. The Nimbus 7 orbit is sun-synchronous, so that the SMMR views any point on the Antarctic coast at most twice per day, near local noon and midnight. Because of spacecraft power limitations, the SMMR is on a 50% duty cycle and hence is operational only every other day, which corresponds in our study to even Julian days. Conical cross-track scanning provides a constant angle of incidence and a swath width of 780 km such that complete coverage at our 60°-70° S latitude region of interest requires a minimum of 2 days of SMMR data or 4 calendar days. As we show below, this temporal resolution is adequate to observe the openings and closings of the coastal polynyas.

The spatial resolution of the SMMR is frequency-dependent, with the highest resolution obtained from the 0.81-cm wavelength radiometer. This radiometer, which has an integrated field of view of 27 x 32 km^2, measures both horizontal and vertical polarizations on each scan, while the longer-wavelength radiometers measure alternate polarizations on each half scan. Thus, the 0.81-cm radiometer has the added advantage of sampling a given polarization at twice the rate of the longer wavelengths. This permits the 0.81-cm radiances to be mapped on a 15 x 15 km^2 grid, but with some loss of observational independence resulting from the overlap of adjacent 27 x 32 km^2 footprints.

The sharp contrast in thermal microwave emission (usually expressed in terms of a radiometric brightness temperature) at 0.81 cm between ice-free and ice-covered ocean allows

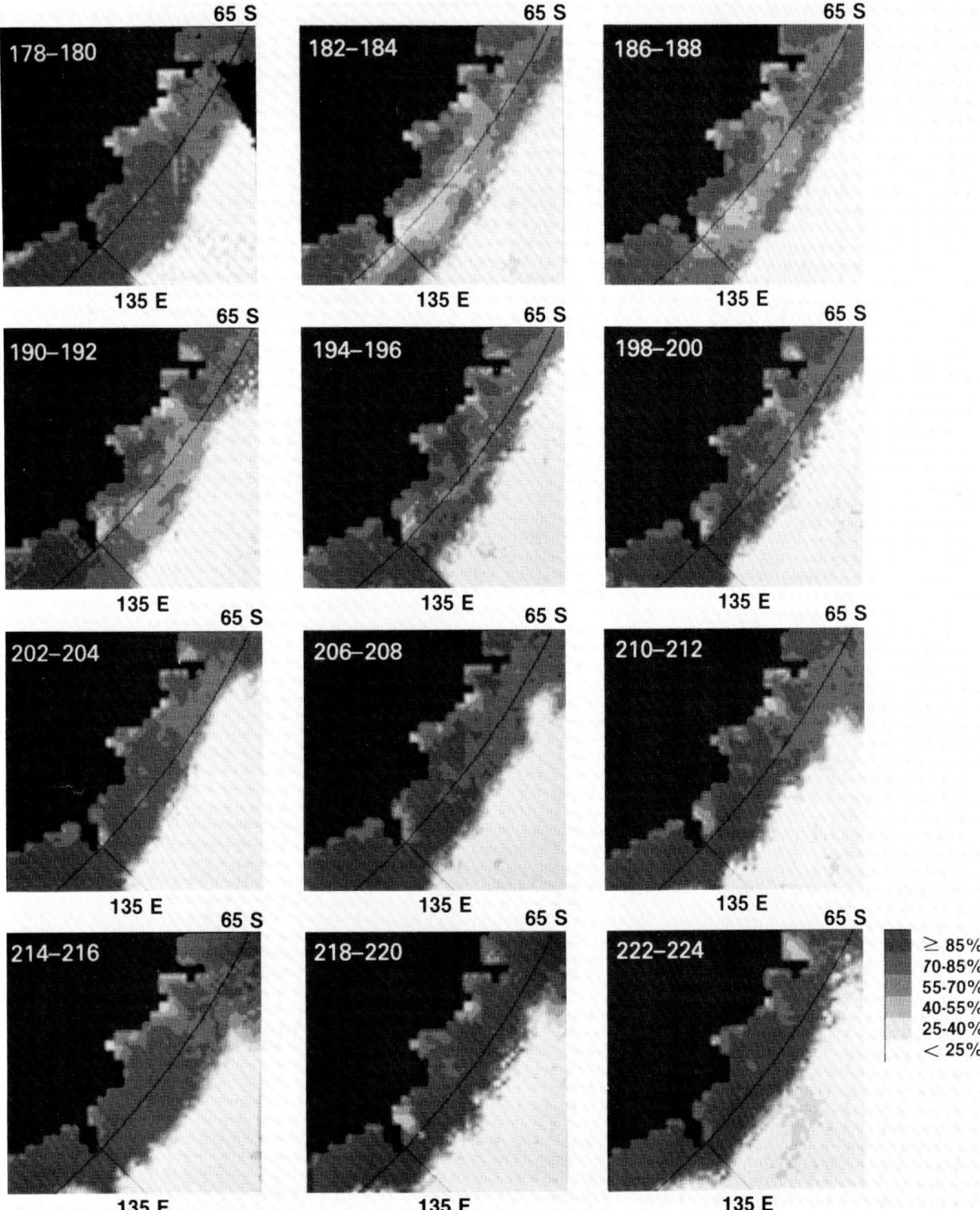

Fig. 4a. Sea ice concentration derived from the vertically polarized 0.81-cm SMMR brightness temperatures for a portion of the Wilkes Land coast: days 178-222.

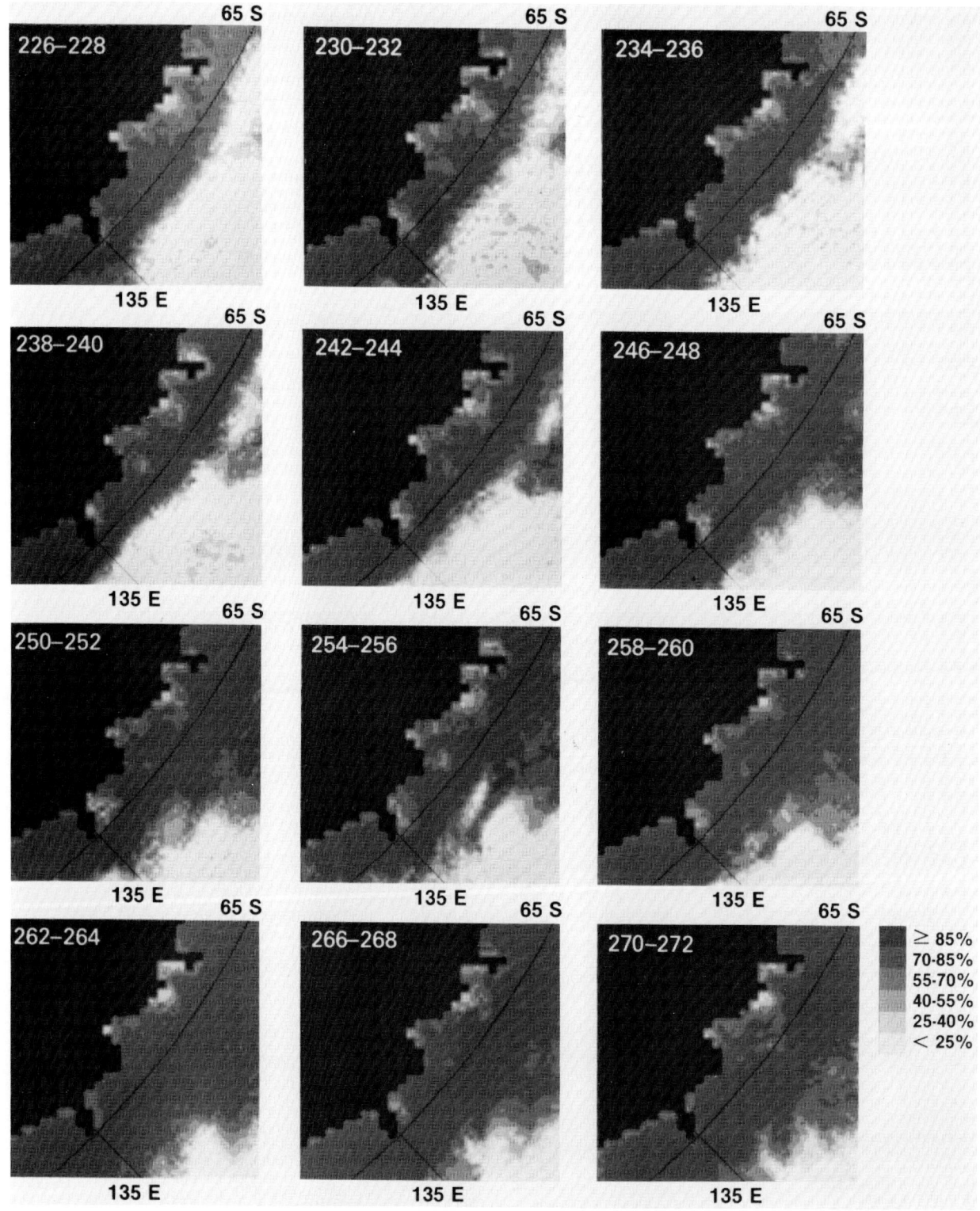

Fig. 4b. Sea ice concentration derived from the vertically polarized 0.81-cm SMMR brightness temperatures for a portion of the Wilkes Land coast: days 226-270.

TABLE 1. Meteorological Stations Nearest the Six Geographic Features

Station	WMO Station Number	Location
Mirny	89592	66.55°S, 93.02°E
Casey (Wilkes)	89611	66.25°S, 110.53°E
Dumont d'Urville	95502	66.67°S, 140.02°E

calculation of an ice concentration for each 15 x 15 km² pixel. In this calculation, even though both horizontal and vertical polarizations are available, we use only the vertical polarization, since Cavalieri et al. [1983] show that the horizontally polarized brightness temperatures are more sensitive to the presence of thin ice and thus would bias the computed ice concentration toward lower values.

Calculation of the ice concentration proceeds from a linear interpolation between two vertically polarized 0.81-cm-wavelength brightness temperatures, one corresponding to ice-free ocean and the other to consolidated sea ice. The expression for the concentration C is given by

$$C = (T_B - T_{BO})/(T_{BI} - T_{BO}) \quad (1)$$

where T_B is the 4-day-averaged observed brightness temperature within the pixel and T_{BO} and T_{BI} are the ice-free ocean and the consolidated sea ice average brightness temperatures, respectively. The values of T_{BO} and T_{BI} chosen for this 3-month period are 200 K and 245 K. These are average values based on an analysis of brightness temperature histograms generated for each 4-day period (2 SMMR data days).

A histogram of the 0.81-cm vertically polarized SMMR brightness temperatures obtained over an area from approximately 50°S to the Wilkes Land coast and from 90°E to 180°E for the 4-day period beginning with Julian day 242 is shown in Figure 3. The histogram has two peaks; one corresponds to the area of ice-free ocean, and the other to the Antarctic pack ice. For this 4-day period, the minimum and peak ocean brightness temperatures are 195 K and 204 K, respectively. The maximum and peak ice brightness temperatures are 248 K and 241 K. Variations in this distribution from causes other than real ice concentration changes are sources of error in the calculated ice concentration. These variations may result from changes in the physical temperature of the ice, in the ice surface radiometric properties, in ocean surface roughness, and in atmospheric attenuation and emission. While it is difficult to estimate the errors associated with each of these changes, it is possible to place limits on the relative error of the calculated ice concentration by using the total variation in observed brightness temperature.

Over the 3-month period, the mean ocean minimum brightness temperature is 196 K with a standard deviation of 0.7 K; the mean ice maximum brightness temperature is 248.4 K with a standard deviation of 1.6 K. The maximum difference between the chosen ice-free ocean value of 200 K and the minimum brightness temperature is 5 K. This difference translates into a maximum relative error in concentration of 11%. Likewise, at the ice end, the maximum difference between the chosen ice point of 245 K and the maximum ice brightness temperature is 6 K. This corresponds to a relative error of 13% in ice concentration. The error associated with either one of these maximum differences in a region of 50% ice concentration will be about half the cited values. It should be emphasized that these estimates are limits on the relative error in the computed concentration and do not necessarily represent the actual error of the calculated concentration, which may be considerably less.

The sequence of gray-scale ice concentration images shown in Figures 4a and 4b for the 3-month period illustrates the spatial variation of the polynyas along the Antarctic coast from the Dibble Iceberg Tongue to the Dalton Iceberg Tongue. Image pixels which are filled or partially filled by land or shelf ice are masked and appear black in the images. The lightest gray-scale level corresponds to ocean pixels having ice concentrations of less than 25%. Darker grays correspond to higher concentrations in increments of 15%.

Interestingly, in addition to the polynyas associated with the Dibble and Dalton Iceberg Tongues, Figure 4 shows another polynya located off the Voyeykov Ice Shelf and a smaller polynya in Porpoise Bay (see map in Figure 1). An examination of the sequence of images reveals that the Dibble polynya is more variable than either the Dalton polynya, which tends to have less open water, or the Voyeykov polynya, which maintains an open water area of about 50% throughout the period.

3.2. <u>Meteorological Data</u>

The meteorological data for this study were obtained from the World Data Center A (WDC-A) for Meteorology, National Climatic Center, Asheville, North Carolina. The data include 6-hour southern hemisphere surface analysis charts and 6-hour wind speeds, wind directions, and air temperatures for the 3 operating weather stations along the Wilkes Land coast (Table 1).

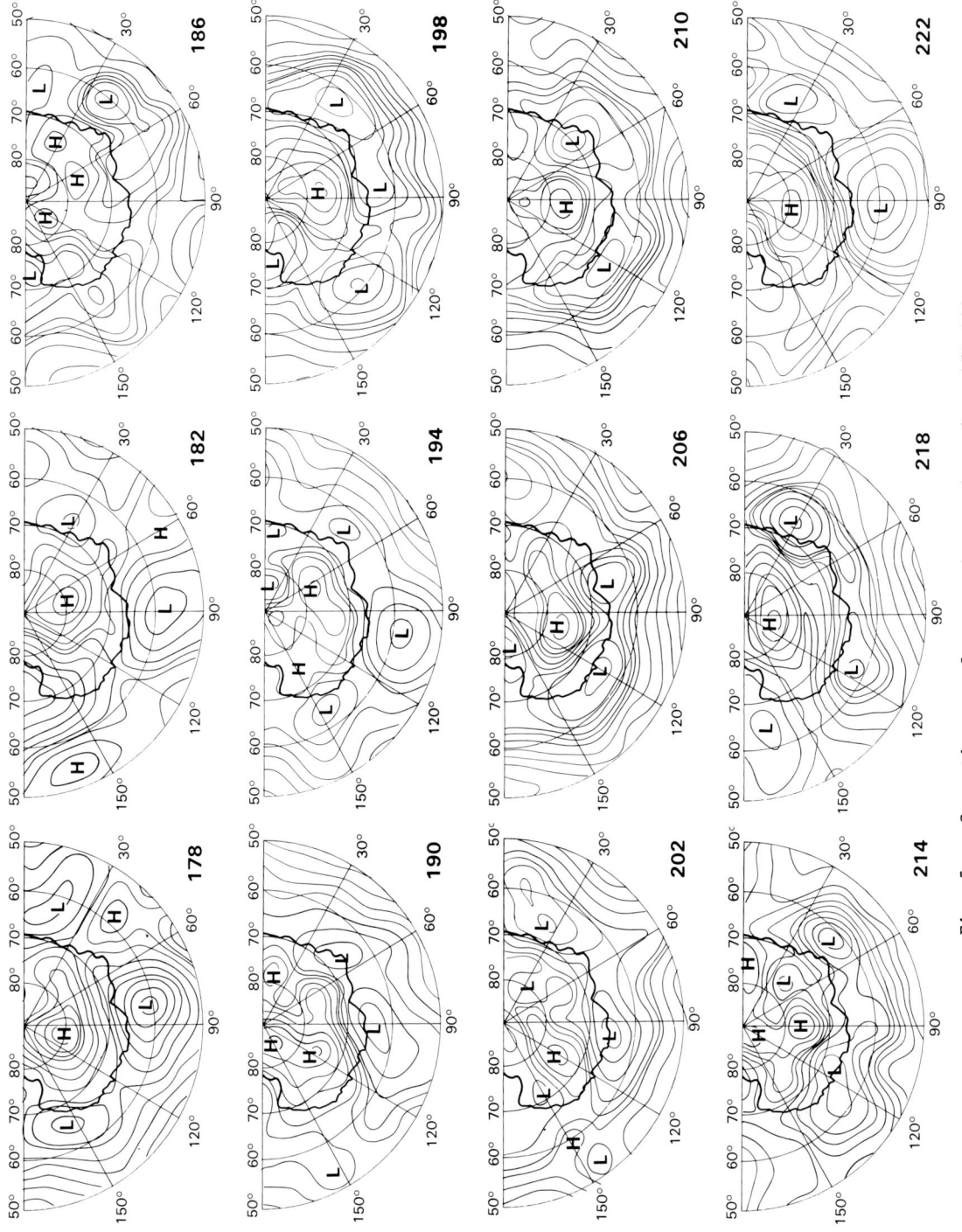

Fig. 5a. Synoptic maps for east Antarctica: days 178-222.

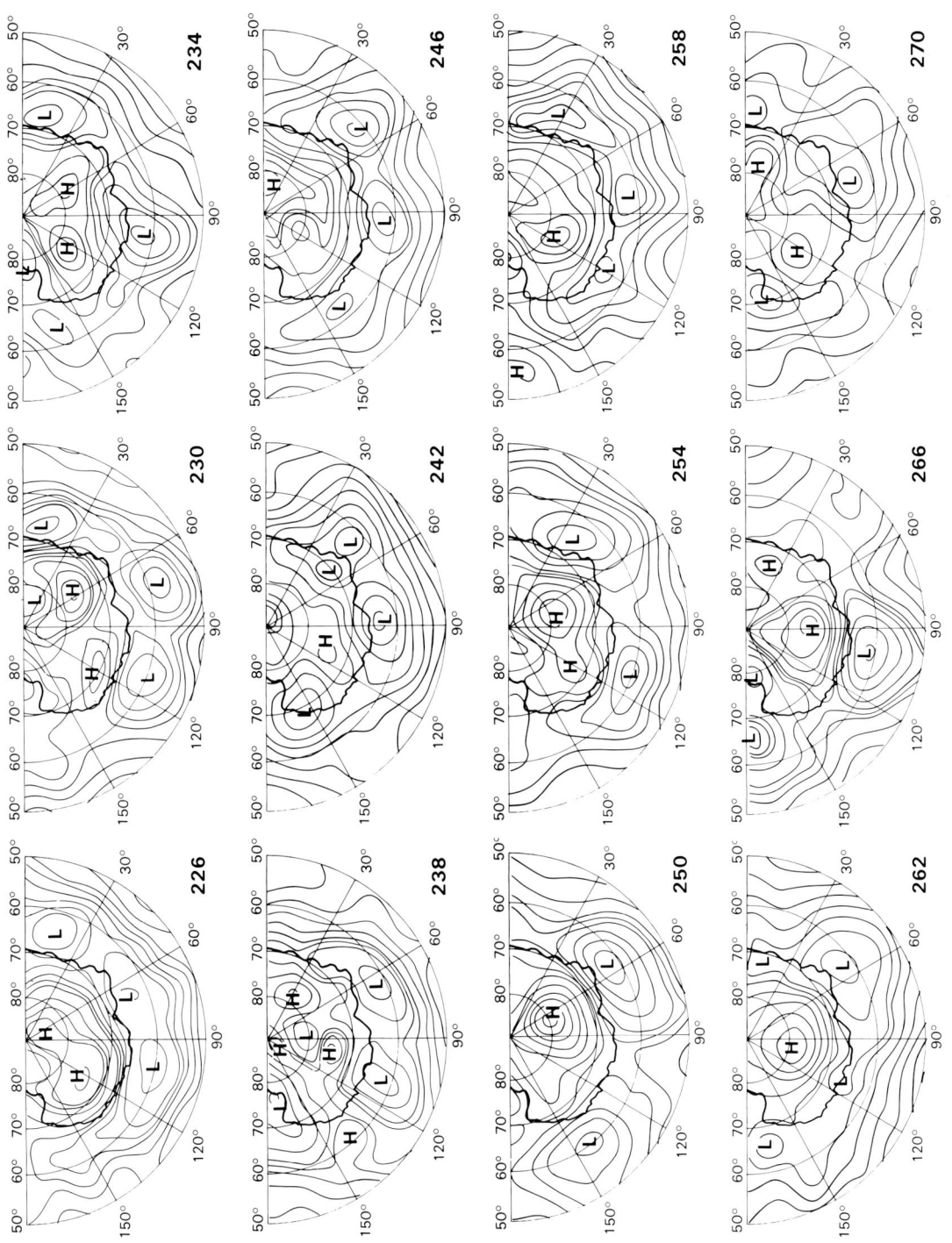

Fig. 5b. Synoptic maps for east Antarctica: days 226-270.

The sequence of surface analysis charts covering the 3-month period is presented in Figures 5a and 5b. Each chart corresponds to the analysis for 0000 UT of the first SMMR day for each 4-day period. Since we show only one chart for each 4-day period of averaged SMMR data, the synoptic charts are not truly representative of the conditions over the 4-day period but give only an indication of the relative positions of the high and low pressure systems. Examination of these figures shows that over the 3-month period, there is a generally eastward drift of cyclonic centers which have preferred positions centered at the 90°E and 160°E longitudes. These observations are consistent with climatological data which show three to five low-pressure centers encircling the continent, defining the position of the circumpolar trough for June through September, with lows in the eastern hemisphere centered at about 60°S, 90°E and 70°S, 160°E [Taljaard et al., 1969].

The 6-hour wind speeds and directions for each of the three stations are presented in Figures 6a and 6b. Wind roses for each station are given in Figure 7, and the surface air temperatures are presented in Figure 8. As many authors show [Ball, 1960; Kurtz and Bromwich, 1983; Parish, 1980], the surface winds measured at each of the coastal stations are a composite of the regional, topographically induced katabatic winds and of the winds associated with the synoptic weather conditions. The station position relative to the coastal topographic features determines the strength of the katabatic component, which is characteristically of short duration and of constant direction. For each of our three stations, the wind roses in Figure 7 show that in spite of the large variability in wind speed, the majority of the winds were from the southeast at about 135° on average.

In the southern hemisphere, the geostrophic turning angle is to the left of the wind, so that for steady winds, we would expect, following McPhee [1979], that the wind stress on the ice will be 20° to the left of the wind. This predominant wind direction supports our observation and that of earlier authors cited by Kurtz and Bromwich [1983] that the polynyas lie to the west of the major geographic features.

Casey (formerly Wilkes), one of the three stations used in this study, is located at 66°S, 110°E on a small ice cap called Law Dome. Here the local topography favors easterly winds, since the ice cap slopes are the steepest to the east, and the prevailing easterlies associated with the circumpolar winds tend to reinforce the katabatic winds [Zimmerman, 1960]. More recent work, though, indicates that the surface wind regime at Casey is largely influenced by synoptic conditions [D.H. Bromwich, personal communication, 1984].

Indeed, there seems to be a strong 5 to 6 day period in the wind speeds shown in Figure 6a consistent with synoptic-scale variability. The wind direction is fairly constant (90°-135°) for only those winds exceeding about 10 m s^{-1}. While some of these may be associated with katabatic flow, many of the events are associated with warm air advection, as a comparison of the 6-hour station temperatures (Figure 8) with the synoptic weather charts (Figure 5) shows.

Winds at Dumont d'Urville, our second station, are considerably stronger and more directional than those at Casey, even though the station is located on an island approximately 5 km off the coast. In a study of the topographic factors affecting the surface wind in Antarctica, Mather and Miller [1967] report that the most frequent wind direction at Dumont d'Urville is about 130°. They attribute the strong easterly component to the offshore station location, which results in a greater susceptibility to circumpolar than to katabatic flow. The wind rose (Figure 7) for this station shows that the winds are predominantly within the southeast quadrant.

A comparison of the wind roses of Figure 7 shows that the Mirny winds are on the average more intense and more directional than the winds from the other two stations. The direction of the Mirny winds spans the range 100°-170°, an observation consistent with Tauber's [1960] earlier results. Further, Parish [1980] shows that the yearly resultant wind direction of 127° at Mirny is 50° from the direction of the fall line. As noted by Parish [1980], the latter result underscores the important effect of synoptic events on the station wind data. In fact, earlier studies have stressed the important effect of synoptic events on the katabatic regime [Ball, 1960; Dzerdzeevskii, 1960; Tauber, 1960]. Tauber [1960] argues that the highest intensity of katabatic wind occurs at the rear of cyclones, while the lowest occurs at the front of the system due to the penetration of warm air onto the continent. Ball [1960] further shows that the migration of depressions along the coast causes considerable variability in the strength of the katabatic wind.

4. The Polynya Time Series

The open water area for each of the six polynyas was calculated by first defining a rectangle within which we computed the ice concentration using equation (1) for each of the enclosed image pixels. The open water area for each pixel is calculated from $AP(1 - C)$ where AP is the pixel area and C the computed ice concentration. Figures 9a-9c show the coastline and rectangle for each of the six polynyas, and Table 2 lists the area of and the number of pixels within each rect-

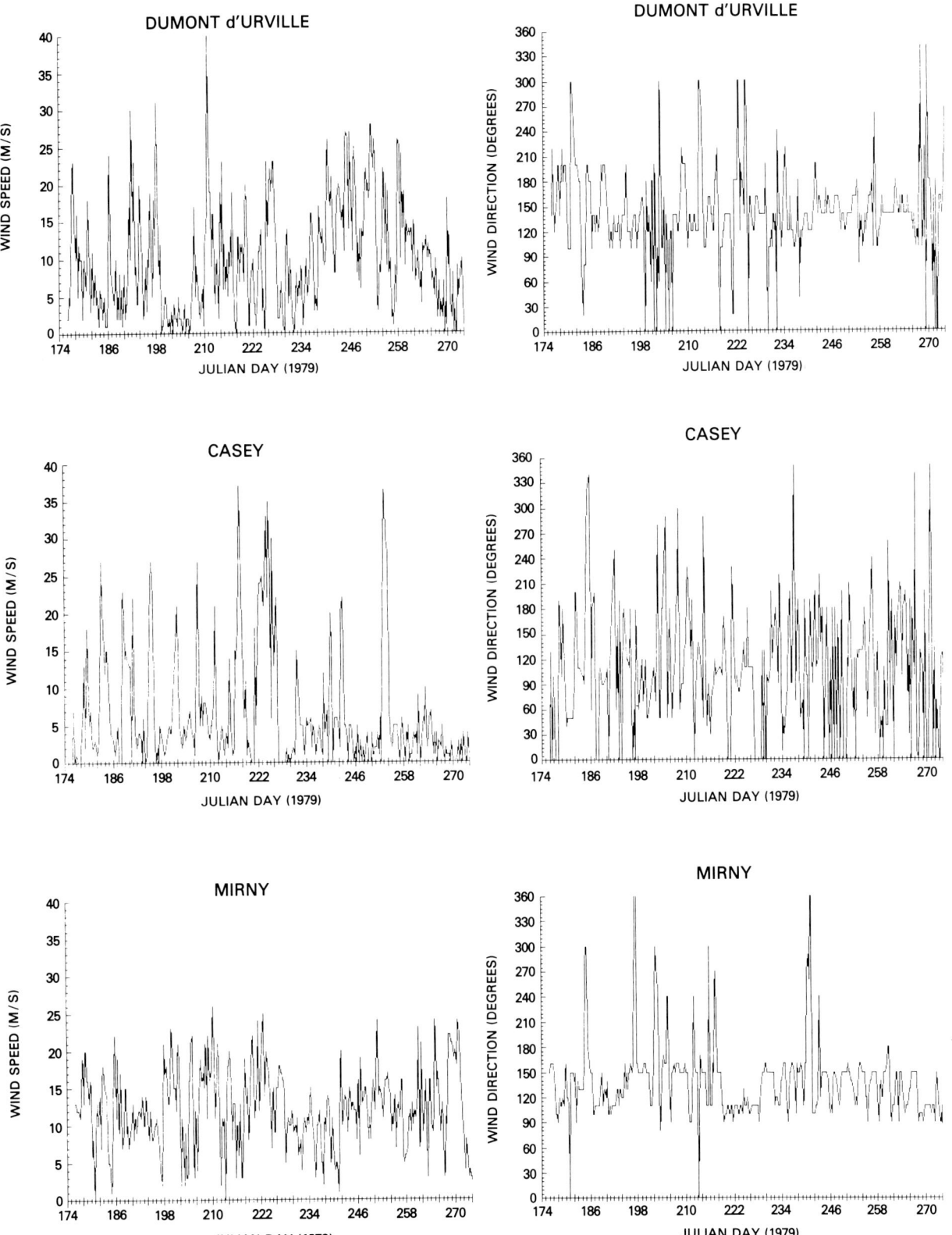

Fig. 6a. Six-hour station wind speeds for Dumont d'Urville, Casey, and Mirny.

Fig. 6b. Six-hour station wind directions for Dumont d'Urville, Casey, and Mirny.

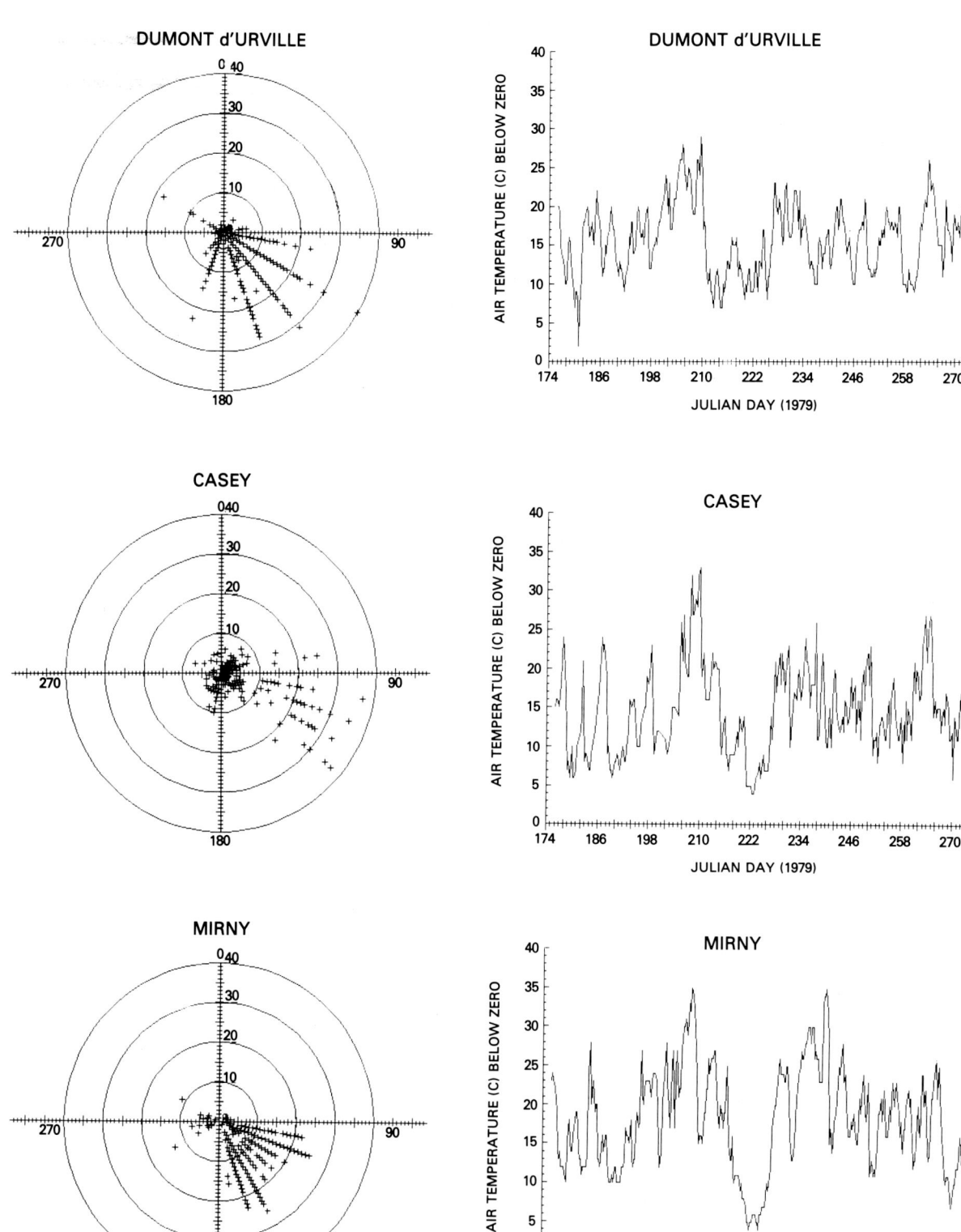

Fig. 7. Wind roses for Dumont d'Urville, Casey, and Mirny.

Fig. 8. Six-hour station air temperatures for Dumont d'Urville, Casey, and Mirny.

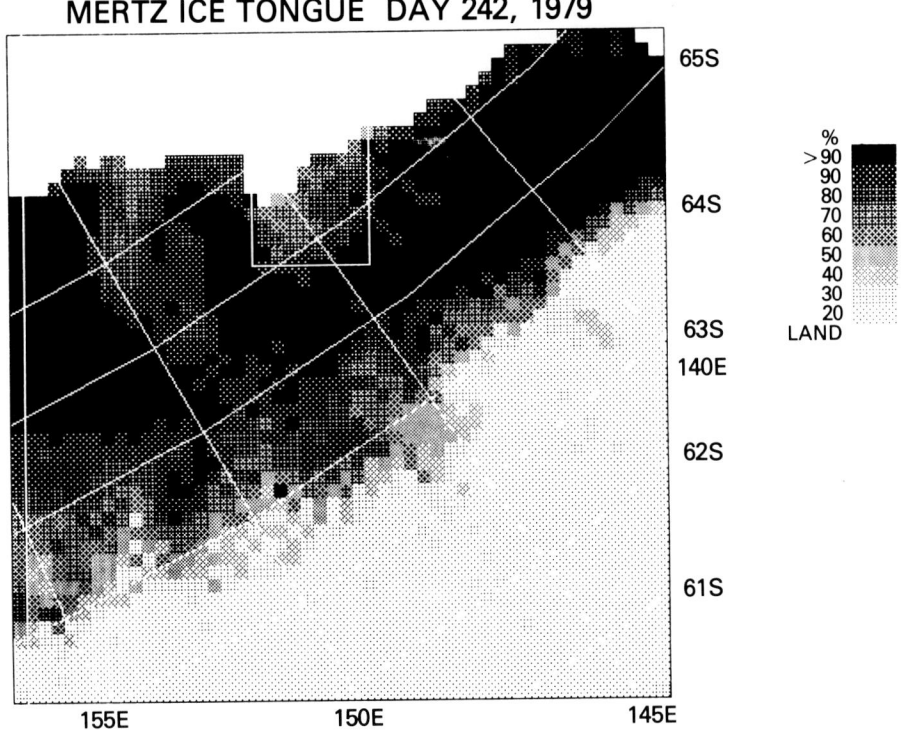

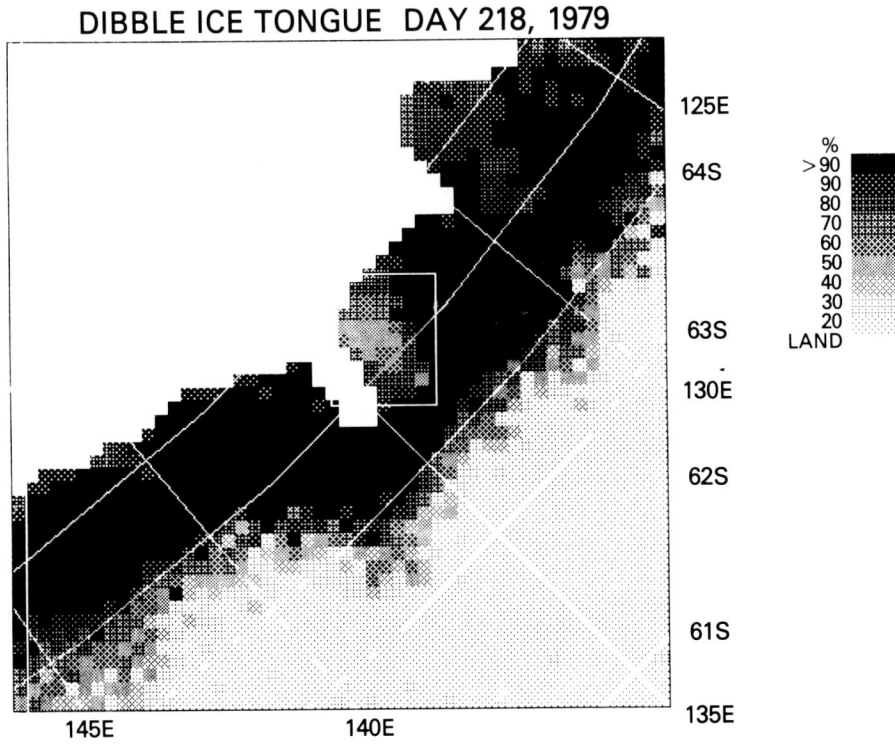

Fig. 9a. Rectangle for defining open water areas and coastline for the Mertz and Dibble polynyas. The gray scale gives percent ice concentration.

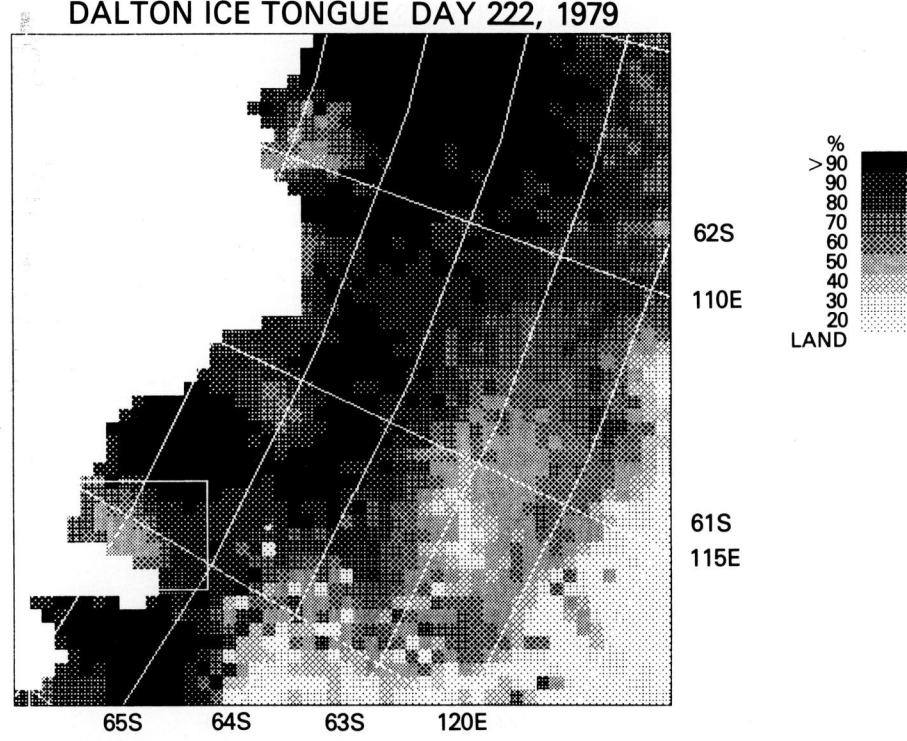

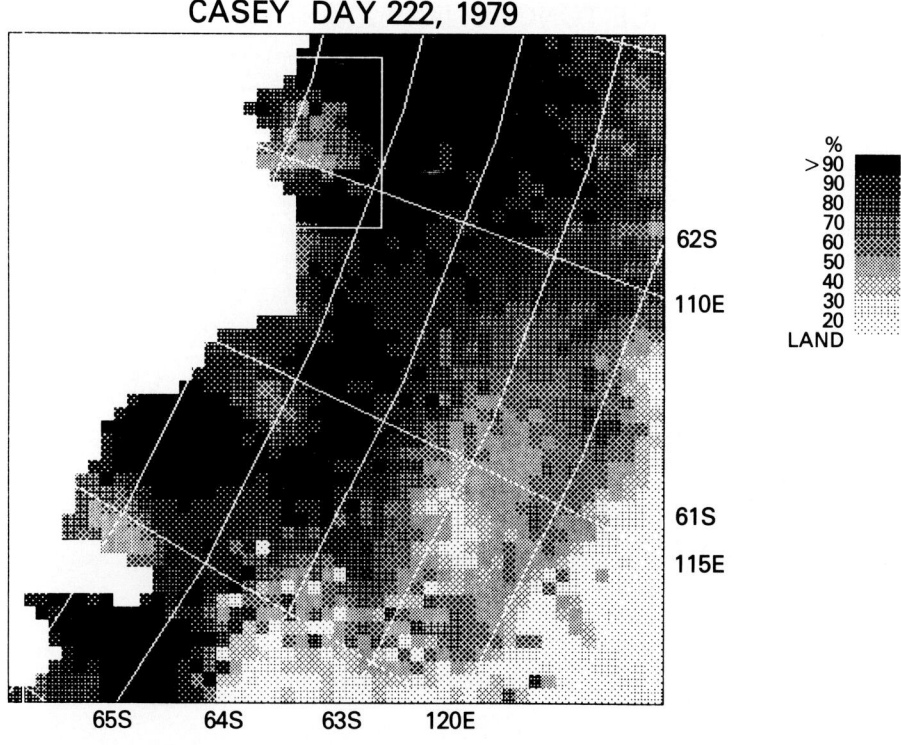

Fig. 9b. Rectangle for defining open water areas and coastline for the Dalton and Casey polynyas. The gray scale gives percent ice concentration.

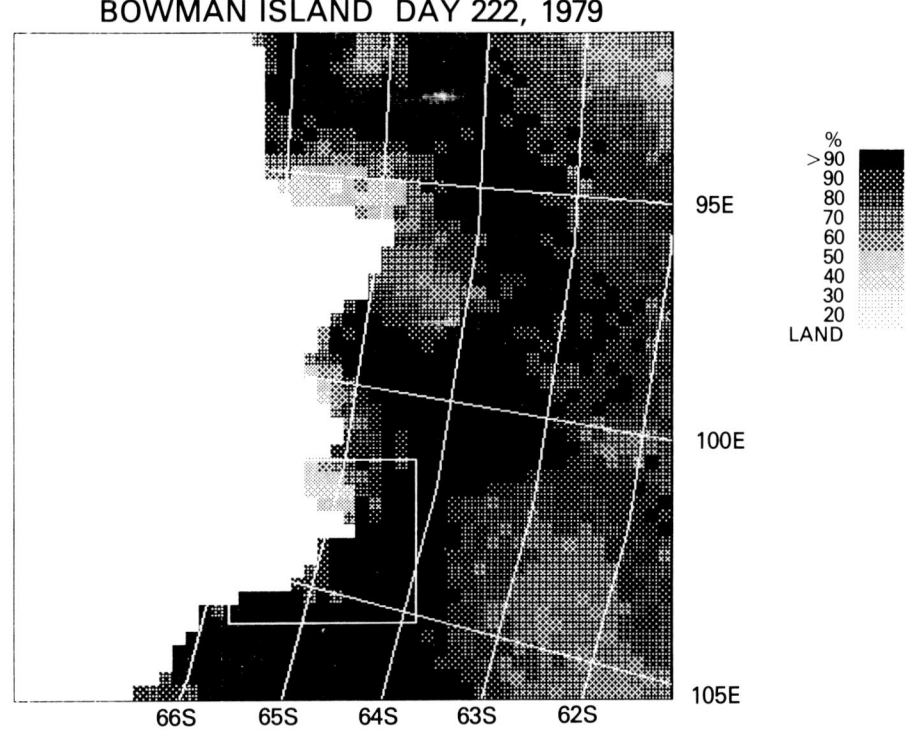

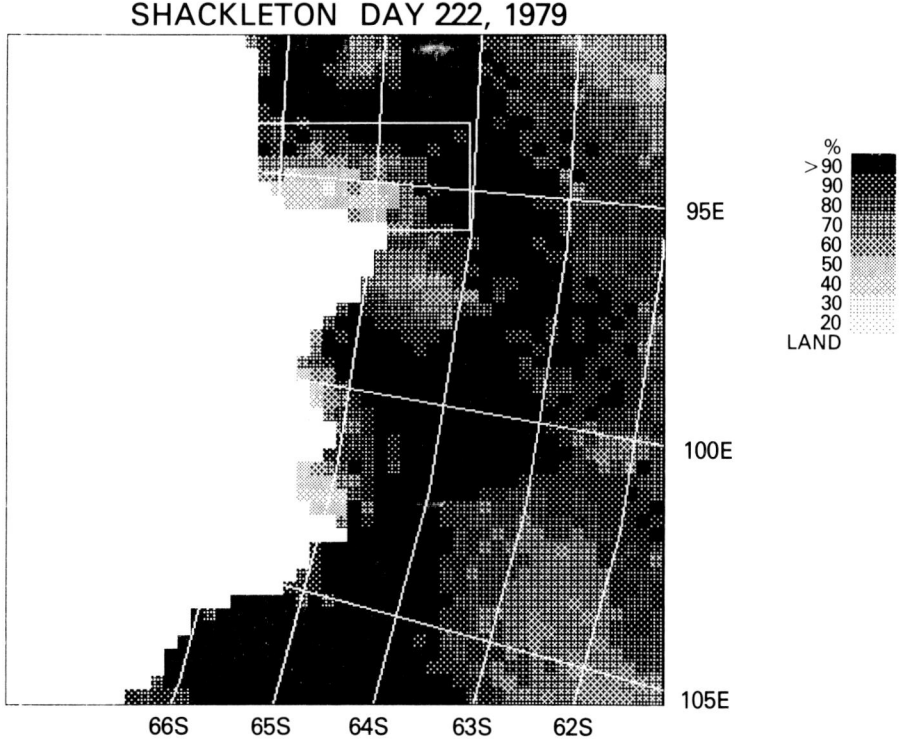

Fig. 9c. Rectangle for defining open water areas and coastline for the Bowman and Shackleton polynyas. The gray scale gives percent ice concentration.

TABLE 2. The Area and Number of Pixels for Each Polynya Rectangle

Polynya	Area, km^2	Number of Pixels
Mertz	18,346	79
Dibble	17,052	74
Dalton	17,125	74
Casey	27,738	120
Bowman	28,687	125
Shackleton	30,580	133

angle. In Figure 9, as in Figure 4, any pixel filled or partially filled with either land or glacier ice is treated as land and is masked. In these polynyas, the low ice concentration is located either near or adjacent to the coast. Examination of the images shows that the Dibble and Dalton Iceberg Tongues yield particularly compact polynyas, while the Shackleton Ice Shelf yields a large, low ice concentration polynya.

Within each rectangle, we use the ice concentration data not only to calculate the open water area but also to discriminate between true polynyas and regions of dense pack ice with leads. The opening and closing of leads within the ice pack may result in extensive areas of ice cover with up to 15% open water, in contrast to polynyas which tend to be localized regions of low ice concentration. For example, Table 2 shows for the Mertz rectangle that a 90% ice cover yields a spurious polynya area of 2×10^3 km^2. To discriminate against such regions, the open water area within each rectangle was calculated in three ways. First, we calculated the open water area from all pixels with concentrations less than 100%. Second, we considered any concentration greater than 75% as solid ice and calculated the open water area only from those pixels with a 75% or less ice concentration. Third, we followed the same procedure, treating anything greater than 50% ice concentration as solid ice.

The open water area time series are presented in Figures 10a-10c and tabulated in Martin and Cavalieri [1984] for the three polynya definitions and the six locations. With the exception of the Dibble and Dalton polynyas, the area time series run from day 177 through 273. The reason that the Dibble and Dalton time series start late at day 185 and 181, respectively, is as follows: for the Dibble time series, Figure 4a shows on days 182-184 that the large low ice concentration region to the west of the Dibble Iceberg Tongue is not a polynya, but rather is part of the general diffuse nature of the ice edge.

For the Dalton time series, Figure 4a also shows on days 178-180 a region of missing SMMR data indicated by a blacked-out wedge extending toward the Dalton Iceberg Tongue. To avoid both the diffuse ice edge and the lack of SMMR coverage, we started the Dibble and Dalton time series later than the other polynyas.

In Figure 10, the top of the white bars shows the open water area based on the 100% concentration definition; the top of the gray bars, the 75% definition; and the top of the black bars, the 50% definition. Associated with each pair of polynya areas in Figure 10 are the 4-day-averaged winds from the closest weather station. The 6-hour data from each station were averaged over a 4-day period to correspond with the SMMR observations. The 4-day averaging period begins on the day preceding a SMMR data day, so that for a given 4-day interval, the SMMR days fall on the second and fourth day. The following geographic features were paired with the closest weather station (the numbers in parentheses give the distance of the feature from the weather station): the Mertz Glacier Tongue (250 km) and Dibble Iceberg Tongue (275 km) with the Dumont d'Urville Station; the Dalton Iceberg Tongue (475 km) and the Cape Poinsett peninsula (0 km) on which Casey is located with the Casey Station; Bowman Island (475 km) and the west end of the Shackleton Ice Shelf (150 km) with the Mirny Station.

Before discussion of the mathematical correlation between the station winds and area time series, we relate the temporal variability of the polynyas to the synoptic events over the 3-month period. First, from a visual examination of the polynya area and average winds presented in Figure 10, we see that for the case of the Dalton and Casey polynyas there is a striking increase in open water at the 50% and 75% level during the strong wind peak which occurs near day 222. Similarly, for the Mertz and Dibble polynyas, the strong wind event between days 234 and 270 is also correlated with the change in open water area. For the Bowman and Shackleton polynyas, however, the large variability at Shackleton at all levels does not follow the more uniform Mirny winds, even though the relative wind peak at day 222 is again followed by an open water increase.

As previously noted, the winds measured at each station are a composite of both a topographically induced katabatic component and a synoptic component. While we cannot divide the station wind data into these components, a comparison of the open water time series in Figure 10 with the synoptic maps in Figure 5 suggests that synoptic forcing plays a dominant role in the variability of the polynyas. For example, the increase in open water centered on day 194 for the Dibble Iceberg Tongue polynya shown in Figure 10a is associated with a low-pressure system centered at about 65°S, 148°E on day 194 (Figure 5a). A second event, centered on day 250 and apparent in both the Mertz and Dibble time series, is associated with the cyclonic system centered at 60°S, 145°E on that day. In both these cases, the orientation of the low-pressure systems supports an off-continent circulation.

By far the strongest event of the entire 3-month period is centered on day 222 and is observed in the Dalton, Casey, Bowman, and Shackleton time series. This event is strongly associated with the intense low-pressure system centered off the coast at 60°S, 95°E and is reinforced by the high-pressure ridge just to the south over the continent. The strong warm air advection associated with this event is observed in the 6-hour air temperature station data of Figure 8. Interestingly, a comparison of the temperature time series for Casey and Mirny shows that the Casey temperatures started increasing from 8 to 12 days before those at Mirny, reflecting the effect of the low-pressure system as it approached the Mirny Station from the east. Further, the maximum open water area for the Dibble Iceberg Tongue polynya occurred on day 218, which also illustrates the influence of the westward migration of the system shown in Figures 4a and 5a.

Finally, the Shackleton polynya has the largest open water area of the six polynyas. The reason is that the Shackleton Ice Shelf provides the requisite north-south projection just east of the mean position of a quasi-stationary low-pressure system, which serves as a source of almost constant easterly winds. The transient weather systems then tend to modulate the intensity of the winds and to pump warm maritime air onto the continent. Tauber [1960] reports that warm air associated with these systems penetrates onto the continent for hundreds of kilometers. Each of the five warmest periods (days 178, 182, 190, 222, and 270) observed at Mirny in Figure 8 are associated with intense weather systems as shown in Figure 5.

Because the interaction of the synoptic and katabatic winds strongly depends on the local topographical conditions, winds acting on the polynyas may differ from the station winds. For example, Parish [1980] shows that the Mertz Glacier Tongue is at the focus of a region of strong katabatic winds, which may not be reflected in the measured winds at Dumont d'Urville. Furthermore, as discussed also by Bromwich and Kurtz [1984], when a katabatic wind flows off the continent out over the ice, an internal hydraulic jump occurs, yielding slower wind speeds. Thus, even if the wind measurements at a single point can be applied to the polynya, the winds over the polynya may not be uniform.

Proceeding to the correlation of the 4-day averaged wind speed with the open water area, we correlate the wind and areas in two ways: first, with the 10-m wind speed (U) and then with wind speed squared (U^2). Our reason for use of the U correlation follows the argument used by Kurtz and Bromwich [1983], that the ice motion induced by the wind is approximately proportional to the geostrophic wind; the use of the U^2 correlation follows McPhee [1979], who shows that the drag on the ice is proportional to U^2.

Table 3 lists the correlation coefficients of open water area with U and U^2 for the six polynyas, and the three different open water definitions. The results in Table 3 show that neither the U nor U^2 correlation is higher than the other. In the calculation, we found the largest correlations at zero lag; also, using Student's T test, we found that all correlations greater than 0.48 are significant at the 95% confidence level. Examination of the table shows that Dalton and Casey have the highest correlations with wind speed, with the largest correlations occurring at the 75% open water definition. Mertz and Shackleton show the next highest correlations, which for the U correlation lie just above the significant level; Bowman and Dibble show the worst, with the possible exception of the 0.47 correlation at the 50% open water definition for Bowman. The coefficient magnitudes, then, are not correlated with the distance of the polynya from the station. Further, for most of these cases, the correlations are much better than the 0.30 value observed by Kurtz and Bromwich [1983] for the Terra Nova Bay correlation of open water with geostrophic winds. The difference between our results and those of Kurtz and Bromwich [1983] may be due to our use of the local weather station data, and to our computer-generated open water areas.

5. Estimation of Heat Flux, Ice Production, and Salt Flux in the Polynyas

In our calculation of the polynya ice and salt production, we make the simplifying assumption that the ice growth only occurs from open water and that the water column under the polynya is at its freezing point. The first assumption neglects the ice growth of the columnar sheet ice surrounding the polynya and thus underestimates the ice production; the second neglects the heat loss necessary to cool the water column down to the freezing point and thus overestimates the production. We justify the first assumption from the laboratory observations of Martin and Kauffman [1981] which show that the growth rate of frazil ice is much greater than that of columnar ice; the second assumption is necessary because of the lack of salinity and tempera-

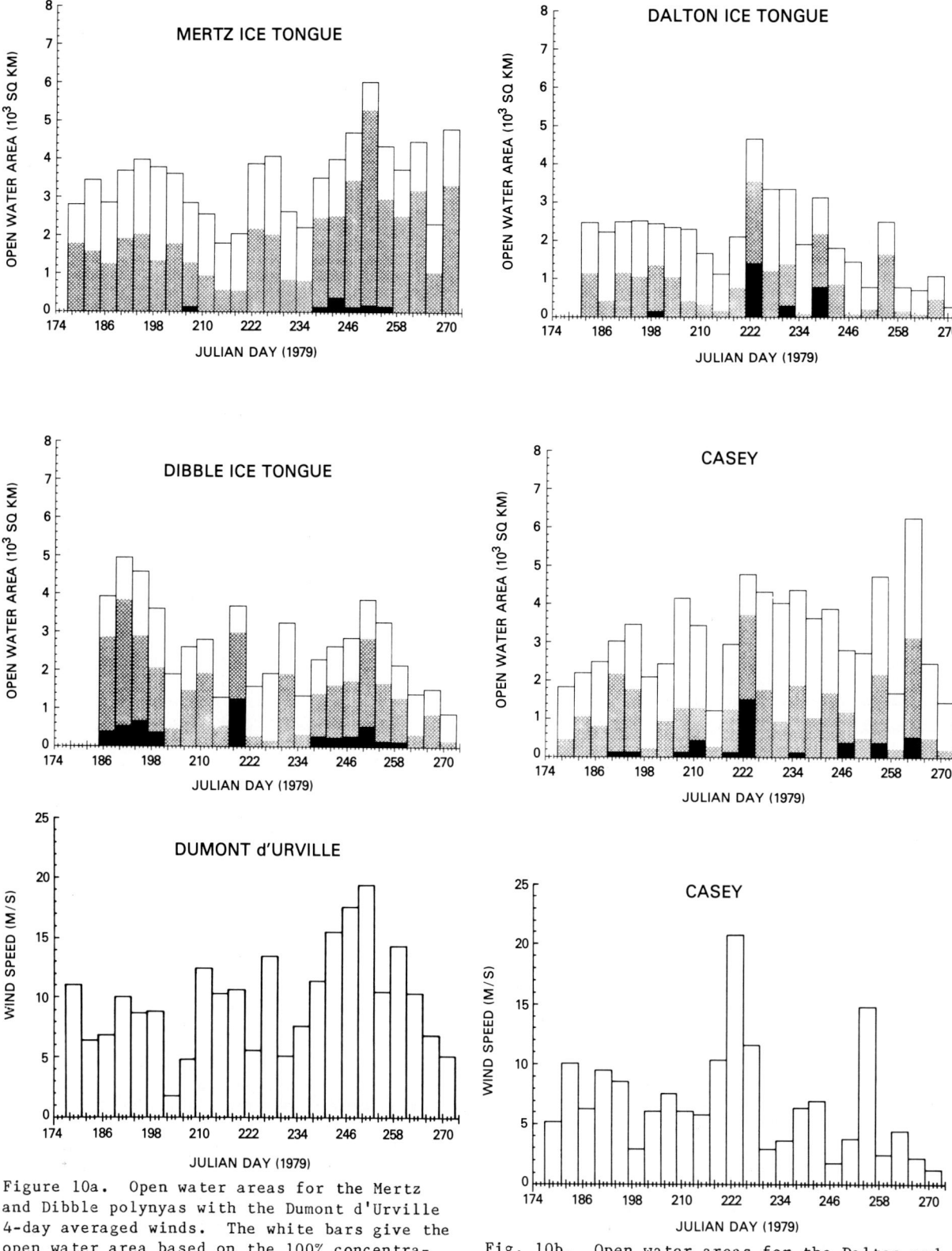

Figure 10a. Open water areas for the Mertz and Dibble polynyas with the Dumont d'Urville 4-day averaged winds. The white bars give the open water area based on the 100% concentration definition; the grey bars, the 75% definition; the black bars, the 50% definition. See text for further description.

Fig. 10b. Open water areas for the Dalton and Casey polynyas with the Casey 4-day-averaged winds. See caption of Figure 10a for further description.

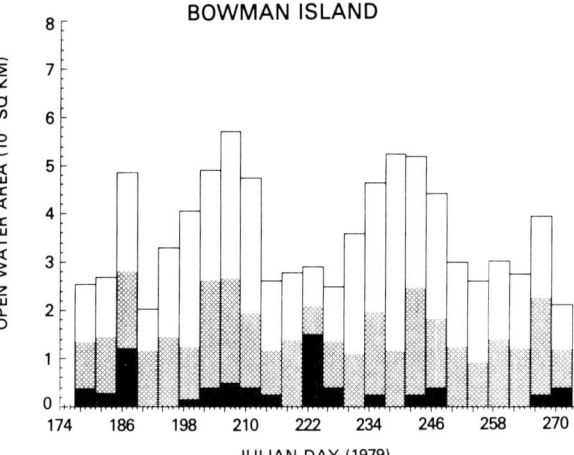

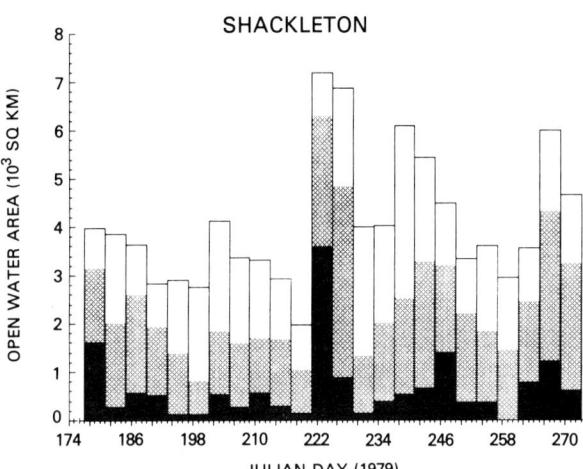

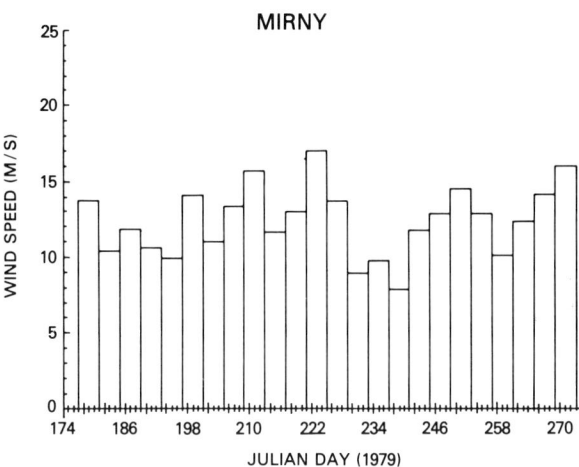

Fig. 10c. Open water areas for the Bowman and Shackleton polynyas with the Mirny 4-day-averaged winds. See caption of Figure 10a for further description.

ture water column data from the winter polynya regions.

The heat flux and ice production rates are calculated following the work of Den Hartog et al. [1983], and Smith et al. [1983], who carried out a field study of the heat loss from the Canadian Dundas Island polynya. We use their heat loss measurements to calculate the ice production rates from the data of each weather station, and then to estimate the total ice produced within each of the polynyas.

Den Hartog et al. [1983] and Smith et al. [1983] measured the advective heat loss from the Dundas polynya during March and April 1980. Their formulation of the sensible heat flux is

$$H = C_T \rho_a c_p U \Delta T \quad (2)$$

where C_T is a heat transfer coefficient, ρ_a is the air density (1.3 kg m^{-3}), the specific heat of air is c_p (1.0 x 10^3 in MKS units), $\Delta T = T_s - T_a$, (where T_s is the surface temperature and T_a is the air temperature), and U is the wind speed. From their measurements at a 4-m height, Smith et al. [1983, Table 3] show for U ≥ 10 m s^{-1} and ΔT ≥ 30 K that

$$C_{T4} = 1.3 \times 10^{-3}$$

Since we are most interested in the heat transfer at large wind speeds and low temperatures, which maximize the ice production by sweeping the water surface free of ice and exposing it to cold air, we choose this relatively small value of C_{T4}. To adjust this C_T to the 10-m station height, we follow J. Overland [personal communication, 1984] and choose $z_0 = 5 \times 10^{-3}$ m, and then reduce C_{T4} by the ratio $\ln(10/z_0)/\ln(4/z_0) = 1.14$, which gives

$$C_{T10} = 1.1 \times 10^{-3}. \quad (3)$$

Calculation of the total heat flux follows Den Hartog et al. [1983, Figure 5], who show that the total flux over 1 month is 1.6 times

TABLE 3. The Correlation Coefficient for Polynya Area with U and U^2 for the Different Open Water Definitions

Polynya	U			U^2		
	50%	75%	100%	50%	75%	100%
Mertz	0.50	0.54	0.42	0.48	0.58	0.49
Dibble	0.28	0.28	0.23	0.26	0.30	0.27
Dalton	0.53	0.75	0.71	0.52	0.74	0.68
Casey	0.63	0.69	0.43	0.64	0.65	0.41
Bowman	0.47	0.14	-0.20	0.51	0.19	-0.13
Shackleton	0.53	0.48	0.20	0.52	0.46	0.22

the advective flux. Combination of the above gives the following equation for the total heat flux H_T:

$$H_T = 2.3 \, U_{10} \Delta T \quad (4)$$

where H_T has units of $W \, m^{-2}$ and U_{10} is the 10-m wind velocity. Even though our heat flux model neglects the radiative and convective losses associated with cold temperatures and no winds, during periods of no winds the sea ice forms as a flat columnar ice sheet, so that the water surface quickly becomes covered by an insulating ice layer, and the ice production falls off rapidly.

To calculate the ice production rate from H_T, we note that the latent heat of freezing for fresh water is $L = 335 \, kJ \, kg^{-1}$ and that the ice density is $\rho_i = 920 \, kg \, m^{-3}$. Because, as Martin [1981] shows, frazil ice consists of freshwater ice crystals coated with a thin saline layer the frazil production rate nearly equals the freshwater ice production rate P, given by

$$P = H_T/\rho_i L = 0.027 \, U_{10} \Delta T \quad (5)$$

where P has units of millimeters per hour of ice growth per unit area of open water.

In the computation of the heat flux and ice production rates from equations (4) and (5), we use the 6-hour recorded winds and air temperatures and take the water surface temperature as $T_s = -2°C$. For all stations, the temperatures presented in Figure 8 show a great deal of variability, with minimums of $-35°$ to $-40°C$. From these data, Figure 11 displays and Martin and Cavalieri [1984] list the 4-day average heat flux calculated from equation (4). Figure 11 shows that Mirny had the highest heat flux with an average of $450 \, W \, m^{-2}$ followed by Dumont d'Urville at $290 \, W \, m^{-2}$, and Casey at $160 \, W \, m^{-2}$. Mirny, then, was by far the coldest of the three stations.

From the heat flux and polynya area data, we next calculate for the six polynyas the 4-day ice production, which is the product of P in equation (5) times the open water area. Because the 50% and 75% open water definitions correlate better with the wind than does the 100% definition, and because the 100% definition does not discriminate against heavy pack ice, we use only the 50% and 75% definitions in these calculations. Figures 12a-12c show, and Martin and Cavalieri [1984] again list, the 4-day ice production for each polynya in cubic kilometers, where the white is the 75% and the black is the 50% level. Examination of this time series shows that the largest single 4-day production equaled 3 km^3, which occurred at the 75% level for Mertz beginning on day 249. The data show that Bowman and Shackleton have the largest average production and that Dalton and Casey have the smallest.

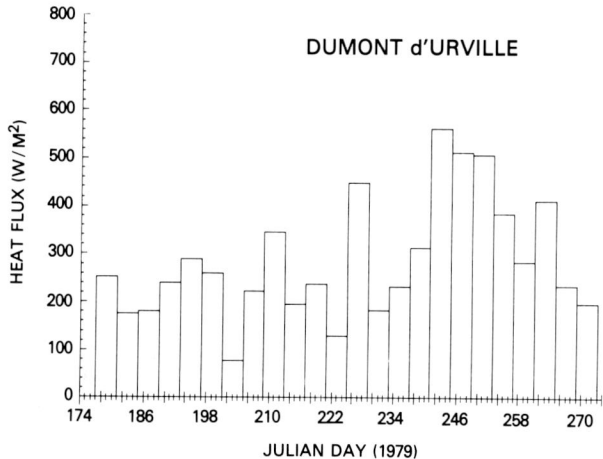

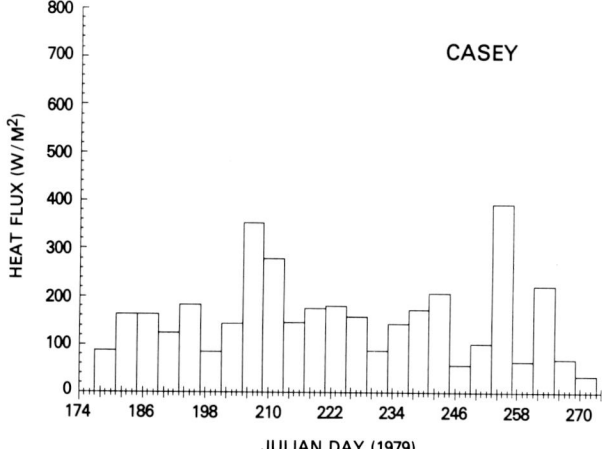

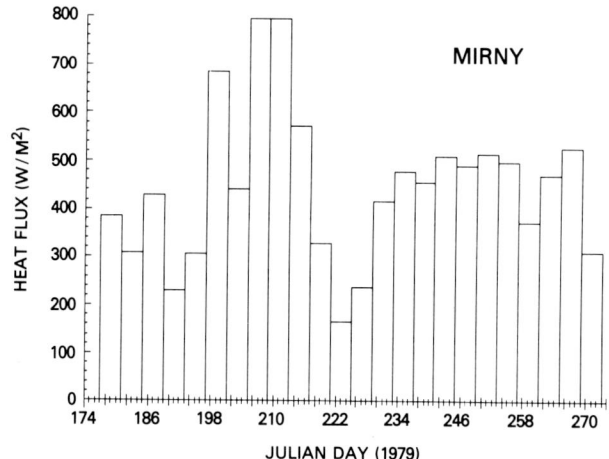

Fig. 11. Heat flux calculated from the 4-day-averaged station data for Dumont d'Urville, Casey and Mirny.

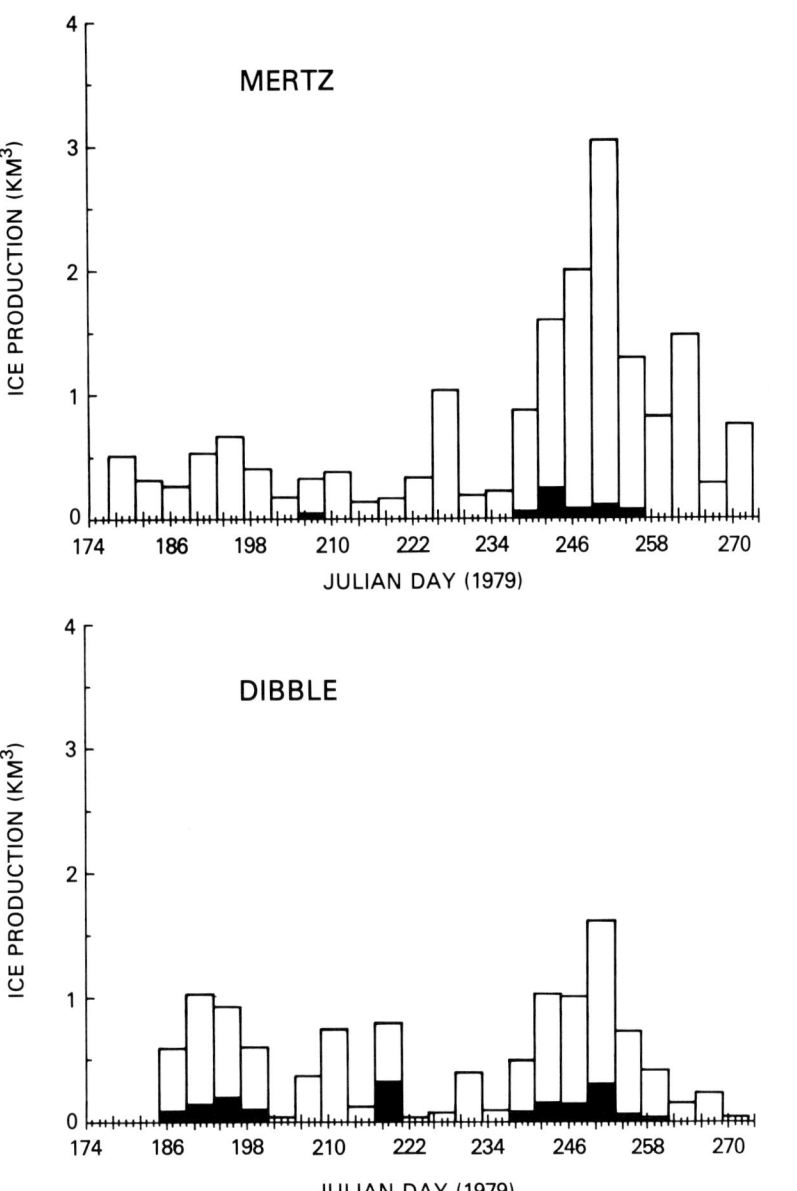

Fig. 12a. Four-day-averaged ice production for the Mertz and Dibble polynyas. The white bars give the ice production for the 75% open water definition; the black bars, the 50% definition.

Finally, Figure 13 shows the total ice production of the six polynyas for the winter period.

For further comparison, Table 4 lists the average and cumulative polynya properties. The first pair of columns lists the average open water area for the winter period; the second pair, the cumulative ice production; the third pair, the cumulative ice thickness, which is the total production divided by the mean area. The table shows the large variation in the open water area, which at the 75% definition varies by a factor of 3 from Dalton to Shackleton, the variation in cumulative ice production, which varies by a factor of 7 for the same stations, and the variation in cumulative ice thickness, which ranges from 4.6 to 12.4 m with an average of 9 m.

For comparison with these cumulative ice thicknesses, Gordon and Tchernia [1974] give 1.5 m as a typical seasonal ice thickness. Further, Anderson [1961] gives a formula for ice growth in still air as a function only of temperature (also described by Bauer and Martin [1983]). Substitution of the temperature data into Anderson's formula, on the assump-

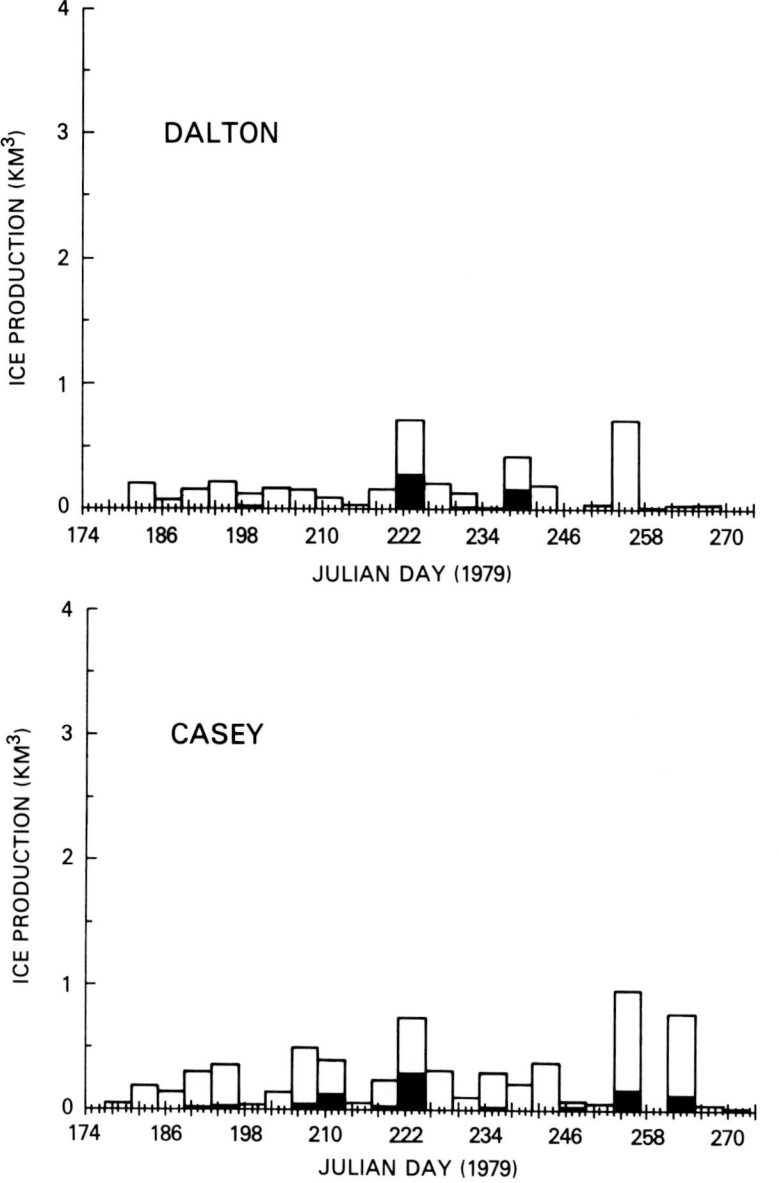

Fig. 12b. Four-day-averaged ice production for the Dalton and Casey polynyas. See caption of Figure 12a for further description.

tion of open water initial conditions at the start of our temperature time series, gives the results shown in Table 5, namely, that the temperature variations at the three weather stations cause between 0.83 and 1.0 m of ice growth over the observational period. Comparison of Tables 4 and 5 shows that the wind-driven polynyas increase the ice growth rate by about a factor of 10 over the still air case.

The estimation of oceanic salt flux from frazil ice growth for each of our six polynyas follows the work of Martin and Kauffman [1981, equation (16)] who studied frazil ice growth in the laboratory. For frazil ice grown in salt water of salinity 34 g kg^{-1}, they found the melted frazil ice, after removal from the water, to have a salinity of 10 g kg^{-1} with a standard deviation of 2 g kg^{-1}. We neglect this residual frazil ice salinity for two reasons: First, the field and laboratory work cited above shows that this salinity takes from a month to a growth season to drain into the underlying ocean. Second, because our implicit assumption is that the wind constantly sweeps the newly formed frazil ice off the

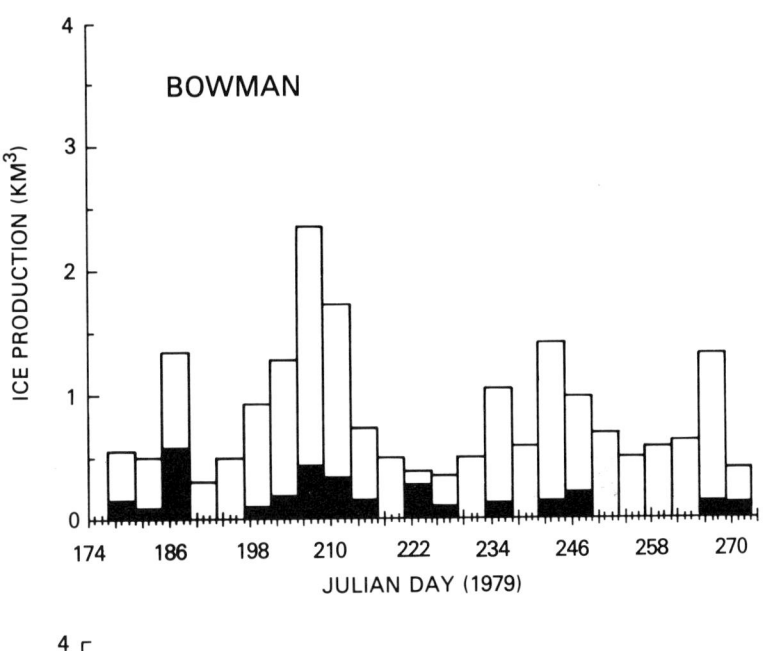

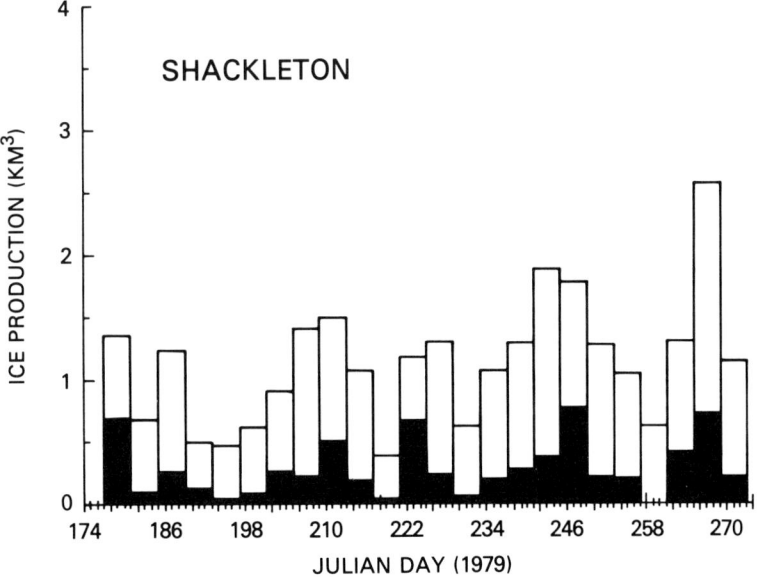

Fig. 12c. Four-day-averaged ice production for the Bowman and Shackleton polynyas. See caption of Figure 12a for further description.

polynya, this additional slow drainage does not enter the polynya. Therefore, within the polynya the formation of 1 kg of frazil ice leads to a salt flux of 24 g. Alternatively, 1 m^3 of frazil ice, where we ignore the weight of the brine coating, corresponds to 920 kg of ice and a salt flux of 22 kg. On an areal basis, the growth of 10 m of ice per 1 m^2 of surface area yields a salt flux of 220 kg m^{-2}.

Finally, Gordon [1974] and Gordon and Tchernia [1974] discuss bottom water formation around the Antarctic coast and speculate on the role of katabatic winds and coastal polynyas in the formation of this water, showing that for bottom water formation to occur, the shelf water salinity must be greater than 34.63 g kg^{-1} (see also Carmack and Killworth, [1978]). Table 6, which is adapted from Gordon [1974, Figure 5] and the results of the present paper, lists the observed salinity elevation above 34 g kg^{-1} of the cold, summer shelf water as well as our calculated polynya areas and cumulative salt production for the six polynyas. We calculated the cumulative

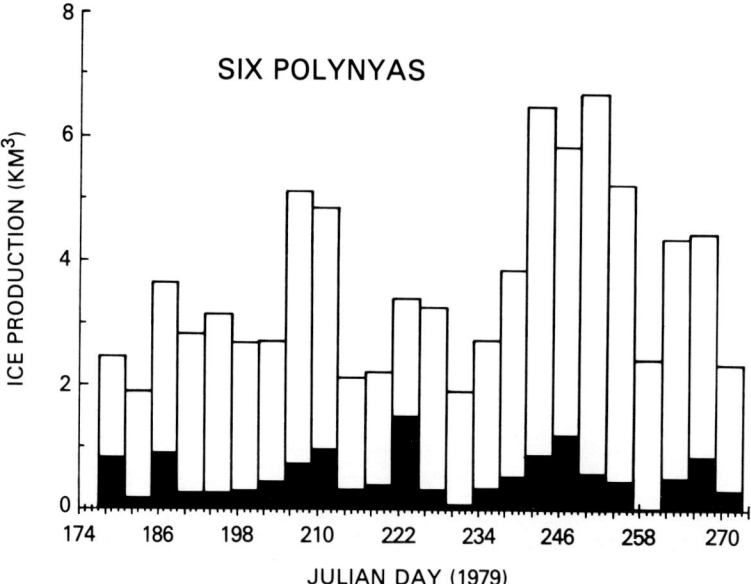

Fig. 13. Four-day-averaged ice production for all six polynyas. See caption of Figure 12a for further description.

salt production listed in Table 6 from the average open water area and the cumulative ice thickness listed in Table 4 and from the salt flux of 22 kg m^{-2} per meter of ice growth. Because the 75% definition of open water yields the highest correlation with the wind data for most of the stations and because it also yields average polynya areas of the order of 10^3 km^2, which is the same order as that observed by Kurtz and Bromwich [1983], we use the results from the 75% definition in Table 6. Examination of Gordon's [1974] numbers in this table shows that high salinity shelf water will only be produced in the vicinity of the Mertz Glacier Tongue and the Shackleton Ice Shelf/Bowman Island region. The table also shows that the high oceanic salinities correspond to relatively large polynya areas and large salt production, whereas the salinity minimum at the Dalton Iceberg Tongue corresponds to a minimum in both polynya area and salt production. The table suggests that our satellite-observed regions of high ice and salt production agree well with the oceanographic regions of increased shelf water salinity.

TABLE 4. Average and Cumulative Properties of the Six Polynyas for the 50% and 75% Open Water Definitions

Polynya	Average Open Water Area, km^2		Cumulative Ice Production, km^3		Cumulative Ice Thickness, m	
	50%	75%	50%	75%	50%	75%
Mertz	43	1973	0.5	17.4	11.6	8.8
Dibble	215	1516	1.7	11.4	7.9	7.5
Dalton	112	845	0.5	3.9	4.5	4.6
Casey	160	1259	0.9	6.4	5.6	5.1
Bowman	286	1625	3.2	20.2	11.3	12.4
Shackleton	661	2435	6.9	27.1	10.4	11.1
All polynyas	1477	9653	13.7	86.4	9.3	9.0

TABLE 5. Cumulative Winter Ice Growth from Temperature Data Only

Station	Period, Julian Days	Thickness, m
Dumont d'Urville	185-269	0.84
Casey	177-269	0.83
Mirny	177-269	1.06

6. Concluding Remarks

The results presented in this paper show that passive microwave observations can be used with confidence in studying the variability of coastal polynyas, since in most cases the correlations of wind speed with open water area are significant at the 95% level. This study also shows that the 75% definition of open water within the polynya yields areas which are consistant with the work of other investigators. From our comparison between the polynya time series and the synoptic maps, we conclude that the polynya variability is largely influenced by the coastal weather systems. Further, the observations of polynya area and the estimates of ice production correlate well with Gordon's [1974] oceanic surveys in that the Shackleton Ice Shelf and the Mertz Glacier Tongue with their high ice and salt production, correspond to a high shelf water salinity, while the Dalton Iceberg Tongue, with its small polynya and low ice and salt production, corresponds to a low shelf water salinity. In summary, the combination of satellite passive microwave observations with local weather data yields a powerful tool for the investigation of ice production in coastal polynyas.

TABLE 6. Observed Range of the Salinity Difference (S-34) [from Gordon, 1974], the Winter Mean Polynya Size and Cumulative Salt Production

Feature	Salinity Elevation, g/kg	Polynya Area, $km^2 \times 10^3$	Cumulative Salt Production, $kg \times 10^{11}$
Shackleton	0.5-0.7	2.4	3.9
Bowman	0.5-0.7	1.6	2.5
Casey	0.5-0.55	1.3	0.9
Dalton	0.3-0.4	0.8	1.5
Dibble	0.3-0.5	1.5	4.4
Mertz	0.6-0.7	2.0	5.9

Acknowledgments. This study was supported in part by the Oceanic Processes Branch of NASA and by the Department of Commerce Spacecraft Oceanography (SPOC) Group under contract MO-A01-00-4335. S. Martin gratefully acknowledges the encouragement and support of John Sherman, Pat S. DeLeonibus, and Paul McClain of SPOC for this work and also acknowledges the support of the Office of Naval Research under task NR083-012 and contract N00014-84-C-0111 for preparation of this manuscript. The authors thank David Bromwich for his many helpful comments in reviewing an early draft of the paper and for directing us to relevant papers on the climatology of East Antarctica. The authors also thank K. Fred Huemmrich of the Science Applications Research Corporation for his programming and graphics support, and Mary G. Reph of the NASA Pilot Climate Data System for providing and reformating the restructured FGGE II data set obtained from the WDC-A for Meteorology, National Climatic Center, Asheville, North Carolina. This is contribution 1406 of the School of Oceanography, University of Washington.

References

Aagaard, K., L.K. Coachman, and E. Carmack, On the halocline of the Arctic Ocean, Deep Sea Res., 28A, 529-45, 1981.

Anderson, D.L., Growth rate of sea ice, J. Glaciol., 3, 1170-1172, 1961.

Ball, F.K., Winds on the ice slopes of Antarctica, in Antarctic Meteorology, Symposium on Antarc. Meteorology, pp. 9-16, Pergamon, New York, 1960.

Bauer, J. and S. Martin, A model of grease ice growth in leads, J. Geophys. Res., 88, 2917-2925, 1983.

Bromwich, D.H. and D.D. Kurtz, Katabatic wind forcing of the Terra Nova Bay polynya, J. Geophys. Res., 89, 3561-3572, 1984.

Carmack, E.C. and P.D. Killworth, Formation and interleaving of abyssal water masses off Wilkes Land, Antarctica, Deep Sea Res., 25, 357-369, 1978.

Cavalieri, D.J., S. Martin, and P. Gloersen, Nimbus-7 SMMR observations of the Bering Sea ice cover during March 1979, J. Geophys. Res., 88, 2743-2754, 1983.

Den Hartog, G., S.D. Smith, R.J. Anderson, D.R. Topham, and R. G. Perkin, An investigation of the polynya in the Canadian Archipelago, 3, Surface heat flux, J. Geophys. Res., 88, 2911-2916, 1983.

Dzerdzeevskii, B.L., Certain features of weather in the coastal area of eastern Antarctica, in Antarctic Meteorology, Symposium on Antarctic Meteorology, pp.37-51, Pergamon, New York, 1960.

Gloersen, P., and F. Barath, A scanning multichannel microwave radiometer for Nimbus-G and Seasat, IEEE, J. Oceanic Eng., OE-2, 172-178, 1977.

Gordon, A.L., Varieties and variability of Antarctic bottom water, Colloq. Int. CNRS, 215, 33-47, 1974.

Gordon, A.L. and P. Tchernia, Waters of the continental margin off Adélie Coast, Antarctica, in Antarctic Oceanography II: The Australian-New Zealand Sector, Antarc. Res. Ser., vol. 19, edited by D.E. Hayes, pp. 59-69, AGU, Washington, D.C., 1974.

Knapp, W.W., Formation, persistance, and disappearance of open water channels related to the meteorological conditions along the coast of the Antarctic continent, Polar Meteorology, Proceedings of WMO/SCAR/ICPM Symposium, WMO Tech. Rep. 87, World Meteorol. Organ., Geneva, 89-104, 1967.

Knapp, W.W., Satellite observations of large polynyas in polar waters, in Sea Ice, Proceedings of an International Conference, May 10-13, 1971, edited by T. Karlsson, pp. 201-212, National Research Council, Reykjavik, 1972.

Kurtz, D.D. and D.H. Bromwich, Satellite observed behavior of the Terra Nova Bay polynya, J. Geophys. Res. 88, 9717-9722, 1983.

Kurtz, D.D., and D.H. Bromwich, A recurring, atmospherically forced polynya in Terra Nova Bay, this volume.

Martin, S., Frazil ice in rivers and oceans, Ann. Rev. Fluid Mech., 13, 379-397, 1981.

Martin, S., and D.J. Cavalieri, A passive microwave study of the Antarctic coastal polynyas between 90°E and 150°E during the austral winter 1979, Spec. Rep. 98, Reference M84-45, 49 pp., School of Oceanography, University of Washington, Seattle, June 1984.

Martin, S., and P. Kauffman, A field and laboratory study of wave damping by grease ice, J. Glaciol., 27, 283-313, 1981.

Mather, K.B., and G.S. Miller, Notes on topographic factors affecting the surface wind in Antarctica, with special reference to katabatic winds; and bibliography, technical report, Geophysical Institute, University of Alaska, June 1967.

McPhee, M.G., The effect of the oceanic boundary layer on the mean drift of pack ice: Application of a simple model, J. Phys. Oceanogr., 9, 388-400, 1979.

Parish, T.R., Surface winds in East Antarctica, Res. Rep., 121 pp., Department of Meteorology, University of Wisconsin, Madison, 1980.

Smith, S.D., R.J. Anderson, G. Den Hartog, D.R. Topham, and R.G. Perkin, An investigation of a polynya in the Canadian Archipelago 2, Structure of turbulence and sensible heat flux, J. Geophys. Res., 88, 2900-2910, 1983.

Streten, N.A., Satellite observations of the summer decay of the Antarctic sea ice, Arch. Meteorol. Geophs. Bioklimatol., Ser. A, 22, 119-134, 1973.

Taljaard, J.J., H. Van Loon, H.L. Crutcher, R.L. Jenne, Climate of the upper air, 1, Southern hemisphere, 1, Sea level pressure and selected heights, temperatures and dew points, Rep. NAVAIR 50-16-55, U.S. Government Printing Office, Washington, D. C., 1969.

Tauber, G.M., Characteristics of Antarctic katabatic winds, in Antarctic Meteorology, Symposium on Antarctic Meteorology, pp. 52-64, Pergamon, New York, 1960.

Zimmerman, J.R., Wilkes climatology, in Antarctic Meteorology, Symposium on Antarctic Meteorology, pp. 415-422, Pergamon, 1960.

Zwally, H.J., J.C. Comiso, and A.L. Gordon, Antarctic offshore leads and polynyas and oceanographic effects, this volume.

(Received November 1, 1984; accepted January 16, 1985.)

SOME EFFECTS OF OCEAN CURRENTS AND WAVE MOTION
ON THE DYNAMICS OF FLOATING GLACIER TONGUES

G. Holdsworth

National Hydrology Research Institute, Environment Canada
Calgary, Alberta, Canada   T3A 0X9

Abstract. A survey is made of several super glacier tongues (SGTs) that have existed in the past or are still in existence along the Antarctic coastline. The dynamics of these SGTs are examined, principally in the context of relationships to ocean currents and wave motion. A discussion of some iceberg calving mechanisms is presented with the aim of attempting to physically explain several cases of documented SGT calving events. It is concluded that both ocean currents and waves, directly or indirectly, play an important role in the dynamics of SGTs.

Introduction

This paper reviews certain aspects of the dynamics of large unconfined ice tongues or streams which flow for long distances, in some cases more than 100 km, beyond the adjacent coastline. The behavior of the super glacier tongues (SGTs) studied seems to be strongly influenced both by ocean currents and the ambient ocean wave field.

Measurements pertinent to complete descriptions of the dynamics of SGTs are singularly lacking. However, measurements made on the smaller scale Erebus Glacier Tongue (EGT) in McMurdo Sound help describe some of the features and behavior of several SGTs.

A large amount of data has been obtained from the Soviet Antarctic Expedition Information Bulletins and the Atlas Antarktiki [Tolstikov, 1966]. These sources are extremely valuable for studying SGT calving, especially for the 1960-1970 decade, when some exceptionally large super tabular icebergs (STI) entered the circumpolar current.

Several SGTs are identified as suitable for studying the interactions between floating ice and the ocean environment. To some extent, selection was influenced by the accessibility of available information and the conciseness of reports, many of which appear in the Antarctic Journal of the U.S., the Soviet Antarctic Expedition Information Bulletin and the Polar Record. Most of these summary accounts are not supported by detailed papers and, as noted, e.g., by Swithinbank [1969] conflicting dimensions have been given for certain SGTs and STIs by different researchers. These discrepancies may be explained by (1) errors in navigation on the ships and aircraft from which many of the SGT surveys were carried out, (2) the use of maximum, mean, and minimum dimensions of features without specifying these qualifiers, or (3) the failure of later researchers to recognize significant scale changes of certain map projections as a function of latitude. Some of these errors have no doubt propagated into the present study, but the information necessary to correct them is not readily available. Such errors will not significantly change the major conclusions reached in this paper. In several cases, the SGT features were transitory and are not now available for modern remeasurement.

Fortunately, modern hydrographic surveys can eliminate one early source of error. Beginning with the British Antarctic Expedition 1907-1909, certain authors arrived at the conclusion that an ice tongue cannot be free floating everywhere, despite insufficient information regarding water depths. A case in point is the Drygalski Ice Tongue. Whereas David and Priestly [1914] concluded that this SGT was probably aground along the central zone, modern hydrographic surveys [Anderson and Kurtz, 1980] have detailed an extensive area of deep water ($\geq$ 1000 m), suggesting that this glacier is afloat almost everywhere beyond the coastline. If this is the case, then a number of simplified theoretical model studies may readily be carried out on this SGT.

That some SGTs and small glacier tongues such as the Erebus Glacier Tongue (EGT), are free floating and can remain intact for several decades is virtually certain. There are also some SGTs for which there is evidence for local support by bedrock rises. Such appeared to be the case for the former Chelyuskintsy Ice Tongue (66°S; 82°E) and may have been the case for the earlier Thwaites Glacier Tongue (75°S; 107°W). However, one may question whether these "pinning points" are completely necessary for the continued existence of SGTs when they are spaced at distances of many tens of kilometers. Accordingly, large sections of an SGT must have been free floating before overriding and gaining support from a bedrock rise. These considerations lead to the exami-

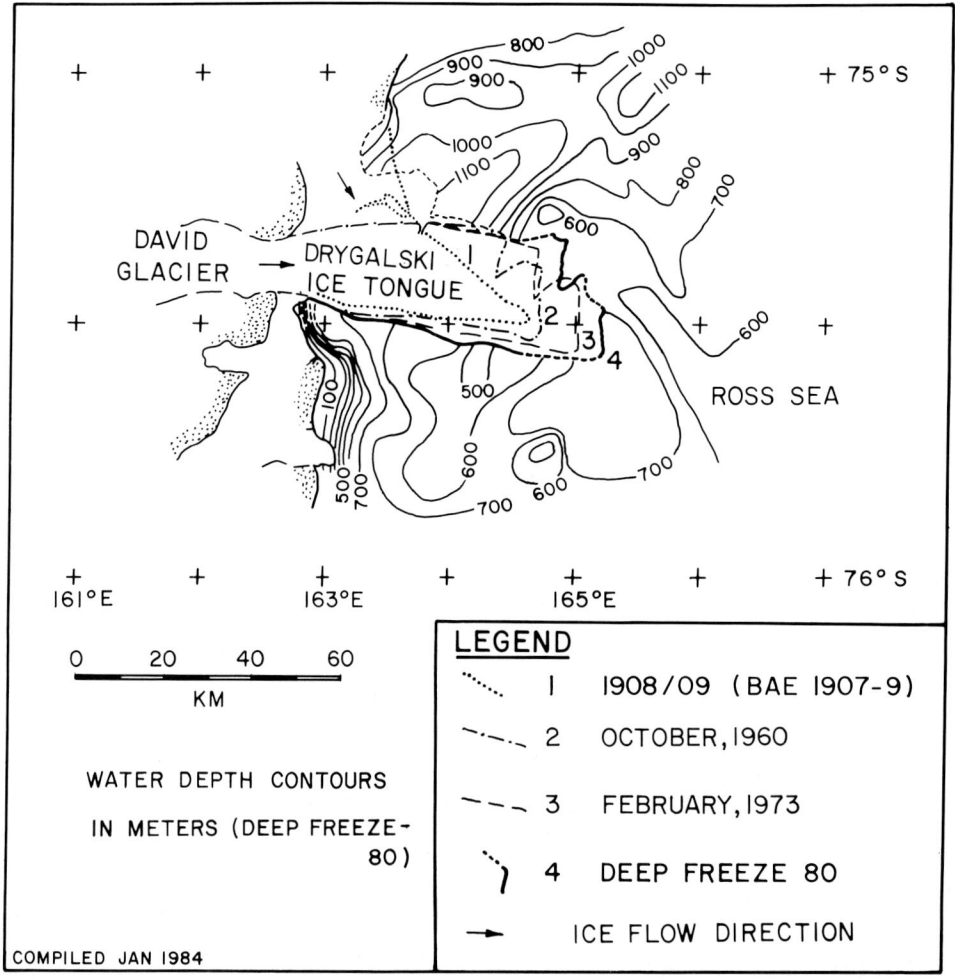

Fig. 1. Drygalski Ice Tongue, Victoria Land coast. The positions of the glacier have been derived from the following sources: (1) the British Antarctic Expedition report by David and Priestly [1914]. The position shown is for the summer 1908-1909, (2) the U.S. Geological Survey Topographical map SS58-60/13 [1968], for the October 1960 position, (3) a February 8, 1973, LANDSAT image, and (4) Anderson and Kurtz [1980, Figure 3]. Bathymetry is also from this reference. The glacier outline is based on a shipborne radar survey (solid line). Unsurveyed sections are shown dashed. A Saltzmann overhead enlarger/reducer and a Bausch and Lomb Zoom Transfer Scope was used for preparing this and subsequent figures.

nation of possible mechanisms for SGT detachment from the parent ice mass. Evidence supporting several possible mechanisms is reviewed and discussed.

Zumberge and Swithinbank [1962] suggested that: "the...problem of calving from ice shelves... [including ice tongues]...needs more attention from glaciologists and oceanographers". The present situation is illustrated by the GEBCO 5.18 map [Vanney and Johnson, this volume] where no bathymetric data seem to be available for several extensive coastal areas where SGTs occur. That map also shows the same outlines of ice shelves and SGTs as the dated [1965] American Geographical Society (AGS) map. The figures in this paper show the known major changes that have taken place on many of the largest SGTs, although the mapping accuracy in many cases is unknown. Mapping errors may cast doubt upon the historical record of some of these glaciers, but some of the apparently drastic changes may be real. A comparison of the AGS [1965] map with the AGS [1970a] or AGS [1970b] editions shows that a new SGT has appeared at 75°S; 22°W). Is this a result of earlier deficiencies in the mapping of that region or evidence for a spectacular glacier surge? Hughes [1977] provides excellent satellite imagery covering several coastal regions in

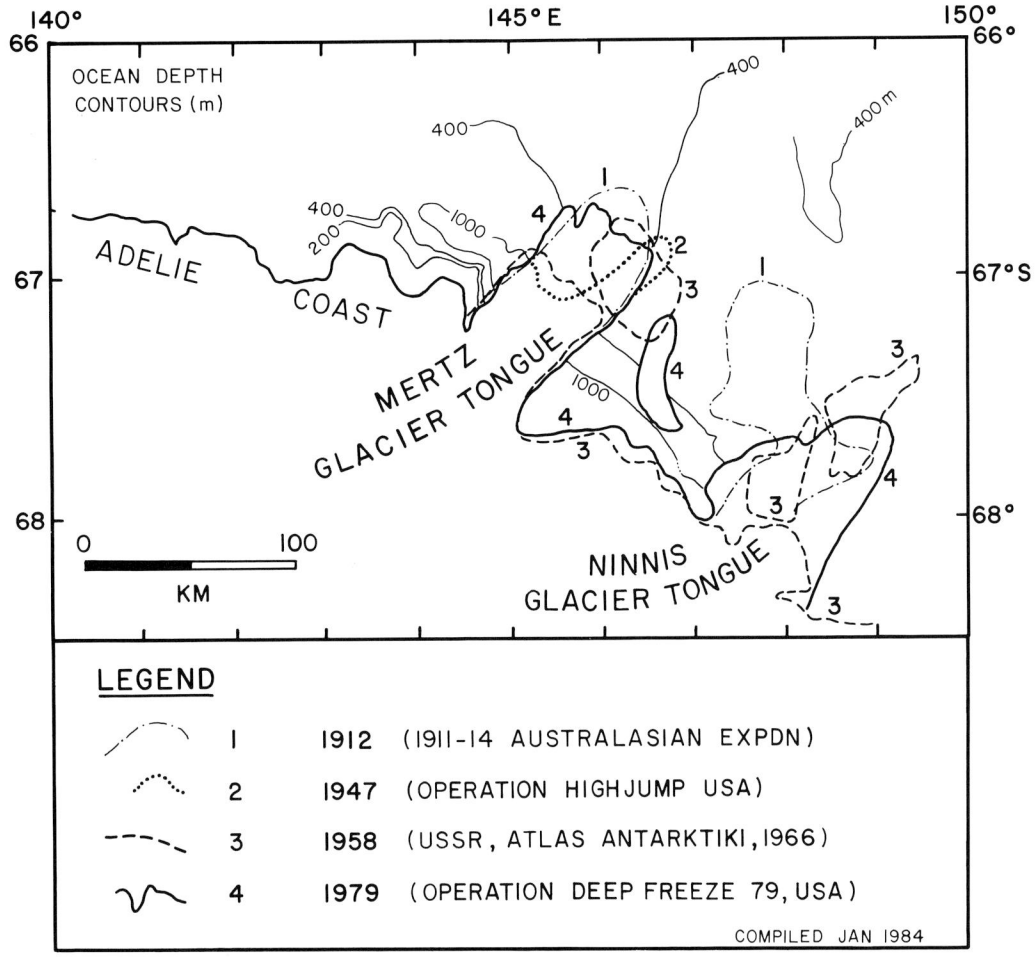

Fig. 2. Mertz and Ninnis Glacier Tongues, Adelie Coast. The glacier positions have been derived from the following sources: (1)-(3) from Atlas Antarktiki [Tolstikov, 1966], (4) Anderson et al. [1979, Figure 3]. Registration for these maps is of low precision. Bathymetry is taken from Jacobs et al. [1979].

which SGTs occur and discusses the interaction between the West Antarctic ice sheet, ice streams, and ice shelves. Many of these images update the available maps.

STIs originating from SGTs have been of some concern to ship operations in the Antarctic [e.g., Nasta and Nawratil, 1969]. They can also provide platforms for experiments in iceberg decay [McClain, 1978; 1985] which might be useful for iceberg towing enterprises, should these ever materialize [Weeks and Campbell, 1973].

Description of SGTs Studied

A number of prominent SGTs were selected for the purpose of examining the relationship between SGT morphology or calving history and oceanographic process such as wave or current action. The SGTs are discussed in order of their westward occurrence along the coastline. Their positional changes are based upon information obtained from multiple sources of greatly varying reliability.

Drygalski Ice Tongue

This SGT was first mapped in 1908-1909 [David and Priestley, 1914] when the glacier was much less extensive than at present (Figure 1). Subsequent mapping shows that the early position along the fixed coastline was very accurate and that the limited soundings of water depth were also reliable. In 1908-1909 the glacier had a very distinctive shape (Figure 1), possibly indicating the nature of a previous calving. This calving may have occurred by the mechanism discussed by Holdsworth and Glynn [1978, 1981] who showed that modal shapes can be dominated by longitudinal bending or by transverse twisting. A complete analysis has not been carried out on Drygalski

Ice Tongue (DIT), but by analogy with the EGT, the diagonal line marking the northern edge in 1908-1909 could have been caused by the glacier vibrating in a predominantly twisting mode, evidently not the fundamental one.

The next available map of DIT is from 1960, although unutilized aerial photographs may exist from the 1947 U.S. Operation Hijump. The entirely new shape in 1960 indicates that at least one calving must have taken place between 1909 and 1960. This shape has been maintained according to Anderson and Kurtz [1980], up to the time of a January 1980, hydrographic survey. The 1973 outline is derived from a Landsat image that was registered along the coastline. Using the change in position of the prominent notch on the north edge, an ice flow rate at about 50 km from the coast is found to be 730 m y^{-1} (± 36 m y^{-1}).

The bathymetry is the result of reasonably accurate soundings also made in January 1980, and prior to that time. From these data and estimates of the DIT thickness from freeboard measurements, it is probably free floating beyond the coast. There is no evidence that DIT is grounded, as hypothesized by David and Priestley [1914].

The Ninnis and Mertz Glacier Tongues

Because of their close proximity, these two SGTs are considered together (Figure 2). The glacier outlines are extremely variable over about 20-year intervals which seem to be of the order of the apparent calving period. Satisfactory registration of the coastal outline and latitude or longitude graticule lines could not be achieved for any of the available maps, so the composite figure is of low positional reliability. However, large-scale relative changes of each glacier appear to be real, and could thus be useful for future study.

The bathymetry [Jacobs et al., 1979] shows that both glaciers traverse deep water, suggesting that the greater part of them is afloat. Shoaling toward the terminal face of Mertz Glacier increases the likelihood of possible grounding there, but this is not established. The extremely irregular outline of Ninnis Glacier Tongue around 1912 (the exact date is uncertain) would seem to indicate the existence of local grounding, but there is also the possibility that the outline represents that of a calved section still in contact with the main glacier. There is a lack of detailed bathymetry in this sector.

Chelyuskintsy Ice Tongue

This former promontory of the West Ice Shelf has been studied principally by Soviet researchers who showed it to have calved prior to March 1965. The existence of Pingvin Island at the former northern tip, shoal areas on the bathymetric map, irregular surface elevations and the stalling of icebergs over areas formerly occupied by the Chelyuskintsy ice tongue (CIT) in Figure 3, indicate that it was locally grounded in several places. CIT is thus not suitable for the simplified dynamic analyses of the type considered in this paper. The existence of CIT prior to 1965 may or may not have been dependent on local grounding, but any grounding did not save it from destruction.

Amery Ice Shelf

Prior to February, 1964, Amery Ice Shelf had an extension of SGT proportions beyond the coastline (Figure 4). While the positions of the front before that time are of uncertain accuracy, those of 1963 and 1964 are probably reliable. The outlines for 1936 and 1937 are certainly not compatible, considering the ice flow rates of ~1 km yr^{-1} given by Law [1967] and Budd et al. [1982]. The 1965 position is evidently in error over the southern extremity of the shelf [see Tret'yakov and Mikheyev, 1970]. Swithinbank [1969] has noted that different values have been given for the dimensions of the Amery Ice Shelf iceberg (designated 1967B). Iceberg 1967B retained its gigantic proportions at least as far as 0° longitude, where it apparently struck Trolltunga (which became 1967A) [Swithinbank et al., 1977]. Therefore, the rift shown in Figure 4, if correctly mapped, would seem to have played no role in the calving of 1967B nor in its subsequent stability.

Some oceanographic information north of the Amery Ice Shelf has been reported by Ledenev and Yevdokimov [1966]. Current speeds in this region in the upper 200 m are of order 0.1 m s^{-1}. Iceberg 1967B traveled with average speeds of 0.14 to 0.16 m s^{-1}.

Trolltunga

This prominent former SGT has been referred to as the Greenwich Meridian Glacier, Ice Shelf D [Swithinbank, 1957], Bellingshausen ice shelf [Savatyugin, 1970] and Trolltunga, which shall be used here (Figure 5).

A map showing the outline of Trolltunga in different years from 1938-1939 is given by Kruchinin and Koblents [1963] but the details are unclear, and the relative accuracy of each outline is unknown. This also applies to the positions shown in Figure 5, where the exact dates which apply to the outlines 1 and 2 were not found. For the purpose of examining the dynamics of a possible collision between iceberg 1967B and this SGT, the outline of the Trolltunga iceberg [1967A] has been superimposed on the outline of the SGT position for

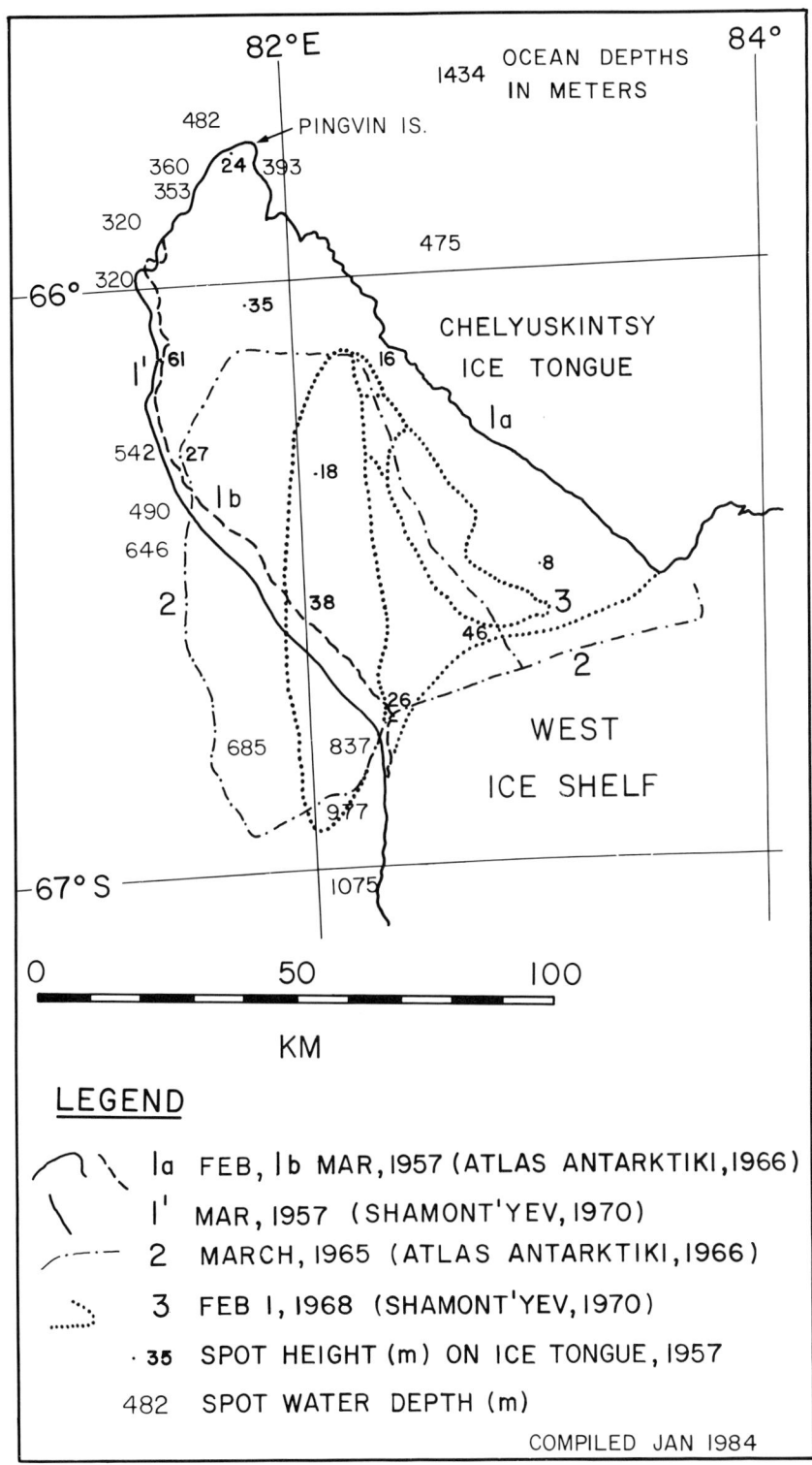

Fig. 3. Chelyuskintsy Ice Tongue, Leopold and Astrid Coast. Outlines of the ice tongue (1a, February 1957; 1b, March 1957), spot heights, and water depths are taken from the Atlas Antarktiki [Tolstikov, 1966] and (1') from Shamont'yev [1970]. The position of the iceberg (2) and the fracture is taken from the Atlas Antarktiki and the outline (3) of the iceberg group is from Shamont'yev [1970].

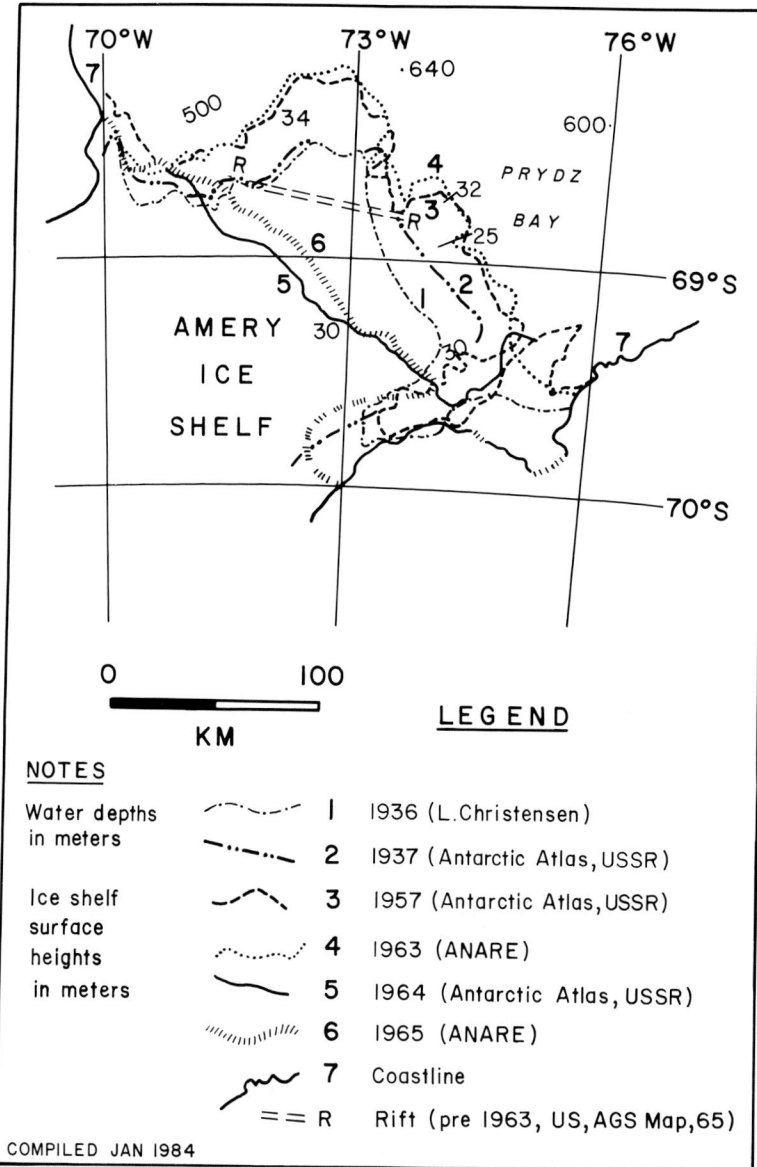

Fig. 4. Amery Ice Shelf. The positions (1), (4), and (6) were taken from Law [1967]. Position (6) appears to be in error at the extreme southeastern end (cf., outline 5). Also Law [1967, p. 439] states that "a major breakout apparently did occur in 1963." Ledenev and Yevdokimov [1966] and the maps in Atlas Antarktiki (outlines 2, 3, 5) indicate that 1964 was the year of this calving. The reason for the large discrepancy between outlines (1) and (2) is unknown. Similarly, the shift between outlines (5) and (6) is far too large to be explained by the (known) ice flow rates there [Budd et al., 1982]. Spot heights are taken from the Atlas Antarktiki and water depths from the AGS [1965] map, from which the position of the rift (R-R) is also taken.

1961 (see figure caption for details). After calving, 1967A spread and suffered minor peripheral attrition, at least until 1973, [McClain, 1978], but it is the relative change in outline that is of interest here.

Trolltunga was surrounded by very deep water according to the Atlas Antarktiki [Tolstikov, 1966]. The GEBCO 5.18 chart [Vanney and Johnson, this volume] shows depth contours of 500-3000 m across the area occupied by Trolltunga. These contours would indicate that the former Trolltunga was everywhere

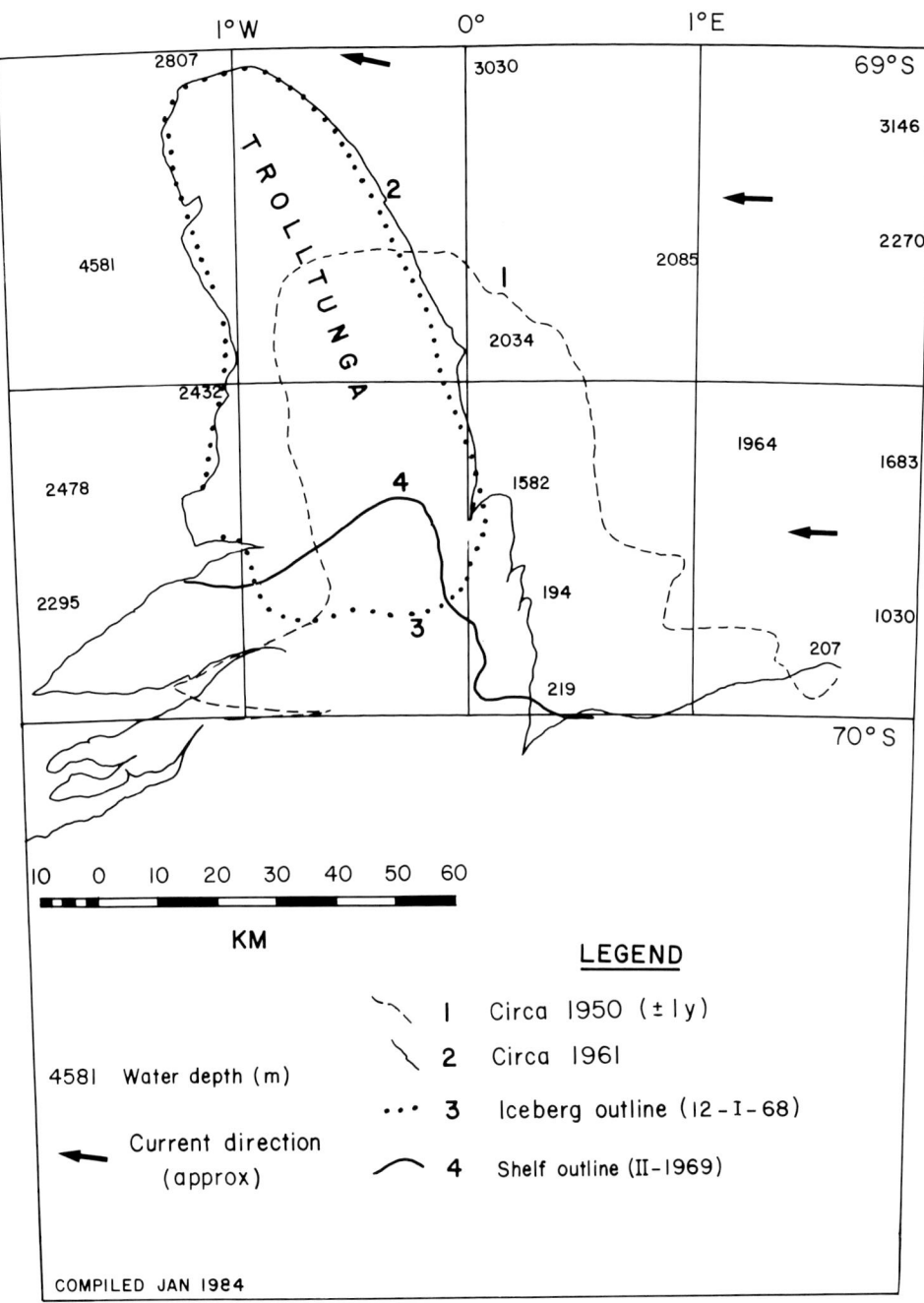

Fig. 5. Trolltunga, Princess Martha Coast. The outlines are derived as follows: (1) from Swithinbank [1957], based on surveys made during 1949-1952. Outline date not specified on the map, but is circa 1950. (2) From Atlas Antarktiki [Tolstikov, 1966], circa 1961. The Trolltunga iceberg outline (3) applies to January 12, 1968, when it was situated near 20°W longitude. It is fitted to the original outline in order to determine if any changes in shape took place as a result of the original calving (ESSA 3, orbit 5869, frame 11). The shelf outline (4) is taken from Savatyugin [1970] and applies to February 1969. Bathymetry is taken from the Atlas Antarktiki, but detailed contours may be obtained from Vanney and Johnson [this volume].

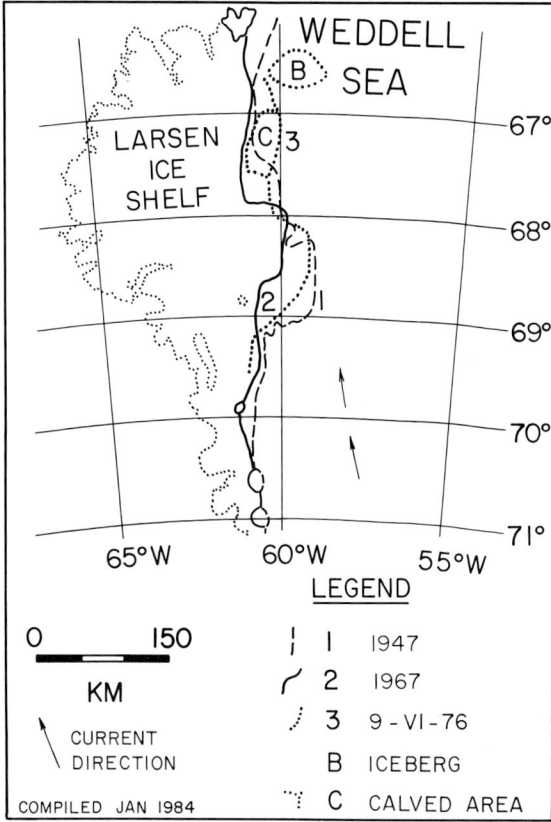

Fig. 6. Larsen Ice Shelf, Antarctic Peninsula. The outline (1) is taken from the AGS [1965] map and applies to 1947 (Operation Hijump). The outline (2) is taken from the AGS [1970a] map and applies to 1967. Outline (3) is from a NOAA-4 IR image of June 9, 1976, and shows the calved section of the shelf (C). The Trolltunga iceberg is to the north (B). No bathymetry is available in this region. Ocean current direction shown is approximate.

afloat since its mean draft was evidently not much greater than about 160 m. Drift speeds of icebergs [Tchernia, 1974; Swithinbank et al., 1977] are of order 0.2 m s^{-1} near Trolltunga, but current speeds against the former SGT may have been quite different (see next section).

Larsen Ice Shelf

This ice shelf has supported an extension 100 km in length, which protrudes out into the Weddell Sea at least 50 km beyond the general edge of the shelf (Figure 6). This feature is shown on the AGS [1965] map of Antarctica based upon 1947 mapping, and on the revised AGS [1970a] map, showing a 1967 position. The shapes are similar for the 2 years but the positions differ significantly. A June 9, 1976 satellite image of Larsen Ice Shelf (LIS) again shows the protuberance, but displaced relative to the previously mapped feature. In March, 1976, or later, iceberg 1967A contacted LIS and this collision may have caused the calving of a section of the shelf [Swithinbank et al., 1977]. Surprisingly, the new iceberg did not derive from the protuberance itself, but from a 30 x 70 km section further north, the removal of which resulted in a corresponding indentation in the body of LIS. Contact between the STI and the LIS extension took place in January 1976, without any noticeable calving.

No hydrographic information is available in this zone according to the GEBCO 5.18 map [Vanney and Johnson, this volume]. From McClain's [1978] data on iceberg drift in 1976-1977, current speeds can be estimated to be of order 0.03 m s^{-1} or less, which may reflect transitory grounding. Currents may be stronger as they are forced to flow around a promontory [Ledenev, 1964]. An apparent northward skewness in the 1947 profile is in the direction of surface currents.

Thwaites Glacier Tongue

According to the AGS [1965] map, the 1961 outline of Thwaites Glacier Tongue (TGT) is that of a lobe about 100 km long and 40 to 80 km wide (Figure 7). This is the shape to be expected from plastic creep spreading of an SGT, where melt rate and other marginal attrition is not significant compared with the spreading rate. It is unknown whether the northern edge of TGT was grounded, as no hydrographic data seem to be available [GEBCO map 5.18, Vanney and Johnson, this volume]. If the former TGT was free floating and the flow rate was known, an order of magnitude estimate could be made for the transverse creep rate. The flow rate of about 2.2 km y^{-1} can be estimated from the displacement of crevasses on aerial photographs but this may have been under surge conditions [T. J. Hughes, personal communication, 1984]. Taking this value as an upper limit, then an upper limit average value for the transverse creep rate over about 40 years, 2.6 x 10^{-2} y^{-1}, implies either very thick ice or an assumed ice speed of surge magnitude. Melting and attrition of the margins were neglected in this calculation, and no relevant water current data were located.

By January 1966, TGT had calved to become Thwaites Iceberg Tongue (TIT), which remained in the vicinity of the parent glacier evidently due to grounding. The shape of TIT bears little resemblance to the 1961 outline of TGT and details of the calving are unknown.

Some Ocean-SGT Interactions

In this section, some results obtained from measurements on EGT [Holdsworth, 1982] will be used to analyze observations made on certain

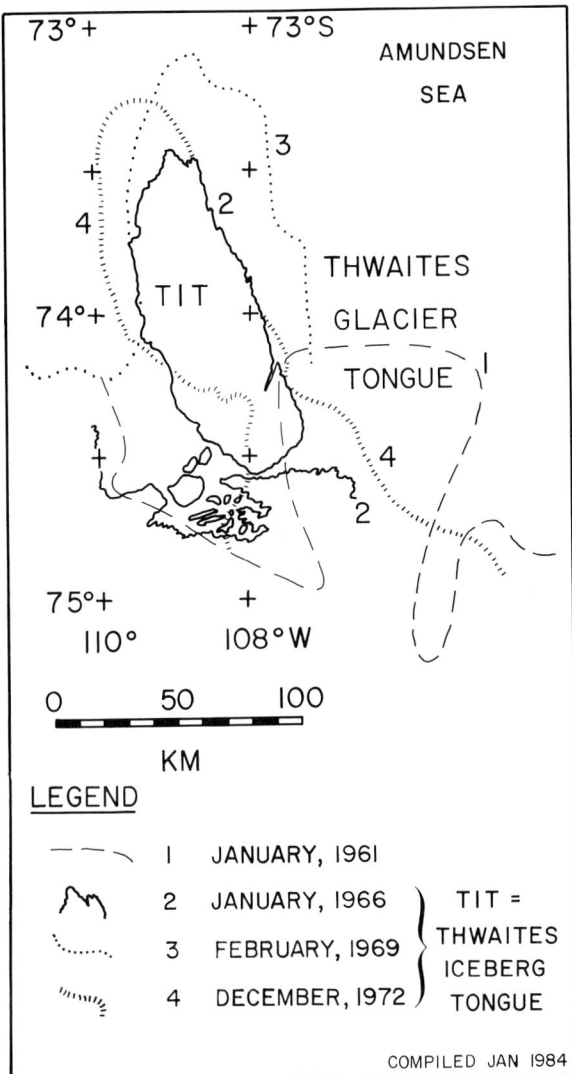

Fig. 7. Thwaites Glacier Tongue, Walgreen coast. Outline (1) is taken from the AGS [1965] map and applies to January 1961. Outline (2) is taken from the AGS [1970a] map. Outline (3) is based on Cosmos-226 data [Savatyugin, 1970] and applies to February 1969, but its reliability is unknown. Outline (4) is based on the U.S. Geological Survey map SS 10-12/12 [1978], and applies to December 1972. No bathymetry is available. Currents flow from east to west [Treshnikov, 1964].

SGTs. These interactions are related to current flow and wave motion. Tidal action will not be covered here, but will be briefly discussed under the section on calving mechanisms.

Current Action on Ice

A water current flowing against an ice surface can cause phase changes and exert pressure to displace the ice. In the first case, melting can contribute importantly in determining the shape of a glacier tongue (GT) both in plan and in vertical section [Holdsworth, 1982]. Irregularities in ice thickness along the length of EGT may play a role in the calving of that glacier. However, for the purposes of this discussion the assumption will be made that melting (or freezing) rates are essentially constant, so that at a given distance from the coastline the thickness of a GT is invariant with time. The shape of a GT in plan is modified by melting, regardless of how a particular geometry evolved, and is taken into account in any mechanical analysis. Thus, only the effects of direct pressure on an ice face will be discussed here.

Holdsworth [1982] showed that current pressure exerted on the edge of EGT is sufficient to explain the observed curvature of that glacier by a process of plastic creep bending. The aspect ratio (length/width) of EGT is approximately 13 and the assumed current speed of 0.5 m s^{-1} appears to be much higher than most other measured or estimated coastal currents. Because such combinations of high aspect ratios and high current speeds are uncommon, this mechanism may have very limited application.

However, some observations made on Trolltunga should be examined for possible current interaction effects. There was an apparent westward displacement of this SGT over about a decade before 1961 (Figure 5). On the basis of maps of Trolltunga since 1938-1939, Kruchinin and Koblents [1963] concluded that the tongue advanced between 1939 and 1955 at least 1.77 km y^{-1}. Ice flow rates would be greater than this if marginal ice was lost by attrition at the edge by collisions or warping [Reeh, 1968]. Assuming a general creep strain rate of about $1.8 \times 10^{-3} \text{ y}^{-1}$, a value applicable to the Maudheim ice shelf of comparable thickness [Swithinbank, 1957], there must be a high initial ice discharge ($\geq 1.6 \text{ km y}^{-1}$) across the root zone of Trolltunga. Kruchinin and Koblents [1963] also noted that the tongue appeared to shift westwards at about 0.8 km y^{-1} between 1939 and 1959 with a somewhat greater shift in the northern part than in the middle and southern parts.

The Kruchinin and Koblents [1963] observations may be used to estimate that it would have taken about 65 years for this SGT to advance 110 km, and that it may not be supported beyond the root zone, but is being deflected by some mechanism in a cumulative way. If this apparent shift is not due to mapping errors, then it could be due to high attrition of ice on the eastern margin. The curvature might also be due to a higher exit speed on the eastern edge, or creep bending due to current pressure from the east, from which direction surface waters are known to be flowing [Treshnikov, 1964, Figure 5]. No exit speed information is available but creep bending may be examined following the analysis carried out on EGT [Holdsworth, 1982]. The problem is to estimate the curvature or cumulative lateral displacement of this SGT as it grows to a length of 110 km while being acted upon by a transverse current. In order to do this, a highly simplified model will be used.

The maximum (anticlockwise) bending stress acting at the root section along the edge of Trolltunga can be specified by equations (5) and (6) of Holdsworth [1982]:

$$\sigma'_{xx}(t) = 1.17\, C_D\, \bar{\rho}\, V_C^2\, W^{-2}\, L^2(t) \qquad (1)$$

where C_D is the form drag coefficient, $\bar{\rho}$ is the mean ice density, V_C is the current speed, W is the width of the glacier at the root section, and $L(t)$ is the length of the glacier as a function of time, the schedule for which is constructed according to

$$t = \sum_{x=0}^{L} [u_o + \dot{\varepsilon}^W_{xx} L]^{-1} \Delta x \qquad (2)$$

where t is the time taken for the ice to travel from the origin ($x=0$) at the root section over a distance L along the centerline, u_o is the exit speed at $x=0$, and $\dot{\varepsilon}^W_{xx}$ the expansive creep rate [Weertman, 1957] of the ice in the x direction, taken here for convenience as constant, but actually varying with x due to thickness changes.

The bending creep rate, actually a deviator value that is positive on the east edge and negative on the west edge, can then be obtained from an assumed flow law in component form

$$\dot{\varepsilon}_{ij} = B^{-3}\tau^2\sigma'_{ij} \qquad (3)$$

where $i,j = x,y$ (x-longitudinal, y-transverse) and $\dot{\varepsilon}_{ij}$ is the creep rate associated with the bending, B is a quantity that depends on material properties of the ice and is strongly temperature dependent, τ is the effective stress defined by

$$2\tau^2 = \sigma'_{ij}\, \sigma'_{ij}$$

where σ'_{ij} is the stress deviator component due only to the bending. Since $\sigma'_{ij} = \sigma'_{ij}(t)$, the corresponding strain rates $\dot{\varepsilon}_{ij}(t)$ from (3) must be used incrementally to compute the creep strains, and hence the elongation or shortening parallel to an edge, given by

$$\Delta L = \dot{\varepsilon}_{xx} \Delta t \cdot \Delta x$$

For simplicity, plane strain conditions in (x,y) were used. In equation (1) the following values were assumed:

$C_D = 1.95 \quad \bar{\rho} = 840\text{ kg m}^{-3} \quad V_c = 0.2\text{ m s}^{-1}$
$W = 3.5 \times 10^4\text{m}$ and $0 < L < 11 \times 10^4\text{m}$

and in equation (2)

$u_o = 1.6 \times 10^3\text{m y}^{-1} \quad \dot{\varepsilon}^W_{xx} = 1.8 \times 10^{-3}\text{y}^{-1}$

The results are that the theoretical curvature due to the anticlockwise bending is insufficient to account for the apparent (observed) westward curvature of Trolltunga. The situation is not significantly improved by increasing V_c to 1 m s^{-1} which may not be unreasonable considering the data of Ledenev [1964] for the Lazarev Ice Shelf. There, a current flowing into an SGT was causing increased flow rates as the current was forced to flow northward around the SGT.

There are two other possibilities for the apparent curvature of Trolltunga. One is differential melting due to current flow against its eastern side. The likelihood of this appears low, since even current speeds as high as 1 m s^{-1} would cause melting of only a few tens of meters per year and an apparent curvature about 10% of that observed. The second possibility is that bombardment by icebergs from the east could cause sections of that edge to calve. C. Swithinbank [personal communication, 1984] noted multiple iceberg collision scars along the edge of the Ronne Ice Shelf front on January 2, 1984. These scars were characterized by a flattening of the cliff with rubble piled up to 15 m above the top surface. The older, outer sections might be expected to be more damaged than the inner sections, resulting in an apparent curvature. The magnitude of possible attrition by this mechanism is unknown. In a later section, collision between the STI and this SGT and bending stresses generated at the root section large enough to cause calving are considered.

Computations of the type outlined above seem to indicate that glacier tongues continuously subjected to lateral ocean current pressure must achieve an aspect ratio of at least 5 and water currents must reach a mean speed of at least 0.5 m s^{-1} before significant, measurable, bending can occur. However, this mechanism might be effective in combination with collisions, and with differential exit speeds on either side of the tongue.

Once calving has occurred, and provided no bottom obstructions interfere with movement of the iceberg, initial acceleration of the mass may be estimated. For the Trolltunga iceberg, neglecting subsequent interaction from the colliding STI, the acceleration would be approximately $0.5\, C_D\, V_C^2\, W^{-1}$ or ~ 10 m s^{-2}, using the values of the constant given above. Assuming a linearly decreasing acceleration it would take about 5 hours before the iceberg reached current speed over a distance of about 40 km. Subsequent collisions with other ice shelf margins and shoals would cause temporary stalling and rotations, resulting in erratic motions. This has perhaps hindered efforts to determine coastal current velocities [Tchernia, 1974] by tracking iceberg drift.

Wave Motion Transmitted into Floating Ice Attached to Land

That thick ice can transmit significant wave energy has been sufficiently demonstrated. Weinman [1958] detected submillimeter oscillations of ~1 min period on the several hundred meter thick Filchner Ice Shelf. Thiel et al. [1960] reported the existence of oscillations with periods ranging from 15 to 50 s on the edge of the Ross Ice Shelf, although the complete wave spectrum was evidently not detected. In winter, the amplitude of the wave motion significantly decreased due to the damping action of the sea ice. On January 4, 1960, C. Swithinbank [personal communication, 1984] measured oscillations of 4 m, 40 s period on the Ross Ice Shelf at 79°00.5'S, 175°45'E, where the ice thickness was about 350 m. The corresponding maximum tilt was estimated to have been 80". According to the wave dispersion curve of Williams and Robinson [1981] the speed of this progressing wave would have been about 50 m s^{-1} at a wavelength of about 14 km. If these figures are of the correct order, then the estimated tilt angle would imply a 1 m amplitude, which appears to be extremely high [cf. Williams and Robinson, 1978]. Holdsworth [1974] measured oscillations of 16s period using a Worden gravity meter at a point of EGT where the ice thickness was over 200 m. Measurements were repeated on EGT during 1977-1978 [Holdsworth and Holdsworth, 1978; Goodman and Holdsworth, 1978] at which time a broad spectrum of oscillations was detected. Significant power occurred at frequencies close to the theoretical free vibration modal harmonic frequencies, suggesting the existence of standing waves or resonance.

Williams and Robinson [1978] reported gravity meter results from the Ross Ice Shelf which indicated that waves with a period of 1-2 min, and amplitudes of up to 8 cm were present where the ice was about 200 m thick. Waves with a period of ~10 min with amplitudes below 2 cm in ≥ 450 m thick ice occurred further to the south. The phase speed of the long, progressive waves was estimated to be close to the theoretical gravity wave speed $\sqrt{gh}$, in water of depth h below the ice shelf [Williams et al., 1978; Williams and Robinson, 1981]. In this case, resonance at the measured frequencies was evidently not taking place. Attenuation of the wave amplitude was estimated to be small so that elastic effects would still appear to predominate. Because there exists a discrete spectrum of waves, the ice at any particular location is undergoing constant loading and unloading at time intervals only a fraction as long as the periods of interest. An expected tendency toward viscous relaxation of stress may thus be quite small.

For unconfined floating glaciers, a simplified numerical model was based on the work of Reeh [1970]. This model [Holdsworth and Glynn, 1978], produced reasonable results for EGT but may not be capable of providing realistic results for very much larger GTs. Since the model constitutive law is purely elastic, for frequencies of oscillation corresponding to periods significantly greater than about 1 min the neglect of a viscous term in the flow law may not be satisfactory. Multiple frequencies were not explicitly considered to occur together, which they do in reality, but the results provide a qualitative picture of the vibrational characteristics of SGTs.

It may be possible to estimate the maximum discrepancy between the results of the purely elastic discrete frequency model and a real glacier by referring to the field results obtained for EGT. There the computed fundamental period of the elastic model at a length of 13 km was about 181 s. Measurements with tilt meters and strain meters on the glacier surface yielded frequency power spectra showing significant peaks at periods of 200 s, 220 s and about 240 s. While these data are not unequivocal, they suggests that the actual fundamental period is above 200 s. Taking the viscous properties of ice into account, an increase in the length of modal periods would be expected. If the maximum increase in the EGT case is taken to be 20-60 s for the fundamental mode, then the percentage error in the model results at 3 min is between 11 and 33%, and could be much less, because there are other influences that might give rise to the discrepancy. Reeh [1970] sought and identified a reasonable value for the fundamental frequency of Jakobshavns Glacier, Greenland. This value was about 6 min, the observed period of seiches in Jakobshavns Fiord. In determining that value, various combinations of unknown glacier length and water depth were used, a procedure which tends to obscure the limitations of the model.

It appears that the period of the fundamental frequency of SGTs such as Drygalski Ice Tongue (Figure 1) is about 10 min according to the elastic model. In reality, the actual value may be as high as 11-14 min. Very long waves could excite this mode. Alternatively, resonance of a section of coastal or shelf water could force oscillations of an SGT at these periods. When the model is run on an SGT (viz., Drygalski Ice Tongue) the existing finite difference grid mesh is too coarse for periods shorter than about 150 s. Until a mesh containing about 100 x 500 nodes (instead of the present 10 x 50) can be accommodated, solutions at high frequencies will not be available. These are the very frequencies which are of interest in terms of storm wave excitation, and for which the elastic model becomes much more realistic. The model shapes are of particular interest because certain glaciers show evidence of a calving mechanism (eg., EGT; Drygalski Ice Tongue, Figure 1; Mertz and Ninnis Glacier Tongues, Figure 2). By matching model antinodal lines along which fracture is assumed to occur with the observed fracture outlines, the approximate period of the oscillation could be determined. With this material as background, it is now feasible to examine the main mechanisms that may be responsible for calving of SGTs.

Major Calving Mechanisms for SGTs

A number of plausible mechanisms have been suggested to account for major ice shelf calvings. The ones that would seem to be applicable to SGT calvings in particular are listed below. Calvings are implicitly connected with extreme events occurring within these mechanisms: (1) tsunami wave interaction [Zumberge and Swithinbank, 1962; Savatyugin, 1974], (2) storm wave interaction (storm surge waves) [Zumberge and Swithinbank 1962; Debenham, 1965], (3) pressure induced long waves with shelf amplification [Maksimov, 1965], (4) collision between an STI and an SGT [Nasta and Nawratil, 1969; Swithinbank et al., 1977], and (5) tidal motion [Zumberge and Swithinbank, 1962].

The references quoted should not be taken to imply that these authors were the first or only workers to suggest these possible iceberg calving mechanisms. Little detailed work has been done to follow up on these hypotheses, perhaps due to the lack or inaccessibility of relevant data.

Sufficient glaciologic and oceanographic data existed for Holdsworth and Glynn [1978] to examine an extension of (2) for EGT. They proposed a vibration calving mechanism in which resonance could be induced in an ice tongue by storm-produced incident wave swell when one of the higher modes of vibration of the tongue is excited by a corresponding component of the wave swell spectrum. Measurements on EGT [Holdsworth and Holdsworth, 1978; Goodman and Holdsworth, 1978] subsequently showed that multiple vibration modes can be excited by the incident ocean wave spectrum. All five mechanisms that have been proposed to explain major calving events will be examined below in the light of available data.

Tsunami Wave Interaction

The propagation of long wavelength, high energy, "tsunami" waves over large distances is well documented. The investigation of Braddock [1970] is of particular interest here because he examined computed tsunami ray paths penetrating into the South Pacific from the great Alaskan earthquake/tsunami of March 28, 1964. Passage of the tsunami wave was recorded at Macquarie Island (55°S; 160°E) and it was assumed that significant energy was transmitted into Antarctic waters.

Examination of the McMurdo Sound tide gauge records did not appear to show any evidence indicating the arrival of a tsunami there at a time consistent with a propagation speed of $C = \sqrt{gh}$, where g is the acceleration of gravity and h is the (variable) water depth along the ray path. Braddock attributed the lack of a tsunami signal there to attenuation of the energy in the pack ice. Whether the tide records were taken at McMurdo Station or Scott Base, the McMurdo Ice Shelf and the physiography of the harbor would also tend to prevent significant wave energy from entering the vicinity. In March the position of the outer edge of the pack ice is quite variable and considerable sections of the coastline may be ice free at that time. (see Atlas Antarktiki, Tolstikov, [1966]; AGS [1970a]). Data in Simpson [1923] indicate the occurrence of intense storms throughout March in the McMurdo Sound region, which could move new sea ice to the north, exposing large areas of open sea in the sound. Long waves may thus penetrate to parts of the coast as late as March of a particular year.

If the wave energy arriving at the coast is sufficiently high, how is the energy transferred to an SGT in order to cause it to fracture and calve? A tsunami wave train may contain waves with periods between several minutes and several tens of minutes. Open ocean tsunami information is difficult to obtain, and at coastal locations the oscillation periods and the amplitude spectra seem to vary considerably. However, the power spectra are similar at one location for different tsunamis, indicating the influence of local resonance [Wiegel, 1964]. If most of the energy of a tsunami is contained in one wave then it might be considered as a solitary wave. Upon meeting a coastline, local hydrography would determine the local wave power spectrum which would be the forcing function applied to an SGT. Local coastal resonant periods are frequently several minutes and greater.

Since the fundamental frequency of free vibration of many SGTs corresponds to periods of this order, tsunami generated waves could well be capable of causing resonance in SGTs. This is the one frequency where the bending stresses are always greatest at the location of complete or partial support, the floatation or hingeline. Sustained resonance could lead to fatigue failure of the ice by cyclic stress reversals [Holdsworth and Glynn, 1981]. Seiche activity or resonance has been observed in water near the Lazarev Ice Shelf [Zakharov, 1967], although the period (35 ± 9 min) there was much longer than the fundamental period of the shelf. The seiche amplitude was 15 cm and no calving activity was associated with the measurements.

The periods of the fundamental frequency for two SGTs (Figures 1 and 5) were estimated numerically employing the model of Holdsworth and Glynn [1981] and using an idealized ice thickness and ocean-bed geometry. Values for the period of the first mode, although ambiguous, were near or above 10 minutes. It thus seems possible that tsunami-wave generated coastal oscillations could excite the fundamental mode of an SGT, apparently necessary for calving of the complete SGT in one piece, of which there were several occurrences in the mid-1960s.

Amery Ice Shelf protruded well beyond the general coastal line of support in 1963 (Figure 4). According to Ledenev and Yevdokimov [1966], calving of the Amery Ice Shelf occurred between February and December 1964, more likely early in the year. Fissures were noted on the ice surface in February, so some dynamic activity had possibly taken place by that time, unless the fissures were related to the rift shown in Figure 1. The tongue had disappeared by December, 1964 (Figure 4). Whether the fissures were related to the main calving is not known, but the prominent rift marked R-R in Figure 4, did not coincide with the fracture line.

Could this calving have been related to the great Alaskan earthquake/tsunami of March 28, 1964? Were there any other calving events in 1964? Ledenev and Yevdokimov [1966] provide two other examples of major calvings prior to March 1965: the huge Chelyuskintsy ice tongue (part of the West Ice Shelf) and the much smaller Polar Record Glacier, also between 70°E and 85°E longitude. Chelyuskintsy ice tongue was previously surveyed in 1957, but it is implied that these two calvings took place not long before March 1965, since Soviet vessels had occupied these waters in previous years making regular observations of iceberg frequency [Ledenev, 1961; Ledenev and Yevdokimov, 1966]. There is some evidence that the former Chelyuskintsy ice tongue (Figure 3) was partly grounded in some areas, a boundary condition that would preclude modeling by simple methods. Still, the resonance of the ice/water system remains a viable calving mechanism, and the shallower the water layer the lower the frequency of oscillation and the more efficient the energy transfer from the water to the ice.

The above mentioned SGTs are not situated in a particularly favorable location with respect to Pacific tsunamis, but it should be noted that the largest SGT in the 60°W to 140°E sector [Thwaites Glacier Tongue, Figure 7] apparently calved between January 1961 and January 1966. No calvings can be definitely associated with the large 1960 Chilean earthquake/tsunami. The seismically active circum-Antarctic plate boundary is another source of local tsunamis, the activity of which may not have been completely documented. Seismic activity and tsunami activity may have been at a maximum during the mid-1960s [Savatyugin, 1974].

Storm Wave Interaction

Short (≤ 1 min) period wave motion appears in some cases to be capable of inducing significant vertical deflections in glacier tongues, which may then calve as a result of fatigue failure. This was postulated for the EGT, which has a calving history that may be explained in terms of resonant excitation to a characteristic frequency [Holdsworth and Glynn, 1981]. To generate high-bending stresses at points not on the floatation line, higher modes of vibration must be excited. Since EGT selectively resonates at multiple harmonic frequencies, as observed in power spectra of surface strain [Goodman and Holdsworth, 1978] and tilt [Holdsworth and Holdsworth, 1978], there is a good case for expecting significant increases in stresses beyond those computed for individual modes.

Observations during the 1910-1913 British Antarctic Expedition established that about 5 km of the EGT broke off during an intense storm in March 1911 [Debenham, 1965]. The meteorological records for Cape Evans [Simpson, 1923] show that intense storms from the southeast or east-southeast occurred on March 4 (maximum wind speed 52 mph (23 m s^{-1}), daily mean 48 mph (21 m s^{-1})) March 12-13 (maximum wind speed 55 mph (25 m s^{-1}), daily mean 35 mph (16 m s^{-1})) and March 21-22 (maximum wind speed 57 mph (26 m s^{-1}), daily mean 42 mph (19 m s^{-1})). Conditions during any one of these approximately 24 hr storm periods would have been capable of generating significant local high amplitude waves in the then ice-free Erebus Bay.

Knowing the fetch, windspeed, and duration of the storm, it is possible to estimate both the dominant period and the amplitude of the generated waves from empirical graphs [Wiegel, 1964]. For a southeast wind, the fetch is about 5 nautical miles (9 km). For fetch controlled waves, the significant wave period would be about 4 s and the amplitude 2 m at EGT. For wind duration controlled waves, the significant wave period is 18 ± 3 s and the significant amplitude 10 ± 1 m. This latter value agrees with an estimate obtained from the empirical formula of Scott [1965]. If the actual significant period lay between these values, EGT could have been induced to oscillate at one of its higher natural frequencies, at which significant bending of the outer half of the tongue would predominate.

Transfer functions relating wave height to glacier tongue deflection amplitudes are not available. However, it appears possible to apply an approximate equation derived by Williams and Robinson [1981] relating the amplitude of the ocean wave to that of the wave in the ice

$$\eta_i/\eta_w \approx \frac{k_i^2 \sinh K_w h_i}{k_w^2 \sinh K_w h_w} \quad (4)$$

where
η_i is the wave amplitude in the ice;
η_w is the wave amplitude in the ocean;
k_i is the wave number of the wave in the ice;
k_w is the wave number of the ocean wave;
h_i is the depth of water beneath the ice;
h_w is the depth of the free ocean water.

Substituting in equation (4) values for the above parameters for a wave period of 16 ± 1 s and mean water depths: $\eta_i/\eta_w \sim 0.06$. The actual ratio is likely to be less than this because of wave diffraction effects along the sides of EGT. There would also be interference caused by diffraction from islands and from damping and filtering of ocean waves at varying distances to sea ice surrounding the glacier. In addition, Holdsworth and Holdsworth [1978] estimated that vertical deflections of the EGT surface in 1978 were of order 1 cm under normal sea state conditions. If wave amplitudes were then of order 0.5 m, when transmission of wave energy was through sea ice surrounding EGT, it is evident that vertical deflections of EGT may have had amplitudes of the order of several centimeters during the 1911 storm. For sustained cyclic stresses at these amplitudes over periods of many hours, fatigue failure of the ice could have occurred.

A calving event might also be associated with multiple mechanisms whose effects are additive. Wave and current pressure on EGT could have increased the horizontal clockwise bending moment acting on the glacier. This moment is a maximum at the floatation line, but corresponding stresses may reach local maxima elsewhere due to significant reductions in glacier width and thickness.

In order to determine whether the vibration mechanism could be applied to SGTs, the program developed for EGT [Holdsworth and Glynn, 1978] was used to analyze the modal character of two SGTs: Drygalski Ice Tongue (Figure 1) and Trolltunga (Figure 5). In the first example a reasonable set of eigen solutions was found between 150 s and 800 s. If account is taken of viscous properties, these periods would increase. Nevertheless, the obtained period values (2-13 min) are still thought to be useful. Several transverse twisting modes were found to exist, corresponding to periods of several minutes. Calving by vibration in one of these modes could explain the shape of the 1908-1909 glacier outline (Figure 1).

In the second example, eigen solutions were found lying in the same general range as in the first example, but, as before, no unambiguous fundamental mode was identified. It would appear that, especially for the shorter period oscillations, the SGTs will have to be analyzed using finer grid mesh than was used for the much smaller EGT. For this to be successful, a computer of large capacity will be needed. For harmonics of period ≤ 1 min, the SGTs would have very complicated modal shapes with the likelihood of multiple bending stress maxima reaching significant levels. If calving did occur under these conditions, it might be expected that the SGT would fracture into numerous ($10-10^2$) pieces. The Dibble (65°-66°S; 135°E) and the Dalton (65°30'-122°E) STIs are shown on the AGS [1965, 1970a] and the Atlas Antarktiki [Tolstikov, 1966] maps. On the latter chart, these STIs are shown explicitly as being comprised of individual pieces. These features may be grounded or copied onto the later charts from their originally mapped positions. Satellite imagery might be useful in resolving this question.

Pressure Induced Long Waves
With Shelf Amplification

Maksimov [1965] advanced the idea that large scale deep cyclonic depressions off the coast of Antarctica could produce traveling gravity waves which could attain significant amplitudes upon entering shallower waters. In this respect, the effects on the SGTs would be the same as for tsunami waves. A solitary wave might be formed by the sudden dispersal of a cyclone in the ocean. Maksimov presents an approximate equation for two-dimensional wave motion and data to demonstrate that an initial wave amplitude could be magnified 6 times to finally exceed 2 m at the coast. If the three-dimensional motion of the wave is considered, a significantly lower wave amplitude would result. If waves of this type are significant, and conditions for their generation are "frequent", as Maksimov states, then they should have been easily identified on the tide gauge records. This does not appear to be the case. These long waves would not be heavily attenuated by the pack ice [Wadhams, 1973], so identification at the coast should be possible.

Collision Between an STI and an SGT

This mechanism was suggested after satellite images revealed that the prominent Trolltunga SGT (Figure 5) had broken off at a time (mid-1967) when the main part of the Amery iceberg [Swithinbank et al. 1977] had drifted into the vicinity of Trolltunga. Later both STIs were seen together. This was strong circumstantial evidence for a collision calving mechanism, which was evidently first suggested by M.H. Fleming [Nasta and Nawratil, 1969]. In tracking the drift of the former Trolltunga, Swithinbank et al. [1977] found circumstances to indicate that it collided with part of the Larsen Ice Shelf (Figure 6), causing the calving of part of that shelf in 1976. Rather than hitting the prominent Larsen SGT (68°30'S; 68°W), a section to the north was impacted, causing fracture lines about 35 and 90 km long into the shelf [Swithinbank et al., 1977]. Later measurements by McClain [1985] showed the STI to be about 30 x 70 km. Thus, there seems to be evidence for two types of collision calving:

1. Lateral impact of a drifting STI with an SGT could cause a horizontal bending stress to be generated at the root of the SGT. In addition, an impact could have induced vibrations in the SGT at the fundamental frequency, in which case the bending stresses would again be a maximum at the root of the SGT, but this time in the vertical plane.

2. Impact of a drifting STI with the edge of an essentially confined ice shelf could induce bending stresses in the shelf. Satellite imagery for 1976 showed STI 1967A to be rapidly rotating, so a current gyre may have caused one end of it to impinge normally or obliquely into the shelf edge. Shearing may have dissipated some of the energy but the remainder could have caused that part of the shelf to vibrate at a very low frequency, giving rise to antinodal lines at very large spacings as reported by Swithinbank et al. [1977].

Referring to Fig. 8, a simplified model may be set up for the first case, whereby a moving plate (the Amery STI or iceberg 1967B) impinges on a stationary plate (Trolltunga). The following treatment ignores the existence of stresses in the stationary plate due to plastic creep spreading [Weertman, 1957] and creep bending due to current pressure [Holdsworth, 1982]. Because there are no data at the time of the assumed collision, some limiting assumptions must be made:

1. The size of the STI from the Amery Ice Shelf was estimated by Swithinbank et al. [1977] to be 110 x 75 km in October 1967, and the thickness, measured during the 1969-1970 field season [Smith, 1972] to be between 200 and 250 m. Therefore, at the time of the assumed collision between May and September 1967, the colliding STI had a volume of at least 1.9×10^{12} m^3. With an average density of 0.846 Mg m^{-3} [Budd, 1966], 1967B had a mass of at least 1.61×10^{12} Mg. Near impact, the average speed of 1967B would have been about 10 nautical miles per day

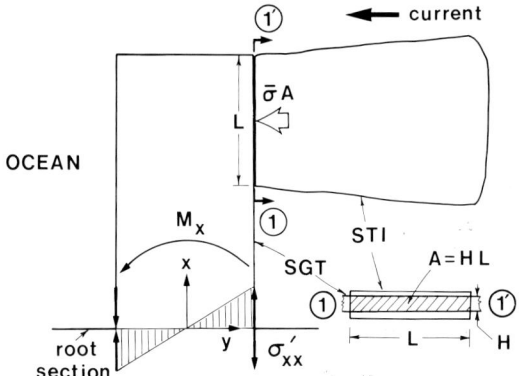

Fig. 8. Model of glacier tongue and colliding iceberg, at the instant of impact. The force applied by the iceberg to the glacier tongue is $\bar{\sigma}A$ where $\bar{\sigma}$ is the mean contact stress and A is the contact area. The bending moment M_x at the root section gives rise to the bending stress $\bar{\sigma}'_{xx}$ as shown.

[Tchernia, 1974]. At impact, therefore, the speed could have been at least 0.21 m s^{-1}. The kinetic energy (E_K) of 1967B could therefore have been at least 35×10^{12} joules.

2. If, at impact, kinetic energy was converted into work done on Trolltunga, then

$$E_K = \bar{\sigma} A X \quad (5)$$

where $\bar{\sigma}$ is the mean stress acting through the cross sectional area of contact A between 1967B and Trolltunga, and X is the distance moved by crushing normal to the plane of A. Suppose, for simplicity, that a complete edge of 1967B contacted Trolltunga. For hydrodynamic reasons, it could be supposed that there was a better than 50% chance that this was the short edge, provided the STI was everywhere free floating at the time. If the end widths were between 50 and 80 km (Figure 4) and the contact depth as the lesser of the two thicknesses (180 m for Trolltunga, Savatyugin [1970]) then A would fall between 9×10^6 and 14×10^6 m^2, say 10^7 m^2.

3. Next, there is the question of the penetration of 1967B into Trolltunga by crushing and possibly by bending failure. Initially, if $A \ll 10^7$ m^2, the value would rise on penetration, which would cause significant variations in the stresses acting on the edge of Trolltunga. An attempt was made to superimpose the image of the Trolltunga iceberg (1967A of Swithinbank et al. [1977]) over its original outline (Figure 5) to determine whether significant damage had been done to Trolltunga by the collision. The satellite image resolution of this era [McClain, 1978] and the quality of the 1966 outline make such an estimate tenuous, but an upper limit of 4 km for the mean penetration distance can be

established. Adopting the above model, where it is implied that $\bar{\sigma} = \sigma_c$, the crushing strength of shelf ice, and assuming that all kinetic energy absorbed in the collision, i.e., 1967B was stopped by the impact, then

$$X = E_K/A\sigma_c \qquad (6)$$

The crushing strength of shelf ice is expected to be much less that of sea ice because of its lower density. Thus, it is estimated that $\sigma_c \leq 1.3$ MPa [Michel, 1970; Croasdale, 1982]. Again using order estimates only, a value of $\sigma_c \sim 1$ MPa yields $X \sim 4$ m, two orders of magnitude less than the probable resolution of the shelf contact outline. If A were only 10^6 m^2, corresponding to a contact length of 6 km, then X would be 35 m, still small compared to any other dimension. Regardless of the values of A or X, $\bar{\sigma} = \sigma_c$, so the average force acting on Trolltunga would be $\bar{\sigma}A$, or about 10^7 MN.

4. If it were known where the contact occurred, then the impact-generated bending moment at the root section of Trolltunga could be computed (Figure 8). If contact were along the outer 60 km of Trolltunga then the lever arm would have been at least 84 km, assuming a length from the tip to the fracture line of at least 104 km [Swithinbank et al. 1977]. If the inner 60 km of shelf were contacted, the lever arm would have been only 30 km. If only one tenth of the contacting edge of 1967B touched the outer part of Trolltunga, then the lever arm would have been about 101 km. These figures may be used to establish approximate limits on the root bending moment, M_x: $10^{11} < M_x < 8 \times 10^{11}$ MNm.

5. The elastic bending stress, $\sigma^e_{xx})_b$, for the root zone may now be computed by the equation for elastic bending. Since the failure was probably catastrophic, an elastic constitutive law would seem to be appropriate. The beam equation gives

$$\sigma^e_{xx})_b = 6 M'_x W^{-2} \qquad (7)$$

where M'_x is the moment per unit ice thickness generated by the collision and W is the width of the plate at the root section [Holdsworth, 1982]. From Figure 5, W is approximately 35 km, so from (7), $\sigma^e_{xx})_b$ takes values from 2.7 to 21.8 MPa.

These values certainly exceed the bending strength of pure ice, which is only of order 1 MPa [Lavrov, 1969]. $\sigma^e_{xx})_b$ would not need to reach even this value due to the likely existence of a plastic creep bending stress, σ'_{xx}, discussed earlier (equation 1), in addition to the ubiquitous plastic spreading creep stress.

From the above, certain implications are possible:

1. The collision did not completely stop 1967B. A contact pressure of only 5 to 37% of σ_c would have been necessary to produce a bending stress $\sigma^e_{xx})_b \sim 1$ MPa and a probable fracture at the root. Alternatively, the area of contact was only 5 to 37% of the initially assumed area (10^7 m^2) if the crushing stress was reached. Propagation of the crack at the root would have proceeded according to a stress, similar to (7), expressed in the form:

$$\sigma^e_{xx})_b \propto M'_x(t)/W^2(t), \qquad (8)$$

where $\sigma^e_{xx})_b$ acts normal to the crack length. If contact was maintained after the initial fracture, $M'_x(t)$ is the transient bending moment and $W(t)$ is the effective width of the SGT at the root. W decreases with time, so even if M'_x decreases with time, $\sigma_{xx})_b$ may be kept at a level sufficient to maintain crack propagation across the width.

2. The collision temporarily stopped 1967B, until fracture of Trolltunga was complete. This implies that the kinetic energy, E_K, of 1967B may have been less than estimated, that is, the speed was substantially less than 0.2 m s^{-1}. This would reduce the stress, according to an impulse-momentum equation

$$\bar{\sigma} = mA^{-1} \, dv/dt \qquad (9)$$

where m is the mass of 1967B and dv/dt is the instantaneous deceleration. The speed v is reduced in this case to zero in time dt, which is unknown. However, a crude estimate for this time may be made. If $\bar{\sigma}$ was 50% of the crushing strength, or 0.50 MPa (to ensure a value of $\sigma^e_{xx})_b$ of 1 MPa was reached), and

$$dv = 0.1 \text{ m s}^{-1} \quad (m = 1.61 \times 10^{12} \text{Mg};$$

$$A = 10^7 \text{ m}^2)$$

then from equation (9), dt is of order half a minute. A further time would be required before the crack propagated across the width. The bending stress during this period would follow an equation of the type (8). Continued current drag pressure along the eastern edge of the ice masses would maintain some value of M'_x, as W decreased. Stress concentration effects at the crack tip would further aid crack propagation. It is not possible to quantitatively examine here the true nature of the failure at the root section of Trolltunga, but a vertical buckling instability could have developed along the potential failure line in analogy to the prefailure buckling of an overloaded slender beam.

Tidal Motion

The possibility of tidally induced calving of SGTs might be considered in terms of extreme tidal ranges generated, e.g., by transient tidal resonance. Tidally produced "hingeline" cracks may be found at the junction of most ice shelves with land [Zumberge and Swithinbank, 1962] but the continuous transport of ice across the floatation zone ensures that cyclic, stress-induced fatigue failure, if at all possible, does not occur there. This is because the bending stresses decay very rapidly away from the grounding line [Holdsworth, 1977]. However, surface crevasses and sets of basal crevasses [Jezek and Bentley, 1983] may act as lines of weakness which might be significant in terms of other mechanisms of calving [Holdsworth, 1977].

If one or more of the other calving mechanisms was operating with its maximum effect focused at the floatation line, then the existence of a concurrent, tidally induced transient stress could increase the effective stress and strain rates. Thus, tidally induced stresses appear to be of minor importance, but may indirectly be a contributing factor to a particular calving event.

Conclusion

A survey of past and present super glacier tongues (SGTs) indicates that many of them have shown very dynamic behavior over the last several decades. In examining the possible effects of currents and wave motion on these SGTs, it was found that these oceanographic parameters probably exert a significant influence on their stability. Investigations of this kind are limited by the availability of oceanographic and glaciologic data and the reliability of topographic maps. The use of satellite imagery to plan the surveillance of a particular SGT, aided by ground control, would benefit from repetitive SAR SEASAT type imagery, with a potential resolution of 25 m [E. P. McClain, 1985]. The Drygalski Ice Tongue (Figure 1) would be a logical SGT for future study since it is near the regular air and sea traffic lanes, and ice thickness data is available [D. Drewry, unpublished data, 1984], along with oceanographic and meteorological observations [Jacobs and Haines, 1982; Jacobs et al., this volume; Kurtz and Bromwich, this volume].

Acknowledgments. I thank E. Paul McClain (NOAA) for providing satellite images of the Weddell Sea region, and Clifton Fry (USGS) for providing topographic maps of certain coastal areas. Bruce Smith helped with some of the numerical model simulations. Movement data for Thwaites Glacier Tongue was provided by R. Allen (USGS). Discussion with M. Metge helped clarify the section on STI-SGT collisions. T.J. Hughes, S. S. Jacobs, D. H. Lennox, and C. W. M. Swithinbank significantly improved the manuscript. This paper is published with the permission of the Director, National Hydrology Research Institute, Ottawa.

References

AGS, American Geographical Society, Map of Antarctica, scale 1:5,000,000, 1965.

AGS, American Geographical Society, Map of Antarctica, scale 1:5,000,000, 1970a.

AGS, American Geographical Society, Antarctic Map Folio Series, 1970b.

Anderson, J.B., and D.D. Kurtz, USCGC Glacier Deep Freeze 80, Antarct. J. U.S., 15, 114-117, 1980.

Anderson, J.B., K. Balshaw, E. Domack, D. Kurtz, R. Milam, and R. Wright, Geological survey of east antarctic continental margin aboard USCGC Glacier, Antarct. J. U.S., 14, 142-144, 1979.

Braddock, R.D., Tsunami propagation over large distances, in Proc. of the International Symposium Tsunamis in the Pacific Ocean, pp. 285-303, East-West Center Press, Honolulu, Hawaii, 1970.

Budd, W.F., The dynamics of the Amery Ice Shelf, J. Glaciol., 6, 335-358, 1966.

Budd, W.F., M.J. Corry, and T.H. Jacka, Results from the Amery Ice Shelf Project, Ann. Glaciol., 3, 36-41, 1982.

Croasdale, K.R., Some implications of ice ridges and rubble fields on the design of Arctic offshore structures, in Proc. of workshop on sea ice ridging and pile up, Tech. Memo., 134, edited by R.M.W. Frederking and G.R. Pilkington, Natl. Res. Counc., pp. 157-180, 1982.

David, T.W.E., and R.E. Priestly, Geology I, Glaciology, Physiology, Stratigraphy, and Tectonic Geology of South Victoria Land, Br. Antarct. Exped. 1907-1909, pp. 48-63, W. Heinemann, London, 1914.

Debenham, F., The glacier tongues of McMurdo Sound, Geogr. J., 131, 369-371, 1965.

Goodman, D.J., and R. Holdsworth, Continuous surface strain measurements on sea ice and on Erebus Glacier Tongue, McMurdo Sound, Antarctica, Antarct. J. U.S., 13, 67-70, 1978.

Holdsworth, G., Erebus Glacier Tongue, McMurdo Sound, Antarctica, J. Glaciol., 13, 27-35, 1974.

Holdsworth, G., Tidal interaction with ice shelves, Ann. Geophys., 33, 133-146, 1977.

Holdsworth, G., Dynamics of Erebus Glacier Tongue, Ann. Glaciol., 3, 131-137, 1982.

Holdsworth, G. and J.E. Glynn, Iceberg calving from floating glaciers by a vibration mechanism, Nature, 274, 464-468, 1978.

Holdsworth, G. and J.E. Glynn, A mechanism for the formation of large icebergs, J. Geophys. Res., 86, C4, 3210-3222, 1981.

Holdsworth, G. and R. Holdsworth, Erebus Glacier Tongue movement, Antarct. J. U.S., 13, 61-63, 1978.

Hughes, T.J., West Antarctic ice streams, Rev. Geophys. Space Phys., 15, 1-46, 1977.

Jacobs, S.S., and W.E. Haines, Ross Ice Shelf Project, Oceanographic Data 1976-1979, Rep. LDGO-82-1, 505 pp., Lamont-Doherty Geol. Observ., Palisades, N.Y., 1982.

Jacobs, S.S., J.J. Szelag, S.M. Patla and P. Bruchhausen, Oceanographic observations near the Antarctic continental Margin and Ross Ice Shelf, Antarct. J. U.S., 14, 116-117, 1979.

Jacobs, S.S., R.G. Fairbanks, and Y. Horibe, Origin and evolution of water masses near the Antarctic continental margin: Evidence from $H_2^{18}O/H_2^{16}O$ ratios in seawater, this volume.

Jezek, K.C., and C.R. Bentley, Field studies of bottom crevasses in the Ross Ice Shelf, Antarctica, J. Glaciol., 29, 118-126, 1983.

Kruchinin, Yu. A., and Ya. P. Koblents, Dynamics of the Trolltunga Ice Shelf, Sov. Antarct. Exped. Info. Bull., Engl. Transl., 5, 58-60, 1963.

Kurtz, D.D., and D.H. Bromwich, A recurring, atmospherically forced polynya in Terra Nova Bay, this volume.

Lavrov, V.V., Deformation and strength of ice, Leningrad, Engl. Transl., Israel Program for Scientific Translations, Jerusalem, 1969.

Law, P., Movement of the Amery Ice Shelf, Polar Rec., 13, 439-441, 1967.

Ledenev, V.G., Study of iceberg discharge from the coast of Antarctica, Sov. Antarct. Exped. Info. Bull. Engl. Transl., 4, 146-147, 1961.

Ledenev, V.G., Currents in the Lazarev Ice Shelf Region, Sov. Antarct. Exped. Info. Bull. Engl. Transl., 5, 95-98, 1964.

Ledenev, V.G. and A.P. Yevdokimov, Changes in the West and Amery Ice Shelves, Sov. Antarct. Exped. Info. Bull. Engl. Transl., 6, 4-7, 1966.

Maksimov, I.V., Causes of sudden break of Antarctic Ice Shelves, Sov. Antarct. Exped. Info. Bull. Engl. Transl., 5, 423-424, 1965.

McClain, E.P., Eleven year chronicle of one of the world's most gigantic icebergs, Mariners Weather Log, 22, 328-333, 1978.

McClain, E.P., Giant tabular icebergs, in Satellite Image Atlas of Glaciers, U.S. Geol. Surv. Prof. Pap., edited by R.S. Williams and J. Ferrigno, in press, 1985.

Michel, B., Ice pressure on engineering structures, U.S. Army Cold Reg. Res. Eng. Lab Monogr., III-B1b, 1970.

Nasta, R. and Nawratil, R., Giant icebergs in the Weddell Sea, Mon. Weather Rev., 98, 774-775, 1969.

Reeh, N., On the calving of ice from floating glaciers and ice shelves, J. Glaciol., 7, 215-232, 1968.

Reeh, N., Natural frequency of the system of a heavy elastic plate covering shallow water, Bygning. Medd., 41, 1967-1987, 1970.

Savatyugin, L.M., Change in the configuration of the front of the Bellingshausen, Thwaites and Lazarev Ice Shelves (according to Cosmos-226 data), Sov. Antarct. Exped. Info. Bull. Engl. Transl., 7, 28-29, 1970.

Savatyugin, L.M., Possibility of tsunamis in the southern ocean, Sov. Antarct. Exped. Info. Bull. Engl. Transl., 8, 621-622, 1974.

Scott, J.R., A sea spectrum for model tests and long-term ship prediction, J. Ship Res., 9, 145-152, 1965.

Shamont'yev, V.A., Change in the shape of the West Ice Shelf, Sov. Antarct. Exped. Info. Bull. Engl. Transl., 7, 575-576, 1970.

Simpson, G.C., Meteorology, III, Br. Antarct. Exped. 1910-1913, London, 1923.

Smith, B.M.E., Airborne radio echo sounding of glaciers in the Antarctic Peninsula, Br. Antarct. Sur. Sci. Rep., 72, 1972.

Swithinbank, C., Glaciology I, Norwegian British Swedish Antarct. Exped. 1949-1952, Scientific Results, III, Norsk Polarinstitutt, Oslo, 1957.

Swithinbank, C., Giant icebergs in the Weddell Sea, Polar Rec., 14, 477-478, 1969.

Swithinbank, C., P. McClain, and P. Little, Drift tracks of Antarctic icebergs, Polar Rec., 18, 495-501, 1977.

Tchernia, M.P., Etude de la derive antarctique Est-Ouest au moyen d'icebergs suivis par le satellite Eole, C. R. Hebd. Seances Acad. Sci., Ser. B, 278, 667-670, 1974.

Thiel, E., A.P. Crary, R.A. Haubrich, and J.C. Behrendt, Gravimetric determination of ocean tide, Weddell and Ross Seas, Antarctica, J. Geophys. Res., 65, 629-636, 1960.

Tolstikov, Ye. I. (Ed.), Atlas Antarktiki Moscow, 1966.

Treshnikov, A.F., Surface water circulation in the Antarctic Ocean, Sov. Antarct. Exped. Info. Bull. Engl. Transl., 5, 81-83, 1964.

Tret'yakov, N.F. and O.F. Mikheyev, Change in the front of ice shelves of east Antarctica, Sov. Antarct. Exped. Info. Bull. Engl. Transl., 8, 30-31, 1970.

U.S. Geol. Surv., Topographical Map, Relief Inlet, SS58-60/13, 1:250,000, 1968.

U.S. Geol. Surv., Topographical Map, Thwaites Glacier Tongue, SS10-12/12, 1:250,000, 1978.

Vanney, J.R., and G.L. Johnson, GEBCO Bathymetric Sheet 5.18 (Circum-Antarctic), this volume.

Wadhams, P., Attenuation of swell by sea ice, J. Geophys. Res., 78, 3552-3563, 1973.

Weeks, W.F., and W.J. Campbell, Icebergs as a fresh water source: An appraisal, J. Glaciol., 12, 207-233, 1973.

Weertman, J., Deformation of floating ice shelves, J. Glaciol., 3, 38-42, 1957.

Weinman, J.A., Ice shelf oscillations, J. Glaciol., 3, 187, 1958.

Wiegel, R.L., *Oceanographical Engineering*, Prentice-Hall, Englewood Cliffs, N.J., 532 p., 1964.

Williams, R.T. and E.S. Robinson, The ocean tide and waves beneath the Ross Ice Shelf (abstract), *Eos Trans. AGU*, 59, 308, 1978.

Williams, R.T., and E.S. Robinson, Flexural waves in the Ross Ice Shelf, *J. Geophys. Res.*, 86, 6643-6648, 1981.

Williams, R.T. and E.S. Robinson, and H.A.C. Neuberg, Ocean wave measurements at the RISP J9 camp, *Antarct. J. U.S.*, 13, 91-92, 1978.

Zakharov, V.F., Seiches in the vicinity of Novolazarevskaya station, *Sov. Antarct. Exped. Info. Bull. Engl. Transl.*, 6, 268-269, 1967.

Zumberge, J.H. and C. Swithinbank, The dynamics of ice shelves, in *Antarct. Res. Geophys. Monogr. Ser.*, vol. 7, edited by H. Wexler et al., pp. 197-208, AGU, Washington, D.C., 1962.

(Received March 13, 1984;
accepted April 27, 1984.)

TIDAL MEASUREMENTS ALONG THE ANTARCTIC COASTLINE

J. R. E. Lutjeharms and C. C. Stavropoulos

National Research Institute for Oceanology, Stellenbosch 7600, South Africa

K. P. Koltermann

Deutsches Hydrographisches Institut, 2000 Hamburg 4, Federal Republic of Germany

Abstract. Some recent tidal measurements along the Queen Maud Land coast of Antarctica have brought into focus the dearth of high-quality tidal information along most parts of the Antarctic coastline. A review of Antarctic tidal measurements is given, and the results of new tidal measurements at SANAE, the South African Antarctic research station, and at the Georg von Neumayer Station of the Federal Republic of Germany are presented. It is concluded that there exists an urgent need for modern, high-quality tidal data along large sectors of the Antarctic coastline.

Introduction

Of all coastal regions in the world ocean, the Antarctic continent is particularly poorly served with sea level measurements, especially so in the South Atlantic and South Indian Ocean sectors of the "Southern Ocean". Only two continuously recording gauges, both situated on the Antarctic Peninsula [Intergovernmental Oceanographic Commission, 1983], are known to be presently in operation in Antarctica.

The geographic distribution of all tidal measurements taken along the Antarctic coastline up to 1980 is portrayed in Figure 1. There are only 75 stations at which sea level records were obtained on the Antarctic continent itself and within the adjacent coastal region (Figure 1). Stations in Chile, Argentina, South Georgia, Bouvetøya Iles, Kerguelen, Tasmania, New Zealand, Macquarie Island, Auckland Island, Campbell Island, etc. have not been included here, although they lie south of 40°S and thus form part of the data set for the Southern Ocean [Lutjeharms, 1976, 1980]. Of the tidal stations in Antarctica, 43% are on the Antarctic Peninsula, and a further 18, or 24%, around the Ross Sea. The remaining 25 measurements are sparsely spread along the whole coast of the continent. No records could be found indicating that any measurements have ever been undertaken between the Ross Sea (160°W) and the Antarctic Peninsula (70°W). Only five records seem to exist of measurements along the Queen Maud Land (Dronning Maud Land) coast, while only three are available for the coastline between the stations Mawson (67°S, 63°E) and Mirnyy (67°S, 93°E) in the South Indian Ocean sector of the Southern Ocean.

This geographic distribution includes all tidal measurements up to 1980 for which records are available or about which mention has been made in the literature. It is instructive to investigate the periods during which these measurements were made and their duration. The distribution in time of all known tidal measurements along six sectors of the Antarctic coast are portrayed in Figure 2. If a measurement of any duration was made during a particular year at a certain station, that year has been marked. A marking for a year does not imply that a full 12-month record for that year exists for that station. In certain cases, such as in the Ross Sea, regular sea level measurements were made during the austral summer, but not in winter. A more detailed record of all long-term sea level measurements, i.e., exceeding 5 years in duration, has been compiled by Lutjeharms and Alheit [1982].

A few interesting facts emerge from the analysis on which Figure 2 is based. First, there are very few measurements of long duration. Lutjeharms and Alheit [1982] could find only 12 records exceeding 5 years duration. Of these, seven are to be found on the Ross Ice Shelf; the rest on the Antarctic Peninsula. Of those on the Antarctic Peninsula, three exceed 9 years in duration. These include records from gauges at the Argentine bases Hope Bay (Bahia Esperanza) and Almirante Brown which have been operating with a few interruptions since 1958 [Association d'Oceanographique Physique, 1961, 1963; Instituto Antarctico Argentino, 1961-1968, 1970, 1971-1973; Permanent Service for Mean Sea-Level, 1968] and at the British base "F" on the Argentine Islands where a measurement was made from May to December in 1935 [Roberts and Corkan, 1941]

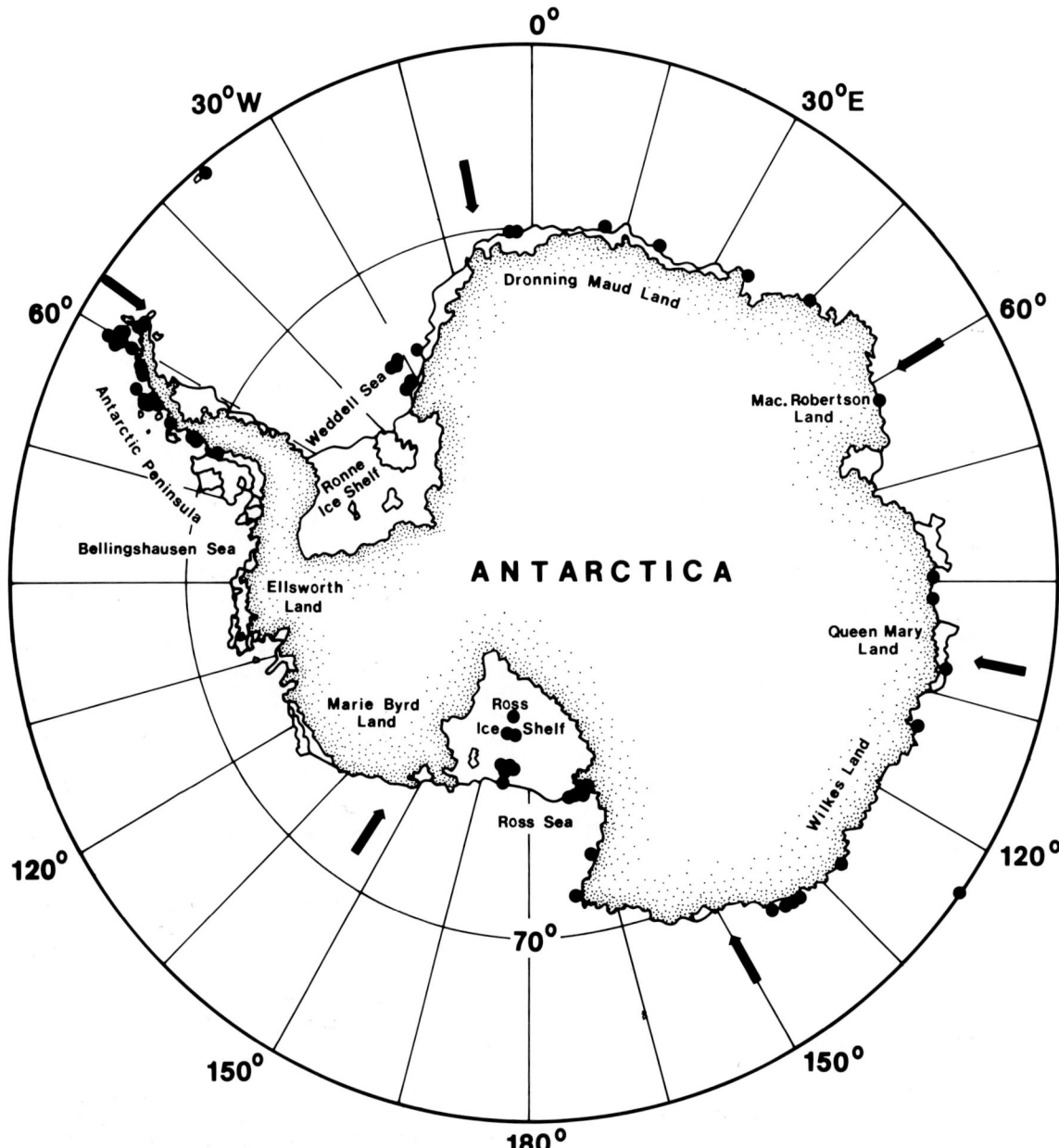

Fig. 1. The geographic locations of all tidal or sea level measurements along the Antarctic coastline up to 1980, according to Lutjeharms [1980]. The concentration of stations on the Antarctic Peninsula and in the vicinity of the Ross Sea is immediately apparent. Arrows demarcate coastline sectors discussed in the text. Dronning Maud Land is also referred to as Queen Maud Land.

and then, with interruptions, since 1958 [Royal Society 1960-1961, 1962, 1963, 1964-1965, 1966-1968, 1969, 1970, 1971-1974].

The upper panel of Figure 2 shows the number of tidal stations at which some measurement was made during that year. In the first decade of this century a number of measurements were made at places such as Scotia Bay [Royal Society, 1908], Gauss Station, Cape Armitage, Port Circumcision, and Port Foster [International Hydrographic Bureau, 1966]. The locations of these early measurements are fairly evenly distributed around Antarctica and were carried out by investigators from France, the United Kingdom, Argentina, and the United States.

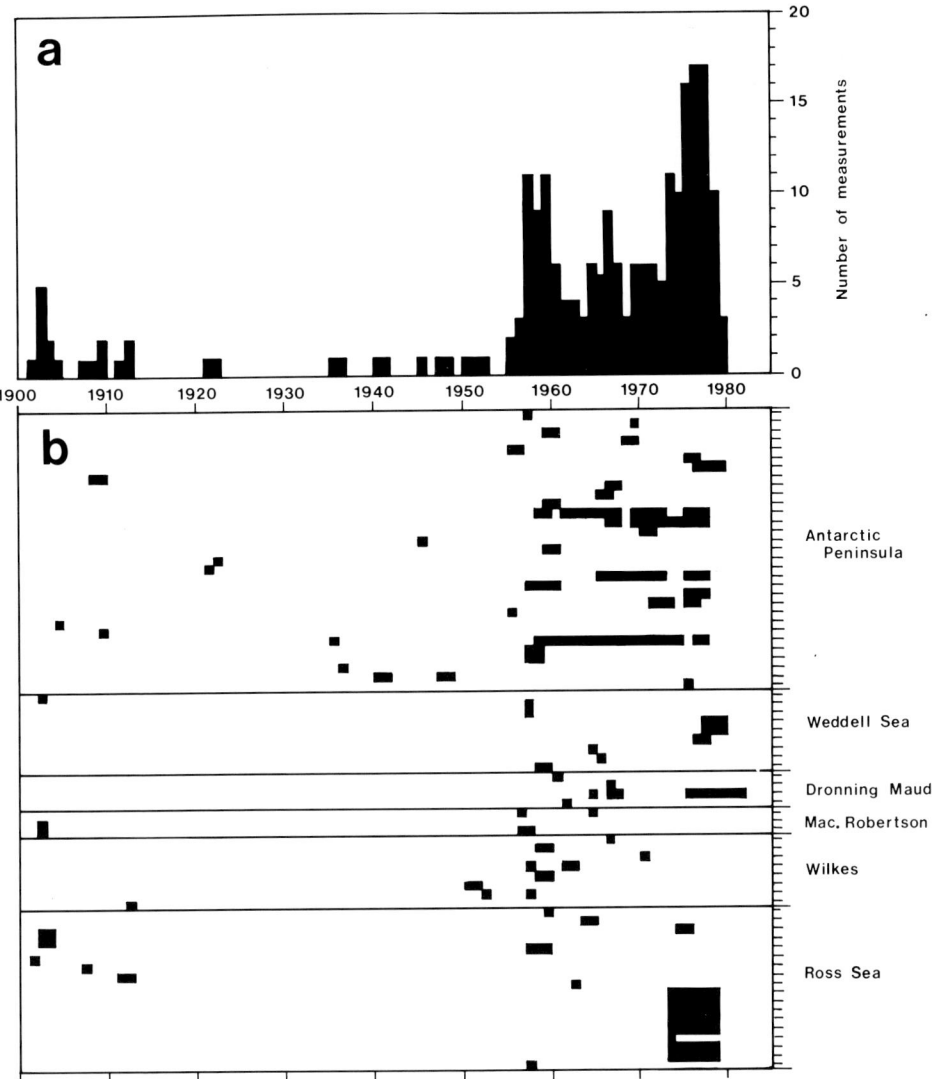

Fig. 2. The temporal distribution of tidal measurements along the Antarctic coast from 1900 to 1980. (a) The total number of tidal measurements undertaken during each year. (b) The distribution with time of tidal records at individual stations in six sectors of the coastline of Antarctica.

Very few measurements were made for the period between about 1912 and the advent of the International Geophysical Year (IGY) in 1957-1958. The IGY years were notable for a tenfold increase in the number of observations [Permanent Service for Mean Sea-Level, 1959]. In a number of cases, a measuring program at stations with very long records was also initiated at that time. The number of operational stations then declined to an average of about five per year until the late 1970s, when a number of instruments were in operation for several years as part of the International Southern Ocean Studies Program [Nowlin et al., 1975; Nowlin and Pillsbury, 1979] or as part of the Ross Ice Shelf Project [Williams and Robinson, 1976, 1979].

It is thus clear that the geographic distribution of tidal measurements is very inhomogeneous around Antarctica and that there are very few records exceeding a few years in duration anywhere on the Antarctic coast. It should also be recognized that the quality of the data may be suspect [Bishop and Walton, 1977; MacDonald and Burrows, 1959] in a significant number of cases. It is therefore important to briefly review the measurements, results, and data quality for the entire coastline. It is not the aim of this discussion to enumerate tidal constants determined for each location. These have been collected and published in detail elsewhere [e.g., International Hydrographic Bureau, 1966] and are thus generally available for many stations. Where known, the availability of tidal constants is mentioned.

Results of Tidal Measurements

Antarctic Peninsula

The number of tidal measurement stations for the Antarctic coast is greatest on the western side of the Antarctic Peninsula. This coverage extends approximately from the northernmost tip of the peninsula to 70°S (Figure 1). For the purposes of this discussion the Antarctic Peninsula has been taken to cover that part of the coast between the islands near the tip of the Antarctic Peninsula and the eastern edge of the Ross Sea. Again, there are no known tidal measurements along the coastline west of the peninsula proper, i.e., from the Bellingshausen Sea to the Ross Sea (Figure 1).

Of the 32 recognized tidal measurements, a large number are known from activities reported to the Scientific Committee for Antarctic Research (SCAR). These include the Argentine bases Jubany, Teniente Camara, Petrel, Caillet Bois, Almirante Brown, Destacamento Melchior, Caleta Potter, and Bahia Esperanza [Instituto Antarctico Argentino, 1961-1968, 1970, 1971-1973], the Chilean bases Presidente Aguirre Cerda, Bernardo O'Higgins [e.g., Comite Chileno de Investigaciones Antarcticas, 1967] and Bahia Ardley [Comite Chileno de Investigaciones Antarcticas, 1972], and some U.S. and British bases such as Palmer Station [National Research Council, 1973] and Base "F" at Stella Creek on the Argentine Islands, respectively [Royal Society, 1960-1961]. At some of these stations, such as Jubany, Teniente Camara, and Petrel, tidal staffs were used, while at others, such as Bahia Esperanza and Ballant, float marigraphs were employed. A number of these records extend over a long time. The tide gauge at Bahia Esperanza has been operating since 1958, while that at the base Bernardo O'Higgins since 1966. The Servicio de Hidrografia Naval publishes yearly tide tables for the Argentine-claimed sector of Antarctica [Servicio de Hidrografia Naval, 1962] based on the records of some of these stations. The tidal constants for a number of these stations have been published by the International Hydrographic Bureau [1966]. The tidal constants published here are in certain cases the only records in the literature that measurements have been made at a certain location. This is true for Admiralty Bay and Puerto Soberania (62°29'S, 59°38'W), both on the South Shetland Islands, Port Charcot, Ferin Head, Lent Island (66°53'S, 66°48'W), and Neko Harbor.

Soviet investigators [Kudryatsev, 1973; Vorobyev, 1973] have reported measurements on King George Island in the South Shetland Islands. From a yearlong record they established that the long-period tides, the semimonthly and the monthly tides, agree with the static tide in this area.

During the British Graham Land Expedition of 1920-1922, tidal observations were taken at Waterboat Point and Foyn Harbor with a tide pole of local construction, the longest record being for 2 months. During an expedition in 1934-1937, observations were made at the two base stations, Stella Creek and Barry Island, for a period of more than 4 months [Roberts and Corkan, 1941]. Tidal constants were calculated for these and all other known records on the coasts of Graham Land. One of these is Port Foster on the South Shetland Islands where the Second French Antarctic Expedition made measurements in 1908-1909 [Godfroy, 1912].

Alvarez and Cobas [1966] obtained readings at Matienzo Station by using a ship as a float and favorably compared the tidal constants they calculated to those determined for Hope Bay [International Hydrographic Bureau, 1966]. Bishop and Walton [1977] constructed a novel tide measurement device to study tidal movement through the ice. They successfully made readings in lake ice abutting the George VI Ice Shelf near the small Antarctic base of Fossil Bluff.

Some of the longer tidal records in the Drake Passage area, such as those collected at Bahia Esperanza, Almirante Brown, Destacamento Melchior, and Base "F" were used by McKee [1971] to study long-term sea level oscillations. He concluded that the variations were such that they suggested large fluctuations in the barotropic volume transport of the Antarctic Circumpolar Current. Van Loon [1972] used the long-term measurements at Melchior, at Port Circumcision, and at Marguerite Bay to study half-yearly oscillations in sea level across the Drake Passage. Both the sea surface slope and the wind records suggest such an oscillation, but are not in phase.

These tantalizing results were to some degree responsible for the fact that very long sea level records were subsequently obtained as part of the International South Ocean Studies program [e.g., Nowlin et al., 1975]. Pressure gauges placed off Livingston Island at the Antarctic Peninsula and off Cape Horn were deployed simultaneously for 1-year intervals and renewed annually. Data from 3 years of measurements were interpreted to show that transport fluctuations in the Antarctic Circumpolar Current exceeding 30 days are highly correlated with fluctuations in wind stress integrated over the Southern Ocean [Wearn and Baker, 1980].

Although a number of measurements on the Antarctic Peninsula are of a rather preliminary nature, are of very short duration, or have been obtained with unconventional or even crude gauges, the general coverage of data is good, and a number of long records exist. The distribution of stations is poor, however, and measurements on the coast of Ellsworth Land and Marie Byrd Land are needed [Lutjeharms, 1985].

Weddell Sea

For the purposes of this discussion the Weddell Sea coast is defined as the eastern coast of the Antarctic Peninsula and along the Ronne and Filchner ice shelves and the coast of Coats Land as far east as 10°W longitude, where Queen Maud Land is considered to start (Figure 1). In this sector, tidal measurements have been made in three areas, namely, on the South Orkney Islands, on the Ronne Ice Shelf, and at three semicoastal stations: Shackleton (37°10'W), Ellsworth (41°08'W), and Halley Bay (26°43'W). The latter two were on moving ice, with Ellsworth being on the Filchner Ice Shelf.

The tidal measurements on the South Orkney Islands were taken in Scotia Bay of Laurie Island from May to October 1903 [Royal Society, 1908], and a month-long record was obtained somewhere in the period 1934-1937 [Roberts and Corkan, 1941]. A tidal pole was used in the former and a recording tide gauge in the latter case. The tidal constants were published by the International Hydrographic Bureau [1966].

Investigators from Norway undertook readings with a TG-3A water level recorder at three sites centered at about 74°S, 38°W [Lutjeharms, 1980] on the Weddell Sea pack ice in the period 1976-1979 [Norwegian Academy of Science and Letters, 1977-1978] (see also Foldvik et al. [this volume]). Thiel et al. [1960] used a Frost gravity meter to estimate tidal components at Ellsworth on the edge of the Filchner Ice Shelf. Although the observations were complicated by high-frequency oscillations of the ice due to oceanic factors (see also Holdsworth [this volume]), they were able to calculate the amplitude and phase for the more significant tidal components. They established that the Weddell Sea tide has both diurnal and semidiurnal components. Tidal measurements were made at the neighboring British base, Shackleton, with a Worden gravimeter during two periods in 1957 [Pratt, 1960]. The tidal constituents based on these readings were also published by the International Hydrographic Bureau [1966]. One further reading, obtained in December 1964 at Base Z in Halley Bay on moving ice, has been reported by British investigators [Royal Society, 1964-1965].

The measurements taken at most of these stations were of 1 month or longer duration and can be assumed to give full suites of tidal constants. No records seem to exist for the eastern side of the Antarctic Peninsula or for the Weddell Sea itself with the exception of the Norwegian Stations mentioned. Our overall knowledge of the tides in the Weddell Sea can thus be considered to be poor.

Queen Maud Land

That sector of the Antarctic coastline lying astride the Greenwich meridian (10°W-60°E) is known as Queen Maud Land (labeled on the GEBCO map [Vanney and Johnson, this volume] as Dronning Maud Land (Figure 1)). According to Lutjeharms [1980] there exist sea level records for only five locations in this sector. These include measurements made at SANAE, the South African Antarctic base at 2°45'W (previously known as Norway Station), Lazarev Station at 12°58'E, Polarhavbay at 24°36'E, the Japanese station Syowa at 39°35'E and measurements from the Alasheyev Bight at 45°47'E by Soviet investigators. Sverdrup [1954] made measurements of tidal currents off Maudheim, roughly at 12°W.

Tidal measurements at Norway Station during a period of about 3 months in the austral summer of 1958-1959 were taken by an Ott-Schreibpegel instrument and were reported by Hisdal [1965]. Two very short records were obtained at the edge of the ice shelf, the longest lasting 3 days. The measurements at SANAE [Pollak and Sharwood, 1971] used a gravimeter to determine the vertical motion of the Fimbul Ice Shelf, and no estimates were made of harmonic constants. The range of shelf motion was found to lie between 0.7 and 1.7 m. The measurements made at Lazarev Station in September 1960 consisted of a jury-rigged apparatus when there was a fortuitous grounding of an iceberg near the base [Dubrovin, 1962]. These data were subsequently reanalyzed by Shesterikov and Dubrovin [1963] but questioned by Hisdal [1965] due to the primitive instrument used and because of the short duration of the record. The measurement at Polarhavbay, in Breid Bay, reported by Van Grondelle [1967] is one of the few in which the tide gauge was not attached to the ice but was moored in 150 m of water away from the ice shelf. A 7-day record was obtained with a modified Smitt tide gauge.

The record at Syowa is by Antarctic standards a long one. A 1-month record was made in 1964 [Science Council of Japan, 1966], and in 1965 a pressure-type tide gauge was installed [Oka and Fuchinoue, 1984]. This has undergone some modifications but has run continuously and produced useful results. Oura and Fujino [1965] and Hori and Inbe [1968] have tabulated harmonic constants of several principal component tides at Syowa, Norway and Molodezhnaya stations. The Syowa record is one of half a dozen high-quality records taken over a period exceeding 5 years [Suzuki and Kurano, 1982].

At Alasheyev Bight, in the proximity of the Soviet Molodezhnaya Station, two sets of measurements were made. The one record of 25 days duration was made in the austral summer of 1962 [Shamontyev, 1963] with a gauge consisting of a tide staff and marigraph. The harmonic constants that were subsequently calculated exhibit the characteristics of a mixed, predominantly diurnal tide [Shamontyev, 1965; Tarasov, 1964]. Further measurements were made for a 6-month period in 1976 [Vorobyev and Uranov, 1983]. It was determined that the amplitude and angle of the eight principal winter tidal waves show little variability and that the maximum range of overall sea level fluctuations amounted to 1.5 m, while the difference between extreme average daily levels was 56 cm.

With the exception of the good Syowa record, the other records are very short and in some cases of doubtful quality. Bearing in mind the length of coastline, Queen Maud Land is very much undersampled for tides.

Mac. Robertson Land to Queen Mary Coast

This sector of the Antarctic coastline lies between Queen Maud Land and Wilkes Land and stretches from 60° to 100°E (Figure 1). Records from tidal measurements at only three places are known, namely, at Mawson (62°53'E), Gauss Station (89°38'E), and Mirnyy Station (93°01'E).

The measurement at Gauss Station was made in 1902 with an instrument of unknown origin. The harmonic constants derived from this measurement were published by the International Hydrographic Bureau [1966]. Two month-long measurements were made at Mawson in 1956, and the harmonic constants were published by the International Hydrographic Bureau [1966]. Further readings were made in 1964 with a tide recorder of unknown description [Australian Academy of Science, 1964-1965]. The measurements that were made at the Soviet station Mirnyy came from a registering tide gauge of their own construction and from a Valdai tide gauge which made measurements from November 1956 to January 1957 [Samov and Kopteva, 1958; Shesterikov, 1959].

No records exist for the Amery basin or Amery Ice Shelf or for some long stretches of this coastline.

Wilkes Land

Wilkes Land lies to the west of the Ross Sea on the Antarctic coast, with tidal stations from 100°E to 150°E (Figure 1). Of the nine known measurements along this coastline and its adjacent waters [Lutjeharms, 1980], six are clustered about the French base Dumont d'Urville, which lies at about 140°E.

The westernmost station in this sector is in the Zaliv Rybiy Khvost (Paz Cove) of the Bunger Hills at the Soviet Dobrovolskaya Station. The record is for 1 day only, in January 1966, when that base closed, and so is not likely to be repeated in the near future [Dubrovin and Zalewski, 1969]. A simple tide staff was used, and it was established that the tide was very similar to that in the Mirnyy area [Samov and Kopteva, 1958] with a difference in tidal range of only 7 cm. The times of high and low water between these stations differed by 30-40 minutes. The only other station west of the Dumont d'Urville base was at the Wilkes Station of the United States at 110°31'E [National Academy of Sciences, 1959]. Measurements were made during a few months of 1958 and 1959 with a portable tide gauge, and the resulting tidal harmonics were reported by the International Hydrographic Bureau [1966].

The only pelagic tidal measurement off any part of the Antarctic continental shelf was made off Wilkes Land at 60°S [Cartwright et al., 1979]. A deep-sea instrument capsule was placed on the ocean bottom for a month in 1970, and various parameters were recorded [Snodgrass, 1971]. Tidal currents were less than 1 cm/s. The observed tide was predominantly a diurnal one that could be interpreted as a Kelvin wave moving westward along the Antarctic coast. A Kelvin model did not give a good fit to the semidiurnal tide [Irish and Snodgrass, 1972].

The measurement stations clustered about Dumont d'Urville are with one exception all of French origin. Tidal constants at Rocher X (136°42'E) have been published by the International Hydrographic Bureau [1966] while those established in 1952, 1957 and 1961-1962 at Ile des Petrels (Petrel Island) (140°01'E) with a recording tide gauge were published by the Association d'Oceanographie Physique [1963], the Permanent Service for Mean Sea-Level [1968], and the International Hydrographic Bureau [1966]. At Dumont d'Urville, tidal measurements were made from May 1958 to January 1959 [Sous-Comite Antarctique Francais, 1959] and again in 1961 [Academie des Sciences, 1962]. A number of measurements of relatively short duration were made in 1950 and 1951 at Cape Margerie (141°25'E) with a tide pole [Imbert, 1953, 1956] and the tidal harmonic constants were published by the International Hydrographic Bureau [1966].

An old Australian record from 1912 exists for Cape Denison (142°40'E) [Doodson, 1939]. The gauge ran for over 2 months. The distribution of measuring points along this long part of coastline is thus not very uniform. A few good records exist but are clustered mostly around one location.

Ross Sea

For the purposes of this discussion, the Ross Sea is considered to be that area of the

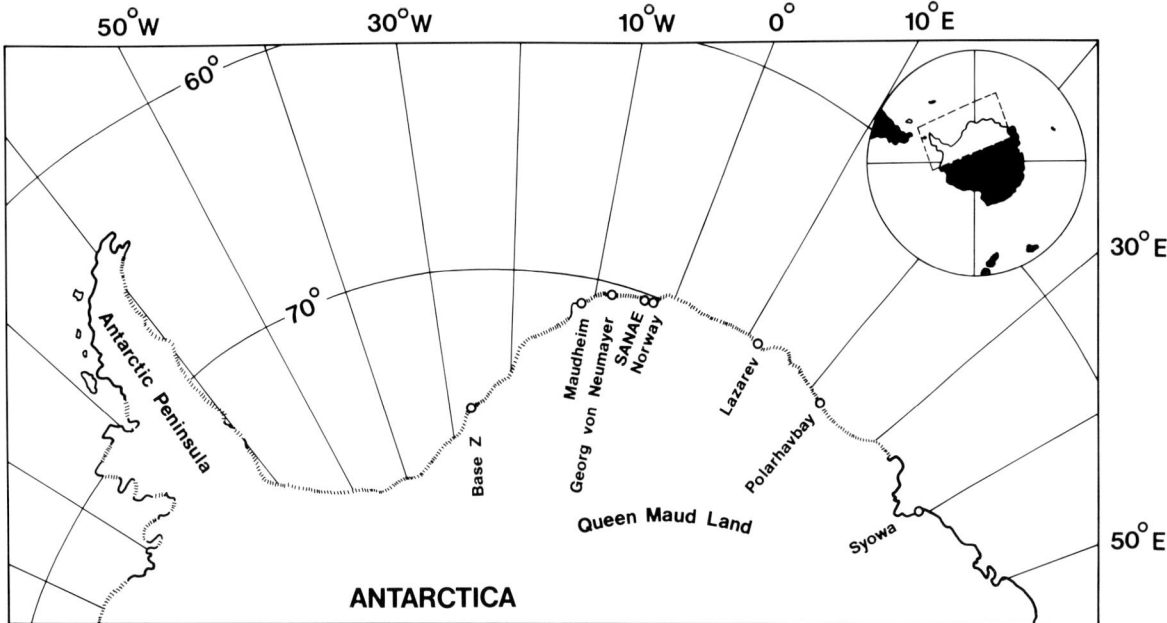

Fig. 3. The geographic location of the known sea level stations on Queen Maud Land, Antarctica. Dashed coastline denotes an ice shelf edge; a solid coastline denotes the Antarctic landmass [after Lutjeharms, 1980].

Antarctic coastline lying between 150°E and 140°W (Figure 1). Tidal measurements in this Ross Sea area fall within three identifiable periods. Between about 1901 and 1912 a number of stations were occupied. During and after the International Geophysical Year, 1957-1958, there were some further measurements (see Figure 2), while a major research effort to measure the tidal effect on the Ross Ice Shelf was initiated in the 1970s.

The earliest tidal record known for this area was taken at the McMurdo Station on Ross Island in 1901 [Lutjeharms, 1980] with a wireline tide gauge. Further readings at Ross Island were undertaken in 1902-1903 with a homemade tide gauge [Royal Society, 1908]. Other measurements of this nature were made at Cape Armitage (1902-1903), Cape Royds (1907), and Cape Evans (1911-1912). The tidal constants derived from these have all been reported by the International Hydrographic Bureau [1966].

During the International Geophysical Year, tidal readings were taken at Hallett Station [National Science Foundation, 1960], the Little America base [National Academy of Sciences, 1959], and Scott Base [Royal Society of New Zealand, 1960]. Using the data collected at the Little America base (163°W), Thiel et al. [1960] showed that the Ross Sea tide is diurnal, with the solar component predominating. The recording methods used at Scott Base with a modified Foxboro liquid level recorder were reported by MacDonald and Burrows [1959]. Harmonic constants for the tides were presented by Gilmour et al. [1962]. Further measurements were made in McMurdo Sound to the north of Scott Base by Australian investigators from February 1963 to September 1964. The data have been reported by the Flinders University [1969]. Meaurements were also made in 1962 at Cape Crozier on Ross Island [Royal Society of New Zealand, 1963].

Heath [1971] and Venzke [1971] have reported tidal measurements made at McMurdo Station in McMurdo Sound. A Foxboro tide gauge was employed, and the first 30-day record showed that the main components of the tide are the diurnal constituents O_1 and K_1.

The Ross Sea area is unique for the Southern Ocean in that it has seen a determined modern effort to study the tides in this region over a period of a number of years. Measurements were made at eight sites during the austral summer starting in 1973 to determine the ocean tide beneath the Ross Ice Shelf [Robinson et al., 1975; Williams and Robinson, 1976] as part of the Ross Ice Shelf Project (RISP). Characteristics of tidal constituents were calculated from tidal fluctuations of gravity measured on the floating ice shelf surface [Robinson et al., 1977a]. Robinson et al. [1977b] and Williams and Robinson [1979] established that diurnal constituents are predominant and constructed provisional cotidal-corange charts for the Ross Sea. It was also shown that the amplitudes of the diurnal tidal constituents in the Ross Sea are larger than in the adjacent sector of the Southern Ocean,

TABLE 1. Information Concerning Sea Level Measurements Along the Coast of Queen Maud Land, Antarctica [from Lutjeharms, 1980]

Point	Location	Latitude	Longitude	Period	Instrument	Tidal Constituents	Reference
Base 'Z'	Halley Bay	75°31'S	26°43'W	Dec. 1964	Tide gauge	---	BNCAR [1965]
SANAE	Fimbul Ice Shelf	70°15'S	02°45'W	Aug. 1965	Worden gravi-meter	---	Pollak and Sharwood [1971]
Norway Station	Queen Maud Land	70°30'S	02°32'W	Nov.-Dec. 1958 Jan. 1959	OTT-Schreibpegel	M_2, S_2, K_1, O_1	Hisdal [1965]
Lazarev Station	Queen Maud Land	69°55'S	12°58'E	Sept. 1960	Primitive apparatus	M_2, S_2, Q_1, K_1	Shesterikov and Dubrovin [1963]
Polarhavbay	Breid Bay	70°18'S	24°36'E	Jan. 1966	Modified Smitt	M_2, S_2, K_1, O_1 etc.	Van Grondelle [1967]
Syowa Station	Enderby Land	69°00'S	39°35'E	Feb.-March 1964 Feb. 1966-Jan. 1967	LPT-2 pressure-type tide gauge	M_2, S_2, K_1, O_1	Oura and Fujino [1965]
				Feb. 1975-Jan. 1978	SWL-7 tidal gauge		Hisdal [1965]

indicating the existence of a diurnal resonance related to the shape and depth of the Ross Sea. The detailed analysis of the tidal wave behavior in this sea (see also MacAyeal [this volume]) cannot at present be undertaken for any similar area of the Antarctic coast due to the lack of data.

Tidal Measurement Availability

From the above it is clear that a good tidal data set exists for very restricted areas of the Antarctic coastline. On the western side of the Antarctic Peninsula, good coverage exists, and at least two gauges are presently in operation [Intergovernmental Oceanographic Commission, 1983]. In the Ross Sea a specifically directed research project has established the progression of the tidal wave under the shelf ice (see also Pillsbury and Jacobs [this volume]). In most other sectors there is a dearth of data. A number of good, long-running records exist at a few stations, but in most cases the records are of short duration, with all the limitations this entails, or are of doubtful quality due to the nature of the gauges used [Lutjeharms, 1985].

Tidal models are very dependent on sufficient tidal measurements for a particular area. Titov and Shesterikov [1965] state that the lack of publications on the fundamentals of tidal movement and behavior in the Southern Ocean may be related directly to the limited number of measurements. They have shown that the difference between their global tide model, which does not have an amphidromic point southeast of Madagascar, and that of Dietrich [1963], which does, can only be resolved by critical data from Iles Crozet. Even a few strategic data points may thus be of cardinal importance for a proper understanding of the tide, particularly its coastal behavior. The inhomogeneous distribution of historic readings along the Antarctic coast is therefore a serious problem [Lutjeharms, 1985].

A few attempts have been made to use the available data in cartographic descriptions of the tide in the Southern Ocean [Dietrich, 1963; Villain, 1952]. The tidal maps in these documents and in atlases such as that of the U.S. Navy Hydrographic Office [1957] are therefore rather hypothetical in coastal areas of Antarctica. The tidal components with long time constants are particularly poorly served by present measurements in the vicinity of Antarctica [Vorobyev, 1973].

The Tidal Measurements in Queen Maud Land

The above overview of the history of tidal and sea level measurements along the coast of Antarctica identifies certain coastal areas as particularly poor in data. Of all the data-sparse areas, Queen Maud Land is certainly one

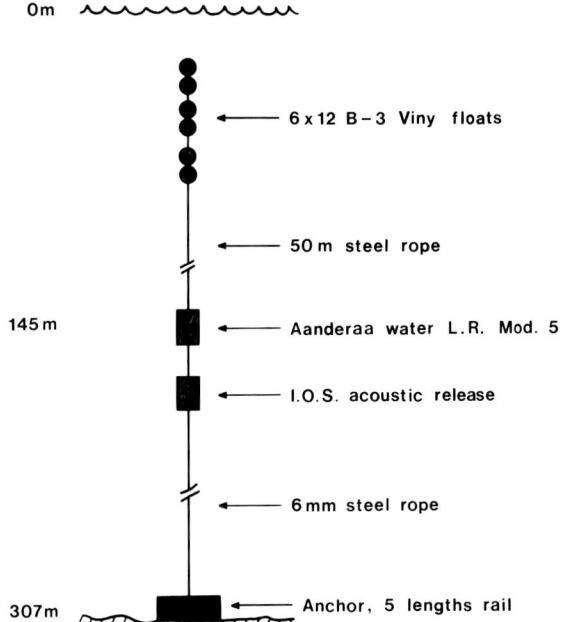

Fig. 4. Mooring scheme for a precision pressure gauge (Aanderaa water level recorder, model 5) deployment at SANAE in January 1982. The diagram is not to scale. Depths are according to corrected echo sounding measurements.

which is in great need of further measurement projects.

It was noted above that Lutjeharms [1980] had established the existence of sea level records for only eight locations in Queen Maud Land. The details of these records are set out in Figure 3 and Table 1. Tidal constituents, to our knowledge, have been published for only four of these stations. Only the Syowa Station record exceeds 7 days duration; the one at Norway Station is for 3 days [Hisdal, 1965], and the one at Polarhavbay for 7 days [Van Grondelle, 1967]. The need for accurate and longer records in this geographic area is thus evident.

We made tidal measurements at the South African Antarctic base SANAE (70°12.5'S, 02°43.5'W) from January 9 to February 7, 1982, and at the research station Georg von Neumayer of the Federal Republic of Germany (17°37'S, 8°22'W) from February 4 to March 2, 1983. We present here the results of these measurements.

Equipment, Methods, and Data

SANAE

Since all previous sea level recordings in the direct vicinity of SANAE were taken on the ice shelf [Hisdal, 1965; Pollak and Sharwood, 1971], thus measuring the movement of the ice shelf relative to the ocean floor, it was decided to place a gauge in the water column on this occasion. A precision pressure gauge, namely, an Aanderaa water level recorder, model 5, was used.

Because of practical problems the gauge could not be placed on the seafloor itself, but formed part of a taut-wire mooring portrayed in Figure 4. The corrected depth at the mooring was 307 m, the gauge itself thus being supported 160 m above the bottom. The mooring was placed in Polarbjornbukta, a sheltered enclave in the ice shelf at 70°15'S, 2°45'W, where no severe currents were expected to occur. Sverdrup [1954] found average tidal currents of 6.2 cm/s with maxima of only 11.1 cm/s at Maudheim (Figure 3). The mooring system was designed so that, under static conditions, a net upward tension of 550 N would be exerted on the anchor. It was assumed that this tension would be sufficient to overcome any significant current-induced lateral movements in the mooring and changes in the depth of the gauge below sea level. The very low noise level in the data suggests that this assumption was justified at least as far as non-tidal currents are concerned.

The instrument was deployed for 29 days, with data recorded at 10-minute intervals (Figure 5). The occasional data point lacking in the record of the first 20 days was easily replaced by interpolation from the 10-minute values. However, record quality deteriorated to such a degree after day 20 that data gaps became too large to carry out further interpolation. The cause of this equipment malfunction has not been determined. Although a 29-day observation period is normally considered a minimum requirement for calculating a full suite of tidal constants, it was possible to carry out a reliable analysis for a maximum of 12 tidal constituents using only the first 20 days of record. The number of tidal constituents that could be addressed was a function of the short recording interval of 10 minutes. The analysis method is described by Shipley [1967] and follows closely that of Schureman [1958]. It has been described in greater detail elsewhere [Shipley, 1977, 1980]. Since the record is so short, it is not possible to separate with sufficient accuracy close constituents such as K_2 and S_2 or P_1 and K_1 [Shipley, 1967]. No corrections were made to the data. The results are presented in Table 2. Residuals between observed and predicted values (Figure 6) are small.

Georg von Neumayer

An Aanderaa water level recorder model 5 was used on a subsurface mooring at a site 7.8 km northeast of the shelf ice edge in Atka Iceport at a water depth of 252 m. The in-

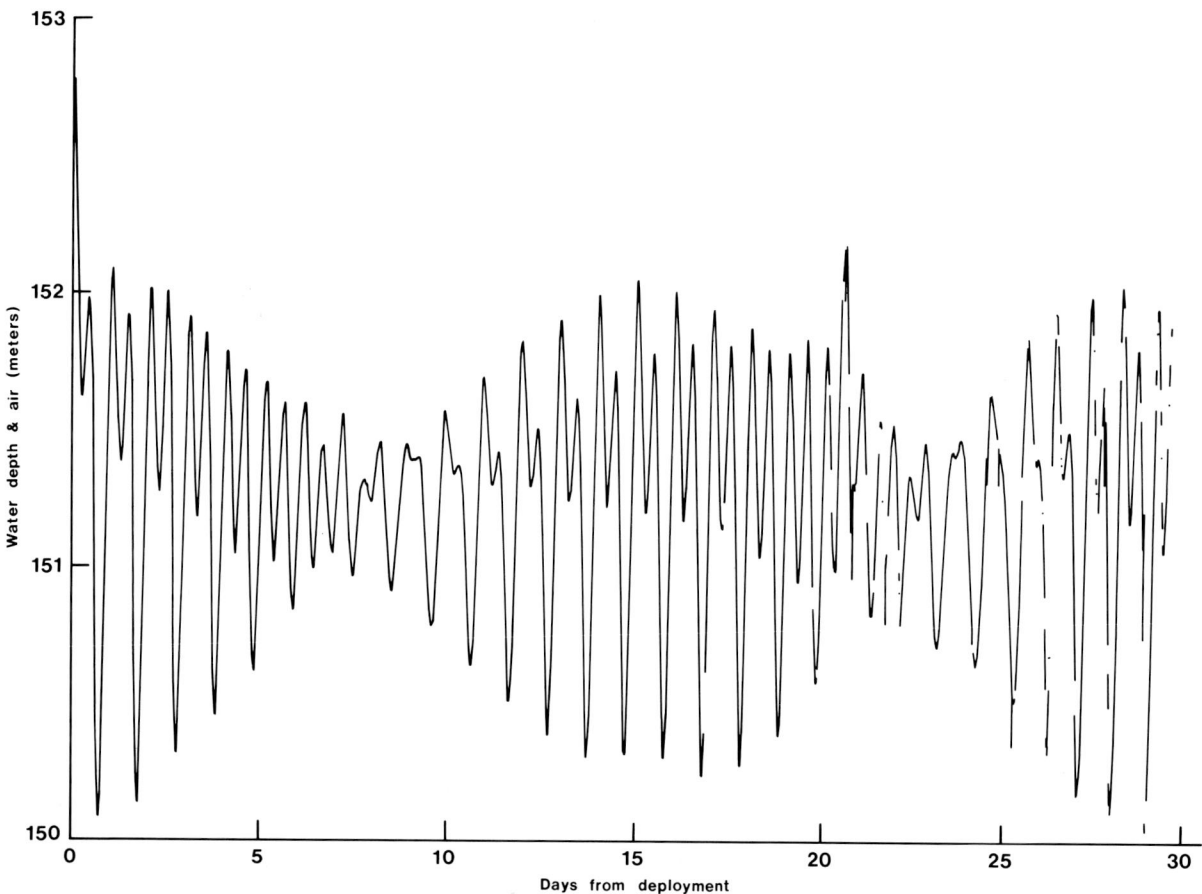

Fig. 5. Sea level record for a 29-day measurement at SANAE. The first deployment day was January 9, 1982. The mixed, predominantly semidiurnal characteristics of the tidal signal are clear, as is the rapid deterioration of the record after day 20.

strument was cradled on the anchor stone to give it a stable position relative to the seafloor. Current meter data from a site 5.7 km east of the tide gauge during the austral summer 1980-1981 show bottom current speeds of less than 20 cm/s, too small to influence the pressure sensor opening. The raw data were of perfect quality, with a time increment of 5.0 minutes. They have been calibrated with pre- and post-cruise calibration data and corrected for air pressure variations. The air pressure record from the German Antarctic Station Georg von Neumayer for the entire deployment period on 3-hourly intervals was splined and sampled at 5-minute-intervals to yield synchronous data points for the tide gauge record. No density corrections have been applied, so the data presented are seabed pressures, corrected for local air pressure effects (Figure 7).

A routine harmonic analysis has been run on the resultant time series. Besides the expected main tidal constituents K_1, O_1, M_2, and S_2, the suite of tidal constituents was se-

lected so as to minimize the rms deviation (to 7.41 ± 0.52 cm) of predicted from observed time series. With respect to the amplitude

TABLE 2. Tidal Constituents for SANAE (70°12.5'S, 02°43.5'W)

Constituent	Radial Speed, deg/h	Amplitude, m	Phase, deg
M_2	28.98	0.35	174.3
S_2	30.00	0.38	168.6
N_2	28.44	0.05	128.7
K_2	30.00	0.27	176.2
MU_2	27.97	0.02	147.7
M_4	57.97	0.01	193.2
MS_4	58.98	0.00	240.5
J_1	15.59	0.02	2.9
Q_1	13.40	0.06	329.6
P_1	14.96	0.06	81.8
O_1	13.94	0.27	343.0
K_1	15.04	0.36	359.3

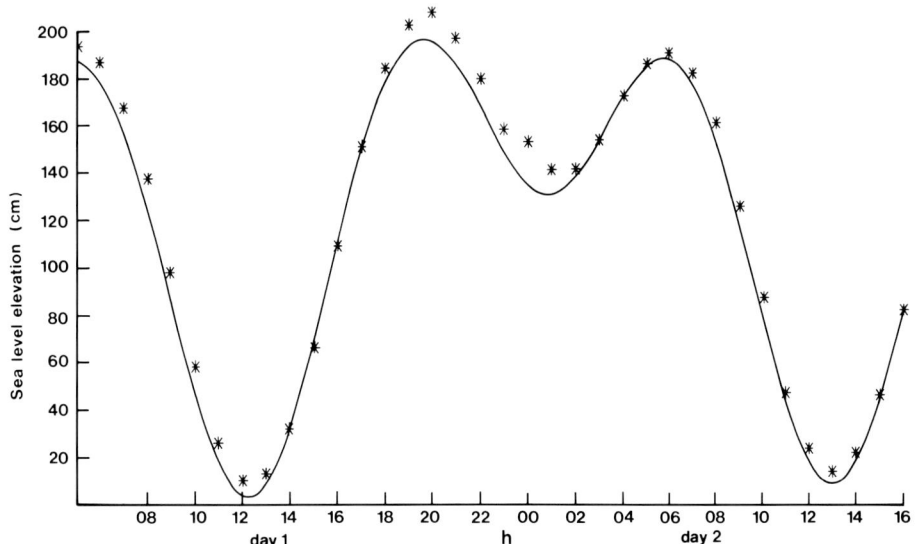

Fig. 6. The observed (solid curve) and predicted (stars) values for parts of the first 2 days of the tidal record at SANAE.

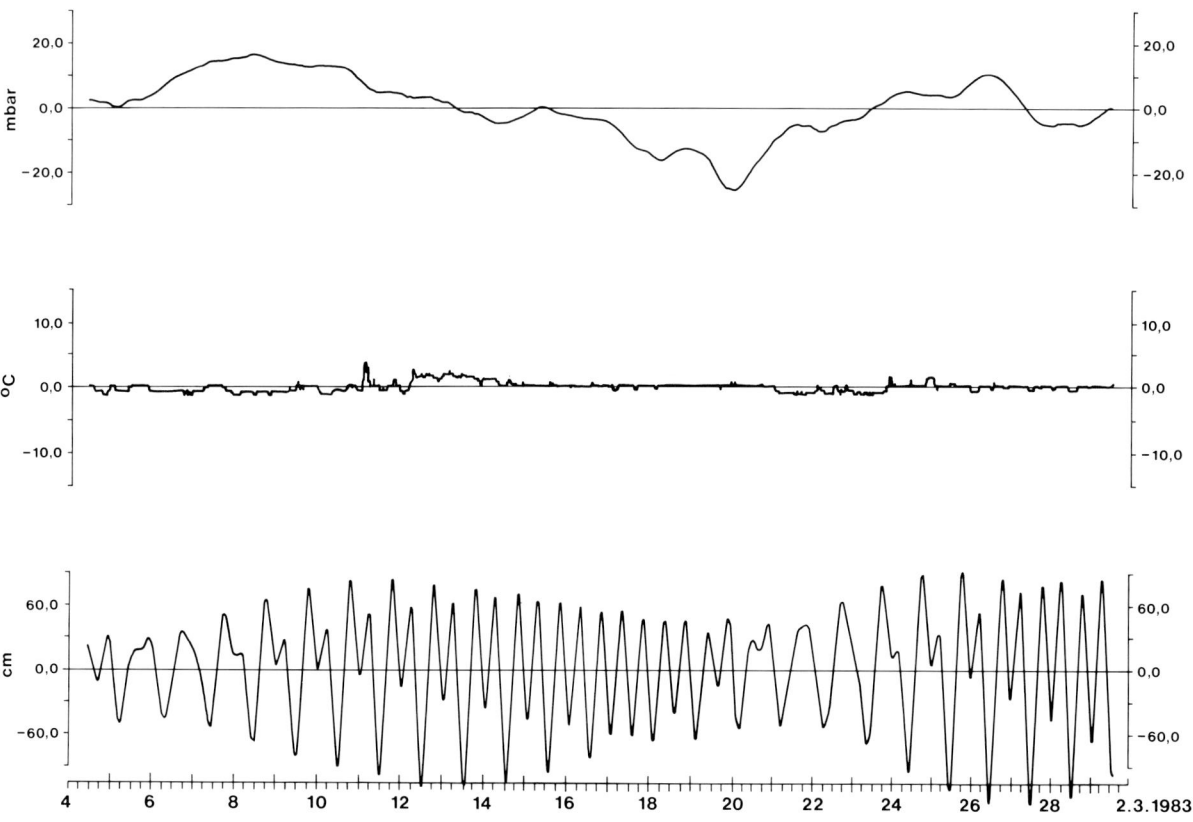

Fig. 7. Sea level record for a 27-day measurement at the Georg von Neumayer Station at Atka Iceport. The first deployment day was February 4, 1983. The upper panel shows the air pressure with a middle reading of 993.58 mbar; the middle panel shows the water temperature with a mean reading of -1.95°C; the lower panel shows sea level, corrected for air pressure. The mixed, predominantly semidiurnal characteristic of the tidal signal is evident.

TABLE 3. Tidal Constituents for Georg von Neumayer Station (70°37'S, 08°22'W)

Constituent	Radial Speed, deg/h	Amplitude, m	Phase, deg
$2Q_1$	12.85_4	0.02_3	245.1_2
Q_1	13.39_9	0.07_8	314.7_5
O_1	13.94_3	0.27_0	336.0_8
K_1	15.04_1	0.25_1	2.1_8
SO_1	16.05_7	0.01_0	326.5_1
$\mu 2$	27.96_8	0.01_6	348.6_2
M_2	28.98_4	0.39_1	162.9_7
S_2	30.00_0	0.35_2	201.0_3
$\eta 2$	30.62_6	0.00_8	173.6_5
M_3	43.47_6	0.00_3	148.3_0
SK_3	45.04_1	0.00_7	357.8_2
M_4	57.96_8	0.00_5	124.3_3
MS_4	58.98_4	0.00_2	200.7_2
S_4	60.00_0	0.00_3	176.5_3
M_6	86.95_2	0.00_0	296.5_2
$2MSN_8$	116.40_8	0.00_0	223.0_3
$2(MS)_8$	117.96_8	0.00_0	288.6_8

ratio of M_2 and S_2 to N_2 and K_2 in the tidal potential, the M_2 and S_2 contributions are of more importance, and they have been used here. The rms deviation between observed and predicted series was greater using K_2 and N_2.

Results and Discussion

The results presented in Table 2 for the measurements made at SANAE show an amplitude of 35 cm for the M_2, 38 cm for the S_2, 36 cm for the K_1 and 27 cm for the O_1 constituent. The corresponding results for the measurements made at the Georg von Neumayer Station, presented in Table 3, show an amplitude of 39 cm for the M_2, 35 cm for the S_2, 25 cm for the K_1 and 27 cm for the O_1 constituent. The values for these two stations are very similar and compare well with the amplitudes for Norway Station, based on a much shorter record, as well as with amplitudes of other stations on Queen Maud Land (Table 4). Following Dietrich [1963], the following may be calculated for SANAE and the Georg von Neumayer Station:

	SANAE	Georg von Neumayer
form factor $F = \dfrac{K_1 + O_1}{M_2 + S_2}$	0.86	0.70
mean spring tide $2(M_2 + S_2)$	146 cm	149 cm
mean neap tide $2(K_1 + O_1)$	126 cm	104 cm

According to Dietrich [1963] a tide with a form factor between 0.25 and 1.5 may be described as mixed, predominantly semidiurnal. The tidal characteristics at both stations are of this form.

The maximum range of 173 cm observed at SANAE and of 214 cm observed at George von Neumayer (Figure 7) may be compared to the amplitude of between 70 cm and 165 cm for vertical ice movement established at SANAE by Pollak and Sharwood [1971], the 150-cm maximum range found by Hisdal [1965] at Norway Station, and the 110-cm value given by Dubrovin [1962] for Lazarev Station. Dubrovin and Simonov [1965] have claimed a maximum tidal amplitude of between 58 cm and 310 cm for the Novolazarevskaya Station area. Novolazarevskaya Station is about 100 km south of Lazarev in Figure 3.

Shesterikov and Dubrovin [1963] found a form factor of 1.35 for Lazarev, confirming a mixed, predominantly semidiurnal nature for

TABLE 4. Harmonic Constants for Stations Along the Coast of Queen Maud Land

	M_2		S_2		N_2		K_2		O_1		K_1		Q_1		P_1	
	H	κ	H	κ	H	κ	H	κ	H	κ	H	κ	H	κ	H	κ
Norway Station	33	169	23	194					31	331	25	345				
Lazarev	25	119	12	128							42	321	30	308		
Polarhavbay	27	164	23	193	5	154	6	193	24	351	21	003			7	003
Syowa	18	154	20	171					22	337	20	351				
SANAE	35	174	38	169	5	129	27	176	27	343	36	355	6	330	6	82
Georg von Neumayer	39	163	35	201					27	336	25	002	8	315		

Amplitudes (H) are in centimeters while phase lags (κ) refer to the local meridian.

the tide at this station also. Sverdrup [1954] calculated a form factor of 2.5 for Maudheim Station, typifying the tidal character as mixed, predominantly diurnal. These two different characteristics agree with the predictions of Titov and Shesterikov [1965], whose model, based on data collected during the International Geophysical Year, predicted that the tide along the Antarctic coast should be chiefly mixed, irregular diurnal, or semidiurnal.

It is interesting to compare the results of the measurements made at SANAE and at Georg von Neumayer Station to those predicted by numerical models. Schwiderski [1983] has presented such a model for the semidiurnal principal lunar tide M_2 which gives an amplitude between 30 cm and 40 cm for SANAE (actual amplitude 35 cm). It predicts an amplitude of 40 cm at Georg von Neumayer Station, where the observed amplitude was 39 cm. Our measurements thus establish the accuracy of this model along this part of the Antarctic coast.

For the Atka tide gauge site near the Georg von Neumayer Station, the Schwiderski [1983] model gives an O_1 amplitude of 29 cm and a phase of 344°, versus the observed 27 cm and 336°. Similar deviations in amplitude and phase occur for the K_1 and the O_1, but, in general, the phases in Table 4 show a large scatter compared with the Schwiderski [1983] data. The amphidromic point in the vicinity of Maud Rise and the clockwise progression of the tidal wave along the Antarctic coast are evident from the M_2 tide gauge data. The only station which shows larger deviations is Lazarev, where all phases seem to be lagging by about 1 hour.

To estimate the quality of the Atka tide gauge analysis, the predicted times of high and low waters for the gauge site were compared with observed times at the ice edge. For this purpose a specially tuned echo sounder was operated continuously on RV Polarstern during the austral summer 1983-1984. A comparison during the deployment period of the tide gauge in 1983 gave a mean time lag of +22 ± 7 minutes between the tide gauge and the ship's berth for 13 consecutive times of high water. During 1983-1984 the same observations for nine consecutive highwater periods gave differences between predicted and observed high water from +1 hour 46 minutes at the beginning of the period to +25 minutes at the end. An explanation for this strong decline in time difference is apparent from simultaneous ice observations in the bay. In 1983 the bay was covered with large ice floes only during high water, but was completely free of ice during low-water periods. In 1984 the bay was covered with large ice floes even at low water, the ice cover decreasing considerably toward the end of the period, when again the ice floes came in with the tide. Apparently, the strongly varying ice cover takes up energy from the tidal wave, causing the lag between the observed and theoretically determined phases. This may also explain why the observed phases and amplitudes along the coast show a large scatter in comparison with the Schwiderski [1983] model. Longer time series would be needed to fully establish a relationship with the ice cover. Foldvik and Kvinge [1974], using the results of yearlong current meter records taken in the southern Weddell Sea, show that the amplitude of the diurnal tidal current component K_1 is about 3 times larger than the amplitude of the semidiurnal tidal current component M_2. In winter, when ice cover extended over the measuring point, the diurnal tidal component broke down. They have attributed this phenomenon to a weakening of the local stratification and the subsequent breakdown of baroclinic tidal forcing [Foldvik et al., this volume].

Conclusions

From a review of historic and contemporary tidal measurements along the coast of the Antarctic continent it is evident that large sectors of the coastline are not well enough sampled to fully determine the tidal constituent. This is particularly true for the longer-period constituents. We have calculated a full suite of tidal constituents from records exceeding 20 days in duration for the SANAE and Georg von Neumayer bases on the data-sparse Queen Maud Land coast. The results agree well with those predicted by a modern numerical model [Schwiderski, 1983]. Further measurements of this nature may be an effective way of calibrating such models in other coastal areas of Antarctica. Longer records are required at most locations, including Queen Maud Land, in order to separate the tidal components more accurately. The observed strong tidal retardation by ice floes also requires further investigation.

Acknowledgments. The deployment of the gauge at SANAE was undertaken by D. Pim of NRIO in cooperation with R. Wonnacoot and G. I. Hudson of the Surveys and Mapping Branch of the Department of Community Development. Tidal constituents were calculated by A. M. Shipley from a tape prepared by R. L. Neilson. We thank these colleagues as well as the ship's complement of the RV S.A. Agulhas for their support. The Atka Iceport tide gauge was deployed during the maiden voyage of the German RV Polarstern. The assistance of the ship's crew is gratefully acknowledged. Participation in the cruise was made possible through the Alfred-Wegener-Institut fur Polarforschung, Bremerhaven.

References

Academie des Sciences, Sommaire des activites scientifiques de l'annee 1961 et des programmes de recherches scientifiques pour l'annee 1963, Rapp. 4, Comite Scientifique Pour la Recherche Antarctique, Conseil International des Unions Scientifiques, Paris, 1962.

Alvarez, J.A., and F.J. Cobas, Rapid determination of tides at a point to the east of the Antarctic Peninsula, Argentine Republic, Bol. Argent. Serv. Hidrogr. Nav., 3(3), 217-223, 1966.

Association d'Oceanographie Physique, Catalogue of published mean sea level data (1807-1958), Publ. Sci. 23, 64 pp. Union Geod. et Geophys. Int., Permanent Service for Mean Sea-Level, Paris, 1961.

Association d'Oceanographie Physique, Monthly and annual mean heights of sea-level 1959 to 1961 and unpublished data for earlier years, Publ. Sci. 24, 59 pp., Union Geod. et Geophys. Int., Paris, 1963.

Australian Academy of Science, Summary of Antarctic research programmes, 1963 (1964) and proposed Antarctic research programmes, 1965 (1966), Natl. Comm. for Antarctic Res., Melbourne?, 1964-1965.

Bishop, J.F., and J.L.W. Walton, Problems encountered when monitoring tidal movement in extremely cold conditions, Polar Rec., 18(116), 502-505, 1977.

BNCAR (British National Committee on Antarctic Research), United Kingdom national report no. 7 to SCAR, Royal Society, London, 1965.

Cartwright, D.E., B.D. Zetler, and B.V. Hamon, Pelagic tidal constants, compiled by IAPSO advisory committee on tides and mean sea-level, IAPSO Publ. Sci. 30, 65 pp., Union Geod. et Geophys. Int., Paris, 1979.

Comite Chileno de Investigaciones Antarcticas, Informe Nacional al SCAR, Ano 1967, Actividades 1966, Programas 1968, Santiago, 1967.

Comite Chileno de Investigaciones Antarcticas, Informe Nacional al SCAR, Ano 1971, Actividades 1970-71, Programas 1972, Santiago, 1972.

Dietrich, G., General Oceanography: An Introduction, 588 pp., Interscience, New York, 1963.

Doodson, A.T., Australasian Antarctic Expedition, 1911-1914, Scientific Reports, Series A, Vol. II, Oceanography, 2, Tidal Observations, 65-85, Govt. Printer, Sydney, 1939.

Dubrovin, L.I., Sea level fluctuations in the Lazarev Station area, Sov. Antarct. Exped. Inf. Bull., Engl. Transl., 4(4), 213-214, 1962.

Dubrovin, L.I., and I.M. Simonov, Tides in the Novolazarevskaya station area, Sov. Antarct. Exped. Inf. Bull., Engl. Transl., 5(4), 272-275, 1965.

Dubrovin, L.I., and M. Zalewski, Tides in the inlets of Bunger Hills, Sov. Antarct. Exped. Inf. Bull., Engl. Transl., 7(1), 20-33, 1969.

Flinders University, Tidal data, Data Register, 2, 11 pp., Horace Lamb Centre for Oceanogr. Res., Adelaide, South Australia, 1969.

Foldvik, A., and T. Kvinge, Bottom currents in the Weddell Sea: Results of long time current meter moorings at 74°S, 40°W, during IWSOE 1968-1973, Rep. 37, 43 pp., Univ. of Bergen, Geophys. Inst., Bergen, Norway, 1974.

Foldvik, A., T. Kvinge, and T. Tørresen, Bottom currents near the continental shelf break in the Weddell Sea, this volume.

Gilmour, A.E., W.J.P. MacDonald, and F.G. van der Hoeven, Winter measurements of sea currents in McMurdo Sound, N.Z.J. Geol. Geophys., 5(5), 778-789, 1962.

Godfroy, R.E., Etude sur les Marees, Deuxieme Expedition Antarctique Francais (1908-1910), commandee par le Dr. Jean Charcot, 74 pp., Masson et Cie, Paris, 1912.

Heath, R.A., Tidal constants for McMurdo Sound, Antarctica, N.Z.J. Mar. Freshwater Res., 5(2), 376-380, 1971.

Hisdal, V., On the tides at Norway Station, and adjacent coastal areas of Antarctica, Norwegian Antarctic Expedition, 1956-60, Sci. Res. 9, Skr. 133, 21 pp., Nor. Polar-Inst., Oslo, 1965.

Holdsworth, G., Some effects of ocean currents and wave motion on the dynamics of floating glacier tongues, this volume.

Hori, S., and E. Inbe, Tides at Syowa Station, Antarct. Rec., 32, 48-54, 1968.

Imbert, B., Nouveaux enregistremernts de maree en Terre Adelie, Bull. Inf. Comite Central Oceanogr. Etude Cotes, 5(7), Paris, 1953.

Imbert, B., Terre Adelie 1950-1952, Marees, 2, Expeditions Polaires Francaises, 87 pp., Paris, 1956.

Instituto Antarctico Argentino, Report to SCAR on Antarctic scientific activities for the year(s) 1959 and 1960 (to 1967) and planned program for 1962 (to 1969), Buenos Aires, 1961-1968.

Instituto Antarctico Argentino, National Report to SCAR on Antarctic scientific activities for the year 1969 and planned program for 1971, Direccion Nacional del Antarctico, Buenos Aires, 1970.

Instituto Antarctico Argentino, National report to SCAR on Antarctic scientific activities for the years 1970-71 (to 1972-73) and planned program for 1972 (to 1974), Prog. Rep. 12-15, Direccion Nacional del Antarctico, Buenos Aires, 1971-1973.

Intergovernmental Oceanographic Commission, Operational sea-level stations, Tech. Ser. 23, 40 pp., UNESCO, Paris, 1983.

International Hydrographic Bureau, Tides--Harmonic constants, Spec. Publ. 26, Monte Carlo, Monaco, 1966.

Irish, J.D., and F.E. Snodgrass, Australian-Antarctic tides, in Antarctic Oceanology II: The Australian-New Zealand Sector, Antarct. Res. Ser., vol. 19, edited by D.E. Hayes, pp. 101-116, AGU, Washington, D.C., 1972.

Kudryatsev, N.F., A periodic sea-level fluctuation in Ardley Inlet, Sov. Antarct. Exped. Inf. Bull., Engl. Transl., 8(3), 152-154, 1973.

Lutjeharms, J.R.E., A catalogue of sea level measurements in the Southern Ocean, Spec. Rep. 63, 141 pp., Dep. Oceanogr., Univ. of Wash., Seattle, 1976.

Lutjeharms, J.R.E., Sea level in the Southern Ocean: A catalogue of measurements, CSIR Res. Rep. 365, 206 pp., Nat. Res. Inst. for Oceanol., Stellenbosch, South Africa, 1980.

Lutjeharms, J.R.E., and M.M. Alheit, Long-term sea level mesurements: A global catalogue, CSIR Tech. Rep. T/SEA 8210, 99 pp., Nat. Res. Inst. for Oceanol., Stellenbosch, South Africa, 1982.

Lutjeharms, J.R.E., Die bepaling van seevlak in die Suidelike Oseaan, S. Afr. Tydskr. Natuurwet. Tegnol., 4,(1), 29-34, 1985.

MacAyeal, D.R., Tidal rectification below the Ross Ice Shelf, Antarctica, this volume.

MacDonald, W.J.P., and A.L. Burrows, Sea-level recordings at Scott Base, Antarctica, 1957, N.Z.J. Geol. Geophys., 2(2), 297-314, 1959.

McKee, W.D., A note on the sea-level oscillations in the neighbourhood of the Drake Passage, Deep Sea Res., 18(5), 547-549, 1971.

National Academy of Sciences, Report to SCAR on United States Antarctic research activities for the International Geophysical Year 1957-58, Rep. 1, Committee on Polar Research, Washington, D.C., 1959.

National Research Council, Report to SCAR on United States Antarctic research activities, 1972-73, United States research activities planned for 1973-74, Rep. 15, Committee on Polar Research, National Academy of Sciences, Washington, D.C., 1973.

National Science Foundation, Report to SCAR on United States Antarctic research activities 1959, Rep. 2, Committee on Polar Research, Washington, D.C., 1960.

Norwegian Academy of Science and Letters, Report to SCAR on Norway's Antarctic research activities, October 1976 (1977)-September 1977 (1978), Norway's Antarctic research activities planned for October 1977 (1978)-September 1978 (1979), Norwegian National Committee on Polar Research, Oslo, 1977-1978.

Nowlin, W.D., and R.D. Pillsbury, F DRAKE 1979 operations aboard R.V. Melville, Antarct. J. U.S., 14, 121-123, 1979.

Nowlin, W.D., S.L. Patterson, R.D. Pillsbury, and G.C. Anderson, Contributions of R.V. Melville to F DRAKE, 1975, Antarct. J. U.S., 19, 144-146, 1975.

Oka, K., and S. Fuchinoue, Oceanographic data of the 23rd Japanese Antarctic Research Expedition from November 1981 to April 1982, JARE Data Rep. 91 (Oceanography 3), 5 pp., Natl. Inst. of Polar Res., Tokyo, 1984.

Oura, H., and K. Fujino, Tides of Syowa Station, Antarct. Rec., 24, 14-17, 1965.

Permanent Service for Mean Sea-Level, Monthly and annual mean heights of sea-level for the period of the International Geophysical year (1957 to 1958) and unpublished data for earlier years, Publ. Sci. 20, 65 pp., Ass. Oceanogr. Phys., Union Geod. et Geophys. Int., Paris, 1959.

Permanent Service for Mean Sea-Level, Monthly and annual mean heights of sea-level, 1962 to 1964, Publ. Sci. 26 Monogr. 30, 109 pp., Ass. Int. des Sci. Phys.-des Oceans, Union Geod. et Geophys. Int., Paris, 1968.

Pillsbury, R.D., and S.S. Jacobs, Preliminary observations from long-term current meter moorings near the Ross Ice Shelf, Antarctica, this volume.

Pollak, W.H., and D. Sharwood, Gravimetric determination of ocean tidal effect on the Fimbul Ice Shelf, Princess Martha Coast, Queen Maud Land, S. Afr. J. Antarct. Res., 1, 30-32, 1971.

Pratt, J.G.D., Tides at Shackleton, Weddell Sea, Trans-Antarctic Expedition, 1955-58, Sci. Rep. 4, 21 pp., American Philosophical Society, Philadelphia, 1960.

Roberts, B., and R.H. Corkan, Tidal observations in Graham Land, British Graham Land Expedition, 1934-37, Sci. Rep., 1(8), 327-335, British Museum (Natural History), London, 1941.

Robinson, E.S., R.T. Williams, H.A.C. Neuberg, C.S. Rohrer, and R.L. Ayers, Southern Ross Sea tides, Antarct. J. U.S., 10(4), 155-157, 1975.

Robinson, E.S., R.T. Williams, H.A.C. Neuberg, C.S. Rohrer, and R.L. Ayers, Interaction of the ocean tide and the solid earth gravity tide in the Ross Sea area of Antarctica, preliminary results, Ann. Geophys., 33(1/2), 147-150, 1977a.

Robinson, E.S., H.A.C. Neuberg, R.T. Williams, B.B. Whitehurst, and G.E. Moss, Provisional cotidal charts for the southern Ross Sea, Antarct. J. U.S., 7(4), 48, 1977b.

Royal Society, National Antarctic Expedition 1901-1904, iii + 192 pp., + map of folios, London, 1908.

Royal Society, Scientific work completed by the Falkland Islands Dependencies Survey in Antarctica during 1959 (1960) and provisional plans for 1961 (1962), U.K. Prog. Rep., 2(3), submitted to the Fourth (Fifth) Meeting of SCAR, Br. Natl. Comm. on Antarct. Res., London, 1960-1961.

Royal Society, 1) Scientific work completed by the British Antarctic Survey in Antarctica during 1961 and provisional plans for 1963,

2) The Royal Society expedition to Tristan da Cunha 1962, U.K. Prog. Rep. 4, submitted to the Sixth Meeting of SCAR, Br. Natl. Comm. on Antarct. Res., London, 1962.

Royal Society, Scientific work completed by the British Antarctic Survey in Antarctica during 1962 and provisional plans for 1964, U.K. Prog. Rep. 5, submitted to the Seventh Meeting of SCAR, Br. Natl. Comm. on Antarct. Res., London, 1963.

Royal Society, Report to SCAR, U.K. Natl. Rep. 6(7), Br. Natl. Comm. on Antarct. Res., London, 1964-1965.

Royal Society, United Kingdom report on Antarctic research 1965/1967 (1966/1968, 1967/1969), Br. Natl. Comm. on Antarct. Res., London, 1966-1968.

Royal Society, Summary of scientific work completed in the Antarctic during 1968 and provisional plans for 1970, United Kingdom report on Antarctic research 1969/1970, Br. Natl. Comm. on Antarct. Res., London, 1969.

Royal Society, United Kingdom Antarctic research 1970 report; April 1969-March 1970: record of activities; October 1970-September 1971: planned activities, Br. Natl. Comm. on Antarct. Res., London, 1970.

Royal Society, United Kingdom Antarctic research 1971 report; April 1970 (to 1973)-March 1971 (to 1974): records of activities; winter 1971 (to 1974): current activities; October 1971 (to 1974)-September 1972 (to 1975): planned activities, Br. Natl. Comm. on Antarct. Res., London, 1971-1974.

Royal Society of New Zealand, Summary of research programme 1959, Report to SCAR, Rep. 2, Natl. Comm. for Antarct. Res., Christchurch, 1960.

Royal Society of New Zealand, Summary of research programmes, 1962 and proposed research programmes, 1964, Report to SCAR, Rep. 5, Natl. Comm. for Antarct. Res., Christchurch, 1963.

Samov, M.M., and A.V. Kopteva, Tides in the Mirnyy observatory region, (in Russian), Inf. Byull. Sov. Antarkt. Eksped., 1, 73-78, 1958. (Engl. Transl., Sov. Antarct. Exped. Inf. Bull., 1, 46-51, 1964.)

Schureman, P., Manual of harmonic analysis and prediction of tides, Spec. Publ. 98, U.S. Coast and Geod. Surv., 1958.

Schwiderski, E.W., Atlas of ocean tidal charts and maps, I, The semidiurnal principal lunar tide M_2, Mar. Geod., 6(3-4), 219-265, 1983.

Science Council of Japan, National report of the Japanese Antarctic research expedition, compiled by the National Antarctic Committee, 1966.

Servicio de Hidrografia Naval, Tide tables for the year 1963, 1, Coasts of America including Argentine Antarctica, Argentine Republic, Publ. H.698, 610 pp., 1962.

Shamontyev, V.A., Tides in Zaliv Alasheyev Bight (in Russian), Inf. Byull. Sov. Antarkt. Eksped., 43, 31-32, 1963. (Engl. Transl., Sov. Antarct. Exped. Inf. Bull., 5(1), 15-16, 1965.)

Shamontyev, V.A., Results of observations of tidal phenomena in the Molodeshnaya Station area (in Russian), Tr. Sov. Antarkt. Eksped., 44, 115-123, 1965.

Shesterikov, N.P., Sea level variations in the Mirnyy observatory area (in Russian) Inf. Byull. Sov. Antarkt. Eksped., 11, 29-32, 1959. (Engl. Transl. Sov. Antarct. Exped. Inf. Bull., 2, 18-21, 1964.)

Shesterikov, N.P., and L.I. Dubrovin, Tides in the Lazarev Station region, Sov. Antarct. Exped. Inf. Bull., (Engl. Transl.) 1(5), 52-54, 1963.

Shipley, A.M., Recent developments in tidal analysis in South Africa, in Proceedings of the Symposium on Tides, Organized by the International Hydrographic Bureau, Monaco, 28-29 April 1967, pp. 59-74, UNESCO, Paris, 1967.

Shipley, A.M., Tidal analysis, CSIR Rep. T/SEA 7911, 26 pp., Natl. Res. Inst. for Oceanol., Stellenbosch, South Africa, 1977.

Shipley, A.M., Tidal prediction program, CSIR Rep. SEA 8001, 4 pp., + addenda, Natl. Res. Inst. for Oceanol., Stellenbosch, South Africa, 1980.

Snodgrass, F.E., Eltanin cruise 41, Antarct. J. U.S., 6(1), 12-14, 1971.

Sous-Comite Antarctique Francais, Rapport d'activite des expeditions Antarctiques Francaises 1957-1959, Annee Geophisique Internationale, Paris, 1959.

Suzuki, M., and T. Kurano, Report on tidal observation of the 19th Japanese Antarctic Research Expedition (from January to December 1978), Antarct. Rec., 74, 290-299, 1982.

Sverdrup, H.U., Tidal currents off the Antarctic Ice Barrier, Queen Maud Land, Arch. Meteorol. Geophys. Bioklimatol., Ser. A,7, 385-390, 1954.

Tarasov, A.N., Sea level variations in Alasheev Bay in Enderby Land (in Russian), Okeanologiia, 4(2), 265-266, 1964.

Thiel, E., A.P. Crary, R.A. Haubrich, and J.C. Behrendt, Gravimetric determination of ocean tide, Weddell and Ross seas, Antarctica, J. Geophys. Res., 65(2), 629-636, 1960.

Titov, V.B., and N.P. Shesterikov, Propagation and character of the tidal wave in the Antarctic ocean, Sov. Antarct. Exped. Inf. Bull., (Engl. Transl.) 5(3), 170-172, 1965.

U.S. Navy Hydrographic Office, Oceanographic atlas of the polar seas, I, Antarctica, Washington, D.C., 1957.

Van Grondelle, A., New tidal measurements in Antarctica Queen Maud Land, Breid Bay, Hydrogr. Newsl., 1(5), 295-302, 1967.

Van Loon, H., Half-yearly oscillations in the Drake Passage, Deep Sea Res., 19(7), 525-527, 1972.

Vanney, J.R., and G.L. Johnson, GEBCO bathymetric sheet 5:18 (circum-Antarctic), this volume.

Venzke, N.C., Tidal measurements at McMurdo Sound, Antarct. J. U.S., 6(5), 231-232, 1971.

Villain, C.M., Cartes des lignes cotidales dans les oceans, Ann. Hydrogr., 4, 3, 1952.

Vorobyev, V.N., Contribution to the study of long-period tides in the Antarctic, Sov. Antarct. Exped. Inf. Bull., (Engl. Transl.) 8(8), 433-435, 1973.

Vorobyev, V.N., and E.N. Uranov, Statistical structure of sea level fluctuations in Alasheyev Bight in winter (in Russian), Inf. Byull. Sov. Antarkt. Eksped., 103, 54-60, 1983.

Wearn, R.B., and D.J. Baker, Bottom pressure measurements across the Antarctic Circumpolar current and their relation to wind, Deep Sea Res., 27(11A), 875-888, 1980.

Williams, R.T., and E.S. Robinson, Tidal currents in the sea beneath the Ross Ice Shelf, Antarct. J. U.S., 11(3), 159-161, 1976.

Williams, R.T. and E.S. Robinson, Ocean tide and waves beneath the Ross Ice Shelf, Antarctica, Science, 203(4379), 443-445, 1979.

(Received June 18, 1984;
accepted May 13, 1985.)

OCEANOGRAPHIC INFLUENCES ON SEDIMENTATION ALONG THE ANTARCTIC CONTINENTAL SHELF

Robert B. Dunbar, John B. Anderson, and Eugene W. Domack[1]

Department of Geology, Rice University, Houston, Texas 77251

Stanley S. Jacobs

Lamont-Doherty Geological Observatory of Columbia University, Palisades, New York 10964

Abstract. Continental shelf surface sediment samples from the Ross Sea west to the George V Coast are examined in relation to the ocean circulation. The relative concentrations of ice-rafted debris, biogenic phases, and current-reworked material in modern surface sediments are controlled by the interaction of glacial, biological, and oceanographic processes. Ice rafting is presently an important sedimentary process only within a few tens of kilometers of the coast where outlet glaciers drain. Ice-rafted debris is a minor component of surface sediments near the front of the Ross Ice Shelf, consistent with basal melting landward of the northern edge of the shelf ice. Siliceous biogenic material is a significant component of surface sediments in shelf basins. Its distribution is strongly influenced by marine currents, which generally decrease with depth and transport material onshore and east to west on the inner shelf. Marine currents rework relict glacial and glacial-marine deposits, leaving residual glacial marine sediments on shallow banks and on the outer shelf. Reworking of these very cohesive relict deposits is facilitated by biological mixing of surface sediments. Maximum sustained current velocities (1 meter off the bottom) indirectly derived from grain-size data range from < 8 cm/s in shelf basins and troughs to 20-25 cm/s on the outer shelf/upper slope. Pelagic sedimentation and reworking by marine currents appear to be the prime contributors of sediment to the shelf today, a situation that differs dramatically from that of the last glacial maximum when glacial and glacial-marine sedimentation were predominant on the shelf.

Introduction

The Antarctic continental shelf is unique in its great depth, rugged topography, and glacial setting. Glacial loading and erosion have produced an average shelf depth of about 500 m [Vanney and Johnson, this volume]. In many areas the shelf deepens in the onshore direction. Most of the coastline and nearshore region is covered by floating or grounded glacial ice, so that beaches are uncommon. This limits the potential roles of pack ice and anchor ice in supplying ice-rafted debris to deeper regions of the shelf. Antarctic glaciers supply sediment directly to the sea by subglacial sedimentation or by ice rafting of material entrained in icebergs or floating ice shelves and ice tongues. In contrast to the Arctic seas, where meltwater streams supply large amounts of sediment to the coastal zone, modern sediment input by meltwater is considered to be minimal in the Antarctic. Biogenic sedimentation rates are high on the deeper areas of the shelf, and in some areas aeolian deposition may be significant.

Given its great depth, absence of meltwater run-off, and lack of a wave-dominated coastal zone, marine currents and mass flow processes are likely to be key sedimentary agents on the continental shelf. Reworking of shelf sediments by glacial ice, an important process in the Arctic, is of minor importance in the Antarctic because relatively little of the shelf is sufficiently shallow (< 150-200 m) to be scoured by sea ice and icebergs. Because of these unique features, models for sediment accumulation on other continental shelves, including other high latitude areas, do not adequately depict sedimentation on the Antarctic shelf.

In this paper we describe the distribution patterns of surface sediments on the continental shelf from 150°W west to 140°E. This sector includes those portions of the Antarctic coastline where a sufficient sample density exists for such studies. Both compositional and textural sediment properties are examined and interpreted, along with the distribution patterns, in terms of present-day sedimentary processes.

Modern sediment supply to the Antarctic shelf is dominated by ice rafting, pelagic

---

[1] Now at Department of Geology, Hamilton College, Clinton, New York 13323.

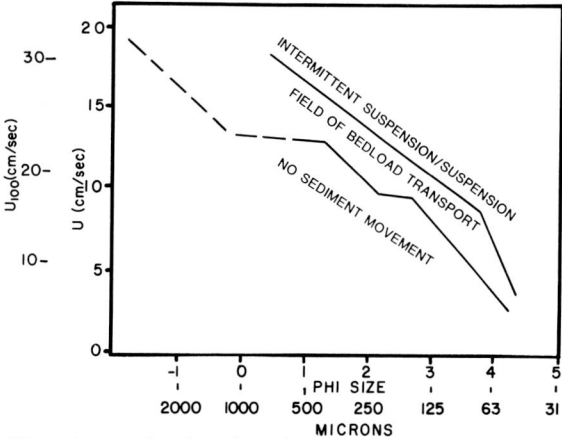

Fig. 1. Relationship between bottom current velocity and sediment transport by traction and saltation (bed load field), and intermittent suspension and suspension for particles of varying size. Field boundaries were derived from flume experiments [Singer and Anderson, 1984] utilizing poorly sorted sediments, mechanically mixed to simulate the effects of bioturbation. Velocities (U) measured between 1 and 3 cm above the bed are equivalent to those measured at higher levels in the flume and are thus taken as mean flow velocities. The current velocity 1 meter off the bottom in an oceanic setting (U_{100}) is calculated from mean flow velocity in the flume experiment following the procedure of Southard et al. [1971].

production and deposition, and the redistribution of material reworked from relict units on the seafloor. These processes involve vertical and/or horizontal transport of sediment through the shelf water column. Physical processes that control the structure and chemistry of the shelf water column will influence the nature of sedimentation. Therefore, we briefly discuss what is known about shelf circulation and currents in relation to sedimentology. For the Pennell Coast region, we show how current directions may be inferred from mineralogical patterns in surface sediments. Our ultimate goal is to integrate knowledge of shelf water column processes with surface sediment distribution patterns in order to formulate accurate models of sediment deposition on the Antarctic shelf.

Although links between the oceanography and sedimentology of the Antarctic shelf are not well defined at this time, the sediments provide a potential proxy record of shelf water column conditions. Sediment textural and compositional data can be used to derive bottom current intensity and direction as well as information about biological productivity in surface waters. Once the credibility of such proxy techniques is more firmly established, they will become an additional tool for examining the response of the Antarctic marginal seas to past global climatic perturbations.

Estimation of Relative Bottom Current Intensity

A variety of interrelated physical parameters influence the supply of terrigenous and biogenic sediments and their distribution on the Antarctic shelf. The current and wind regimes, light intensity, water column stability, seasonal ice coverage, and wave energy are among the most important. One principal variable which will be discussed in this paper is bottom current intensity, which can be determined at least qualitatively from textural and compositional analysis of surface sediments. Sediment grain-size distribution can provide important information about suspension versus bottom transport mechanisms and maximum velocities of bottom currents. Our approach to grain size/current speed relationships is based upon results from flume experiments by Southard et al. [1971] and Singer and Anderson [1984], so some discussion of the method is warranted.

Shields [1936] and Hjulstrom [1939] experimented with particle transport versus flow conditions, and published curves that are still widely cited. Those experimenters measured the bottom shear stress or velocity at which particles of a given size experienced initial transport. Their studies have limited application to sedimentology, in part because threshold velocities were defined by the initial movement of uniform sediment grains in equigranular (well-sorted) beds. In the marine environment, sediments are often poorly sorted, and complex sediment transport mechanisms include traction, saltation, intermittent suspension and suspension [e.g., Middleton and Southard, 1977]. These transport mechanisms operate simultaneously on grains of different sizes, with each mechanism involving a quantum increase in transport rate. At any given velocity, different size particles are thus transported and deposited by various mechanisms, and different size populations may be segregated through time. The velocity required to transport particles of a given size can only be determined by knowing how those particles were transported.

Visher [1969] and Middleton [1976] recognized that grain-size distributions of sand samples consist of discrete modes which represent the different transport populations. Although the polymodal nature of many marine sediments renders a strict Visher-type analysis difficult, useful information on transport mode can be acquired by examining specific regions of the grain-size spectrum that show evidence of current influence, e.g., near truncation points. Assuming that transport modes

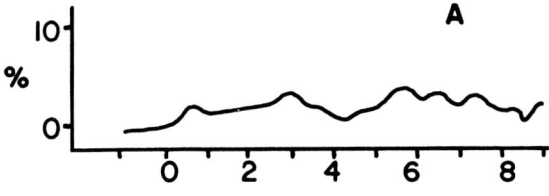
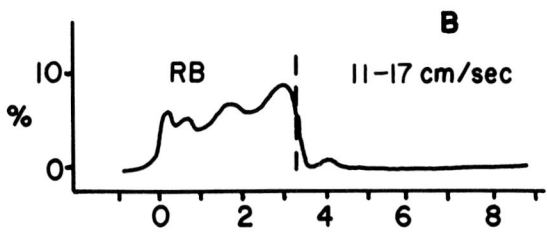
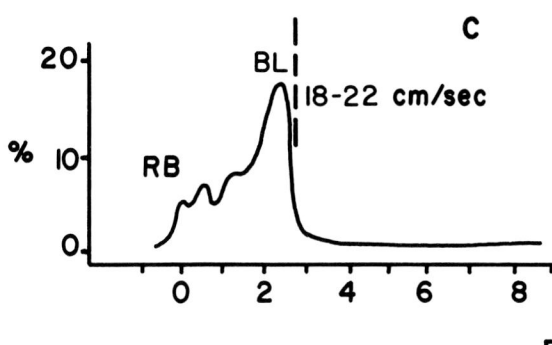
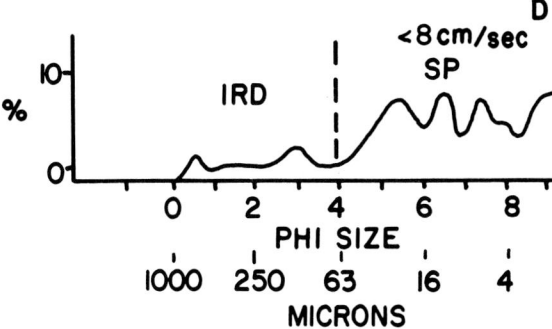

Fig. 2. Examples of typical grain-size curves for Antarctic surface sediment samples and the inferred current velocites. Sample A is a typical unsorted glacial-marine sediment similar in size distribution to starting material used in flume experiments of Singer and Anderson [1984]. Sample B is a residual bed (RB) resulting from winnowing of material finer than 90 μm (3.5 φ). Sample C shows a bed load mode (BL) resulting from winnowing of material finer than 150 μm (2.75 φ) and associated residual bed. Sample D consists of unsorted ice-rafted debris (IRD) in a typical fine-grained basin deposit where most material has settled out of suspension (SP).

could be discerned by grain size analysis, Singer and Anderson [1984] conducted a series of experiments to measure critical transport velocities for different particle sizes, by various transport mechanisms. Their experiments differed from previous flume studies in that they used an initially unsorted bed, designed to resemble glacial and glacial-marine sediments of the Antarctic continental shelf.

In poorly sorted, cohesive sediment beds only the very surface of the bed is eroded in the absence of sediment mixing. After fine grains are removed from the top of the bed, a surficial lag of coarser particles protects (armors) the bed from further erosion. In the marine environment armoring is prevented by biological mixing, thus rendering bottom sediments more susceptible to current erosion. The high standing stock of benthic organisms on most areas of the Antarctic continental shelf [e.g., Bullivant and Dearborn, 1967; Everson, 1977] and analyses of box cores [Nittrouer et al., 1984] suggest that sediment mixing by bioturbation is important. Singer and Anderson [1984] simulated the effects of bioturbation by mechanically mixing the sediment bed. They justify using a much faster mixing rate than occurs in the marine environment on the grounds that the rate of mixing only influences the efficiency of bed erosion; shear velocity (or shear stress) influences the size of particles eroded from the bed. Their results demonstrate that without thorough mixing of the bed, unsorted sediments are not sufficiently eroded to produce measurable grain size differences at mean velocities less than 15 cm/s (measured to within 1 to 3 cm of the bed). With sediment mixing, silts and clays were resuspended at velocities less than 5 cm/s to produce a residual bed consisting of unsorted sand and gravel. These results are consistent with theoretical treatments of the subject of sediment suspension processes [Blatt et al., 1980; Middleton and Southard, 1977; Anderson and Kurtz, 1985]. Singer and Anderson also measured the mean current speeds at which different size sand grains moved along the floor of the flume by either traction or saltation, using sampling trays placed at various levels on or above the floor of the flume, one meter downstream of the bed. Their field for bed load transport is shown in Figure 1. The critical transport velocities derived through their experiments are lower than those obtained in previous studies [e.g. Shields, 1936]. These differences are ascribed to the fact that Singer and Anderson used a rough unsorted bed (made to resemble glacial marine sediment of the Antarctic shelf) whereas previous workers have dealt mainly with smooth, well sorted beds. The actual shear velocity acting on the bed will be greater for a current of given velocity passing over a rough (unsorted) bed than over

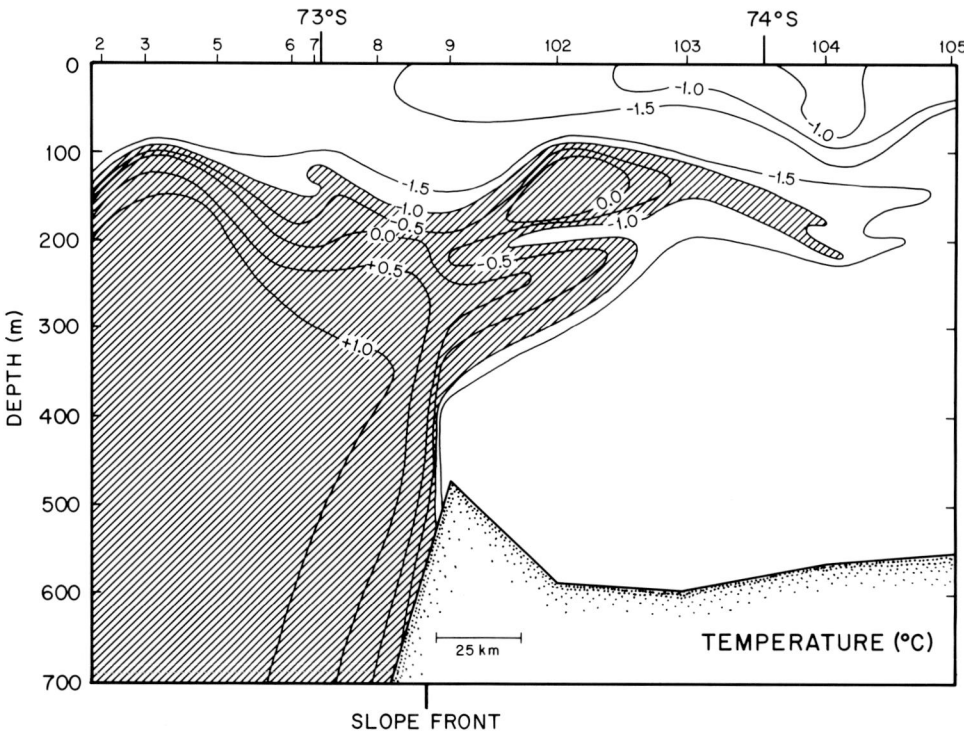

Fig. 3. Temperature section across the continental shelf break in the northwest Ross Sea near 178°E, December 19-20, 1976. Temperatures were measured by a salinity-temperature-depth instrument (STD 2-9) and by expendable bathythermographs (XBT 102-105) from the U.S. Coast Guard icebreaker <u>Northwind</u> [Jacobs and Haines, 1982]. Relatively warm water (T > -1.0°C) intrudes onto the continental shelf above the slope front that separates very cold (-1.9°C) shelf water from deep water (>+1.0°C) north of the shelf.

a smooth bed [Middleton and Southard, 1977]. Because of these differences, the Singer and Anderson curve was used to derive current speed for samples containing bed load transport fractions. Singer and Anderson measured mean flow velocity which, in the case of flume experiments involving shallow uniform flow, can be converted to the current velocity one meter off the bed using

$$\frac{U_y}{U_*} = 5.6 \log_{10} \frac{\rho U_* y}{\mu} + 4.9$$

[Daily and Harleman, 1966], where U_y is the velocity at distance y above the bed, U_* is shear velocity, ρ is fluid density, and μ is fluid viscosity. The shear velocity (U_*) is derived from the mean flow velocity using

$$\frac{U_{mean}}{U_*} = 5.75 \log_{10} \frac{\rho U_* h}{\mu} - 4.72$$

[Rouse, 1949], where h is flow depth. See Southard et al. [1971] for a justification of these computations when dealing with shallow uniform flow in flume experiments. The method used in deriving current speed from sediment grain size data is illustrated in the following set of examples. For a more detailed description of this method see MacDonald and Anderson [1985].

Sample A in Figure 2 is a typical unsorted glacial sediment. Material finer than 90 μm (3.5 ∅) has been eroded from sample B relative to sample A, resulting in the formation of a residual bed (RB). This implies either sediment mixing and removal of fines via bed load transport by currents with velocities (U_{100}) of at least 11 cm/s, the lowermost line in Figure 1, or winnowing via intermittent suspension at current velocities of at least 17 cm/s, the boundary between the bed load and intermittent suspension fields. Sediment finer than 150 μm (2.75 ∅) has been removed from sample C, and a sorted bed load (BL) population (177 μm, 2.5 ∅) is clearly illustrated. Traction is a slow process, and the traction mode coexists with the residual bed material (1000 to 250 μm, 0 to 2 ϕ) in sample C. Here, implied current speeds 1 meter off

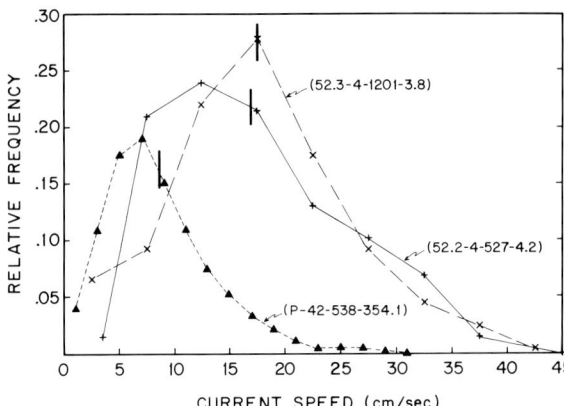

Fig. 4. Current speed frequency records from three instruments moored near bottom in the Ross Sea. Vertical bars denote mean currents. Within the brackets, read (mooring designator-height above bottom (m)-water depth (m)-record length (days)). The moorings were located at 72°55'S, 177°20.2'E on the continental slope (52.3), 73°21'S, 176°57.9'E on the outer shelf (52.2), and 78°05.5'S, 175°30'W near the Ross Ice Shelf (P). Data from Jacobs et al. [1974] and Pillsbury and Jacobs [this volume].

the bottom are between 18 and 22 cm/s, the range of the bed load field for particles 177 μm in diameter. Sample D shows a typical fine grain size population (SP) associated with unsorted ice-rafted debris (IRD), and reflects sedimentation under quiescent bottom conditions (< 8 cm/s).

There are several limitations to this technique. The original grain size distribution of the source material influences the distribution of respective transport modes. If the input material were well sorted, e.g. recycled beach sands, this kind of analysis would be of little use. However, the diverse sediment sources and unique supply mechanisms on the Antarctic continental shelf make it difficult to model the formation of an extensive well-sorted modern deposit, except via bottom current influences and aeolian transport. The latter has been shown to be important in McMurdo Sound [Barrett et al., 1983]. Regional variations may exist in the grain-size distributions of terrigenous sediments supplied from the continent, but our underlying assumption is that all of this source material is poorly sorted. This assumption is supported by the poorly sorted nature of glacial sediments in the area (e.g., sample A in Figure 2).

Following the procedure of Southard et al. [1971] we have calculated ocean bottom velocities at one meter above the seabed (U_{100}) using the flume data of Singer and Anderson [1984]. The transformation is sensitive to a number of factors that influence erosion velocity (bed cohesion, roughness, and fluid turbulence) which we are unable to precisely evaluate in the field. Calibration of sediment textural data to absolute velocity within the water column is hampered by the absence of boundary layer studies and the scarcity of direct current measurements near the seafloor on the Antarctic continental shelf.

Methods

Samples used in this study include trigger core tops, Dietz-La Fonde grab samples, and samples taken from the tops of box cores. Grain-size analyses were performed using an automated settling tube for sand-sized material (> 63 μm, 4 φ), and a hydrophotometer [Jordan et al., 1971] for fine-grained (< 63 μm) sediments. For samples collected on the George V shelf, the amount of terrigenous material within the very fine sand and coarse silt fraction of siliceous oozes (31 μm to 125 μm, 8.5 φ to 3 φ) was based on 300 point counts on each of 15 processed slides.

For our studies of biogenic silica and organic carbon distribution, we utilized 230 sediment grab samples that we collected and another 25 samples taken from the tops of box cores collected in the southern Ross Sea by DeMaster et al. [1983]. The sediment samples were dried, gently ground with mortar and pestle, and sieved through a 425 μm (1.25 φ) screen to remove coarse crystalline particles. Biogenic silica was determined by a kinetic dissolution experiment in 0.1 molar NaOH at 85°C following the procedures of DeMaster [1979, 1981]. We modified his technique by measuring silica concentrations at 4 hours and using data from the 3-, 4-, and 5-hour samplings to determine the clay mineral influence. As described by DeMaster [1981], grinding produces a slight increase in weight percent biogenic silica calculated for sediment samples with < 4% biogenic silica. Organic carbon contents were determined using a LECO analyzer following standard techniques [Boyce and Bode, 1972]. Weight percent biogenic silica and organic carbon values reported here are not corrected for sea salt content of the dried samples, which ranges from 1 to 5%.

Continental Shelf Circulation and Currents

Ocean circulation near the Antarctic continental margin is marked by seasonal waxing and waning of sea ice, prevailing winds off the ice sheet, large lateral property gradients combined with upwelling and sinking over the continental slope, net melting of the large floating masses of glacial ice, and currents dominated by a westward drift and diurnal tides. The continental shelf is largely ice covered from April through October, and portions of it remain beneath the sea ice and glacial ice year-round [Naval Polar Oceanography Center, 1974-1985]. Even during the winter period, however, leads and polynyas along the coastline constitute more than 15% of the shelf area north of the ice shelves, and sea ice that forms in these open water areas is rapidly moved offshore [Kurtz and Bromwich, this volume; Zwalley et al., this volume]. Winter brine drainage increases the salinity of shelf waters, the densest of which accumulate on the western sides of the wider shelves and in depressions along the coastline [Jacobs et al., 1970, this volume; Gordon and Tchernia, 1972; Killworth, 1974]. In combination with surface winds, the resulting density field leads to a cyclonic (clockwise) circulation pattern in the large embayments, with onshelf flows at shallow to intermediate depths east of and above the high density regions (Figure 3).

Near the continental shelf break, some of the denser shelf waters flow along and down the slope, mixing with, entraining and modifying the deep water or producing Antarctic Bottom Water [Carmack and Killworth, 1978; Foldvik et al., this volume]. At some locations, deep water or its derivatives upwell onto the continental shelf and subsequently provide heat for melting glacial ice [Jacobs et al., 1979]. Over the upper continental slope, a well-developed subsurface frontal zone (Figure 3) is characterized by sharp horizontal thermohaline gradients, evidence of deep vertical mixing and relatively high levels of biological productivity [Ainley and Jacobs, 1981].

Few current measurements have been made very near bottom on the Antarctic continental shelf. From a number of brief (5-10 min) direct current observations about 1 m above the seafloor in the Ross Sea, Jacobs et al. [1970] reported currents in excess of 15 cm/s on the outer shelf and generally less than 10 cm/s on the inner shelf. More recent measurements, some for a full year, also show stronger near-bottom currents near the continental shelf break. Two of the curves in Figure 4 show current speed distributions for 4-day moorings near 1200 m on the continental slope and 525 m on the outer shelf in the Ross Sea. The left-hand curve in Figure 4 displays similar data from a 354-day record at 508 m near the Ross Ice Shelf. Mean current speed at the inner shelf site, 42 m above the seafloor, is only half that at the outer shelf and slope sites, 4 m above bottom. The 4-day records were obtained during the austral summer, a time when both higher and lower energy levels have been noted in yearlong records on the Antarctic continental shelf [Foldvik et al., this volume; Pillsbury and Jacobs, this volume]. Bimodal current speed distributions could result in the formation of fine laminations in some shelf sediments, even without seasonal variations in productivity. More importantly, the frequent incidence of high current speeds, particularly on the outer shelf (Figure 4) raises the prospect that high-energy events may play as important a sedimentologic role as sustained, slower drifts much as described by Hollister and McCave [1984].

Most current measurements at intermediate depths have been taken beneath the sea ice, either in McMurdo Sound [Tressler and Ommundsen, 1962; Heath, 1977; Carter et al., 1981; Lewis and Perkin, this volume] or along and under the ice shelves [Jacobs et al., 1979; Jacobs and Haines, 1982; Pillsbury and Jacobs, this volume; Potter and Paren, this volume]. These measurements revealed currents that are dominated by diurnal periods, and commonly range up to half a knot (≈ 25 cm/s). The mean directions for several year-long observations near the ice shelf show significant southward and westward components. Currents associated with vertical overturning during winter sea ice formation events are likely to reach the seafloor at some locations, particularly beneath coastline leads associated with katabatic winds off the ice sheet [Kurtz and Bromwich, this volume; Cavalieri and Martin, this volume].

The near-surface circulation, seasonal sea ice distribution, and water column stratification all influence biological productivity and the resulting concentrations of biogenic components in shelf sediments. Surface currents are variable, but appear to be strongest (in excess of 1 knot (≈ 50 cm/s)) near the coastline or ice shelf [U.S. Naval Oceanographic Office, 1960]. Substantial currents have also been measured directly beneath the sea ice [Mitchell and Bye, this volume]. Geostrophic currents are not large, but there are indications from the density field and from ship drift that surface flows set to the east on the south side of the shelf break/slope front, and to the west on its north side [Ainley and Jacobs, 1981].

Nutrient levels are of minor concern in surface waters on the shelf, since abundances seem to be rarely depleted below relevant thresholds by phytoplankton blooms. The most stable surface layers will be those longest exposed to summer insolation, specifically in the southwest Ross Sea. Sea ice usually persists throughout the austral summer over the slope and outer shelf in the eastern Ross Sea, along the Victoria Land coast, the Pennell Coast west of 167°E, and on the George V Coast east of 147°E [Naval Polar Oceanography Center, 1974-1985].

Ross Sea (150°W to 163°E)

Physiography

The bathymetry of the Ross Sea north of the Ross Ice Shelf has been described and mapped by Lepley [1966], Hayes and Davey [1975], Vanney et al. [1981] and Vanney and Johnson [this volume]. The shelf has an average depth around 500 m and slopes toward the continent (Figure 5). It is bounded on the east by Sulzberger Bay, which contains several small basins with depths as great as 900 m. The central Ross Sea shelf is characterized by a series of elongate, north to northeast trending banks ($\approx$ 300 m) and basins (> 500 m). The western shelf is more rugged and deeper, particularly near the Victoria Land coast where outlet glaciers have eroded the seafloor to create narrow transverse troughs, down to 1200 m in the case of Drygalski basin. The irregular topography of the western Ross Sea shelf is further accentuated by volcanic islands and seamounts of the McMurdo Sound complex, which extends along a roughly north-south line from McMurdo Sound to Cape Adare.

Sediments

The Ross Sea was sampled as part of Eltanin cruises 27, 32, and 52 and during "Deep Freeze" cruises on U.S. Coast Guard icebreakers (DF 76, 78, 80, 83, and 84). Sedimentological studies have been conducted by Stetson and Upson [1937], Kennett [1966], Chriss and Frakes [1971], Glasby et al. [1975], Kellogg et al. [1979], Barrett et al. [1983], and Anderson et al. [1984]. Fifty-nine surface grab samples collected in the Ross Sea during DF 84 provide an important basis for this discussion.

Surface sediments in the Ross Sea are composed of mixtures of unsorted ice-rafted debris, siliceous biogenic material (mainly diatom frustules), calcareous shell debris, and terrigenous silts and clays that have been transported in suspension by marine currents. The concentrations and distributions of these various components (Figure 6) reflect the relative influence of glacial, oceanographic, and biological processes. Sediments consisting of greater than 20% ice-rafted debris and associated with fine-grained current-derived terrigenous material are referred to as Compound Glacial Marine (CGM) after Anderson et al. [1980a]. These are common in Sulzberger Bay and in a relatively narrow zone along the Victoria Land coast. Compound glacial marine sediments with > 10% diatoms (dCGM) occur on the central shelf east of 180°. Current-winnowed sediments containing abundant calcareous shell debris are referred to as Residual Glacial Marine (RGM), also after Anderson et al. [1980a]. These sediments blanket most of the outer continental shelf and upper continental slope and the tops of banks in the western Ross Sea. Terrigenous silts and clays (cZ and zC) make up more than 80% of the surface sediments collected along the front of the Ross Ice Shelf east of approximately 180°.

To the west of 180°, the silts and clays contain 10-50% by weight biogenic silica and are labeled SiM (siliceous mud) and SiO (siliceous ooze). Sands and muddy sands were sampled in western McMurdo Sound and off the Edward VII Peninsula.

The concentration of coarse (larger than 63 μm, 4 φ) ice-rafted debris (IRD) in surface sediments increases in an offshore direction from the Ross Ice Shelf (Figure 7). The low concentrations near the calving line of the ice shelf suggest that melt-out of most basal debris occurs landward of the barrier, consistent with the oceanographic data [Jacobs et al., this volume; MacAyeal, this volume].

Grain-size data from central and outer shelf samples exhibit a prominent mode between 125 and 177 μm (3 φ to 2.5 φ, Figure 7). At this time we are unable to evaluate the contribution of aeolian debris to this mode. Aeolian material is known to be of importance in some coastal areas [Barrett et al., 1983], especially in regions affected by katabatic winds. We would expect some size sorting to occur during the aeolian transport process. We note that, although the absolute concentration of ice-rafted material (aeolian and glacial) varies across the shelf, sands from different areas exhibit consistent truncation points at 125 μm (3 φ, Figure 7). This implies that the maximum sustained velocity (U_{100}) of bottom currents on the shelf is greater than 18 cm/s, the minimum velocity required to transport this size material by intermittent suspension and suspension (Figure 1), therefore removing it from the bed. At this velocity, sands in the 125 to 177 μm range are transported as bed load (Figure 1) and the occurrence of a sorted mode in many samples implies that bottom currents are indeed acting on these sediments.

An increase in the relative proportion of fine-grained material (finer than 125 μm, 3 φ) in an onshore direction reflects a lower frequency of high velocity (> 18 cm/s) bottom current events. These general trends are supported by bottom current records from the shelf (Figure 4). Also important is the observation that the frequency of low velocity (<8 cm/s) events that will facilitate sedimentation of fine-grained components is significantly greater on the inner shelf (Figure 4).

We attribute the increased concentration of residual ice-rafted and current derived sands along the outer shelf and tops of shallow banks to more efficient winnowing of fine sediments. Increased winnowing efficiency would result from more frequent strong flow events, more extensive sediment mixing by bottom dwellers, and/or reduced sediment input. The winnowed sediments, including diatom frustules, are probably transported both north across the continental shelf break into the deep ocean and south toward the deeper basins on the in-

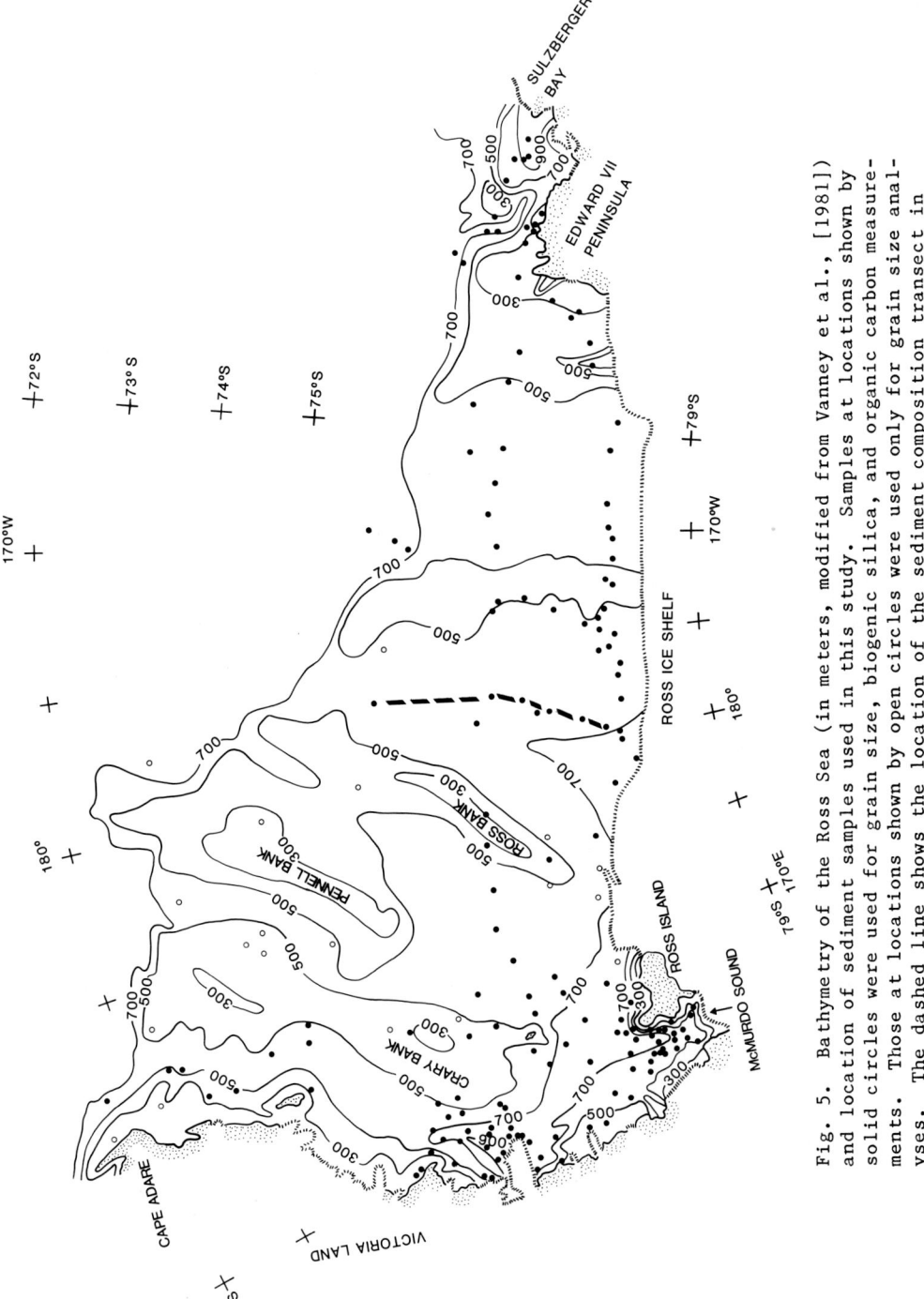

Fig. 5. Bathymetry of the Ross Sea (in meters, modified from Vanney et al., [1981]) and location of sediment samples used in this study. Samples at locations shown by solid circles were used for grain size, biogenic silica, and organic carbon measurements. Those at locations shown by open circles were used only for grain size analyses. The dashed line shows the location of the sediment composition transect in Figure 14.

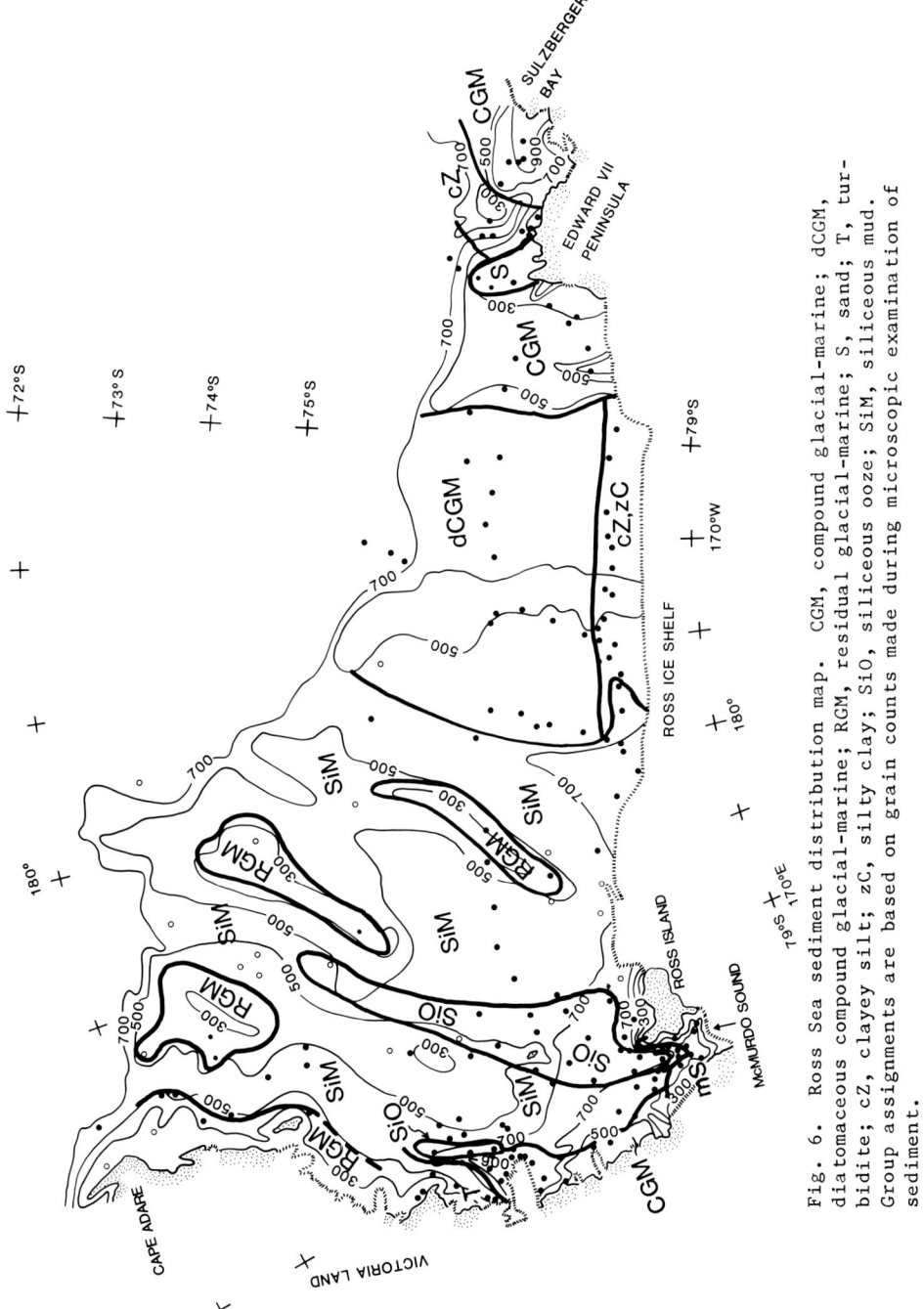

Fig. 6. Ross Sea sediment distribution map. CGM, compound glacial-marine; dCGM, diatomaceous compound glacial-marine; RGM, residual glacial-marine; S, sand; T, turbidite; cZ, clayey silt; zC, silty clay; SiO, siliceous ooze; SiM, siliceous mud. Group assignments are based on grain counts made during microscopic examination of sediment.

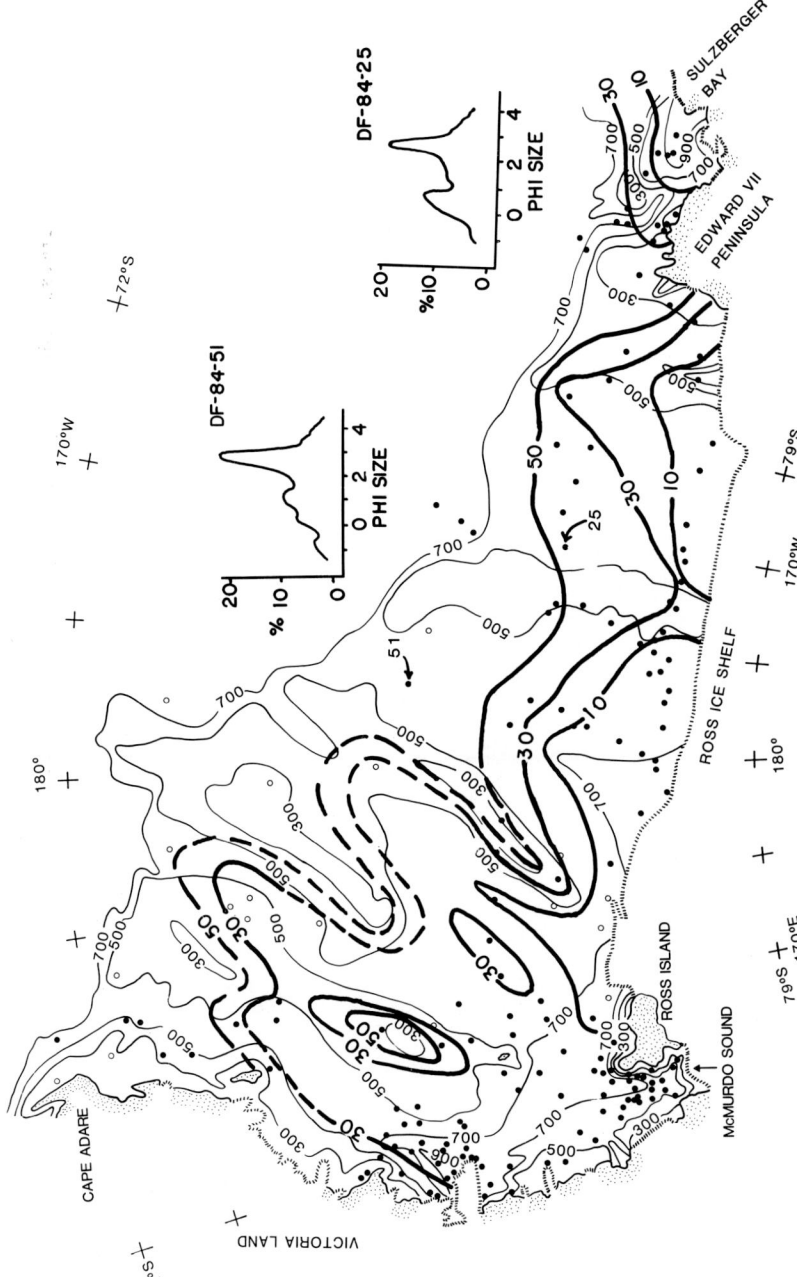

Fig. 7. Bold contours show weight percent of coarse (> 63 μm) ice-rafted debris in Ross Sea surface sediments. Light contours show bathymetry in meters. Also shown are two representative sand fraction grain-size frequency curves for surface sediments from the central Ross Sea. All samples show clear evidence of winnowing, with the finer than 125 μm (3 φ) component partially removed. The sorted 125 μm to 150 μm (3φ-2.5φ) mode is possibly of aeolian origin.

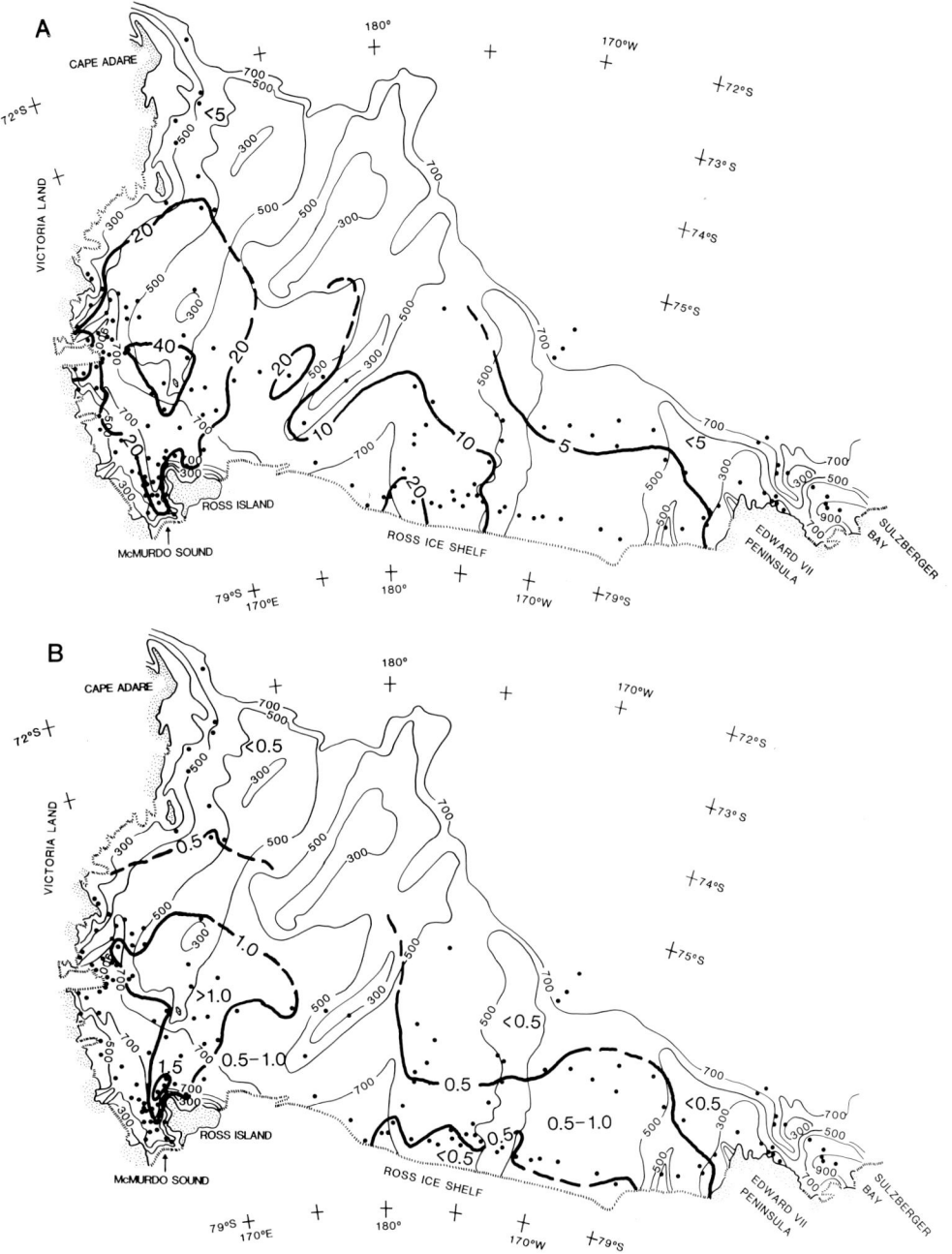

Fig. 8. Bold contours of weight percent concentrations of (a) biogenic silica and (b) organic carbon in surface sediments of the Ross Sea. Light contours show bathymetry in meters.

ner shelf where more quiescent bottom conditions are probable. It is possible that some portion of the IRD maximum along the outer continental shelf results from a higher frequency of iceberg tracks there [e.g., Tchernia and Jeannin, 1983], specifically via deposition from valley glacier icebergs that are more likely than ice shelf icebergs to be sediment laden.

The distribution of biogenic silica in surface sediments increases in the onshelf and east-to-west directions (Figure 8a). This pattern agrees with the observations of Truesdale and Kellogg [1979], who attributed variability in the diatom content of surface sediments to less extensive summer sea ice cover in the western Ross Sea. The highest concentration of biogenic silica (40% by weight) oc-

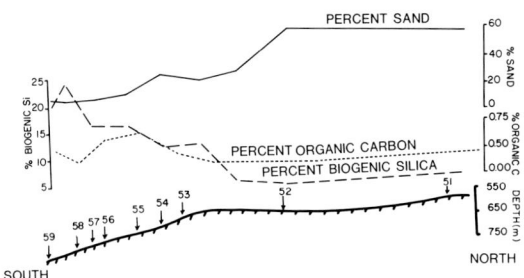

Fig. 9. Concentrations of sand, organic carbon and biogenic silica, along a north-south transect on the Ross Sea continental shelf (dashed line, Figure 5).

curs on the southern and western flanks of Crary bank. This is approximately the region where Smith and Nelson [1985] reported extremely high biogenic silica concentrations (up to nearly 1600 mmol/m^2 integrated from 0 to 150 m) in a surface layer stabilized by meltwater from the receding pack ice edge. The high accumulation rates (1-3 mm/yr) of sediments in the western Ross Sea and their potential importance to the global silica budget are discussed by DeMaster et al. [1983] and Ledford-Hoffman [1984].

Other factors also regulate the distribution of siliceous material on the seafloor. For example, open water is present during the austral summer along much of the Ross Ice Shelf, and some of the highest levels of primary production have been measured in the southeastern Ross Sea [El-Sayed et al., 1983]. High concentrations of opaline material also occur in the surface waters of Sulzberger Bay (R. Dunbar and A. Leventer, unpublished data, 1983), although the sediments of this region are depleted in biogenic components. Organic opaline debris from these locations and from regions of high productivity near the continental slope is likely to be transported westward and onshelf by the mean current flow. Diatom oozes on the Ross Sea shelf are mainly in the 16 μm to 63 μm (6 φ to 4 φ) range, a size that can be maintained in suspension by weak currents (a few centimeters per second). The opal distribution is thus strongly influenced by the hydrographic regime, especially for diatom frustules that settle through the water column as discrete particles rather than in fecal pellets. Fecal pellets accounted for only a minor portion of the total opal flux collected in sediment traps from the central Ross Sea and McMurdo Sound [Dunbar, 1984; Dunbar et al., 1984]; most material was transported as small aggregates of organic opaline debris. The westward drift of surface and subsurface water on the southern half of the shelf will serve to concentrate biogenic silica in that direction.

The mean circulation, the general decrease in current speed with depth, and the deeper water in the southwest Ross Sea will together concentrate fine grained sediments there. Samples from a transect taken across the central Ross Sea (Figure 9) illustrate the effects of winnowing and redeposition. From the outer shelf to the inner shelf there is a threefold decrease in sand content, while the opal content increases from less than 10% to greater than 20% by weight.

The organic carbon content of sediments also increases from east to west and, to a lesser degree, from north to south in the Ross Sea (Figure 8b). Despite relatively high opaline silica content (> 20%), sediments from depressions in the central Ross Sea tend to be depleted (< 0.50%) in organic carbon (Figure 9). We believe this reflects a more rapid degradation of organic carbon relative to dissolution of silica in the cold well-oxygenated waters of the Antarctic shelf. Thus, as winnowing and transport continue, the organic carbon/opal ratio of a reworked sediment is decreased. Sediments with > 1% organic carbon by weight blanket approximately 50,000 km^2 of the western Ross Sea. Higher organic carbon/opal ratios in these biogenic sediments indicate a lesser degree of reworking via resuspension in the water column there than in the central Ross Sea.

A persistent polynya is present in Terra Nova Bay [Kurtz and Bromwich, this volume] and might be expected to be accompanied by higher levels of illumination, surface warming, and higher annual productivity. However, the greatest enrichment in opal content of these sediments occurs not beneath the polynya but immediately to the east on the western flank of Crary bank (Figure 10a). The region of greatest organic carbon enrichment appears slightly west of the area of maximum opal content (Figure 10b). Surface productivity in the bay may in fact be low because of frequent deep mixing by strong winds and as a result of brine drainage from newly formed sea ice. Alternatively, freshly produced biogenic debris may be rapidly removed to the north and east by surface currents generated by offshore winds. In addition, biogenic sediments in the Drygalski basin may be diluted by terrigenous material arriving as wind-blown debris [Barrett et al., 1983], or rafted from the Drygalski Ice Tongue. It is not known to what extent the transport of terrigenous components into the basin might be facilitated by high density plumes sinking as a result of sea ice formation. Detailed analyses of biogenic components and sediment accumulation rates are needed to further elucidate controls on sedimentation in the region.

Pennell Coast (166°E to 171°E)

Physiography

The continental shelf of the Pennell Coast region of north Victoria Land has an average

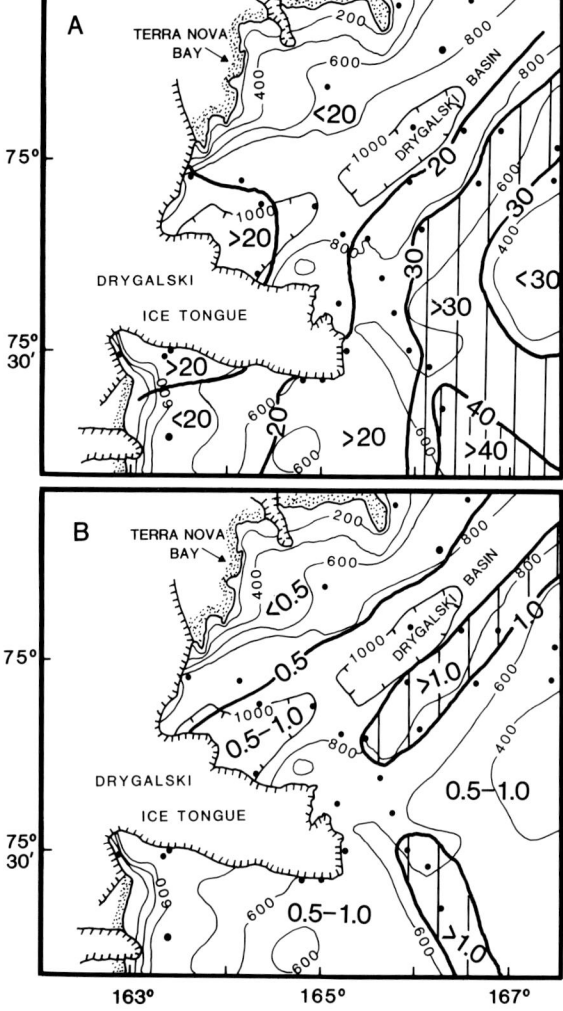

quired during DF 80. Sands, muddy sands, and residual glacial marine sediments blanket the outer shelf, and finer-grained sediments occupy the inner shelf (Figure 11). Diatomaceous oozes, muds, and sandy muds are accumulating below about 350 m in glacial troughs. These deposits consist of a mixture of diatom frustules, terrigenous silts and clays, and very fine volcanic sand. Trough sediments contain less than 10% by weight ice-rafted sand and gravel. Perennial sea ice may limit the source of biogenic material west of about 167°E, assuming ice-edge productivity enhancement does not outweigh the effect of ice cover.

Poorly sorted glacial-marine sediments (CGM) are confined to a narrow (<20 km) coastal zone between Cape Adare and Dennistoun Glacier (71°11'S, 168°00'E), and to the region west of Yule Bay (70°44'S, 166°40'E). Residual glacial-marine sediments (RGM) occur on the shallower (<250 m) portions of the shelf between Dennistoun Glacier and Yule Bay. The distribution of sands and muddy sands on the outer shelf suggest a continuation of the sand unit toward the northwest Ross Sea (Figure 6).

There are two distinctly different sources of terrigenous sediment along the Pennell Coast of north Victoria Land, the quartz-rich Robertson Bay Group which outcrops along the entire coastline west of Robertson Bay, and

Fig. 10. Bold contours show weight percent concentrations of (a) biogenic silica and (b) organic carbon in surface sediments of the Terra Nova Bay region of the western Ross Sea. Light contours show bathymetry in meters. The polynya in Terra Nova Bay [Kurtz and Bromwich, this volume] lies directly north of the Drygalski Ice Tongue and commonly occupies a 1000-km^2 area, roughly equivalent to 1° longitude by 20' latitude.

depth just over 200 m, which makes it one of the shallowest areas of the Antarctic continental margin (Figure 11) [Vanney and Johnson, this volume]. The inner shelf is rugged, while the outer shelf exhibits relatively smooth topography [Brake and Anderson, 1983]. Transverse troughs are associated with large outlet glaciers.

Sediments

The following discussion is based upon analyses of grab samples and piston cores ac-

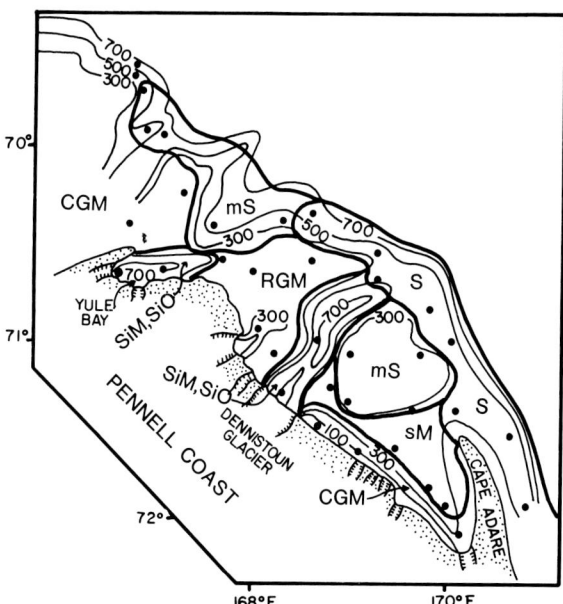

Figure 11. Bathymetry (in meters, from Brake and Anderson [1983]) and surface sediment distribution map for the Pennell Coast continental shelf. RGM, residual glacial-marine sediment; CGM, compound glacial-marine sediment; S, sand; mS, muddy Sand; sM, sandy Mud; SiM, siliceous mud; and SiO, siliceous ooze. Dots show station locations.

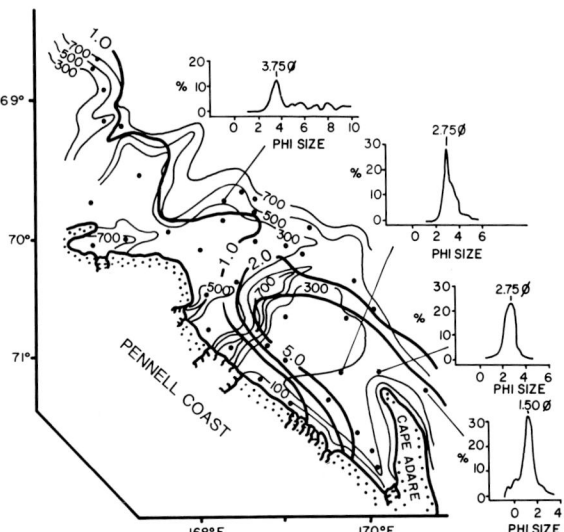

Fig. 12. Bold contours show volcanic sand/quartz sand ratios for sands and muddy sands of the Pennell Coast continental shelf [from Brake, 1982]. Light contours show bathymetry in meters. Also shown are representative grain-size frequency curves.

the McMurdo Sound Volcanic Group, restricted to the Cape Adare region [Brake, 1982]. Sediment transport paths can thus be determined through mineralogical analyses, such as the volcanic sand/quartz sand ratio (Figure 12). Volcanic sands comprise the major fraction of terrigenous sediment accumulating on the central and outer shelf, and are transported as far as 60 km west of Cape Adare. Bottom current velocities may be estimated from the grain size distributions in Figure 12. East of Cape Adare, bottom current velocities (U_{100}) of up to 25 cm/s are indicated by relatively coarse bed load fractions (175 μm to 350 μm, 2.5 ϕ to 1.5 ϕ). Current speeds of 16 to 22 cm/s to depths of 350 m would be needed to transport the volcanic sands to the west of Cape Adare. The deep basins adjacent to the coast are accumulating diatomaceous muds with opal contents up to 35% and are therefore characterized by more sluggish circulation.

George V Continental Shelf (140°E to 150°E)

Physiography

The bathymetry of the George V Coast is typical of other portions of the East Antarctic continental shelf in that it is characterized by a rugged and deep inner shelf (Figure 13) [Vanney and Johnson, this volume]. The inner shelf is dominated by the George V Basin, which attains depths greater than 1000 m parallel to the coastline and apparently extends beneath the Mertz and Ninnis Glacier tongues. The rest of the shelf is smoother, with alternating banks (≈ 200 m) and depressions (≈ 500-700 m). The continental shelf break occurs abruptly at 400-500 m. The continental slope exhibits two distinct morphologies. The area west of 143°30'E is characterized by highly irregular topography and a gentle slope. The eastern region is very steep along the upper slope, but marked by several large (> 300 km^2) isolated banks and knolls at depths of 2000 to 3000 meters on the lower slope, not shown in Figure 13.

Sediments

Fifty-six grab samples and piston cores were acquired from the George V continental shelf in early 1979 from the USCGC Glacier (Figure 13). Initial accounts of the surface sediments and fauna were given by Domack [1980, 1982], Milam and Anderson [1981], and Anderson et al. [1983]. A detailed description of relict glacial and glacial-marine deposits of the area can be found in the works by Anderson et al. [1980a] and Domack [1982]. Surface sediments obtained during the 1911-1914 Australasian Expedition to this region were examined for composition [Chapman, 1922], mineralogy [Von der Borch and Oliver, 1968], and palynology [Trueswell, 1982].

Surface sediments in this sector vary widely in texture and composition (Figure 14). Siliceous muds and oozes predominate below 500 m. Relict glacial and glacial-marine sediments occur on the westernmost shelf and along the flanks of Mertz and Ninnis banks. Residual glacial-marine sediments, sands, and muddy sands, often containing calcareous debris, blanket the upper slope and outer shelf and the easternmost shelf along the Ninnis bank. Calcareous biogenic sediments are found nearshore at depths of less than 200 m.

Siliceous muds and oozes are poorly to very poorly sorted, are typically laminated, and contain a low diversity arenaceous foraminiferal assemblage [Milam and Anderson, 1981]. The concentration of coarse ice-rafted material is low in both sediment types, but tends to be more concentrated in oozes (Figure 14).

Sediment-laden icebergs are rarely sighted in Antarctic waters, and this form of ice rafting probably supplies only a small amount of material to the modern sediments of the shelf. However, during Deep Freeze 79, several sediment laden icebergs were observed and sampled along the George V Coast (Figure 15). The material recovered from these icebergs was very poorly sorted, ranging in size from 1000 to < 4 μm (0 to 8 ϕ) [Anderson et al., 1980b]. Approximately 50% of the entrained debris was finer than 63 μm (4 ϕ).

Sediments with biogenic silica content greater than 10% and organic carbon content greater than 0.5% are generally restricted to

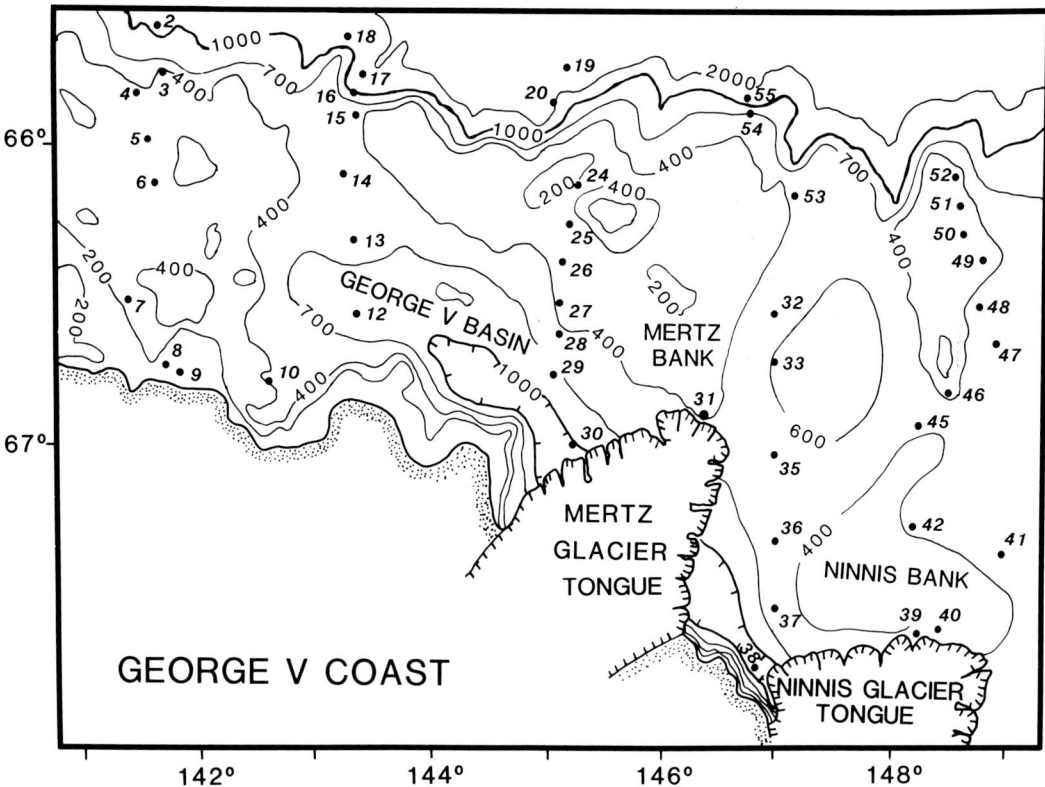

Fig. 13. Bathymetry (in meters, modified from Vanney and Johnson [1979] and Domack [1980]) of the George V continental shelf and locations of Deep Freeze 79 sediment sample stations (black dots).

depths below 500 m (Figure 16). There is an east to west increase in the thickness of biogenic units as well as in the biogenic silica and organic carbon concentrations of sediments within the George V basin. Up to 40 meters of diatomaceous ooze with a biogenic silica content greater than 30% has accumulated in the western end of the George V basin (Figure 16a). From 12 kHz subbottom profiler data, Domack and Anderson [1983] estimated the average thickness of these deposits to be 6 m and the areal extent to exceed 625 km^2. The accumulation rate for these sediments is about 0.3 cm/yr, based on ^{210}Pb analyses of the piston core at site 12 in Figure 13 (K. Cochran and D. DeMaster, personal communication, 1981).

The biogenic sediment distribution pattern results in part from the east to west transport of freshly produced opaline debris by surface water moving along the coastline. Sediment enrichment in opal and organic carbon is lower in the central and eastern portions of the George V basin, perhaps due to lower productivity related to the persistent sea ice cover east of 147°E. In addition, some sediment accumulating at deeper levels, such as in the western end of the George V basin, must be derived from the adjacent shallow banks where opaline sediments are absent. Information about winnowing on the shelf and upper slope is provided by grain-size data.

Residual glacial-marine sediments, sands, and muddy sands occur on the upper slope and shallow regions of the shelf and only differ in relative concentration of fine-grained matrix, i.e., mean grain size and sorting (Figure 14). These sediment types tend to be highly bioturbated and are in gradational contact with underlying glacial sediments. These stratigraphic and grain-size relationships indicate that these modern sediments are the by-products of reworked glacial sediments.

The efficiency with which fine-grained sediment is winnowed from a bed is primarily a function of mixing of the bed, whereas the size material eroded from the bed is velocity dependent, as demonstrated by Singer and Anderson [1984]. The size material being winnowed from these modern surface sediments is 125 μm (3 φ) and finer, which implies that they are subject to currents in the range of 18 to 22 cm/s. An offshore transition from residual glacial marine sediment to muddy sand and finally to well-sorted sands of the shelf break/upper slope is attributed to more efficient mixing by benthic organisms, lower sedi-

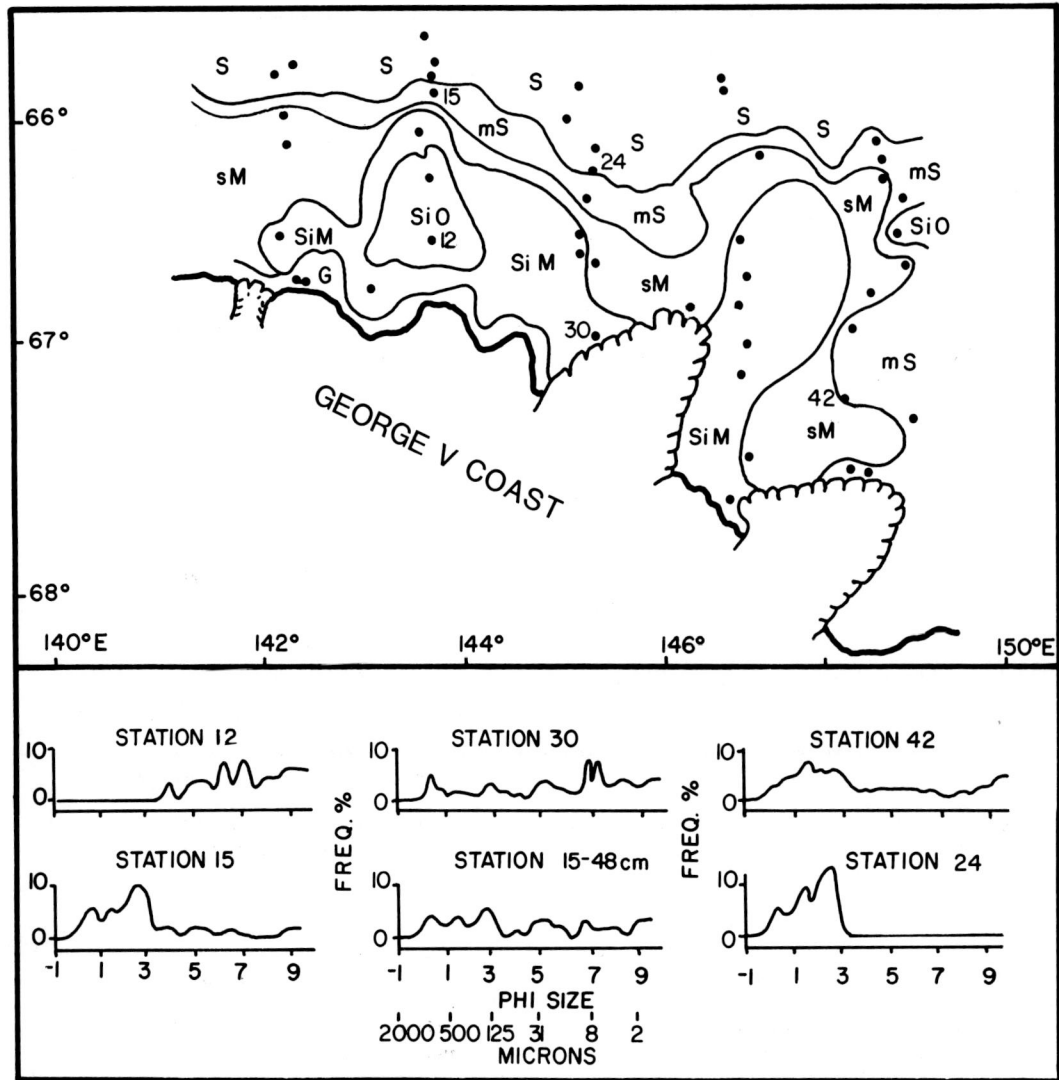

Fig. 14. Surface sediment distribution map for the George V continental shelf and representative grain size frequency curves for major sediment types. Grain-size data from an ice-rafted glacial marine unit (station 15 at 48 cm) illustrates the unsorted nature of sediment deposited on the shelf during the Pleistocene. SiM and SiO, siliceous mud and ooze, respectively. These two differ only in the relative concentration of siliceous biogenic material. The frequency curve for station 12 is representative of these poorly sorted, polymodal sediments. Sand (S), includes both well sorted saltation modes and poorly sorted residual sands with associated traction modes (cf. station 24). Poorly sorted sandy muds (SM) resemble compound glacial marine sediments but are more the product of reworking of relict glacial sediments than modern ice rafting and settling from suspension. The frequency curve for station 30 is representative of this group. Muddy sands (mS) exhibit evidence of winnowing (e.g., the frequency curves for stations 15 and 42) and therefore resemble residual glacial marine sediments, except that they lack significant amounts of gravel. Gravel (G) occurs in a shallow (< 150 m) nearshore zone.

mentation rates, or to more persistent, but not necessarily stronger, currents in that direction. Any of these factors would result in more efficient winnowing. In contrast, coarser sands on the continental slope imply stronger currents (U_{100} up to 25 cm/s).

The fine-grained fraction eroded from relict glacial sediments is redeposited in shelf basins along with freshly produced and reworked diatomaceous material to comprise siliceous muds and oozes. This locally derived fine-grained terrigenous material (very fine

Fig. 15. Large iceberg with thick debris zones observed during Deep Freeze 79 off the Ninnis Glacier, George V Coast. After Anderson et al. [1980b].

sand and coarse silt) comprises no more than 30% of the total sample weight of siliceous oozes. This would represent a mass of approximately 5×10^{14} g within our estimated volume for the present distribution of ooze (625 km^2 and 6 m thick). Reworking during the Holocene of a 20-cm surface layer of relict glacial and glacial-marine sediment over that portion of the shelf shallower than 500 m ($\approx$ 8000 km^2) would provide 4 to 10×10^{14} g of fine-grained terrigenous sand and silt. Therefore, the terrigenous component of fine-grained basin sediments can be derived via scouring of relict glacial and glacial-marine sediments on the adjacent shallower portions of the shelf. Ice rafting and meltwater run-off may also supply material to the shelf, but significant sediment input by these processes is not needed to produce a balanced sediment budget.

Conclusions

Surface sediments of the Ross Sea, Pennell, and George V continental shelves are similar in composition and texture and have distribution patterns that indicate similar processes are active in these areas. Sedimentation on the shelf is presently dominated by the production and settling of biogenic material and reworking by bottom currents. An aeolian-derived component may be important in some areas.

Compound glacial-marine sediments, which consist of unsorted ice-rafted debris and fine-grained terrigenous and biogenic material, occur in nearshore areas drained by outlet glaciers and across much of the central and eastern Ross Sea shelf. There appears to be a relatively limited input of terrigenous material by glaciers and meltwater at the present time. By contrast, the Antarctic continental shelf was blanketed with glacial and glacial-marine sediments during the last glacial maximum [Anderson et al., 1980a].

Residual glacial-marine sediments are very poorly sorted glacial-marine deposits from which fine-grained components have been winnowed by marine currents. They occur on shallow banks and on the outer continental shelf and typically display physical and textural evidence of bottom current reworking. Biological mixing of relict glacial and glacial-marine sediment is believed to decrease the cohesive properties of these deposits and ren-

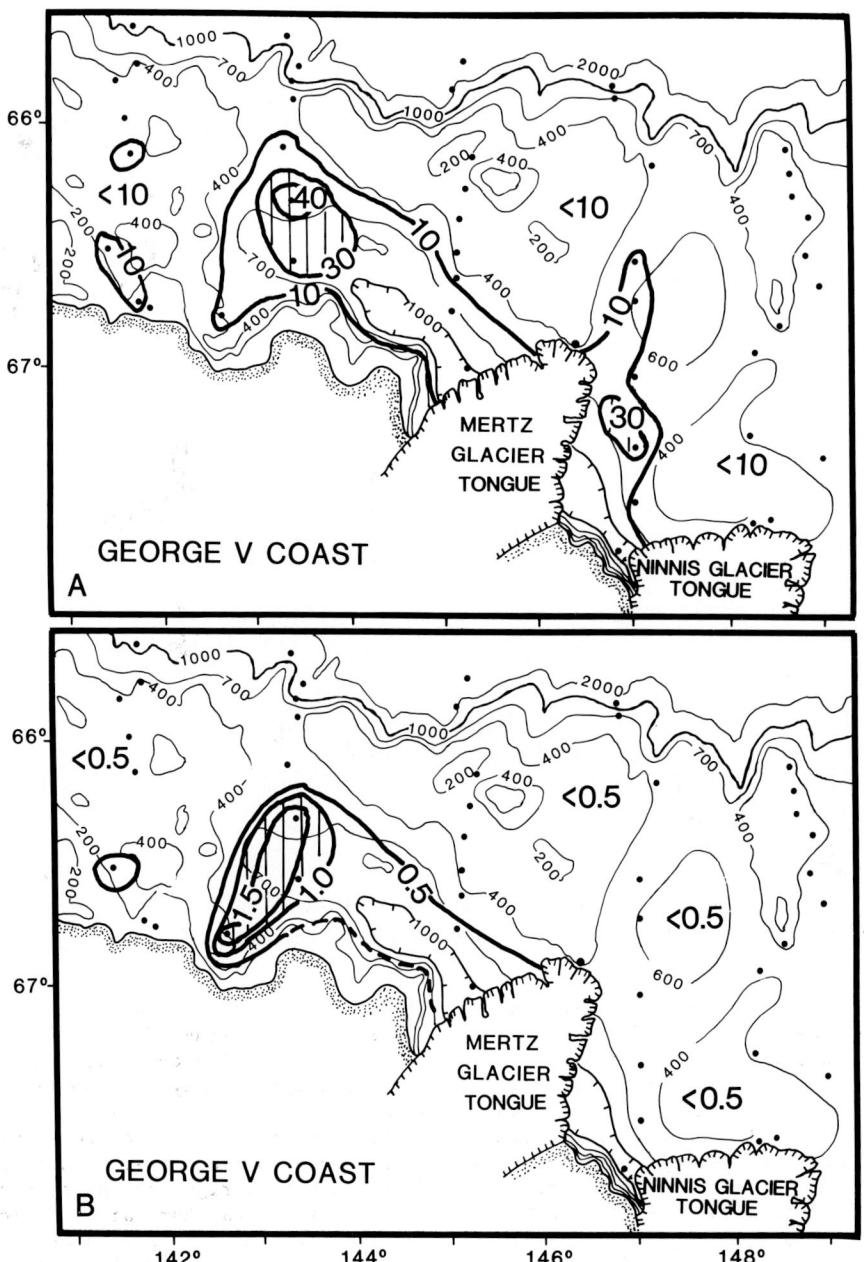

Fig. 16. Bold contours show weight percent concentrations of (a) biogenic silica and (b) organic carbon in surface sediments of the George V continental shelf. Light contours show bathymetry in meters. The concentrations of biogenic opal and organic carbon are high at the western end of George V trough and low on banks and ridges.

der them more susceptible to marine current influence.

Sediments containing up to 45% biogenic silica and up to 2% organic carbon are found in shelf basins. They frequently contain only minor amounts of ice-rafted debris and coarse terrigenous silt, due to rapid accumulation rates and/or limited input of terrigenous material via ice rafting and meltwater streams. These biogenic sediments are redistributed by marine currents, which tend to erode fine-grained sediments from shallow portions of the shelf and redeposit some portion of this material in shelf basins. These biogenic components are especially enriched in areas where reduced sea ice conditions prevail during summer months. Siliceous sediments of the Antarctic have anomalously low organic carbon/opal ratios compared to other continental shelf deposits. This may result from more rapid oxidation of organic carbon relative to dissolution of biogenic silica in the cold, oxygenated waters of the Antarctic, abetted by reworking of the shelf sediments.

In all of the areas studied, bottom currents are effectively reworking relict glacial and glacial-marine deposits on unobstructed regions of the outer and central shelf to depths of about 500 m, i.e., to about the depth of the continental shelf break. The degree of winnowing decreases in an onshore direction and with increasing depth. Grain size data suggest that sustained current speeds one meter off the bottom range from less than 8 to 25 cm/s. Maximum sustained velocities appear to be highest (20-25 cm/s) on the outer shelf and somewhat lower and more uniform on the central shelf. Seabed current velocities of only a few centimeters per second are inferred from the characteristics of material being deposited in shelf basins and troughs. These inferred velocities are in agreement with available current meter data in the Ross Sea.

Most sediments winnowed from Antarctic continental shelf deposits fall within the very fine sand to clay size range. Some experiments with flumes and observations of fluvial systems have indicated that cohesive beds comprised of material in this size range can only be winnowed and transported by currents with speeds in excess of 20 to 30 cm/s, measured or calculated 1 m above bottom [Hjulstrom, 1939; Sundborg, 1967]. We infer that lower threshold speeds (U_{100} < 20 cm/sec) will transport this size material on the Antarctic shelf, based, in part, upon the flume experiments of Singer and Anderson [1984]. Those experiments suggest that low velocity currents (< 8 cm/s) may be competent for erosion and transport of silts and clays, consistent with the work of Rees [1966]. Heezen and Hollister [1971] suggested that many previously determined relationships between erosion velocity and grain-size were not valid for abyssal sediments consisting of grains finer than medium sand (350 µm, 1.5 φ), because of variability in the degree of sediment cohesion. On the Antarctic shelf, sediment cohesion is reduced by bioturbation, and fine-grained sediment may thus be winnowed by relatively low-velocity currents.

In all three study areas, we infer onshore and east to west transport of surface sediments from compositional and textural variations. Sediment transport across the shelf break and down the continental slope also occurs, as evidenced by light-scattering observations and by shelf water on the slope and rise [Jacobs et al., 1970; Eittreim et al., 1972; Foldvik et al., this volume]. Estimates of current velocity based on grain-size analysis are reasonably consistent with the limited current meter data available in the Ross Sea. A more precise calibration of grain size data to bottom water current velocity must await detailed benthic boundary layer studies on the Antarctic continental shelf.

Acknowledgements. Financial and logistic support for this work was provided by the National Science Foundation, Division of Polar Programs, grants DPP-83-12486 (R.B.D.), DPP-83-15555 (J.B.A.) and DPP-81-19863 (S.S.J.). We thank D. DeMaster, W. Gardner, and other reviewers for their many helpful comments during three cycles of the review process. M. Smith collected the Deep Freeze 84 samples, and he and A. Leventer ably carried out essential laboratory work. Word processing by B. Hautau and D. Criscione. The officers and crews of U.S. Coast Guard icebreakers Glacier and Polar Sea facilitated our field work. Lamont-Doherty Geological Observatory contribution 3807.

REFERENCES

Ainley, D.G., and S.S. Jacobs, Sea-bird affinities for ocean and ice boundaries in the Antarctic, Deep Sea Res., 28A, 10, 1173-1185, 1981.

Anderson, J.B., and D.D. Kurtz, The use of silt grain size parameters as a paleo-velocity gauge: A critical review and case study, Geo. Mar. Lett., 4, in press, 1985.

Anderson, J.B., D.D. Kurtz, E.W. Domack, and K.M. Balshaw, Antarctic glacial-marine sediments, J. Geol., 88, p. 399-414, 1980a.

Anderson, J.B., E.W. Domack, and D.D. Kurtz, Observations of sediment-laden icebergs in Antarctic waters, Implications to glacial erosion and transport, J. Glaciol., 25, 387-396, 1980b.

Anderson, J.B., C. Brake, E.W. Domack, N. Myers, and J. Singer, Sedimentary dynamics of the Antarctic continental shelf, in Antarctic Earth Science, edited by R.L. Oliver, P.R. James, and J.B. Jago, pp. 387-389, Australian Academy of Science, Canberra, 1983.

Anderson, J.B., C.F. Brake, and N.C. Myers, Sedimentation on the Ross Sea continental shelf, Antarctica, Mar. Geol., 57, 295-333, 1984.

Barrett, P.J., A.R. Pyne, and B.L. Ward, Modern sedimentation in McMurdo Sound, Antarctica, in Antarctic Earth Science, edited by R.L. Oliver, P.R. James, and J.B. Jago, pp. 550-554, Australian Academy of Science, Canberra, 1983.

Blatt, H., G. Middleton, and R. Murray, Origin of sedimentary rocks, 782 pp., Prentice-Hall, Englewood Cliffs, N.J., 1980.

Boyce, R.E., and G.W. Bode, Carbon and carbonate analyses, Leg 9, in Initial Rep. Deep Sea Drill. Proj., 9, 797-816, 1972.

Brake, C.F., Sedimentology of the north Victoria Land continental margin, Antarctica, M.S. thesis, Rice Univ., Houston, Tex., 175 pp., 1982.

Brake, C.F., and J.B. Anderson, The bathymetry of the north Victoria Land continental margin, Mar. Geod., 6, 139-147, 1983.

Bullivant, J.S., and J.H. Dearborn, The fauna of the Ross Sea, 5, General accounts, station lists and benthic ecology, N. Z. Oceanogr. Inst. Mem., 32, 77 pp., 1967.

Carmack, E.C., and P.D. Killworth, Formation and interleaving of abyssal water masses off Wilkes Land, Antarctica, Deep Sea Res., 25, 357-370, 1978.

Carter, L., J.S. Mitchell, and N.J. Day, Suspended sediment beneath permanent and seasonal ice, Ross Ice Shelf, Antarctica, N. Z. J. Geol. Geophys., 24, 249-262, 1981.

Cavalieri, D.J., and S. Martin, A passive microwave study of polynyas along the Antarctic Wilkes Land Coast, this volume.

Chapman, F., Sea floor deposits from soundings, Australasian Antarct. Exped. 1911-1914, Sci. Rep., Ser. A, 2, 1-60, 1922.

Chriss, T.M., and L.A. Frakes, Glacial marine sedimentation in the Ross Sea, in Antarctic Geology and Geophysics, edited by R.J. Adie, Oslo, Universitetsforlaget, pp. 747-762, 1971.

Daily, J.W., and D.R.F. Harleman, Fluid Dynamics, Addison-Wesley, Reading, Mass. 154pp., 1966.

DeMaster, D.J., The marine budgets of silica and ^{32}Si, Ph.D. thesis, Yale Univ., New Haven, Conn., 1979.

DeMaster, D.J., The supply and accumulation of silica in the marine environment, Geochim. Cosmochim. Acta, 45, 1715-1732, 1981.

DeMaster, D.J., C.A. Nittrouer, and P.A. Ledford-Hoffman, Biogenic silica accumulation on the Antarctic continental shelf, Antarct. J. U.S., 18, 132-134, 1983.

Domack, E.W., Glacial marine geology of the George V-Adelie continental shelf, M.S. thesis, 142 pp., Rice Univ., Houston, Tex., 1980.

Domack, E.W., Sedimentology of glacial and glacial-marine deposits on the George V-Adelie continental shelf, East Antarctica, Boreas, 11, 79-97, 1982.

Domack, E.W., and J.B. Anderson, Marine geology of the George V continental margin: Combined results of Deep Freeze 1979 and the 1911-1914 Australasian Expedition, in Antarctic Earth Science, edited by R.L. Oliver, P.R. James, and J.B. Jago, Australian Academy of Science, Canberra, pp.402-406, 1983.

Dunbar, R.B., Sediment trap experiments on the Antarctic continental margin, Antarct. J. U.S., 19(5), 70-71, 1984.

Dunbar, R.B., A.J. MacPherson, G. Wefer, and D.C. Biggs, Biogenic fluxes from sediment trap experiments on the Antarctic margin, EOS Trans. AGU, 65, 917, 1984.

Eittreim, S., E.M. Thorndike, and M. Ewing, Vertical distribution of turbidity in the South Indian and South Australian basins, in Antarct. Oceanology II, The Australian-New Zealand Sector, Antarct. Res. Ser., vol. 19, edited by D. Hayes, pp. 51-58, AGU, Washington, D.C., 1972.

El-Sayed, S.Z., D.C. Biggs, and O. Holm-Hansen, Phytoplankton standing crop, primary productivity and near surface nitrogenous nutrient fields in the Ross Sea, Antarctica, Deep Sea Res., 30, 871-886, 1983.

Everson, I., The Southern Ocean: The living resources of the Southern Ocean, FAO Rep., GLO/SO/77/1, 156 pp., South. Ocean Fish. Surv. Prog., Rome, 1977.

Foldvik, A., T. Gammelsrød, and T. Torresen, Circulation and water masses on the southern Weddell Sea shelf, this volume.

Glasby, G.P., P.J. Barrett, J.C. McDougall, and D.G. McKnight, Localized variations in sedimentation characteristics of the Ross Sea and McMurdo Sound regions, Antarctica, N.Z. J. Geol. and Geophys., 18, 605-621, 1975.

Gordon, A.L., and P. Tchernia, Waters of the continental margin off Adelie Coast Antarctica, in Antarctic Oceanology II, The Australian-New Zealand Sector, Antarct. Res. Ser., vol. 19, edited by D. Hayes, pp. 59-69, AGU, Washington, D.C., 1972.

Hayes, D.E., and F.J. Davey, A geophysical study of the Ross Sea, Antarctica, Initial Rep. Deep Sea Drill. Proj., 28, 887-907, 1975.

Heath, R.A., Circulation across the ice shelf edge in McMurdo Sound, Antarctica, in Polar Oceans, edited by M.J. Dunbar, pp. 129-149, Arctic Institute of North America, Calgary, Alta., 1977.

Heezen, B.C., and C.D. Hollister, The Face of the Deep, 659 pp., Oxford University Press, New York, 1971.

Hjulstrom, E., Transportation of detritus by moving water, in Recent Marine Sediments, American Association of Petroleum Geologists, Tulsa, Okla., edited by P.D. Trask, 5-31, 1939.

Hollister, C.D., and I.N. McCave, Sedimentation under deep-sea storms, Nature, 309, 220-225, 1984.

Jacobs, S.S., and W.E. Haines, Ross Ice Shelf Project, Oceanographic Stations, 1976-1979, Tech. Rep. 82-1, 505 pp., Lamont-Doherty Geol. Obs., Palisades, N.Y., 1982.

Jacobs, S.S., A.F. Amos, and P.M. Bruchhausen, Ross Sea oceanography and Antarctic Bottom Water formation, Deep Sea Res., 17, 935-962, 1970.

Jacobs, S.S., E.B. Bauer, P.M. Bruchhausen, A.L. Gordon, T.F. Root, and F.L. Rosselot, Eltanin Reports, Cruises 47-51, 1971; 52-55, 1972, Tech. Rept. CU-2-74, 502 pp., Lamont-Doherty Geol. Obs., 1974.

Jacobs, S.S., A.L. Gordon, and J.L. Ardai,

Circulation and melting beneath the Ross Ice Shelf, Science, 203, 439-442, 1979.

Jacobs, S.S., R.G. Fairbanks, and Y. Horibe, Origin and evolution of water masses near the Antarctic continental margin: Evidence from $H_2^{18}O/H_2^{16}O$ ratios in sea water, this volume.

Jordan, C.F., Jr., G.E. Fryer, and E. Hemmen, Size analysis of silt and clay by hydrophotometer, J. Sediment. Petrol., 41, 489-496, 1971.

Kellogg, T.B., R.S. Truesdale, and L.E. Osterman, Late Quaternary extent of the West Antarctic ice sheet: New evidence from Ross Sea cores, Geology, 7, 249-253, 1979.

Kennett, J.P., Foraminiferal evidence of a shallow calcium carbonate solution boundary, Ross Sea, Antarctica, Science, 153, 191-193, 1966.

Killworth, P.D., A baroclinic model of motions on the Antarctic continental shelves, Deep Sea Res., 21, 815-838, 1974.

Kurtz, D.D., and D.H. Bromwich, A recurring atmospherically forced polynya in Terra Nova Bay, this volume.

Ledford-Hoffman, P.A., Biogenic-silica accumulation in the Ross Sea and the importance of Antarctic continental shelf deposits in the marine silica budget, M.S. thesis, 60 pp., North Carolina State Univ., Raleigh, 1984.

Lepley, L.K., The submarine geomorphology of Sulzberger Bay and the eastern Ross Sea, Antarctica, Tech. Rep. TR 172, pp. 1-34, U.S. Naval Oceanogr. Office, Suitland, Md., 1966.

Lewis, E.L. and R.G. Perkin, The winter oceanography of McMurdo Sound, Antarctica, this volume.

MacAyeal, D.R., Tidal rectification below the Ross Ice Shelf, Antarctica, this volume.

MacDonald, S.E., and J.B. Anderson, Paleoceanographic implications of terrigenous deposits on the Maurice Ewing Bank, Southwest Atlantic Ocean, Marine Geology, in press, 1985.

Middleton, G.V., Hydraulic interpretation of sand size distributions, J. Geol., 84, 405-426, 1976.

Middleton, G.V., and J.B. Southard, Mechanics of sediment movement, Soc. Econ. Paleontol. Mineral., Short Course Notes, No. 3, 191 pp., 1977.

Milam, R.M., and J.B. Anderson, Distribution and ecology of recent benthonic foraminifera of the Adelie-George V continental shelf and slope, Antarctica, Mar. Micropaleontol., 6, 297-325, 1981.

Mitchell, W.M., and J.A.T. Bye, Observations in the boundary layer under the sea ice in McMurdo Sound, this volume.

Naval Polar Oceanography Center, Weekly Antarctic Ice Charts, Navy-NOAA Joint Ice Center, Suitland, Md., 1974-1985.

Nittrouer, C.A., P.A. Ledford-Hoffman, D.J. DeMaster, and K.A. Dadey, Accumulation of modern sediment in the Ross Sea, Antarctica, EOS Trans. AGU, 65, 916, 1984.

Pillsbury, R.D., and S.S. Jacobs, Preliminary observations from long-term current meter moorings near the Ross Ice Shelf, this volume.

Rees, A.I., Some flume experiments with fine silt, Sedimentology, 6, 209-239, 1966.

Rouse, H., Elementary Mechanics of Fluids, Wiley, New York, 376 pp., 1946.

Shields, A., Anwendung der Ahnlichkeitsmechanik und der Turbulenzforschung auf die Geschiebebewegung, Mitt. Preuss Versuchanst. Wasserbau Schiffbau, Berlin, 26, 26 pp., 1936. (English translation by W.P. Ott and J.C. van Uchelen, U.S. Dep. of Agric., Soil Conserv. Serv. Coop. Lab., Calif., Inst. of Technol., Pasadena, 1936.

Singer, J.K., and J.B. Anderson, Use of total grain-size distributions to define bed erosion and transport for poorly sorted sediment undergoing simulated bioturbation, Mar. Geol., 57, 335-359, 1984.

Smith, W.O., and D.M. Nelson, Phytoplankton bloom produced by a receding ice edge in the Ross Sea: Spatial coherence with the density field, Science, 227, 163-166, 1985.

Southard, J.B., R.A. Young, and C.D. Hollister, Experimental erosion of calcareous ooze, J. Geophys. Res., 76, 5903-5909. 1971.

Stetson, H.C., and J.F. Upson, Bottom deposits of the Ross Sea, J. Sediment. Petrol., 7, 55-56, 1937.

Sundborg, A., Some aspects of fluvial sediments and fluvial morphology, I. General views and graphic methods, Geograf. Ann., 49a, 333-343, 1967.

Tchernia, P., and P.F. Jeannin, Quelques aspects de la circulation oceanique Antarctique reveles par l'observation de la derive d'icebergs (1972-1983), 92 pp., Centre National d'Etudes Spatiales, Museum National d'Histoire Naturelle, Paris, 1983.

Tressler, W.L., and A.M. Ommundsen, Seasonal oceanographic studies in McMurdo Sound, Antarctica, Tech. Rept. 125, 141 pp., U.S. Navy Hydrographic Office, Washington, D.C., 1962.

Truesdale, R.S., and T.B. Kellogg, Ross Sea diatoms: modern assemblage distributions and their relationship to ecologic oceanographic and sedimentary conditons, Mar. Micropaleo., 4, 13-31, 1979.

Truswell, E.M., Palynology of sea floor samples collected by the 1911-1914 Australasian-Antarctic Expedition: Implications for the geology of coastal East Antarctica, J. Geol. Soc. Aust., 29, 343-356, 1982.

U.S. Naval Oceanographic Office, Sailing directions for Antarctica, H.O. Pub. 27, 2nd ed., 431 pp., U.S. Government Printing Office, Washington, D.C., 1960.

Vanney, J.R., and G.L. Johnson, Wilkes Land continental margin physiography, East Antarctica, Polarforschung, 49, 20-29, 1979.

Vanney, J.R. and G.L. Johnson, GEBCO Bathymetric Sheet 5.18 (Circum-Antarctic), this volume.

Vanney, J.R., R.K.H. Falconer, and G.L. Johnson, Geomorphology of the Ross Sea and adjacent oceanic provinces, Mar. Geol., 41, 73-102, 1981.

Visher, G.S., Grain size distributions and depositional processes, J. Sediment. Petrol., 39, 1074-1106, 1969.

Von der Borch, C.C., and R.L., Oliver, Comparison of heavy minerals in marine sediments with mainland outcrops along the coast in Antarctica, between longitudes 40°E and 150°E, Sediment Geol., 2, 77-80, 1968.

Zwally, H.J., J.C. Comiso, and A.L. Gordon, Antarctic offshore leads and polynyas and oceanographic effects, this volume.

(Received April 3, 1984;
accepted March 12, 1985).

551.4 Oc2j c.1
150101 000
Oceanology of the Antarctic co

3 9315 00107640 2
MANCHESTER COLLEGE LIBRARY

FUNDERBURG LIBRARY
MANCHESTER COLLEGE

WITHDRAWN
from
Funderburg Library

551.4 Oc2j

Oceanology of the Antarctic
continental shelf